AF588080

Texts in Applied Mathematics

Volume 77

The mathematization of all sciences, the fading of traditional scientific boundaries, the impact of computer technology, the growing importance of computer modelling and the necessity of scientific planning all create the need both in education and research for books that are introductory to and abreast of these developments. The aim of this series is to provide such textbooks in applied mathematics for the student scientist. Books should be well illustrated and have clear exposition and sound pedagogy. Large number of examples and exercises at varying levels are recommended. TAM publishes textbooks suitable for advanced undergraduate and beginning graduate courses, and complements the Applied Mathematical Sciences (AMS) series, which focuses on advanced textbooks and research-level monographs.

Pierre Brémaud

An Introduction to Applied Probability

Pierre Brémaud
Paris, France

ISSN 0939-2475 ISSN 2196-9949 (electronic)
Texts in Applied Mathematics
ISBN 978-3-031-49305-8 ISBN 978-3-031-49306-5 (eBook)
https://doi.org/10.1007/978-3-031-49306-5

Mathematics Subject Classification (2020): 60-01, 60J10, 60G55, 60F05, 60G15, 28-01, 62, 94

This Springer imprint is published by the registered company Springer Nature Switzerland AG
The registered company address is: Gewerbestrasse 11, 6330 Cham, Switzerland

Pour Marion

Preface

The purpose of this book is

(a): to provide the elements of probability and stochastic processes of direct interest to the applied sciences where probabilistic models play an important role, most notably the information and communications sciences, the computer sciences, operations research and electrical engineering, but also epidemiology, biology, ecology, physics and the earth sciences,

(b): to introduce very progressively the basic notions of probability and to help the reader to acquire the computational skills necessary for the manipulation of random variables and vectors (the elementary "calculus of probability"), and

(c): to give the essentials of the mathematical theory that will bring the reader to the level that is indispensable for a profitable application of probability in the fields mentioned above.

The treatment is mathematical yet not unnecessarily abstract. It maintains the balance between depth and width that is adequate for the efficient manipulation, based on solid theoretical foundations, of the most popular probabilistic models. The theoretical tools are presented gradually in such a way as to not deter the reader with a wall of technicalities before having the opportunity to understand their relevance in simple situations. In particular, the use of the so-called modern integration theory (that is, the Lebesgue integral) is postponed until the fifth chapter, where it is reviewed in sufficient detail for a rigorous treatment of the topics of interest in the various domains of application listed above. All the results are proved, except in the rare situations where the tools needed for the proof require a deeper immersion into the foundations of measure and integration theory and only when their content is intuitive. They are then accompanied by meaningful examples of application.

The contents are organized in three parts.

Part I: The Elementary Calculus

In this part (Chapters 1 to 3), the beginner is acquainted with the vocabulary of probability theory and with the methods and tricks of the trade that suffice to treat simple, yet significant, examples. The first two chapters are devoted to *discrete probability* models and the third chapter to *continuous random variables and vectors.* This part features, among other topics, generating functions, the Gaussian vectors, linear regression and the elementary theory of conditional expectation.

The focus there is on practical computations and only a working knowledge of series, of the Riemann integral and of matrices is required.

Part II: The Essential Theory

The book then proceeds to the basic *theory* of probability, starting with a brief survey of *integration theory*, that is then used to revisit, formalize and generalize the results of the previous chapters that were either admitted or proved in the special framework of discrete probability and continuous random vectors. It introduces the various types of *convergence of sequences of random variables*: almost-sure, in probability, in distribution, in variation and in the quadratic mean, featuring in particular the strong law of large numbers and the central limit theorem, and gives the intermediate, and then the advanced, theory of *conditional expectation*. Chapter 8 is an introduction to *martingales*, one of the fundamental tools of probability. It may be considered to be a continuation of the theme "Convergence of sequences", especially of the chapter on almost-sure convergence.

The first and second parts provide the material for a basic course in the theory of probability.

Part III: The Important Models

The results gathered at this point are then applied to the four most important and ubiquitous categories of probabilistic models:

- *Markov chains*, an omnipresent and most versatile model of applied probability,
- *Poisson processes* (on the line and in space), which occur in a number of applications, ranging from ecology to queuing and mobile communications networks,
- *Brownian motion*, which models fluctuations of the stock market and the "white noise" of physics, and
- *Wide-sense stationary processes*, which are of special importance in signal analysis and design, and also in the earth sciences.

An appendix on Hilbert spaces is given for easy reference and self-containedness.

Each chapter contains a final section with exercises. In the important transition chapters 4 and 5, the solutions are given.

This book can be used as a text in a variety of ways and at various levels of study. Essentially, it provides the material for a two-semester graduate course on probability and stochastic processes in a department of applied mathematics, or for students in departments where stochastic models play an essential role.

The progressive introduction of the concepts and of the tools, together with the inclusion of numerous examples, also make this book well-adapted to self-study.

Paris, October 15, 2023

Pierre Brémaud

Contents

Chapter 1

Basic Notions

Probability theory aims at quantifying randomness. It concerns "experiments" (performed by man or Nature, or both) whose outcome is uncertain, and evaluates the probability of the resulting events. The meaning of these terms (outcomes, events and probability) is given in the so-called *axiomatic framework* embodied in the trinity $(\Omega, \mathcal{F}, P)$, called the probability space, that will be progressively introduced in this chapter.

1.1 Outcomes and Events

We first recall the notation concerning the basic set operations: *union*, *intersection*, and *complementation*.

If A and B are subsets of some set Ω, $A \cup B$ denotes their union and $A \cap B$ their intersection. In this book, $\overline{A}$ denotes the complement of A in Ω. The notation $A + B$ (the *sum* of A and B) implies by convention that A and B are *disjoint*, in which case it stands for the union $A \cup B$. Similarly, the notation $\sum_{k=1}^{\infty} A_k$ is used for $\cup_{k=1}^{\infty} A_k$ only when the A_k's are pairwise disjoint. The notation $A - B$ is used only if $B \subseteq A$, and it stands for $A \cap \overline{B}$. In particular, if $B \subset A$, then $A = B + (A - B)$.

A subset of Ω consisting of just one element $a \in \Omega$ is called a *singleton* and is denoted by $\{a\}$. Similar notation is used for sets with a finite number of elements. For instance $\{a, b, c\}$ represents the set consisting of the three distinct elements a, b and c in Ω.

The *indicator function* of the subset $A \subseteq \Omega$ is the function $1_A : \Omega \to \{0, 1\}$ defined by

$$1_A(\omega) := \begin{cases} 1 & \text{if } \omega \in A\,, \\ 0 & \text{if } \omega \notin A. \end{cases}$$

P. Brémaud, *An Introduction to Applied Probability*, Texts in Applied Mathematics 77,
https://doi.org/10.1007/978-3-031-49306-5_1

Let now $\mathcal{P}$ be a property that an element of some set X may or may not possess. The notations

$$1_{\mathcal{P}}(x) \text{ or } 1_{\{x \text{ satisfies } \mathcal{P}\}}$$

stand for $f(x)$, where $f(x) = 1$ if x satisfies property $\mathcal{P}$, $= 0$ otherwise. A variety of similar notations will be used and should be self-explanatory in a given context. For instance, f and g being real-valued functions defined on a set X, $1_{\{f \geq g\}}(x)$ is equal to 1 if $f(x) \geq g(x)$, and to 0 otherwise.

Random phenomena are observed by means of experiments. Each experiment results in an *outcome*. The collection of all possible outcomes ω is called the *sample space* Ω. Any subset A of the sample space Ω can be regarded as a representation of some *event*[1].

EXAMPLE 1.1.1: TOSSING A DIE, TAKE 1. The experiment consists in tossing a die once. The possible outcomes are $\omega = 1, 2, \ldots, 6$ and the sample space is the set $\Omega = \{1, 2, 3, 4, 5, 6\}$. The subset $A = \{1, 3, 5\}$ is the event "result is odd."

EXAMPLE 1.1.2: THROWING A DART. The experiment consists in throwing a dart at a wall. The sample space can be chosen to be the plane $\mathbb{R}^2$. An outcome is the position $\omega = (x, y)$ hit by the dart. The subset $A = \{(x, y); x^2 + y^2 > 1\}$ is an event that could be named "you missed the dartboard" (the disk of radius 1 centered at 0).

EXAMPLE 1.1.3: HEADS OR TAILS, TAKE 1. The experiment is an infinite succession of coin tosses. One can take for the sample space the collection of all sequences $\omega := \{x_n\}_{n \geq 1}$, where $x_n = 1$ or 0, depending on whether the n-th toss results in heads or tails. The subset $A = \{\omega;\ x_k = 1 \text{ for } k = 1 \text{ to } 1{,}000\}$ is a lucky event for anyone betting on heads!

Probability theory was born out of the study of practical problems (mostly gambling, but not exclusively) and this has led the probabilists to develop their own dialect which connects their science to reality and favors intuition.

One says that outcome ω *realizes* event A if $\omega \in A$. For instance, in the die model of Example 1.1.1, the outcome $\omega = 1$ realizes the event "result is odd", since $1 \in A = \{1, 3, 5\}$. Obviously, if ω does not realize A, it realizes $\overline{A}$. Event $A \cap B$ is realized by outcome ω if and only if ω realizes both A and B. Similarly, $A \cup B$ is

[1] However, in general, the appellation "event" will be reserved to a more restricted class of subsets. See Definition 1.1.4.

realized by ω if and only if ω realizes *at least* one event among A and B (both can be realized). Two events A and B are called *incompatible* when $A \cap B = \varnothing$. In other words, event $A \cap B$ is impossible: no outcome ω can realize both A and B. For this reason one refers to the empty set $\varnothing$ as the *impossible* event. Naturally, Ω is called the *certain* event.

Recall now that the notation $\sum_{k=1}^{\infty} A_k$ is used for $\cup_{k=1}^{\infty} A_k$ only when the subsets A_k are pairwise disjoint. In the terminology of sets, the sets A_1, $A_2, \dots$ form a *partition* of Ω if $\sum_{k=0}^{\infty} A_k = \Omega$. One then calls events A_1, $A_2, \dots$ *mutually exclusive and exhaustive*. They are exhaustive in the sense that any outcome ω realizes at least one among them. They are mutually exclusive in the sense that any two distinct events among them are incompatible. Therefore, any ω realizes *one and only one* of the events $A_1, \dots, A_n$.

If $B \subseteq A$, event B is said to *imply* event A, because ω realizes A whenever it realizes B.

Probability theory associates with each event a number, the *probability* of the said event. The collection $\mathcal{F}$ of events to which a probability is assigned is not always identical to the collection of all subsets of Ω. The requirement on $\mathcal{F}$ is that it should be a σ-field, whose definition follows.

Definition 1.1.4 *Let $\mathcal{F}$ be a collection of subsets of Ω, such that*

(i) *the certain event Ω is in $\mathcal{F}$,*

(ii) *if A belongs to $\mathcal{F}$, then so does its complement $\overline{A}$, and*

(iii) *if $A_1, A_2, \dots$ belong to $\mathcal{F}$, then so does their union $\cup_{k=1}^{\infty} A_k$.*

One then calls $\mathcal{F}$ a σ-field *on Ω, here the σ-field of* events.

The requirements in the definition are in a sense minimal if you want the σ-field $\mathcal{F}$ to contain the "interesting" events (those for which you are eager to compute the probability). Indeed, the complement, the unions and intersections of interesting events are most likely interesting events. A natural question at this point is: why not accept in general as an event the union of an arbitrary (not just countable) collection of events. The answer is given in the next section.

Note that the impossible event $\varnothing$, being the complement of the certain event Ω, is in $\mathcal{F}$. Note also that if $A_1, A_2, \dots$ belong to $\mathcal{F}$, then so does their intersection $\cap_{k=1}^{\infty} A_k$ (see Exercise 1.5.27).

The collection $\mathcal{P}(\Omega)$ of all subsets of Ω and $\mathcal{F} = \{\Omega, \varnothing\}$ are called respectively the *trivial* σ-field and the *gross* σ-field.

If the sample space Ω is finite or countable, one usually (but not always and not necessarily) considers any subset of Ω to be an event. That is $\mathcal{F} = \mathcal{P}(\Omega)$. But this is not true in the general case, both for technical reasons that will be of little concern in this course.[2] Even in the discrete case, this is not necessarily true. For instance, suppose that you wish to play heads or tails and have no coin (as an inveterate gambler, you are probably broke), but you keep handy a precious die in your pocket. You can use the die model of Example 1.1.1, calling "even" heads and "odd" tails. That is, you will use the σ-field $\{\Omega, \varnothing, \{1,3,5\}, \{2,4,6\}\}$ instead of the trivial σ-field.

Definition 1.1.5 *Let Ω be an arbitrary set, and let $\mathcal{C}$ be a non-empty collection of its subsets. The* σ-field generated by $\mathcal{C}$*, denoted by $\sigma(\mathcal{C})$, is by definition the smallest σ-field containing all the subsets in $\mathcal{C}$.*

Let us now agree to call an interval of $\mathbb{R}$ any convex subset of $\mathbb{R}$: $[a,b]$, $[a,b)$, $(a,b]$, (a,b), $(-\infty,b]$, $(-\infty,b)$, $(a,+\infty)$, $[a,+\infty)$, $(-\infty,+\infty)$.

Definition 1.1.6 *The σ-field on $\mathbb{R}^n$, denoted by $\mathcal{B}(\mathbb{R}^n)$ and called the* Borel σ-field *on $\mathbb{R}^n$ is, by definition, the smallest σ-field on $\mathbb{R}^n$ that contains all rectangles, that is, all sets of the form $\prod_{j=1}^{n} I_j$, where the I_j's are arbitrary intervals of $\mathbb{R}$.*

In other words, $\mathcal{B}(\mathbb{R}^n)$ is the σ-field on $\mathbb{R}^n$ generated by the rectangles.

The above definition of the Borel σ-field is not constructive and therefore one may wonder if there exist sets that are not Borel sets. The theory tells us that there are indeed such sets, but they are in a sense "exotic" and never met in applications. At this stage, you just have to know that any set for which you have once computed the n-volume is in $\mathcal{B}(\mathbb{R}^n)$.

EXAMPLE 1.1.7: HEADS OR TAILS, TAKE 2. Let $\mathcal{F}$ be the smallest σ-field that contains all the sets $\{\omega\,;\, x_k = 1\}$ $(k \geq 1)$. It also contains the sets $\{\omega\,;\, x_k = 1\}$, $k \geq 1$ (pass to the complements), and therefore (take intersections) all the sets of the form $\{\omega\,;\, x_1 = a_1, \ldots, x_n = a_n\}$ $(n \geq 1,\, a_1, \ldots, a_n \in \{0,1\})$.

1.2 Probability of Events

The probability $P(A)$ of an event A measures the likeliness of its occurrence. As a function defined on $\mathcal{F}$, it is required to satisfy a few properties, the *axioms*

[2] See however the comment in the next subsection.

of probability. These are motivated by the following heuristic interpretation of $P(A)$ as the empirical frequency of occurrence of event A. If n "independent" experiments are performed, among which n_A result in the realization of A, then the empirical frequency of occurrences of event A,

$$F(A) = \frac{n_A}{n},$$

should be close to $P(A)$ if n is "sufficiently large". (This statement will be clarified later on by the law of large numbers.) Clearly, the empirical frequency function F satisfies the axioms listed in the following definition.

Definition 1.2.1 *A* probability *on* $(\Omega, \mathcal{F})$ *is a mapping* $P : \mathcal{F} \to \mathbb{R}$ *such that*

(i) $0 \leq P(A) \leq 1$,

(ii) $P(\Omega) = 1$, *and*

(iii) $P(\cup_{k=1}^{\infty} A_k) = \sum_{k=1}^{\infty} P(A_k)$ *whenever the sets* $A_k \in \mathcal{F}$ $(k \geq 1)$ *are mutually disjoint.*

Property (iii) is called *σ-additivity.* The triple $(\Omega, \mathcal{F}, P)$ is called a *probability space*, or an *abstract probability model.*

EXAMPLE 1.2.2: TOSSING A DIE, TAKE 2. An event A is a subset of $\Omega = \{1, 2, 3, 4, 5, 6\}$. The formula

$$P(A) = \frac{|A|}{6},$$

where $|A|$ is the *cardinality* of A (the number of elements in A), defines a probability P.

EXAMPLE 1.2.3: HEADS OR TAILS, TAKE 3. Choose probability P such that for any event of the form $A = \{x_1 = a_1, \ldots, x_n = a_n\}$, where $a_1, \ldots, a_n$ are arbitrary in $\{0, 1\}$,

$$P(A) = \frac{1}{2^n}.$$

Note that this does not define the probability of all events of $\mathcal{F}$. But the theory tells us that there exists such a probability satisfying the above requirement and that it is unique.[3]

---

[3] In this book, all the results concerning the existence and uniqueness of probabilities will be assumed, as they require a deeper immersion in the theory and are in fact easily admitted. The interested reader will find the proofs in [1] or [3] for instance.

EXAMPLE 1.2.4: RANDOM POINT IN THE SQUARE. The following is a possible model of a random point in the unit square: $\Omega = [0, 1]^2$, $\mathcal{F}$ is the collection of sets in the Borel σ-field $\mathcal{B}(\mathbb{R}^2)$ that are contained in $[0, 1]^2$. The theory tells us that there indeed exists one and only one probability P assigning to rectangles therein their area in the usual sense, called the Lebesgue measure on $[0, 1]^2$, which formalizes the intuitive notion of area.

The probability of Example 1.2.2 suggests an *unbiased die*, where the outcomes 1, 2, 3, 4, 5 and 6 are equiprobable. As we shall see later on, the probability P of Example 1.2.3 implies an *unbiased coin* and *independent tosses* (the emphasized terms will be defined later).

We now answer a question that the reader may have in mind. Why, for instance in Example 1.2.4 above, don't we take for the σ-field of events the trivial σ-field? The answer is easy, although its proof is not immediate and belongs to an advanced course on measure theory: there exists no probability P on the trivial σ-field on the square $[0, 1]^2$ that assigns to rectangles therein their area in the usual sense.

Another question is: why impose only σ-additivity, and not unrestricted additivity
$(P(\cup_{i\in I} A_i) = \sum_{i\in I} P(A_i)$ where the index set I is arbitrary and the A_i's are mutually disjoint)? In fact, if unrestricted additivity was part of the definition of probability, there would exist no such "probability" on the Borel σ-field on the square $[0, 1]^2$ assigning to rectangles therein their area in the usual sense (see Exercise 1.5.5).

We now list some properties that follow directly from the axioms:

Theorem 1.2.5 *For any event A*

$$P(\overline{A}) = 1 - P(A), \tag{1.1}$$

and

$$P(\varnothing) = 0. \tag{1.2}$$

Proof. For a proof of (1.1), use additivity:

$$1 = P(\Omega) = P(A + \overline{A}) = P(A) + P(\overline{A}).$$

Applying (1.1) with $A = \Omega$ gives (1.2). □

Theorem 1.2.6 Monotonicity*:*

$$A \subseteq B \Longrightarrow P(A) \le P(B). \tag{1.3}$$

Proof. Observe that $B = A + (B - A)$ when $A \subseteq B$, and therefore

$$P(B) = P(A) + P(B - A) \geq P(A).$$

□

Theorem 1.2.7 Sub-σ-additivity:

$$P(\cup_{k=1}^{\infty} A_k) \leq \sum_{k=1}^{\infty} P(A_k). \tag{1.4}$$

See Exercise 1.5.7.

The next property, the *sequential continuity* of probability, is close to a tautology and yet extremely useful.

Theorem 1.2.8 *Let* $\{A_n\}_{n\geq 1}$ *be a* non-decreasing *sequence of events, that is,* $A_{n+1} \supseteq A_n$ $(n \geq 1)$. *Then*

$$P(\cup_{n=1}^{\infty} A_n) = \lim_{n\uparrow\infty} P(A_n). \tag{1.5}$$

Proof. Write

$$A_n = A_1 + (A_2 - A_1) + \cdots + (A_n - A_{n-1})$$

and

$$\cup_{k=1}^{\infty} A_k = A_1 + (A_2 - A_1) + (A_3 - A_2) + \cdots.$$

Therefore,

$$\begin{aligned} P(\cup_{k=1}^{\infty} A_k) &= P(A_1) + \sum_{j=2}^{\infty} P(A_j - A_{j-1}) \\ &= \lim_{n\uparrow\infty} \left\{ P(A_1) + \sum_{j=2}^{n} P(A_j - A_{j-1}) \right\} = \lim_{n\uparrow\infty} P(A_n). \end{aligned}$$

□

Corollary 1.2.9 *Let* $\{B_n\}_{n\geq 1}$ *be a* non-increasing *sequence of events, that is,* $B_{n+1} \subseteq B_n$ $(n \geq 1)$. *Then,*

$$P(\cap_{n=1}^{\infty} B_n) = \lim_{n\uparrow\infty} P(B_n). \tag{1.6}$$

See Exercise 1.5.8.

A central notion of probability is that of a *negligible set*. Its importance is due to the fact that probabilistic calculations bear on the probability of events, not on the events themselves. One will never be able to say that an event such as "the empirical frequency of heads in an infinite sequences of independent tosses of a fair coin is equal to $\frac{1}{2}$" is certain, that is, is *identical* to Ω. One will only be able to prove that the complementary event has null probability.

Definition 1.2.10 *A set $N \subset \Omega$ is called P-*negligible *if it is contained in an event $A \in \mathcal{F}$ of null probability.*

Note that the set N need not be an event (an element of $\mathcal{F}$). An event that is negligible set will of course be called a negligible event.

Theorem 1.2.11 *A countable union of negligible sets is a negligible set.*

Proof. Let N_k $(k \geq 1)$ be P-negligible sets. By definition there exists a sequence A_k $(k \geq 1)$ of events of null probability such that $N_k \subseteq A_k$ $(k \geq 1)$. We have

$$N := \cup_{k\geq 1} N_k \subseteq A := \cup_{k\geq 1} A_k ,$$

and then $P(A) = 0$, by the sub-σ-additivity property of probability. □

EXAMPLE 1.2.12: RANDOM POINT IN THE SQUARE, TAKE 2. Each rational point of the square considered as a set (a singleton) has a null area and therefore null probability. Therefore, in this model, the (countable) set of rational points of the square has null probability. In other words, in this particular model, the probability of drawing a rational point is null.

1.3 Independence and Conditioning

Recall the heuristic frequency interpretation of probability at the beginning of Section 1.2. A situation where

$$\frac{n_{A\cap B}}{n_B} \approx \frac{n_A}{n}$$

(here $\approx$ is a non-mathematical symbol meaning "approximately equal") suggests some kind of "independence" of A and B, in the sense that statistics relative to A do not vary when passing from a neutral sample of population to a selected sample characterized by the property B. For example, the proportion of people with a family name beginning with H is the same among a large population with the usual mix of men and women as it would be among a large all-male population. This prompts us to give the following formal definition of independence, the single most important concept of probability theory.

Definition 1.3.1 *Two events A and B are called* independent *if*

$$P(A \cap B) = P(A)P(B)\,. \tag{1.7}$$

One should be aware that *incompatibility does not mean independence.* As a matter of fact, two incompatible events A and B are independent if and only if at least one of them has null probability. Indeed, if A and B are incompatible, $P(A \cap B) = P(\varnothing) = 0$, and therefore (1.7) holds if and only if $P(A)P(B) = 0$.

The notion of independence carries over straightforwardly to families of events.

Definition 1.3.2 *A sequence $\{A_n\}_{n \in \mathbb{N}}$ of events is called* independent *if for any* finite *set of indices $i_1, \ldots, i_r \in \mathbb{N}$,*

$$P\left(A_{i_1} \cap A_{i_2} \cap \cdots \cap A_{i_r}\right) = P(A_{i_1}) \times P(A_{i_2}) \times \cdots \times P(A_{i_r})\,.$$

One also says that the A_n's ($n \in \mathbb{N}$) are jointly independent.

Definition 1.3.3 *The* conditional probability *of A given B is the number*

$$P(A \mid B) := \tfrac{P(A \cap B)}{P(B)}\,, \tag{1.8}$$

defined when $P(B) > 0$. If $P(B) = 0$, one defines $P(A \mid B)$ arbitrarily between 0 *and* 1.

In particular, if A and B are independent, then $P(A \mid B) = P(A)$.

The quantity $P(A \mid B)$ represents our expectation of A being realized when the only available information is that B is realized. The corresponding heuristic quantity is the *relative frequency* $n_{A \cap B}/n_B$.

Probability theory is primarily concerned with the computation of probabilities of complex events. The following formulas, the so-called *Bayes rules*, are not only useful, but indispensable. They lay the foundations of the *elementary calculus of probability* and give the first opportunity to solve simple yet non-trivial problems.

Theorem 1.3.4 *With $P(A) > 0$, we have the* Bayes rule of retrodiction*:*

$$P(B \mid A) = \tfrac{P(A \mid B)P(B)}{P(A)}\,. \tag{1.9}$$

Proof. Rewrite (1.8) symmetrically in A and B:

$$P(A \cap B) = P(A \mid B)P(B) = P(B \mid A)P(A)\,.$$

□

Theorem 1.3.5 *Let $B_1, B_2, \dots$ be events forming a partition of Ω, that is such that $\sum_{i=1}^{\infty} B_i = \Omega$. Then for any event A, we have the* Bayes rule of total causes*:*

$$P(A) = \sum_{i=1}^{\infty} P(A \mid B_i) P(B_i) . \tag{1.10}$$

Proof. Decompose A as follows:

$$A = A \cap \Omega = A \cap \left(\sum_{i=1}^{\infty} B_i \right) = \sum_{i=1}^{\infty} (A \cap B_i) .$$

Therefore (by σ-additivity and by definition of conditional probability):

$$\begin{aligned} P(A) &= P\left(\sum_{i=1}^{\infty} (A \cap B_i) \right) \\ &= \sum_{i=1}^{\infty} P(A \cap B_i) = \sum_{i=1}^{\infty} P(A \mid B_i) P(B_i) . \end{aligned}$$

□

EXAMPLE 1.3.6: DIPLOIDS AND THE HARDY–WEINBERG LAW. In diploid organisms (you for instance) each hereditary character is carried by a pair of *genes*. Consider the situation in which a given gene can take two forms called *alleles*, denoted a and A. Such was the case in the historical experiments performed in 1865 by the Czech monk Gregory Mendel, who studied the hereditary transmission of the nature of the skin in a species of green pea. The two alleles corresponding to the gene or character "nature of the skin" are a for "wrinkled" and A for "smooth". The genes are grouped into pairs and there are two alleles, thus three *genotypes* are possible for the character under study: aa, Aa (same as aA), and AA ([4]). During the reproduction process, each of the two parents contributes to the genetic heritage of their descendant by providing *one* allele of their pair. This is done by intermediaries of the reproductive cells called *gametes* (in the human species, the spermatozoid and the ovula) which carry only one gene of the pair of genes characteristic of each parent. The gene carried by the gamete is chosen at random among the pair of genes of the parent. The actual process occurring in the reproduction of diploid cells is called *meiosis*.

[4] With each genotype is associated a *phenotype* which is the external appearance corresponding to the genotype. Genotypes aa and AA have different phenotypes —otherwise no character could be isolated—, and the phenotype of Aa lies somewhere between the phenotypes of aa and AA. Sometimes, an allele is *dominant*, that is, A, and the phenotype of Aa is then the same as the phenotype of AA.

A given cell possesses two chromosomes. A chromosome can be viewed as a string of genes, each gene being at a specific location in the chain. A given chromosome duplicates itself and four new cells are formed for every chromosome (see the figure below). One of the four gametes of a "mate" (say, the ovula) chosen at random selects randomly one of the four gametes of the other "partner" (say, the spermatozoid) and this gives "birth" to a pair of alleles.

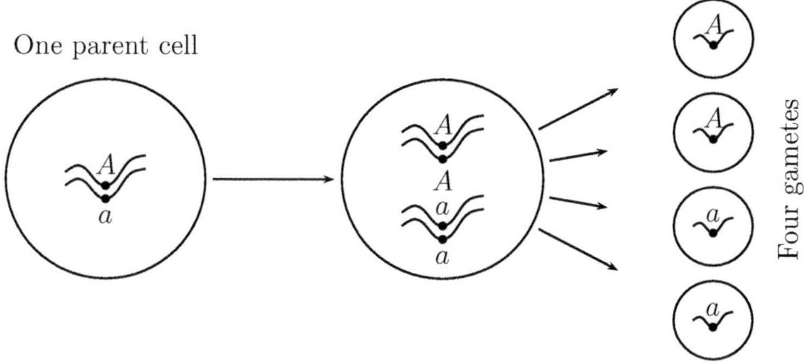

Figure 1.1: Meiosis.

Let us start from an idealistically infinite population where the genotypes are found in the following proportions:

$$AA \; : \; Aa \; : \; aa$$

$$x \; : \; 2z \; : \; y.$$

Here x, y, and z are numbers between 0 and 1, and $x+2z+y=1$. The two parents are chosen independently (random mating), and their gamete chooses an allele at random in the pair carried by the corresponding parent.

We seek the genotype distribution of the second generation. Our first task consists in providing a probabilistic model. We propose the following one. The sample space Ω is the collection of all quadruples $\omega = (x_1, x_2, y_1, y_2)$ where x_1 and x_2 take their values in $\{AA, aA, aa\}$, and y_1 and y_2 take their values in $\{A, a\}$. Four "coordinate functions" X_1, X_2, Y_1, Y_2 are defined by $X_1(\omega) = x_1, X_2(\omega) = x_2, Y_1(\omega) = y_1$ and $Y_2(\omega) = y_2$. We interpret X_1 and X_2 as the pairs of genes in parents 1 and 2 respectively. Y_1 is the allele chosen by gamete 1 among the alleles of X_1, with a similar definition for Y_2. The data available are, for the selection of parents:

$$\begin{aligned} P(X_1 = AA) = P(X_2 = AA) = x\,, \\ P(X_1 = aa) = P(X_2 = aa) = y\,, \\ P(X_1 = Aa) = P(X_2 = Aa) = 2z\,, \end{aligned}$$

and for the choice of allele by gamete 1:

$$\begin{aligned} P(Y_1 = A \mid X_1 = AA) &= 1, \quad P(Y_1 = a \mid X_1 = AA) = 0, \\ P(Y_1 = A \mid X_1 = aa) &= 0, \quad P(Y_1 = a \mid X_1 = aa) = 1, \\ P(Y_1 = A \mid X_1 = Aa) &= \frac{1}{2}, \quad P(Y_1 = a \mid X_1 = Aa) = \frac{1}{2}, \end{aligned}$$

and the similar data for the choice of allele by gamete 2. One must also add the assumptions of independence of X_1 and X_2 and of Y_1 and Y_2. We are required to compute the genotype distribution of the second generation, that is,

$$\begin{aligned} p &= P(Y_1 = A, Y_2 = A)\,, \\ q &= P(Y_1 = a, Y_2 = a)\,, \\ 2r &= P(Y_1 = A, Y_2 = a \text{ or } Y_1 = a, Y_2 = A)\,. \end{aligned}$$

We start with the computation of p. In view of the independence of Y_1 and Y_2, $p = P(Y_1 = A)P(Y_2 = A)$. By the rule of total causes,

$$\begin{aligned} P(Y_1 = A) &= P(Y_1 = A \mid X_1 = AA)P(X_1 = AA) \\ &\quad + P(Y_1 = A \mid X_1 = A_a)P(X_1 = A_a) \\ &\quad + P(Y_1 = A \mid X_1 = aa)P(X_1 = aa) \\ &= 1 \cdot x + \frac{1}{2} \cdot 2z + 0 \cdot y = x + z\,. \end{aligned}$$

Therefore $p = (x+z)^2$ and by symmetry, $q = (y+z)^2$. Now $2r = P(Y_1 = A, Y_2 = a) + P(Y_1 = a, Y_2 = A)$, and therefore by symmetry, $r = P(Y_1 = A, Y_2 = a)$. In view of the independence of Y_1 and $Y_2, r = P(Y_1 = A)P(Y_2 = a)$. Finally, in view of previous computations, $2r = 2(x+z)(y+z)$.

Theorem 1.3.7 *For any sequence of events $A_1, \dots, A_n$, we have the* Bayes sequential formula*:*

$$P\left(\cap_{i=1}^k A_i\right) = P(A_1)P(A_2 \mid A_1)P(A_3 \mid A_1 \cap A_2) \cdots P\left(A_k \mid \cap_{i=1}^{k-1} A_i\right)\,. \tag{1.11}$$

Proof. By induction. First observe that (1.11) is true for $k = 2$ by definition of conditional probability. Suppose that (1.11) is true for k. Write

$$\begin{aligned} P\left(\cap_{i=1}^{k+1} A_i\right) &= P\left(\left(\cap_{i=1}^{k} A_i\right) \cap A_{k+1}\right) \\ &= P\left(A_{k+1} \mid \cap_{i=1}^{k} A_i\right) P\left(\cap_{i=1}^{k} A_i\right), \end{aligned}$$

and replace $P\left(\cap_{i=1}^{k} A_i\right)$ by the assumed equality (1.11) to obtain the same equality with $k + 1$ replacing k. □

EXAMPLE 1.3.8: SHOULD ONE ALWAYS BELIEVE DOCTORS? Doctors apply a test that gives a positive result in 99% of the cases where the patient is affected by the disease. However it happens in 2% of the cases that a healthy patient has a positive test. Statistical data show that one individual out of 1000 has the disease. We compute the probability for a patient with a positive test to be affected by the disease.

Let M be the event "patient is ill," and let $+$ and $-$ be the events "test is positive" and "test is negative" respectively. We have the data

$$P(M) = 0.001,\ P(+ \mid M) = 0.99,\ P(+ \mid \overline{M}) = 0.02,$$

and we must compute $P(M \mid +)$. By the retrodiction formula,

$$P(M \mid +) = \frac{P(+ \mid M)P(M)}{P(+)}.$$

By the formula of total causes,

$$P(+) = P(+ \mid M)P(M) + P(+ \mid \overline{M})P(\overline{M}).$$

Therefore,

$$P(M \mid +) = \frac{(0.99)(0.001)}{(0.99)(0.001) + (0.02)(0.999)},$$

that is, approximately 0.005.

One might have mixed feelings concerning the reliability of the pharmaceutical company that manufactures such a test. There is however a possible explanation. For certain types of illness, it is much preferable to provoke many false alarms than to fail to detect the illness. This is the case in group testing, where the blood samples are mixed. If this mixed sample is positive, the patients are tested individually with a more reliable, in general more expensive, test.

EXAMPLE 1.3.9: THE BALLOT PROBLEM. In an election, candidates I and II have obtained a and b votes respectively. Candidate I won, that is $a > b$. We shall compute the probability that in the course of the vote counting process, candidate I has always had the lead.

Let $p_{a,b}$ be the probability that A is always ahead. By the Bayes rule of total causes, and conditioning on the last vote:

$$\begin{aligned} p_{a,b} &= P(A \text{ always ahead } | A \text{ gets last vote }) P(A \text{ gets last vote }) \\ &\quad + P(A \text{ always ahead } | B \text{ gets last vote }) P(B \text{ gets last vote }) \\ &= p_{a-1,b}\frac{a}{a+b} + p_{a,b-1}\frac{b}{a+b}, \end{aligned}$$

with the convention that for $a = b + 1$, $p_{a-1,b} = p_{b,b} = 0$. The result follows by induction on the total number of votes $a + b$:

$$p_{a,b} = \frac{a-b}{a+b}.$$

Definition 1.3.10 *Let A, B, and C be events, with $P(C) > 0$. One says that A and B are* conditionally independent *given C if*

$$P(A \cap B \mid C) = P(A \mid C)P(B \mid C). \tag{1.12}$$

In other words, A and B are independent with respect to the probability P_C defined by $P_C(A) = P(A \,|\, C)$ (see Exercise 1.5.18).

EXAMPLE 1.3.11: CHEAP WATCHES. Two factories A and B manufacture watches. Factory A produces on average one defective item out of 100, and B produces on average one bad watch out of 200. A retailer receives a container of watches from one of the two above factories, but he does not know which. He checks the first watch. It works!

(a) What is the probability that the second watch he will check is good?

(b) Are the states of the first two watches independent?

Solution: (a) Let X_n be the state of the nth watch in the container, with $X_n = 1$ if it works and $X_n = 0$ if it does not. Let Y be the factory of origin. We express our a priori ignorance of where the case comes from by

$$P(Y = A) = P(Y = B) = \frac{1}{2}.$$

Also, we assume that given $Y = A$ (resp., $Y = B$), the states of the successive watches are independent. For instance,

$$P(X_1 = 1, X_2 = 0 \mid Y = A) = P(X_1 = 1 \mid Y = A)P(X_2 = 0 \mid Y = A).$$

We have the data

$$P(X_n = 0 \mid Y = A) = 0.01\,, \qquad P(X_n = 0 \mid Y = B) = 0.005\,.$$

We are required to compute

$$P(X_2 = 1 \mid X_1 = 1) = \frac{P(X_1 = 1, X_2 = 1)}{P(X_1 = 1)}\,.$$

By the formula of total causes, the numerator of this fraction equals

$$P(X_1 = 1, X_2 = 1 \mid Y = A)P(Y = A) + \cdots$$
$$\cdots + P(X_1 = 1, X_2 = 1 \mid Y = B)P(Y = B)\,,$$

that is, $(0.5)(0.99)^2 + (0.5)(0.995)^2$, and the denominator is

$$P(X_1 = 1 \mid Y = A)P(Y = A) + P(X_1 = 1 \mid Y = B)P(Y = B)\,,$$

that is, $(0.5)(0.99) + (0.5)(0.995)$. Therefore,

$$P(X_2 = 1 \mid X_1 = 1) = \frac{(0.99)^2 + (0.995)^2}{0.99 + 0.995}\,.$$

(b) The states of the two watches are not independent. Indeed, if they were, then

$$P(X_2 = 1 \mid X_1 = 1) = P(X_2 = 1) = (0.5)\,(0.99 + 0.995)\,,$$

a result different from what we obtained.

The example above shows that two events A and B can be conditionally independent given C and conditionally independent given $\overline{C}$, and yet *not* be independent.

1.4 Counting Models

A number of problems in Probability reduce to counting the elements of finite sets. The general setting is the following.

The set of all possible outcomes, Ω, is finite, and for some reason (symmetry for instance) one is led to believe that all the outcomes ω have the same probability. Since the probabilities sum up to one, each outcome has probability $\frac{1}{|\Omega|}$, where $|\Omega|$ denotes the *cardinality* (the number of elements) of the set Ω. Since the probability of an event A is the sum of the probabilities of all outcomes $\omega \in A$, we have

$$P(A) = \frac{|A|}{|\Omega|}. \tag{1.13}$$

Thus, computing $P(A)$ requires *counting* the elements in the sets A and Ω.

There is a whole branch of mathematics devoted mainly to counting, called *combinatorics*. A basic item of combinatorics is the *binomial coefficient*. The binomial coefficient expresses the number of fixed-size subsets of a finite set. Suppose we have a set F containing n elements denoted by $1, 2, \ldots, n$. How many different subsets of p elements of F are there? If we denote this number, called the binomial coefficient, by $\binom{n}{p}$, then we have

$$\binom{n}{p} = \frac{n!}{p!(n-p)!}. \tag{1.14}$$

Proof. To prove this formula, we proceed in two steps. First we shall determine the number of possible *ordered sequences* of p elements taken from F without repetition. (Note the difference between an ordered sequence of p elements without repetition and a subset of p elements: a subset such as $\{1, 2, 3\}$ for instance gives rise to 6 ordered sequences of 3 elements taken from F without repetition:

$$(1,2,3), (2,3,1), (3,1,2), (3,2,1), (1,3,2), (2,1,3).)$$

To make an ordered sequence of p elements taken from F without repetition, we must first select the first element: there are n choices. Having selected the first element, there remains only $n-1$ choices for the second element since we exclude repetitions. We proceed in this way up to the last element, which must be chosen among the $n-p+1$ remaining elements. Thus the number $A(n,p)$ of *ordered* sequences of p elements taken from F without repetition is

$$n(n-1)(n-2)\ldots(n-p+1),$$

that is

$$A(n,p) = \frac{n!}{(n-p)!}.$$

In particular, the number of ordered sequences of length p that one can obtain from a given set of length p is $A(p,p) = p!$, and since each subset of p elements gives rise to exactly $p!$ ordered subsequences of length p, we have

$$\binom{n}{p} = \frac{A(n,p)}{p!},$$

which is formula (1.14). □

Let now F be a finite set with n elements. How many subsets of F are there? One could answer with $\sum_{p=0}^{n} \binom{n}{p}$, and this is true if we use the convention that $\binom{n}{0} = 1$ or equivalently $0! = 1$. (Recall that the empty set $\varnothing$ is a subset of F, and it is the only subset of F with 0 elements. With the above conventions, formula (1.14) also holds for $p = 0$.) Therefore, anticipating the binomial formula (1.15), this number is

$$2^n = \sum_{p=0}^{n} \binom{n}{p}. \qquad (\star)$$

However one can prove directly that the number of subsets of F is 2^n.

Proof. Let $x_1, x_2, \dots, x_n$ be an enumeration of the elements of F. To any subset of F there corresponds a sequence of 0's and 1's of length n, where there is a 1 in the ith position if and only if x_i is included in the subset. Conversely, to any sequence of 0's and 1's of length n, there corresponds a subset of F consisting of all x_i's for which the ith digit of the sequence is 1. Therefore, the number of subsets of F is equal to the number of sequences of length n of 0's and 1's, which is 2^n.

(This method of proof, consisting in establishing a bijection with a set which is easy to count, is fundamental in combinatorics.) □

Formula $(\star)$ is a particular case of the *binomial formula*

$$(x+y)^n = \sum_{p=0}^{n} \binom{n}{p} x^p y^{n-p}. \qquad (1.15)$$

Letting $x = y = 1$ indeed gives $(\star)$.

Proof. (of the binomial formula) Let x_i, y_i $(1 \le i \le n)$ be real numbers. The product $\prod_{i=1}^{n}(x_i + y_i)$ is the sum of all possible products $x_{i_1} x_{i_2} \cdots x_{i_p} y_{j_1} \cdots y_{j_{n-p}}$ where $\{i_1, \dots, i_p\}$ is a subset of $\{1, \dots, n\}$ and $\{j_1, \dots, j_{n-p}\}$ is the complement of $\{i_1, \dots, i_p\}$ in $\{1, \dots, n\}$. Therefore,

$$\prod_{i=1}^{n}(x_i + y_i) = \sum_{p=0}^{n} \sum_{\substack{\{i_1, \cdots i_p\} \\ \{i_1, \dots, i_p\} \subseteq \{1, \cdots, n\}}} x_{i_1} \dots x_{i_p} y_{j_1} \dots y_{j_{n-p}}.$$

The second sum in the right-hand side of this equality contains $\binom{n}{p}$ elements, since there are $\binom{n}{p}$ different subsets $\{i_1, \dots, i_p\}$ of p elements of $\{1, \dots, n\}$. Now, letting $x_i = x, y_i = y$ $(1 \le i \le n)$, we obtain the binomial formula. □

From (1.14) it follows immediately that

$$\binom{n}{p} = \binom{n}{n-p}. \tag{1.16}$$

EXAMPLE 1.4.1: AN URN PROBLEM. ¿From an urn containing N_1 black balls and N_2 red balls, you draw at random, successively and without replacement, n $(n \le N_1 + N_2)$ balls. The probability of having drawn k black balls $(0 \le k \le \inf(N_1, n))$ is:

$$p_k = \frac{\binom{N_1}{k}\binom{N_2}{n-k}}{\binom{N_1+N_2}{n}}. \tag{1.17}$$

Proof. The set of outcomes Ω is the family of all subsets ω of n balls among the $N_1 + N_2$ balls in the urn. Therefore,

$$|\Omega| = \binom{N_1 + N_2}{n}.$$

It is reasonable to suppose that all the outcomes are equiprobable (the urn should be properly shaken before catching the balls with a net). Therefore, formula (1.13) applies and one must count the subsets ω with k black balls and $n-k$ red balls. To form such a set, you first form a set of k black balls among the N_1 black balls, and there are $\binom{N_1}{k}$ possibilities. To each such subset of k black balls, you must associate a subset of $n-k$ red balls. This multiplies the possibilities by $\binom{N_2}{n-k}$. Thus, if A is the number of subsets of n balls among the $N_1 + N_2$ balls in the urn which consist of k black balls and $n-k$ red balls, then

$$|A| = \binom{N_1}{k}\binom{N_2}{n-k}.$$

□

For future reference, we quote here the *negative binomial formula*:

$$(1-z)^{-p} = 1 + \binom{p}{p-1} z + \binom{p+1}{p-1} z^2 + \binom{p+2}{p-1} z^3 + \cdots, \tag{1.18}$$

where $z \in \mathbb{C}$, $|z| \le 1$. (Hint for the proof: For $p \ge 2$, $(1-z)^{-p}$ is the $(p-1)$-th derivative of $(1-z)^{-1}$.)

Poincaré's Formula

Elementary computations give

$$P(A_1 \cup A_2) = P(A_1) + P(A_2) - P(A_1 \cap A_2)$$

and

$$P(A_1 \cup A_2 \cup A_3) = P(A_1) + P(A_2) + P(A_3) \\ - P(A_1 \cap A_2) - P(A_1 \cap A_3) - P(A_2 \cap A_3) \\ + P(A_1 \cap A_2 \cap A_3) .$$

More generally (Exercise 1.5.9):

Theorem 1.4.2 *Let P be a probability on some measurable space $(\Omega, \mathcal{F})$ and let $A_1, \dots, A_r$ be arbitrary events. Then*

$$P(\cup_{i=1}^{r} A_i) = \sum_i P(A_i) - \sum_{i<j} P(A_i \cap A_j) \\ + \sum_{i<j<k} P(A_i \cap A_j \cap A_k) - \cdots + (-1)^{r+1} P(A_1 \cap A_2 \cap \cdots \cap A_r) . \tag{1.19}$$

EXAMPLE 1.4.3: EULER'S FORMULA. Let $\varphi(n)$ denote the number of integers k $(2 \le k \le n)$ that are prime with the integer $n \ge 2$ (the function φ is called *Euler's function*). Euler proved that

$$\frac{\varphi(n)}{n} = \prod_{p|n} \left(1 - \frac{1}{p}\right) ,$$

where the product is over all the prime numbers p that divide n. The proof below uses Poincaré's formula.

The integer $n \ge 2$ has a (unique) representation as

$$n = p_1^{\alpha_1} \cdots p_r^{\alpha_r}$$

where $p_1, \dots, p_r$ are distinct prime numbers > 1. Let $\Omega := \{1, 2, \dots, n\}$. Take for probability the uniform probability on Ω, that is

$$P(A) := \frac{|A|}{n} ,$$

where $|A|$ denotes the cardinality of A. Poincaré's formula then reads:

$$| \cup_{i=1}^{n} A_i| = \sum_i |A_i| - \sum_{i<j} |A_i \cap A_j| \\ + \sum_{i<j<k} |A_i \cap A_j \cap A_k| - \cdots + (-1)^{r+1} |A_1 \cap A_2 \cap \cdots \cap A_r| .$$

We shall apply it to the sets

$$A_k := \{\text{integers divisible by } p_k\} \quad (k = 1, \ldots, r).$$

In particular $A_1 \cup A_2 \cup \cdots \cup A_r$ is the set of integers divisible by at least one of the integers $p_1, \ldots, p_r$ and $\overline{A_1 \cup A_2 \cdots \cup A_r}$ is the set of integers that are not divisible by any of the $p_1, \ldots, p_r$, that is the set of integers that are prime with n. Therefore,

$$|A_1 \cap A_2 \cap \cdots \cap A_r| = n - \varphi(n).$$

Applying Poincaré's formula, and noting that

$$|A_i| = \frac{n}{p_i}, |A_i \cap A_j| = \frac{n}{p_i p_j} \, (i < j), \ldots, |A_1 \cap A_2 \cap \cdots \cap A_r| = \frac{n}{p_1 \cdots p_r},$$

we obtain that

$$n - \varphi(n) = \sum_i \frac{n}{p_i} - \sum_{i<j} \frac{n}{p_i p_j} + \cdots + (-1)^{r-1} \frac{n}{p_1 \cdots p_r}.$$

Therefore

$$\varphi(n) = n - \sum_i \frac{n}{p_i} + \sum_{i<j} \frac{n}{p_i p_j} - \cdots - (-1)^{r-1} \frac{n}{p_1 \cdots p_r},$$

which is Euler's formula.

The next example formalizes the ebriate postman problem, in which letters are randomly distributed in the mailboxes.

EXAMPLE 1.4.4: COINCIDENCES. An urn contains n balls, each one with a different number from 1 to n. One draws the balls, one by one, in succession and without replacement. Each time a ball is drawn, one notes its number. The randomness of the procedure is formalized by a random permutation σ of the set $\{1, 2, \ldots, n\}$, all the permutations being equiprobable, that is, σ_0 being a given permutation,

$$P(\sigma = \sigma_0) = \frac{1}{n!}.$$

One says that there is a coincidence at the i-th sample if $\sigma(i) = i$, and we denote by E_i the corresponding event. We shall compute the probability that there is at least one coincidence occurring in the sequence of successive drawings. This event is

$$A := E_1 \cup E_2 \cup \cdots \cup E_n.$$

We have that,

$$P(E_{i_1} \cap E_{i_2} \cap \cdots \cap E_{i_k}) = \binom{n}{k} \frac{(n-k)!}{n!} \quad (i_1 < i_2 < \cdots < i_k),$$

so that, by Poincaré's formula

$$P(A) = \sum_{k=1}^{n} (-1)^{k-1} \binom{n}{k} \frac{(n-k)!}{n!} = \sum_{k=1}^{n} \frac{(-1)^{k-1}}{k!}.$$

This quantity tends to $1 - \mathrm{e}^{-1} \sim 0.63212$ as $n \uparrow \infty$.

1.5 Exercises

Exercise 1.5.1. De Morgan's rules

(a) Let $\{A_n\}_{n\geq 1}$ be an arbitrary sequence of subsets of Ω. Prove *De Morgan's identities*:

$$\overline{\left(\bigcap_{n=1}^{\infty} A_n\right)} = \bigcup_{n=1}^{\infty} \overline{A}_n \text{ and } \overline{\left(\bigcup_{n=1}^{\infty} A_n\right)} = \bigcap_{n=1}^{\infty} \overline{A}_n.$$

(b) Prove that if $\mathcal{F}$ is a σ-field on Ω, and if $A_1, A_2, \dots$ belong to $\mathcal{F}$, then so does their intersection $\cap_{k=1}^{\infty} A_k$.

Exercise 1.5.2. Finitely often, infinitely often

Let $\{A_n\}_{n\geq 1}$ be an arbitrary sequence of subsets of Ω.

(a) Show that $\omega \in B := \bigcup_{n=1}^{\infty} \bigcap_{k=n}^{\infty} \overline{A_k}$ if and only if there exists at most a *finite* number (depending on ω) of indices k such that $\omega \in A_k$.

(b) Show that $\omega \in D := \bigcap_{n=1}^{\infty} \bigcup_{k=n}^{\infty} A_k$ if and only if there exist an *infinite* number (depending on ω) of indices k such that $\omega \in A_k$.

Exercise 1.5.3. Indicator functions

Prove the following identities for all subsets A, B of a given set Ω, and all sequences $\{A_n\}_{n\geq 1}$ forming a partition of Ω:

$$1_{A\cap B} = 1_A \times 1_B, \qquad 1_{\overline{A}} = 1 - 1_A, \qquad 1 = \sum_{n\geq 1} 1_{A_n}.$$

Exercise 1.5.4. Union of σ-fields

Let $\mathcal{F}_1$ and $\mathcal{F}_2$ be two σ-fields on the set Ω. Give a counterexample contradicting the assertion that $\mathcal{F}_1 \cup \mathcal{F}_2$ is a σ-field.

Exercise 1.5.5. WHY JUST σ-ADDITIVE?
Consider the probability model of Example 1.2.4 (random point on the square). Prove that there exists no totally additive probability P on the Borel σ-field on the square $[0,1]^2$ that assigns to rectangles therein their surface. (By "totally additive", it is meant that the probability of the union of an arbitrary —not necessarily countable— collection of mutually disjoint sets in the Borel σ-field is the sum of the individual probabilities.)

Exercise 1.5.6. IDENTITIES
Let $(\Omega, \mathcal{F}, P)$ be a probability space and let A and B be events ($\in \mathcal{F}$). Prove the identities

$$P(A \cup B) = 1 - P(\overline{A} \cap \overline{B}), \qquad P(A \cup B) = P(A) + P(B) - P(A \cap B).$$

Exercise 1.5.7. SUB-σ-ADDITIVITY
Let $(\Omega, \mathcal{F}, P)$ be a probability space. Prove the *sub-σ-additivity* property: for any sequence $\{A_n\}_{n\geq 1}$ of events,

$$P\left(\bigcup_{n=1}^{\infty} A_n\right) \leq \sum_{n=1}^{\infty} P(A_n).$$

Exercise 1.5.8. SEQUENTIAL CONTINUITY, THE DECREASING CASE
Prove Corollary 1.2.9.

Exercise 1.5.9. POINCARÉ'S FORMULA
Let P be a probability on some measurable space $(\Omega, \mathcal{F})$ and let $A_1, \dots, A_r$ be arbitrary events. Prove that

$$\begin{aligned} P(\cup_{i=1}^r A_i) &= \sum_i P(A_i) - \sum_{i<j} P(A_i \cap A_j) \\ &\quad + \sum_{i<j<k} P(A_i \cap A_j \cap A_k) - \cdots + (-1)^{r+1} P(A_1 \cap A_2 \cap \cdots \cap A_r). \end{aligned} \tag{1.20}$$

Exercise 1.5.10. ROLL IT!
You roll fairly and simultaneously three unbiased dice. What is the probability that one die shows 4, another 2, and another 1?

Exercise 1.5.11. ONE IS THE SUM OF THE TWO OTHERS
You perform three independent tosses of an unbiased die. What is the probability that one of these tosses results in a number that is the sum of the two other numbers?

Exercise 1.5.12. URNS

1. An urn contains 17 red balls and 19 white balls. Balls are drawn in succession at random and without replacement. What is the probability that the first 2 balls are red?

2. An urn contains N balls numbered from 1 to N. Someone draws n balls $(1 \leq n \leq N)$ simultaneously from the urn. What is the probability that the lowest number drawn is k?

Exercise 1.5.13. HEADS OR TAILS AS USUAL
A person, A, tossing an *unbiased* coin N times obtains T_A tails. Another person, B, tossing her own unbiased coin $N+1$ times has T_B tails. What is the probability that $T_A \geq T_B$?

Exercise 1.5.14. EXTENSION OF THE BASIC FORMULA OF INDEPENDENCE
Let $\{C_n\}_{n\geq 1}$ be a sequence of *independent* events. Then

$$P(\cap_{n=1}^{\infty} C_n) = \Pi_{n=1}^{\infty} P(C_n) .$$

This extends formula (1.7) to a countable number of sets.

Exercise 1.5.15. THE SWITCHES
Two nodes A and B in a communications network are connected by three different routes and each route contains a number of links that may fail. These are represented symbolically in Fig. 1.2 by switches that are in the lifted position if the link is in a failure state. In this figure, the number associated with a switch is the probability that the corresponding link is out of order. The links fail independently. What is the probability that A and B are connected?

Exercise 1.5.16. PAIRWISE INDEPENDENCE DOES NOT SUFFICE
Give a simple example of a probability space $(\Omega, \mathcal{F}, P)$ with three events A_1, A_2, A_3 that are pairwise independent, but *not* globally independent (that is, the family $\{A_1, A_2, A_3\}$ is not independent).

Exercise 1.5.17. INDEPENDENT FAMILY OF EVENTS

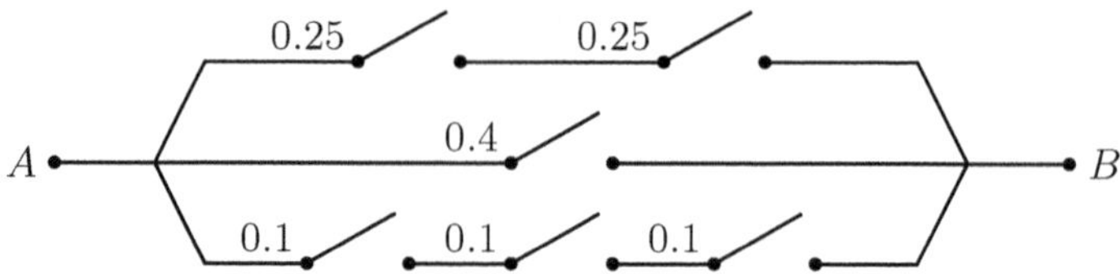

Figure 1.2: All switches up.

If $\{A_i\}_{i\in I}$ is an independent family of events, is it true that $\{\tilde{A}_i\}_{i\in I}$ is also an independent family of events, where for each $i \in I$, $\tilde{A}_i = A_i$ or $\overline{A}_i$ (your choice, for instance, with $I = \mathbb{N}$, $\tilde{A}_0 = A_0, \tilde{A}_1 = \overline{A}_1, \tilde{A}_3 = A_3, \dots$)?

Exercise 1.5.18. Conditional independence and the Markov property

1. Let $(\Omega, \mathcal{F}, P)$ be a probability space. For a fixed event C of positive probability define $P_C(A) := P(A \mid C)$. Show that P_C is a probability on $(\Omega, \mathcal{F})$. (And note that A and B are independent with respect to this probability if and only if they are conditionally independent given C.)

2. Let A_1, A_2, A_3 be three events of positive probability. Show that events A_1 and A_3 are conditionally independent given A_2 if and only if the "Markov property" holds, that is, $P(A_3 \mid A_1 \cap A_2) = P(A_3 \mid A_2)$.

Exercise 1.5.19. Roll it once more!

You roll fairly and simultaneously three unbiased dice. What is the probability that *some* die shows 1, given that the sum of the 3 values equals 5?

Exercise 1.5.20. Social Apartheid University

In the renowned Social Apartheid University, students have been separated into three social groups for "pedagogical" purposes. In group A, one finds students who individually have a probability of passing equal to 0.95. In group B this probability is 0.75, and in group C only 0.65. The three groups are of equal size. What is the probability that a student passing the course comes from group A? B? C?

Exercise 1.5.21. Wise bet

There are three cards. The first one has both faces red, the second one has both faces white, and the third one is white on one face, red on the other. A card is drawn at random, and the color of a randomly selected face of this card is shown to you (the other remains hidden). What is the winning strategy if you must bet on the color of the hidden face?

Exercise 1.5.22. A SEQUENCE OF LIARS
Consider a sequence of n "liars" $L_1, \ldots, L_n$. The first liar L_1 receives information about the occurrence of some event in the form "yes or no" and transmits it to L_2, who transmits it to L_3, etc... Each liar transmits what he hears with probability $p \in (0,1)$, and the contrary with probability $q = 1 - p$. The decision of lying or not is made by each liar independently of the rest of his colleagues. What is the probability x_n of obtaining the correct information from L_n? What is the limit of x_n as n increases to infinity?

Exercise 1.5.23. THE CAMPUS LIBRARY COMPLAINT
You are looking for a book in the campus libraries. Each library has it with probability 0.60 but in each library the book may have been stolen with probability 0.25. If there are three libraries, what are your chances of obtaining the book?

Exercise 1.5.24. SAFARI BUTCHERS
Three tourists participate in a safari in Africa. They encounter an elephant, who is unaware of the rules of the game. The innocent beast is killed, having received two out of the three bullets simultaneously shot by the tourists. The hit probabilities of the tourists are: Tourist A: $\frac{1}{4}$, Tourist B: $\frac{1}{2}$, Tourist C: $\frac{3}{4}$. Give for each tourist the probability that he was the one who missed.

Exercise 1.5.25. THE HARDY–WEINBERG LAW
In Example 1.3.6, show that the genotypic distributions of all generations, starting from the third one, are the same and that the stationary distribution depends only on the proportion c of alleles of type A in the initial population.

Exercise 1.5.26. SLUMBERIDGE UNIVERSITY ALUMNI
A student from the famous Veryhardvard University has with probability 0.25 a bright intelligence. Students from the Slumberidge University have a probability 0.10 of being bright. You find yourself in an assembly with 10 Veryhardvard students and 20 Slumberidge University students. You meet a handsome girl (*resp.* boy) whose intelligence is obviously superior. What is the probability that she (*resp.* he) registered at Slumberidge University?

Exercise 1.5.27. OPERATIONS ON EVENTS
Let $\mathcal{F}$ be a σ-field on some set Ω. Show that if A_1 and A_2 are in $\mathcal{F}$, then so is their *symmetric difference* $A_1 \triangle A_2 := A_1 \cup A_2 - A_1 \cap A_2$.

Exercise 1.5.28. SMALL σ-FIELDS
Is there a σ-field on Ω with 6 elements (including of course Ω and $\varnothing$)?

Exercise 1.5.29. ATOMS
Let the non-empty subsets $A_1, \dots, A_k$ of a set Ω form a partition of the latter.

(a) How many elements are there in the σ-field $\mathcal{F}$ they generate on Ω? (The sets $A_1, \dots, A_k$ are called the *atoms* of $\mathcal{F}$.)

(b) Show that if a σ-field $\mathcal{F}$ on Ω contains a finite number of elements, it is generated by a finite number of sets that form a partition of Ω.

Exercise 1.5.30. LOST UMBRELLA
With probability $p \in (0,1)$ the umbrella that you have lost is, equiprobably, in one of the seven floors of a given building. You have explored without success six floors. What is the probability that you will find your umbrella on the seventh floor?

Exercise 1.5.31. ROLLING DICE
Two (fair) dice are rolled independently in succession. Show that the event "the sum obtained is 7" is independent of the number shown by the first die.

Exercise 1.5.32. THE FIVE COINS
There are five fair coins, two have an A written on both faces, one has a B on both faces, and two have an A on one face and a B on the other face.

(a) Someone picks a coin at random and tosses it. What is the probability that the lower face has an A on it?

(b) Keeping your eyes shut, you pick a coin at random, take this coin into another room and toss it. You open your eyes and see that the upper face shows an A. What is the probability that the lower face has an A on it?

Exercise 1.5.33. PROOF-READING
A book contains four errors. Each time it is proof-read, a so far uncorrected error is corrected with probability $\frac{1}{3}$. The corrections of the different errors are independent. So are the proof-readings. How many proof-readings are necessary for the probability that no error is left to be larger than 0.9? (Hint: take for the probability space the set of 4-tuples (a_1, a_2, a_3, a_4) of positive integers, where a_i is the number of proof-readings necessary to get rid of error i.)

Chapter 2

Discrete Random Variables

The number of heads in a sequence of 10,000 coin tosses, the number of days it takes until the next rain and the size of a genealogical tree are random numbers. All are functions of the outcome of a random experiment performed either by man or nature taking discrete values, that is, *values in a countable set.* In the above examples, the values are numbers, but they can be of a different nature, for instance graphs [1].

2.1 Probability Distribution and Expectation

Definition 2.1.1 *Let E be a countable set. A function $X : \Omega \to E$ such that for all $x \in E$*

$$\{\omega; X(\omega) = x\} \in \mathcal{F}$$

is called a discrete random variable *or* discrete random element.

Being in $\mathcal{F}$, the event $\{X = x\}$ can be assigned a probability.

Definition 2.1.2 *From the probabilistic point of view, a discrete random variable X is described by its* probability distribution function *(or* distribution, *for short) $\{\pi(x)\}_{x\in E}$, where*

$$\pi(x) := P(X = x) .$$

Since E is a countable set, it can always be identified with $\mathbb{N}$ or $\overline{\mathbb{N}} := \mathbb{N} \cup \{\infty\}$, and therefore we shall often assume that either $E = \mathbb{N}$ or $\overline{\mathbb{N}}$.

---

[1] See for instance [6].

P. Brémaud, *An Introduction to Applied Probability*, Texts in Applied Mathematics 77,
https://doi.org/10.1007/978-3-031-49306-5_2

Calling a random variable taking integer values a "random number" is an innocuous habit as long as one is aware that it is not the function X that is random, but the outcome ω. This in turn makes the number $X(\omega)$ random.

One is sometimes faced with the problem of proving that a random variable X, taking its values in $\overline{\mathbb{N}}$ (and therefore for which the value ∞ is *a priori* possible), is in fact almost surely finite. That is, we have to prove that $P(X = \infty) = 0$ or, equivalently, that $P(X < \infty) = 1$. Since

$$\{X < \infty\} = \sum_{n=0}^{\infty}\{X = n\}\,,$$

we have

$$P(X < \infty) = \sum_{n=0}^{\infty} P(X = n)\,.$$

This remark provides an opportunity to recall that in an expression such as $\sum_{n=0}^{\infty}$, the sum is over $\mathbb{N}$ and does not include ∞ as the notation wrongly suggests. A less ambiguous notation would be $\sum_{n\in\mathbb{N}}$. However, we shall stick to the classical notation and in the case where the summation is over all integers plus ∞, we shall *always* use the notation $\sum_{n\in\overline{\mathbb{N}}}$.

In the vein of the above simple rules, let us mention the following often used expression of $P(X < \infty)$ for an integer-valued random variable X:

$$P(X < \infty) = \lim_{n\uparrow\infty} P(X \le n)\,.$$

For the proof (because it requires one), observe that the events $A_n := \{X \ge n\}$ $(n \ge 0)$ form a non-decreasing sequence and that $\cup_n A_n = \{X < \infty\}$. The result then follows by sequential continuity of probability (Theorem 1.2.8).

EXAMPLE 2.1.3: TOSSING A DIE, TAKE 3. In this example, the sample space is $\Omega = \{1, 2, 3, 4, 5, 6\}$. Take for X the identity: $X(\omega) = \omega$. In that sense X is a random number obtained by tossing a die.

EXAMPLE 2.1.4: HEADS OR TAILS, TAKE 4. The sample space Ω is the collection of all sequences $\omega := \{x_n\}_{n\ge1}$, where $x_n = 1$ or 0. Define a random variable X_n by $X_n(\omega) := x_n$. It is the random number obtained at the n-th toss. It is indeed a random variable since for all $a_n \in \{0, 1\}$, $\{\omega\,;\, X_n(\omega) = a_n\} = \{\omega\,;\, x_n = a_n\} \in \mathcal{F}$, by definition of $\mathcal{F}$.

The following are elementary remarks.

Theorem 2.1.5 (a) *Let E and F be countable sets. Let X be a random variable with values in E, and let $f : E \to F$ be an arbitrary function. Then $Y := f(X)$ is a random variable.*

(b) *Let E_1 and E_2 be countable sets. Let X_1 and X_2 be random variable with values in E_1 and E_2 respectively. Then $Y := (X_1, X_2)$ is a random variable with values in $E := E_1 \times E_2$.*

Proof. (a) Let $y \in F$. The set $\{\omega;\ Y(\omega) = y\}$ is in $\mathcal{F}$ since it is a countable union of sets in $\mathcal{F}$, namely:

$$\{Y = y\} = \sum_{x \in E;\, f(x)=y} \{X = x\}\,.$$

(b) Let $x = (x_1, x_2) \in E$. The set $\{\omega;\ X(\omega) = x\}$ is in $\mathcal{F}$ since it is the intersection of sets in $\mathcal{F}$, namely:

$$\{X = x\} = \{X_1 = x_1\} \cap \{X_2 = x_2\}\,.$$

□

Independence and Conditional Independence

Definition 2.1.6 *Two discrete random elements X and Y taking their values in E and F respectively are called* independent *if*

$$P(X = i, Y = j) = P(X = i)P(Y = j) \qquad (i \in E, j \in F)\,. \tag{2.1}$$

The left-hand side of (2.1) is $P(\{X = i\} \cap \{Y = j\})$. This is a general feature of the notational system: commas replace intersection signs. For instance, $P(A, B)$ is the probability that both events A and B occur.

Definition 2.1.7 *The discrete random elements $X_1, \dots, X_k$ taking their values in $E_1, \dots, E_k$ are said to be independent if for all $i_1 \in E_1, \dots, i_k \in E_k$,*

$$P(X_1 = i_1, \dots, X_k = i_k) = P(X_1 = i_1) \cdots P(X_k = i_k)\,. \tag{2.2}$$

Definition 2.1.8 *A sequence $\{X_n\}_{n\geq 1}$ of discrete random elements indexed by the set of positive integers and taking their values in the sets $\{E_n\}_{n\geq 1}$ respectively is called an* independent sequence *if any* finite *collection of* distinct *random elements $X_{i_1}, \dots, X_{i_r}$ extracted from this sequence are independent.*

Definition 2.1.9 *The sequence of discrete random elements $\{X_n\}_{n\geq 1}$ is said to be an* independent and identically distributed sequence *(for short: an* IID *sequence) if*

(a) *the X_ns take their values in the same set E,*

(b) *the family $\{X_n\}_{n\geq 1}$ is independent, and*

(c) *the probability distribution function of X_n does not depend on n.*

EXAMPLE 2.1.10: HEADS OR TAILS, TAKE 5. We show that the sequence $\{X_n\}_{n\geq 1}$ is IID. (Therefore, we have a model for *independent* tosses of an *unbiased* coin.)

Proof. Event $\{X_k = a_k\}$ is the direct sum of the events $\{X_1 = a_1, \ldots, X_{k-1} = a_{k-1}, X_k = a_k\}$ for all possible values of $(a_1, \ldots, a_{k-1})$. Since there are 2^{k-1} such values and each one has probability 2^{-k}, we have $P(X_k = a_k) = 2^{k-1}2^{-k}$, that is,

$$P(X_k = 1) = P(X_k = 0) = \frac{1}{2}.$$

Therefore,

$$P(X_1 = a_1, \ldots, X_k = a_k) = P(X_1 = a_1) \cdots P(X_k = a_k)$$

for all $a_1, \ldots, a_k \in \{0, 1\}$, from which it follows by definition that $X_1, \ldots, X_k$ are independent random variables, and more generally that $\{X_n\}_{n\geq 1}$ is a family of independent random variables. □

EXAMPLE 2.1.11: HEADS OR TAILS, TAKE 6. The number of occurrences of heads in n tosses is $S_n = X_1 + \cdots + X_n$. This random variable is the fortune at time n of a gambler systematically betting on heads. We compute its probability distribution when the X_n's are IID, but $P(X_n = 1) = p \in (0, 1)$ (allowing for a bias of the coin). The sum S_n takes the integer values 0 to n. The event $\{S_n = k\}$ is "k among $X_1, \ldots, X_n$ are equal to 1". There are $\binom{n}{k}$ distinct ways of assigning k values of 1 and $n - k$ values of 0 to $X_1, \ldots, X_n$, and all have the same probability $p^k(1-p)^{n-k}$. Therefore

$$P(S_n = k) = \binom{n}{k} p^k (1-p)^{n-k}.$$

Definition 2.1.12 *Let $\{X_n\}_{n\geq 1}$ and $\{Y_n\}_{n\geq 1}$ be sequences of discrete random elements indexed by the positive integers and taking their values in the sets $\{E_n\}_{n\geq 1}$ and $\{F_n\}_{n\geq 1}$ respectively. They are said to be* independent sequences *if any* finite *collection of random elements $X_{i_1}, \dots, X_{i_r}$ and $Y_{j_1}, \dots, Y_{i_s}$ extracted from their respective sequences are independent, the discrete random elements $(X_{i_1}, \dots, X_{i_r})$ and $(Y_{j_1}, \dots, Y_{i_s})$ are independent.*

This means that

$$P\left(\left(\cap_{\ell=1}^{r}\{X_{i_\ell} = a_\ell\}\right) \cap \left(\cap_{m=1}^{s}\{Y_{j_m} = b_m\}\right)\right)$$
$$= P\left(\cap_{\ell=1}^{r}\{X_{i_\ell} = a_\ell\}\right) P\left(\cap_{m=1}^{s}\{Y_{j_m} = b_m\}\right) \tag{2.3}$$

for all $a_1 \in E_1, \dots, a_r \in E_r$ and all $b_1 \in F_1, \dots, b_s \in F_s$.

The notion of conditional independence for events (Definition 1.3.10) extends naturally to discrete random variables.

Definition 2.1.13 *Let X, Y, Z be random variables taking their values in the denumerable sets E, F, G, respectively. One says that X and Y are* conditionally independent *given Z if for all x, y, z in E, F, G, respectively, events $\{X = x\}$ and $\{Y = y\}$ are conditionally independent given $\{Z = z\}$.*

Theorem 2.1.14 *Let X, Y, and Z be three discrete random variables with values in E, F, and G, respectively. If for some function $g : E \times F \to [0,1]$, $P(X = x \mid Y = y, Z = z) = g(x,y)$ for all x, y, z, then $P(X = x \mid Y = y) = g(x,y)$ for all x, y, and X and Z are conditionally independent given Y.*

Proof. We have

$$\begin{aligned} P(X = x, Y = y) &= \sum_z P(X = x, Y = y, Z = z) \\ &= \sum_z P(X = x \mid Y = y, Z = z) P(Y = y, Z = z) \\ &= g(x,y) \sum_z P(Y = y, Z = z) = g(x,y) P(Y = y)\,. \end{aligned}$$

Therefore,

$$P(X = x \mid Y = y) = g(x,y) = P(X = x \mid Y = y, Z = z)\,.$$

□

Expectation

Definition 2.1.15 *Let X be a discrete random variable taking its values in the countable set E and let $g : E \to \mathbb{R}$ be a function that is either non-negative or such that*

$$\sum_{x \in E} |g(x)| P(X = x) < \infty . \tag{2.4}$$

One then defines $E[g(X)]$, the expectation *of $g(X)$, by the formula*

$$E[g(X)] := \sum_{x \in E} g(x) P(X = x) . \tag{2.5}$$

If the summability condition (2.4) is satisfied, the random variable $g(X)$ is called *integrable*, and in this case the expectation $E[g(X)]$ is a *finite* number. If g is only assumed non-negative, the expectation may be infinite.

EXAMPLE 2.1.16: HEADS OR TAILS, TAKE 7. Consider the random variable $S_n = X_1 + \cdots + X_n$. We compute its expectation.

$$\begin{aligned} E[S_n] &= \sum_{i=1}^{n} i P(S_n = i) = \sum_{i=1}^{n} i \binom{n}{i} p^i (1-p)^{n-i} \\ &= \sum_{i=1}^{n} i \frac{n!}{i!(n-i)!} p^i (1-p)^{n-i} \\ &= \sum_{i=1}^{n} np \frac{(n-1)!}{(i-1)!((n-1)-(i-1))!} p^{i-1} (1-p)^{(n-1)-(i-1)} . \end{aligned}$$

Performing the change of variables $j = i - 1$, we obtain

$$E[S_n] = np \sum_{j=0}^{n-1} \frac{(n-1)!}{j!((n-1)-(j))!} p^j (1-p)^{(n-1)-j} = np .$$

It is important to realize that a discrete random variable taking *finite values* may have an *infinite expectation*:

EXAMPLE 2.1.17: FINITE RANDOM VARIABLES WITH INFINITE EXPECTATIONS. Let X, taking values in $E = \overline{\mathbb{N}}$, have the distribution $P(X = n) = \frac{1}{cn^2}$ ($n \in \overline{\mathbb{N}}$), where the constant c is chosen such that

$$P(X < \infty) = \sum_{n=1}^{\infty} P(X = n) = \sum_{n=1}^{\infty} \frac{1}{cn^2} = 1$$

(that is $c = \sum_{n=1}^{\infty} \frac{1}{n^2} = \frac{\pi^2}{6}$). In fact, the expectation of X is

$$E[X] = \sum_{n=1}^{\infty} nP(X=n) = \sum_{n=1}^{\infty} n\frac{1}{cn^2} = \sum_{n=1}^{\infty} \frac{1}{cn} = \infty .$$

The above example is artificial, but there are more natural occurrences of the phenomenon. Consider for instance Example 2.1.11 ("heads or tails" with an unbiased coin). The quantity $2S_n - n$ is the fortune at time n of a gambler systematically betting one *bitcoin* on heads (and therefore losing one *bitcoin* on tails). Let T be the first integer $n > 0$ (necessarily even) such that $2S_n - n = 0$. Then as it turns out as we shall prove later, in Example 9.1.29, that T is a *finite* random variable with *infinite expectation.*

Theorem 2.1.18 *Let A be some event. The expectation of the indicator random variable $X = 1_A$ is*

$$E[1_A] = P(A) . \tag{2.6}$$

Proof. $X = 1_A$ takes the value 1 with probability $P(X=1) = P(A)$ and the value 0 with probability $P(X=0) = P(\overline{A}) = 1 - P(A)$. Therefore,

$$E[X] = 0 \times P(X=0) + 1 \times P(X=1) = P(X=1) = P(A) .$$

□

Theorem 2.1.19 *Let $g_1 : E \to \overline{\mathbb{R}}$ and $g_2 : E \to \overline{\mathbb{R}}$ be functions such that $g_1(X)$ and $g_2(X)$ are integrable (resp., non-negative), and let $\lambda_1, \lambda_2 \in \mathbb{R}$ (resp., $\in \mathbb{R}_+$). Expectation is linear*

$$E[\lambda_1 g_1(X) + \lambda_2 g_2(X)] = \lambda_1 E[g_1(X)] + \lambda_2 E[g_2(X)] . \tag{2.7}$$

Also, expectation is monotone, in the sense that $g_1(x) \le g_2(x)$ for all x implies

$$E[g_1(X)] \le E[g_2(X)] . \tag{2.8}$$

Also, we have the triangle inequality

$$|E[g(X)]| \le E[|g(X)|] . \tag{2.9}$$

Proof. These properties follow from the corresponding properties of series. □

The next example gives an alternative way of computing the expectation of an integer-valued random variable.

Theorem 2.1.20 *For an integer-valued (that is, taking its values in $\mathbb{N}$) random variable X, we have the* telescope formula*:*

$$E[X] = \sum_{n=1}^{\infty} P(X \geq n) .$$

Proof.

$$\begin{aligned} E[X] &= P(X=1) + 2P(X=2) + 3P(X=3) + \cdots \\ &= P(X=1) + P(X=2) + P(X=3) + \cdots \\ &\qquad\qquad\quad + P(X=2) + P(X=3) + \cdots \\ &\qquad\qquad\qquad\qquad\qquad + P(X=3) + \cdots \end{aligned}$$

□

Theorem 5.2.4 will generalize this formula.

Definition 2.1.21 *Let X be an integer-valued random variable such that $E[|X|] < \infty$. Then X is said to be* integrable. *In this case (only in this case), one defines the mean of X as the (finite) number*

$$\mu = E[X] = \sum_{n=0}^{+\infty} nP(X = n) .$$

From the inequality $|a| \leq 1 + a^2$, true for all $a \in \overline{\mathbb{R}}$, we have that $|X| \leq 1 + X^2$, and therefore, by the monotonicity and linearity properties, $E[|X|] \leq 1 + E[X^2]$ (we also used the fact that $E[1] = 1$). Therefore if $E[X^2] < \infty$ (in which case we say that X is *square-integrable*) then X is integrable. The following definition then makes sense.

Definition 2.1.22 *Let X be a square-integrable random variable. We then define the* variance *σ^2 of X by*

$$\sigma^2 = E[(X - \mu)^2] = \sum_{n=0}^{+\infty} (n - \mu)^2 P(X = n) .$$

The variance is also denoted by Var (X). From the linearity of expectation, it follows that $E[(X - m)^2] = E[X^2] - 2mE[X] + m^2$, that is,

$$\text{Var}\,(X) = E[X^2] - m^2 . \tag{2.10}$$

Markov's Inequality

Theorem 2.1.23 *Let Z be a non-negative discrete random variable and let a be a positive number. Then* (Markov's inequality)*:*

$$P(Z \geq a) \leq \frac{E[Z]}{a}.$$

Proof. Take expectations in the inequality $Z \geq a1_{\{Z\geq a\}}$. □

Taking $Z = (X-m)^2$ in Markov's inequality and $a = \varepsilon^2$, we obtain *Chebyshev's inequality*:

$$P(|X-m| \geq \epsilon) \leq \frac{\text{Var}\,(X)}{\varepsilon^2}.$$

The Markov inequality, its corollary the Chebyshev inequality and the upcoming Jensen's inequality will apply in very general situations, as we shall see in the sequel.

EXAMPLE 2.1.24: BERNSTEIN'S POLYNOMIAL APPROXIMATION A continuous function f from $[0,1]$ into $\mathbb{R}$ can be approximated by a polynomial. More precisely, for all $x \in [0,1]$,

$$f(x) = \lim_{n\uparrow\infty} P_n(x)\,, \tag{$\star$}$$

where

$$P_n(x) := \sum_{k=0}^{n} f\left(\frac{k}{n}\right) \frac{n!}{k!(n-k)]} x^k (1-x)^{n-k}\,,$$

and the convergence of the series in the right-hand side is *uniform* in $[0,1]$. This classical result of analysis will now be proved using probabilistic arguments.

$$E\left[f\left(\frac{S_n}{n}\right)\right] = \sum_{k=0}^{n} f\left(\frac{k}{n}\right) P(S_n = k) = \sum_{k=0}^{n} f\left(\frac{k}{n}\right) \frac{n!}{k!(n-k)!} x^k (1-x)^{n-k}.$$

The function f is continuous on the *bounded* interval $[0,1]$ and therefore *uniformly* continuous on this interval. Therefore to any $\varepsilon > 0$, one can associate a number $\delta(\varepsilon)$ such that if $|y-x| \leq \delta(\varepsilon)$, then $|f(x)-f(y)| \leq \varepsilon$. Being continuous on $[0,1]$, f is bounded on $[0,1]$ by some finite number, say M. Now

$$\begin{aligned}|P_n(x) - f(x)| &= \left|E\left[f\left(\frac{S_n}{n}\right) - f(x)\right]\right| \leq E\left[\left|f\left(\frac{S_n}{n}\right) - f(x)\right|\right] \\ &= E\left[\left|f\left(\frac{S_n}{n}\right) - f(x)\right| 1_A\right] + E\left[\left|f\left(\frac{S_n}{n}\right) - f(x)\right| 1_{\overline{A}}\right],\end{aligned}$$

where $A := \{\omega\,;\, |S_n(\omega)/n) - x| \le \delta(\varepsilon)\}$. Since $|f(S_n/n) - f(x)|1_{\overline{A}} \le 2M1_{\overline{A}}$, we have

$$E\left[\left|f\left(\frac{S_n}{n}\right) - f(x)\right| 1_{\overline{A}}\right] \le 2MP(\overline{A}) = 2MP\left(\left|\frac{S_n}{n} - x\right| \ge \delta(\varepsilon)\right).$$

Also, by definition of A and of $\delta(\varepsilon)$,

$$E\left[\left|f\left(\frac{S_n}{n}\right) - f(x)\right| 1_A\right] \le \varepsilon.$$

Therefore

$$|P_n(x) - f(x)| \le \varepsilon + 2MP\left(\left|\frac{S_n}{n} - x\right| \ge \delta(\varepsilon)\right).$$

But x is the mean of S_n/n, and the variance of S_n/n is $nx(1-x) \le n/4$. Therefore, by Chebyshev's inequality,

$$P\left(\left|\frac{S_n}{n} - x\right| \ge \delta(\varepsilon)\right) \le \frac{4}{n[\delta(\varepsilon)]^2}.$$

Finally

$$|f(x) - P_n(x)| \le \varepsilon + \frac{4}{n[\delta(\varepsilon)]^2},$$

and

$$\lim_{n\uparrow\infty} |f(x) - P_n(x)| \le \varepsilon.$$

Since $\varepsilon > 0$ is otherwise arbitrary, this suffices to prove the convergence in $(\star)$. The convergence is *uniform* since the right-hand side of the latter inequality does not depend on $x \in [0,1]$.

Jensen's Inequality

This inequality concerns the expectation of convex functions of a random variable. We therefore start by recalling the definition of a convex function. Let I be an interval of $\mathbb{R}$ (closed, open, semi-closed, infinite, etc.) with non-empty interior (a,b). The function $\varphi : I \to \mathbb{R}$ is called a convex function if for all $x, y \in I$ and all $0 < \theta < 1$,

$$\varphi(\theta x + (1-\theta)y) \le \theta\varphi(x) + (1-\theta)\varphi(y).$$

If the inequality is strict for all $x \ne y$ and all $0 < \theta < 1$, the function φ is said to be *strictly* convex.

Theorem 2.1.25 *Let I be as above and let $\varphi : I \to \mathbb{R}$ be a convex function. Let X be an integrable discrete real-valued random variable such that $P(X \in I) = 1$. Assume moreover that either φ is non-negative, or that $\varphi(X)$ is integrable. Then* (Jensen's inequality)

$$E[\varphi(X)] \geq \varphi(E[X]).$$

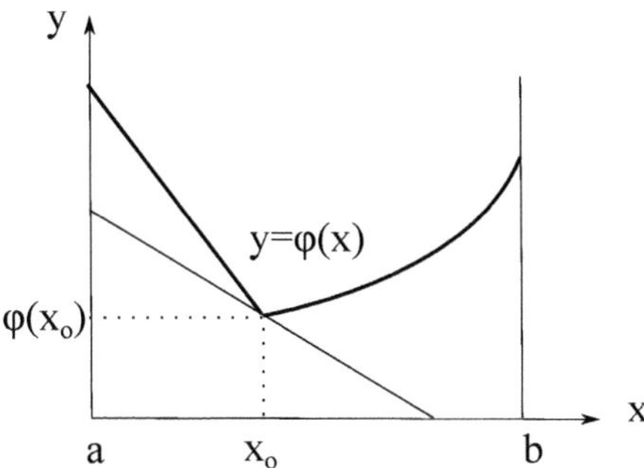

Proof. A convex function φ has the property that for any $x_0 \in (a, b)$, there exists a straight line $y = \alpha x + \beta$ passing through $(x_0, \varphi(x_0))$, that is,

$$\varphi(x_0) = \alpha x_0 + \beta, \qquad (\star)$$

and such that for all $x \in (a, b)$,

$$\varphi(x) \geq \alpha x + \beta, \qquad (\star\star)$$

where the inequality is strict if φ is strictly convex. (The parameters α and β may depend on x_0 and may not be unique.) Take $x_0 = E[X]$. In particular $\varphi(E[X]) = \alpha E[X] + \beta$. By $(\star\star)$, $\varphi(X) \geq \alpha X + \beta$, and taking expectations using $(\star)$,

$$E[\varphi(X)] \geq \alpha E[X] + \beta = \varphi(E[X]).$$

□

Moment Bounds

Theorem 2.1.26 (a) *For any integer-valued random variable X, we have the* first moment bound

$$P(X \neq 0) \leq E[X] \,.$$

(b) *For a square-integrable real-valued discrete random variable X, we have the* second moment bound

$$P(X = 0) \leq \frac{\text{Var}\,(X)}{E[X]^2} \,.$$

Proof. (a) For the first moment bound,

$$\begin{aligned} P(X \neq 0) &= P(X = 1) + P(X = 2) + P(X = 3) + \cdots \\ &\leq P(X = 1) + 2P(X = 2) + 3P(X = 3) + \cdots = E[X] \,. \end{aligned}$$

(b) Since the event $X = 0$ implies the event $|X - E[X]| \geq E[X]$,

$$P(X = 0) \leq P(|X - E[X]| \geq E[X]) \leq \frac{\text{Var}\,(X)}{E[X]^2} \,,$$

where the last inequality is Chebyshev's inequality. □

Product Rule for Expectation

The product formula for expectations featured in the next theorem will be met several times in this book and is in fact very general (see Theorem 5.4.4).

Theorem 2.1.27 *Let Y and Z be two independent discrete random elements with values in the countable sets F and G respectively, and let $v : F \to \overline{\mathbb{R}}$, $w : G \to \overline{\mathbb{R}}$ be functions which are either non-negative, or such that $v(Y)$ and $w(Z)$ are both integrable. Then*

$$E[v(Y)w(Z)] = E[v(Y)]E[w(Z)] \,.$$

Proof. Consider the discrete random element X with values in $E = F \times G$ defined by $X = (Y, Z)$, and let the function $g : E \to \overline{\mathbb{R}}$ be defined by $g(x) = v(y)w(z)$

$(x = (y, z))$. We have, under the prevailing conditions,

$$\begin{aligned}
E[v(Y)w(Z)] &= E[g(X)] = \sum_{x\in E} g(x)P(X = x) \\
&= \sum_{y\in F}\sum_{z\in F} v(y)w(z)P(Y = y, Z = z) \\
&= \sum_{y\in F}\sum_{z\in F} v(y)w(z)P(Y = y)P(Z = z) \\
&= \left(\sum_{y\in F} v(y)P(Y = y)\right)\left(\sum_{z\in F} w(z)P(Z = z)\right) \\
&= E[v(Y)]E[w(Z)]\,.
\end{aligned}$$

□

The following consequence of the product rule is extremely important. It says that for *independent* random variables, "variances add up".

Theorem 2.1.28 *Let $X_1, \dots, X_n$ be independent integrable discrete random variables with real values. Then*

$$\sigma^2_{X_1+\cdots+X_n} = \sigma^2_{X_1} + \cdots + \sigma^2_{X_n}\,. \tag{2.11}$$

Proof. Let $\mu_1, \dots, \mu_n$ be the respective means of $X_1, \dots, X_n$. The mean of the sum $X := X_1 + \cdots + X_n$ is $\mu := \mu_1 + \cdots + \mu_n$. If $i \neq k$, we have, by the product formula for expectations,

$$E\left[(X_i - \mu_i)(X_k - \mu_k)\right] = E\left[(X_i - \mu_i)\right]E\left[(X_k - \mu_k)\right] = 0\,.$$

Therefore

$$\begin{aligned}
\text{Var}\,(X) =& E\left[(X - \mu)^2\right] = E\left[\left(\sum_{i=1}^n (X_i - \mu_i)\right)^2\right] \\
=& E\left[\sum_{i=1}^n\sum_{k=1}^n (X_i - \mu_i)(X_k - \mu_k)\right] \\
=& \sum_{i=1}^n\sum_{k=1}^n E\left[(X_i - \mu_i)(X_k - \mu_k)\right] \\
=& \sum_{i=1}^n E\left[(X_i - \mu_i)^2\right] = \sum_{i=1}^n \text{Var}\,(X_i)\,.
\end{aligned}$$

□

Note that means always add up, even when the random variables are not independent.

Let X be an integrable discrete random variable. Then, clearly, for any $a \in \mathbb{R}$, aX is integrable and its variance is given by the formula

$$\text{Var}\,(aX) = a^2\,\text{Var}\,(X)\,.$$

EXAMPLE 2.1.29: VARIANCE OF THE EMPIRICAL MEAN. ¿From the above remark and Theorem 2.1.28, it follows that if $X_1, \ldots, X_n$ are independent and identically distributed *integrable* random variables with real values and common variance σ^2, then

$$\text{Var}\,\left(\frac{X_1 + \cdots + X_n}{n}\right) = \frac{\sigma^2}{n}\,.$$

EXAMPLE 2.1.30: THE WEAK LAW OF LARGE NUMBERS. Let $\{X_n\}_{n\geq 1}$ be an independent sequence of real-valued discrete random variables with the same probability distribution, mean (supposed well defined) m and variance $\sigma^2 < \infty$. Then, since the variance of the n-th order empirical mean $\overline{X}_n := \frac{X_1+\cdots+X_n}{n}$ is equal to $\frac{\sigma^2}{n}$, we have by Chebyshev's inequality, for all $\varepsilon > 0$,

$$P(|\overline{X}_n - m| \geq \varepsilon) = P\left(\left|\frac{\sum_{i=1}^n (X_i - m)}{n} \geq \varepsilon\right|\right) \leq \frac{\sigma^2}{n^2\varepsilon}\,.$$

Therefore the empirical mean $\overline{X}_n$ converges to the mean m in probability, which means exactly (by definition of the convergence in probability) that, for all $\varepsilon > 0$,

$$\lim_{n\uparrow\infty} P(|\overline{X}_n - m| \geq \varepsilon) = 0\,.$$

This result is called the weak law of large numbers in order to distinguish it from a much more powerful result, the strong law of large numbers in Chapter 6.

2.2 Remarkable Discrete Distributions

Uniform

Let $\mathcal{X}$ be a *finite* set whose cardinality is denoted by $|\mathcal{X}|$. The random variable with values in this set and having the distribution

$$P(X = x) = \frac{1}{|\mathcal{X}|} \quad (x \in \mathcal{X})$$

is said to be uniformly distributed (or to have the *uniform distribution*) on $\mathcal{X}$.

EXAMPLE 2.2.1: IS THIS NUMBER THE LARGER ONE? Let a and b be two numbers in $\{1, 2, \dots, 10,000\}$, with $a > b$. Only one of these numbers is shown to you, chosen at random and equiprobably. Call X this (now random) number. Is there a good strategy for guessing if the number shown to you is the largest of the two? Of course, we would like to have a probability of success strictly larger than $\frac{1}{2}$. Perhaps surprisingly, there is such a strategy, that we now describe. Select at random uniformly on $\{1, 2, \dots, 10,000\}$ a number Y. If $X \geq Y$, say that X is the largest ($= a$), otherwise say that it is the smallest.

Let us compute the probability P_E of a wrong guess. An error occurs when either (i) $X \geq Y$ and $X = b$, or (ii) $X < Y$ and $X = a$. These events are exclusive of one another, and therefore

$$\begin{aligned}
P_E &= P(X \geq Y, X = b) + P(X < Y, X = a) \\
&= P(b \geq Y, X = b) + P(a < Y, X = a) \\
&= P(b \geq Y)P(X = b) + P(a < Y)P(X = a) \\
&= P(b \geq Y)\frac{1}{2} + P(a < Y)\frac{1}{2} = \frac{1}{2}(P(b \geq Y) + P(a < Y)) \\
&= \frac{1}{2}(1 - P(Y \in [b+1, a]) = \frac{1}{2}\left(1 - \frac{a-b}{10,000}\right) < \frac{1}{2}.
\end{aligned}$$

Let $\{X_n\}_{n \geq 1}$ be an IID sequence of random variables taking their values in the set $\{0, 1\}$ and with a common distribution given by

$$P(X_n = 1) = p \qquad (p \in (0, 1)).$$

Define the *Hamming weight* $h(a)$ of the binary vector $a = (a_1, a_2, \dots, a_n) \in \{0, 1\}^n$ by $h(a) := \sum_{j=1}^{k} a_j$. Since $P(X_j = a_j) = p$ or $1 - p$ depending on whether $a_i = 1$ or 0, and since there are exactly $h(a)$ coordinates of a that are equal to 1,

$$P(X_1 = a_1, \dots, X_k = a_k) = p^{h(a)} q^{k-h(a)}, \tag{2.12}$$

where $q := 1 - p$.

Comparing with Examples 1.1.3 and 1.2.3, we see that we have modelized a game of heads or tails, with a biased coin if $p \neq \frac{1}{2}$. The sequence $\{X_n\}_{n\geq 1}$ is called a *Bernoulli sequence* of parameter p.

EXAMPLE 2.2.2: THE GAMBLER'S RUIN. Two players A and B play "heads or tails", where heads occur with probability $p \in (0,1)$ and the successive outcomes form an IID sequence. Calling X_n the fortune in dollars of player A at time n, then $X_{n+1} = X_n + Z_{n+1}$, where $Z_{n+1} = +1$ (resp., -1) with probability p (resp., $q = 1-p$), and $\{Z_n\}_{n\geq 1}$ is IID. In other words, A bets \$1 on heads at each toss, and B bets \$1 on tails. The respective initial fortunes of A and B are a and b (positive integers). The game ends when a player is ruined. The duration of the game is T, the first time n at which $X_n = 0$ or c, and the probability of winning for A is $u(a) = P(X_T = c \mid X_0 = a)$. We shall compute $u(a)$.

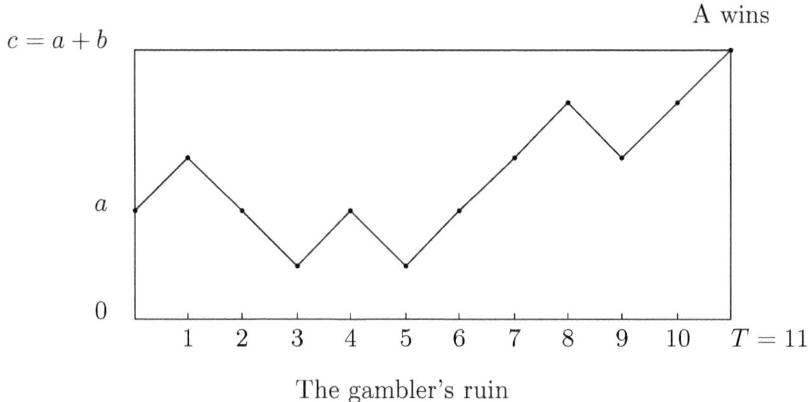

The gambler's ruin

Instead of computing $u(a)$ alone, we shall compute

$$u(i) := P(X_T = c \mid X_0 = i)$$

for all states i ($0 \leq i \leq c$). For this, we first obtain a recurrence equation for $u(i)$ by breaking down event "A wins" according to what can happen after the first step (the first toss) and using the Bayes rule of total causes. If $X_0 = i$ ($1 \leq i \leq c-1$), then $X_1 = i+1$ (resp., $X_1 = i-1$) with probability p (resp., q), and the probability of winning for A with updated initial fortune $i+1$ (resp., $i-1$) is $u(i+1)$ (resp., $u(i-1)$). Therefore, for i ($1 \leq i \leq c-1$),

$$u(i) = pu(i+1) + qu(i-1)\,,$$

with the boundary conditions $u(0) = 0$, $u(c) = 1$. The characteristic equation associated with this linear recurrence equation is $pr^2 - r + q = 0$. It has two distinct roots, $r_1 = 1$ and $r_2 = \frac{q}{p}$, if $p \neq q$, and a double root, $r_1 = 1$, if $p = q = \frac{1}{2}$. Therefore, the general solution is $u(i) = \lambda r_1^i + \mu r_2^i = \lambda + \mu \left(\frac{q}{p}\right)^i$ when $p \neq q$, and $u(i) = \lambda r_1^i + \mu i r_1^i = \lambda + \mu i$ when $p = q = \frac{1}{2}$. Taking into account the boundary conditions, one can determine the values of λ and μ. The result is, for $p \neq q$,

$$u(i) = \frac{1 - (\frac{q}{p})^i}{1 - (\frac{q}{p})^c},$$

and for $p = q = \frac{1}{2}$,

$$u(i) = \frac{i}{c}.$$

In the case $p = q = \frac{1}{2}$, the probability $v(i)$ that B wins when the initial fortune of B is $c-i$ is obtained by replacing i by $c-i$ in the expression for $u(i)$: $v(i) = \frac{c-i}{c} = 1 - \frac{i}{c}$. One checks that $u(i) + v(i) = 1$, which means in particular that the probability that the game lasts forever is null. The reader is invited to check that the same is true in the case $p \neq q$.

The framework of heads or tails shelters the three most common discrete random variables: the binomial, the geometric and the Poisson random variables.

Binomial

Definition 2.2.3 *A random variable X taking its values in the set $E = \{0, 1, \ldots, n\}$ and with the probability distribution*

$$P(X = i) = \binom{n}{i} p^i (1-p)^{n-i} \quad (0 \le i \le n)$$

is called a binomial random variable of size n and parameter $p \in (0, 1)$. *This is denoted by $X \sim \mathcal{B}(n, p)$.*

The mean and the variance of a binomial random variable X of size n and parameter p are given by

$$E[X] = np,$$

$$\text{Var}(X) = np(1-p),$$

Proof. In Example 2.1.11, it was proved that the number of occurrences of heads in n tosses, $S_n = X_1 + \cdots + X_n$, is a binomial random variable $\mathcal{B}(n, p)$. We have

$$E[S_n] = \sum_{i=1}^{n} E[X_i] = nE[X_1]$$

and since the X_i's are IID,

$$\mathrm{Var}\,(S_n) = \sum_{i=1}^{n} \mathrm{Var}\,(X_i) = n\mathrm{Var}\,(X_1)\,.$$

Now

$$E[X_1] = 0 \times P(X_1 = 0) + 1 \times P(X_1 = 1) = P(X_1 = 1) = p\,,$$

and since $X_1^2 = X_1$,

$$E\left[X_1^2\right] = E[X_1] = p\,.$$

Therefore

$$\mathrm{Var}\,(X_1) = E\left[X_1^2\right] - E[X_1]^2 = p - p^2 = p(1-p)$$

and

$$\mathrm{Var}\,(S_n) = np(1-p)\,.$$

□

Geometric

Definition 2.2.4 *A random variable X taking its values in $\mathbb{N}_+$ and with the distribution*

$$P(T = k) = (1-p)^{k-1}p\,,$$

where $0 < p < 1$, is called a geometric random variable with parameter p. *This is denoted by $X \sim \mathcal{G}eo(p)$.*

EXAMPLE 2.2.5: FIRST "HEADS" IN THE SEQUENCE. Define the random variable T to be the first time of occurrence of 1 in the Bernoulli sequence $\{X_n\}_{n\geq 1}$, that is,

$$T = \inf\{n \geq 1; X_n = 1\}\,,$$

with the convention that if $X_n = 0$ for all $n \geq 1$, then $T = \infty$. The event $\{T = k\}$ is exactly $\{X_1 = 0, \ldots, X_{k-1} = 0, X_k = 1\}$, and therefore,

$$P(T = k) = P(X_1 = 0) \cdots P(X_{k-1} = 0)P(X_k = 1)\,,$$

that is, for $k \geq 1$,

$$P(T = k) = (1 - p)^{k-1}p\,.$$

The mean of a geometric random variable X with parameter $p > 0$ is

$$E[X] = \frac{1}{p}\,. \tag{2.13}$$

Proof.

$$E[X] = \sum_{k=1}^{\infty} k\,(1-p)^{k-1}\,p\,.$$

But for $\alpha \in (0, 1)$,

$$\sum_{k=1}^{\infty} k\alpha^{k-1} = \frac{\partial}{\partial \alpha}\left(\sum_{k=1}^{\infty} \alpha^k\right) = \frac{\partial}{d\alpha}\left(\frac{1}{1-\alpha} - 1\right) = \frac{1}{(1-\alpha)^2}\,.$$

Therefore

$$E[X] = \frac{1}{(1-(1-p))^2} \times p = \frac{1}{p^2} \times p\,,$$

that is, finally, (2.13). □

EXAMPLE 2.2.6: THE COUPON COLLECTOR. In a certain brand of chocolate tablets one can find coupons, one for each tablet, randomly and independently chosen among n types. A prize may be claimed once the chocolate amateur has gathered a collection containing a subset with all the types of coupons. We shall compute the average value of the number X of chocolate tablets bought when this happens for the first time. For this, let X_i ($0 \leq i \leq n-1$) be the number of tablets bought while exactly i coupons of different types have been collected, so that

$$X = \sum_{i=0}^{n-1} X_i\,.$$

Each X_i is a geometric random variable with parameter $p_i = 1 - \frac{i}{n}$. In particular ((2.13)),

$$E[X_i] = \frac{1}{p_i} = \frac{n}{n-i}\,,$$

and therefore

$$E[X] = \sum_{i=0}^{n-1} E[X_i] = n \sum_{i=1}^{n} \frac{1}{i}.$$

The sum $H(n) := \sum_{i=1}^{n} \frac{1}{i}$ (called the n-th *harmonic number*) satisfies the inequalities

$$\ln n \le H(n) \le \ln n + 1$$

as can be seen by expressing $\ln n$ as the integral $\int_1^n \frac{1}{x}\,\mathrm{d}x$, cutting the domain of integration into segments of unit length, and using the fact that the integrand is a decreasing function, which gives the inequalities

$$\sum_{i=2}^{n} \frac{1}{i} \le \int_1^n \frac{\mathrm{d}x}{x} \le \sum_{i=1}^{n-1} \frac{1}{i},$$

that is

$$H(n) - 1 \le \ln n \le H(n-1).$$

This gives the inequalities

$$H(n) \le \ln n + 1$$

and

$$H(n) \ge \ln(n+1) = \ln n + \ln \frac{n+1}{n}.$$

Therefore $H(n) = \ln n + O(1)$ [2] and, finally,

$$E[X] = n \ln n + O(n).$$

In fact, observing that $|\sum_{i=1}^{n} 1/i - \ln n| \le 1$, we have that $|E[X] - n \ln n| \le n$.

Poisson

Definition 2.2.7 *A random variable X taking its values in $\mathbb{N}$ and such that*

$$P(X = k) = \mathrm{e}^{-\theta} \frac{\theta^k}{k!} \quad (k \ge 0)$$

is called a Poisson random variable with parameter $\theta > 0$. *This is denoted by $X \sim \mathcal{P}oi(\theta)$.*

[2] $f(n) = O(g(n))$ means that there exists a positive real number M and an integer n_0 such that $|f(n)| \le M|g(n)|$ for all $n \ge n_0$. This notation is part of the so-called Landau notational system.

EXAMPLE 2.2.8: THE POISSON LAW OF RARE EVENTS, TAKE 1. A veterinary surgeon in the Prussian cavalry once collected data concerning accidents due to horse kickbacks among soldiers. He found that the (random) number of accidents of this kind in a given year closely follows a Poisson distribution. The purpose of this example is to explain why.

Suppose that you play "heads or tails" for a large number n of (independent) tosses with a coin such that

$$P(X_i = 1) = \frac{\alpha}{n}.$$

In the Prussian cavalry example, n is the (large) number of soldiers and $X_i = 1$ if the i-th soldier has been hurt by a horse. Let S_n be the total number of *heads* (of wounded soldiers). We show that for all $k \geq 0$,

$$\lim_{n \uparrow \infty} p_n(k) = \mathrm{e}^{-\alpha} \frac{\alpha^k}{k!}, \qquad (\star)$$

with the convention $0! = 1$.

This explains the findings of the veterinary surgeon. The average number of casualties is α and the choice $P(X_i = 1) = \frac{\alpha}{n}$ guarantees this. Letting $n \uparrow \infty$ accounts for n being large but unknown.

Here is the proof of the mathematical statement. As we know, the random variable S_n follows a binomial law:

$$P(S_n = k) = \binom{n}{k} \left(\frac{\alpha}{n}\right)^k \left(1 - \frac{\alpha}{n}\right)^{n-k}$$

of mean $n \times \frac{\alpha}{n} = \alpha$. With $p_n(k) := P(S_n = k)$, we see that

$$p_n(0) = \left(1 - \frac{\alpha}{n}\right)^n \to \mathrm{e}^{-\alpha}$$

and

$$\frac{p_n(k+1)}{p_n(k)} = \frac{\frac{n-k}{k+1}\frac{\alpha}{n}}{1 - \frac{\alpha}{n}} \to \frac{\alpha}{k+1}.$$

Therefore, $(\star)$ holds for all $k \geq 0$, showing that the limit distribution is indeed a Poisson distribution of mean α.

Theorem 2.2.9 *For a Poisson random variable with parameter $\theta > 0$,*

$$E[X] = \theta \text{ and } \operatorname{Var}(X) = \theta.$$

Proof. The following is a direct computation. Later on we shall see a better approach (via generating functions).

$$E[X] = \mathrm{e}^{-\theta} \sum_{k=1}^{\infty} \frac{k\theta^k}{k!} = \mathrm{e}^{-\theta}\theta \sum_{k=1}^{\infty} \frac{\theta^{k-1}}{(k-1)!}$$
$$= \mathrm{e}^{-\theta}\theta \sum_{j=0}^{\infty} \frac{\theta^j}{j!} = \mathrm{e}^{-\theta}\theta \mathrm{e}^{\theta} = \theta\,,$$

$$E\left[X^2 - X\right] = \mathrm{e}^{-\theta} \sum_{k=0}^{\infty} \left(k^2 - k\right) \frac{\theta^k}{k!} = \mathrm{e}^{-\theta} \sum_{k=2}^{\infty} k\,(k-1)\,\frac{\theta^k}{k!}$$
$$= \mathrm{e}^{-\theta}\theta^2 \sum_{k=2}^{\infty} \frac{\theta^{k-2}}{(k-2)!} = \mathrm{e}^{-\theta}\theta^2 \sum_{j=0}^{\infty} \frac{\theta^j}{j!} = \mathrm{e}^{-\theta}\theta^2\mathrm{e}^{\theta} = \theta^2\,,$$

$$\mathrm{Var}\;(X) = E\left[X^2\right] - E\left[X\right]^2$$
$$= E\left[X^2 - X\right] + E\left[X\right] - E\left[X\right]^2 = \theta^2 + \theta - \theta^2 = \theta\,.$$

□

EXAMPLE 2.2.10: SUMS OF INDEPENDENT POISSON VARIABLES. Let X_1 and X_2 be two *independent* Poisson random variables with respective means $\theta_1 > 0$ and $\theta_2 > 0$. Then $X = X_1 + X_2$ is a Poisson random variable with mean $\theta = \theta_1 + \theta_2$.

Proof. For $k \geq 0$,

$$P(X = k) = P(X_1 + X_2 = k) = P\left(\sum_{i=0}^{k} \{X_1 = i, X_2 = k - i\}\right)$$
$$= \sum_{i=0}^{k} P(X_1 = i, X_2 = k - i) = \sum_{i=0}^{k} P(X_1 = i)P(X_2 = k - i)$$
$$= \sum_{i=0}^{k} \mathrm{e}^{-\theta_1}\frac{\theta_1^i}{i!}\mathrm{e}^{-\theta_2}\frac{\theta_2^{k-i}}{(k-i)!} = \frac{\mathrm{e}^{-(\theta_1+\theta_2)}}{k!}\sum_{i=0}^{k}\frac{k!}{i!(k-i)!}\theta_1^i\theta_2^{k-i}$$
$$= \mathrm{e}^{-(\theta_1+\theta_2)}\frac{(\theta_1+\theta_2)^k}{k!}\,.$$

□

Hypergeometric

In Example 1.4.1 we met the following distribution;

$$p_k = \frac{\binom{N_1}{k}\binom{N_2}{n-k}}{\binom{N_1+N_2}{n}} \quad (0 \le k \le \inf(N_1, n)).$$

It is called the *hypergeometric distribution.*

Multinomial

Consider a random vector $X = (X_1, \ldots, X_n)$ where all the random variables X_i take their values in the *same* denumerable space E (this restriction is not essential, but it simplifies the notation). Let $p : E^n \to \mathbb{R}_+$ be a function such that

$$\sum_{x \in E^n} p(x) = 1.$$

Definition 2.2.11 *The discrete random vector X above is said to admit the* probability distribution $p(x)$ $(x \in E^n)$ *if for all sets $C \subseteq E^n$,*

$$P(X \in C) = \sum_{x \in C} p(x).$$

In fact, there is nothing new here since X is a discrete random variable taking its values in the denumerable set $\mathcal{X} := E^n$.

Consider the situation where k balls are placed in n boxes $B_1, \ldots, B_n$, independently of one another, with the probability p_i for a given ball to be assigned to box B_i. Of course,

$$\textstyle\sum_{i=1}^n p_i = 1.$$

After placing all the balls in the boxes, there are X_i balls in box B_i, where

$$\textstyle\sum_{i=1}^n X_i = k.$$

Then

$$P(X_1 = m_1, \ldots, X_n = m_n) = \frac{k!}{\prod_{i=1}^n (m_i)!} \prod_{i=1}^n p_i^{m_i}. \tag{2.14}$$

Proof. Observe that (α): there are $k!/\prod_{i=1}^n (m_i)!$ distinct ways of placing k balls in n boxes in such a manner that m_1 balls are in box B_1, m_2 are in B_2, etc., and (β): each of these distinct ways occurs with the same probability $\prod_{i=1}^n p_i^{m_i}$. □

Definition 2.2.12 *If the random vector $X = (X_1, \dots, X_n)$ admits the probability distribution given by (2.14), it is called a* multinomial (random) vector of size (n, K) and parameters $p_1, \dots, p_n$.

2.3 Generating Functions

Computations relative to discrete probability models often require an enumeration of all the possible outcomes realizing a particular event. Generating functions are very useful for this task and, more generally, for obtaining distribution functions of integer-valued random variables.

In order to introduce this versatile tool, we must first define the expectation of a complex-valued function of an integer-valued variable. Let X be a discrete random variable with values in $\mathbb{N}$, and let $\varphi : \mathbb{N} \to \mathbb{C}$ be a complex function with real and imaginary parts φ_R and φ_I respectively. The expectation $E[\varphi(X)]$ is then defined by

$$E[\varphi(X)] := E[\varphi_R(X)] + iE[\varphi_I(X)]\,,$$

provided that the expectations on the right-hand side are well defined and finite.

Definition 2.3.1 *Let X be an $\mathbb{N}$-valued random variable. Its* generating function (GF) *is the function $g : \overline{D}(0;1) := \{z \in \mathbb{C}; |z| \leq 1\} \to \mathbb{C}$ defined by*

$$g(z) := E[z^X] = \sum_{k=0}^{\infty} P(X = k)z^k\,. \tag{2.15}$$

Since $\sum_{n=0}^{\infty} P(X = n) = 1 < \infty$, the power series associated with the sequence $\{P(X = n)\}_{n \geq 0}$ has a radius of convergence $R \geq 1$. The domain of definition of g could be, in specific cases, larger than the closed unit disk centered at the origin.

In the next two examples below, the domain of absolute convergence is the whole complex plane.

EXAMPLE 2.3.2: GF OF THE BINOMIAL VARIABLE. For the binomial random variable of size n and parameter p,

$$g(z) = \sum_{k=0}^{n} \binom{n}{k} (zp)^k (1-p)^{n-k} = (1 - p + pz)^n\,.$$

EXAMPLE 2.3.3: GF OF THE POISSON VARIABLE. For the *Poisson random variable* of mean θ,

$$g(z) = \mathrm{e}^{-\theta} \sum_{k=0}^{\infty} \frac{(\theta z)^k}{k!} = \mathrm{e}^{\theta(z-1)} .$$

Here is an example where the radius of convergence is finite.

EXAMPLE 2.3.4: GF OF THE GEOMETRIC VARIABLE. For the *geometric random variable* of parameter $p \in (0, 1)$,

$$g(z) = \sum_{k=0}^{\infty} p(1-p)^{k-1} z^k = \frac{pz}{1-(1-p)z} .$$

The radius of convergence of the above power series is $\frac{1}{1-p}$.

Theorem 2.3.5 *The generating function characterizes the distribution of a random variable.*

This means the following. Suppose that, without knowing the distribution of X, you have been able to compute its generating function g, and that, moreover, you are able to give its power series expansion in a neighborhood of the origin:[3]

$$g(z) = \sum_{n=0}^{\infty} a_n z^n .$$

Since g is the generating function of X,

$$g(z) = \sum_{n=0}^{\infty} P(X = n) z^n ,$$

and since the power series expansion around the origin is unique, the distribution of X is identified as

$$P(X = n) = a_n$$

for all $n \geq 0$. Similarly, if two $\mathbb{N}$-valued random variables X and Y have the same generating function, they have the same distribution. Indeed, the identity in a neighborhood of the origin of the power series:

$$\sum_{n=0}^{\infty} P(X = n) z^n = \sum_{n=0}^{\infty} P(Y = n) z^n$$

[3] This is a common situation; see Theorem 2.3.12 for instance.

implies the identity of their coefficients.

Theorem 2.3.6 *Let X and Y be two independent integer-valued random variables with respective generating functions g_X and g_Y respectively. Then the sum $X+Y$ has the* GF

$$g_{X+Y}(z) = g_X(z) \times g_Y(z) . \tag{2.16}$$

Proof. Use the product formula for expectations:

$$g_{X+Y}(z) = E\left[z^{X+Y}\right] = E\left[z^X z^Y\right] = E\left[z^X\right] E\left[z^Y\right] .$$

□

EXAMPLE 2.3.7: SUM OF INDEPENDENT POISSON VARIABLES. Let X and Y be two *independent* Poisson random variables of means α and β respectively. We shall prove that the sum $X+Y$ is a Poisson random variable with mean $\alpha+\beta$. In fact, according to (2.16),

$$g_{X+Y}(z) = g_X(z) \times g_Y(z) = \mathrm{e}^{\alpha(z-1)}\, \mathrm{e}^{\beta(z-1)} = \mathrm{e}^{(\alpha+\beta)(z-1)} ,$$

and the assertion follows directly from Theorem 2.3.5 since g_{X+Y} is the GF of a Poisson random variable with mean $\alpha+\beta$.

Moments from the Generating Function

Generating functions can be used to obtain the moments of a discrete random variable.

Theorem 2.3.8 *We have*

$$g'(1) = E[X] \tag{2.17}$$

and

$$g''(1) = E[X(X-1)] . \tag{2.18}$$

Proof. Inside the open disk centered at the origin and of radius R, the power series defining the generating function g is continuous, and differentiable at any order term by term. In particular, differentiating twice both sides of (2.15) inside the open disk $D(0;R)$ gives

$$g'(z) = \sum_{n=1}^{\infty} nP(X=n)z^{n-1} \tag{2.19}$$

and

$$g''(z) = \sum_{n=2}^{\infty} n(n-1)P(X=n)z^{n-2} . \tag{2.20}$$

When the radius of convergence R is *strictly larger* than 1, we obtain the announced results by letting $z = 1$ in the previous identities.

If $R = 1$, the same is basically true but the mathematical argument is more subtle. The difficulty is not with the right-hand side of (2.19), which is always well defined at $z = 1$, being equal to $\sum_{n=1}^{\infty} nP(X = n)$, a non-negative and possibly infinite quantity. The difficulty is that g may be not differentiable at $z = 1$, a border point of the disk (here of radius 1) on which it is defined. However, by *Abel's theorem*[4], the limit of $\sum_{n=1}^{\infty} nP(X = n)x^{n-1}$ as the *real* variable x increases to 1 is $\sum_{n=1}^{\infty} nP(X = n)$. Therefore g', as a function defined on the real interval $[0, 1)$, can be extended to $[0, 1]$ by (2.17), and this extension preserves continuity. With this *definition* of $g'(1)$, Formula (2.17) holds true. Similarly, when $R = 1$, the function g'' defined on $[0, 1)$ by (2.20) is extended to a continuous function on $[0, 1]$ by *defining* $g''(1)$ by (2.18). □

Another useful result is *Wald's formula* below, which gives the expectation of a random sum of independent and identically distributed integer-valued variables. By taking derivatives in (2.21) of Theorem 2.3.12,

$$E[X] = g'_X(1) = g'_Y(1)g'_T(g_Y(1)) = E[Y_1]E[T].$$

A stronger version of this result is given in Exercise 2.5.18.

The next technical result gives details concerning the shape of the generating function restricted to the interval $[0, 1]$.

Theorem 2.3.9 (α) *Let $g : [0, 1] \to \mathbb{R}$ be defined by $g(x) = E[x^X]$, where X is a non-negative integer-valued random variable. Then g is non-decreasing and convex. Moreover, if $P(X = 0) < 1$, then g is strictly increasing, and if $P(X \le 1) < 1$, it is strictly convex.*

(β) *Suppose $P(X \le 1) < 1$. If $E[X] \le 1$, the equation $x = g(x)$ has a unique solution $x \in [0, 1]$, namely $x = 1$. If $E[X] > 1$, it has two solutions in $[0, 1]$, $x = 1$ and $x = x_0 \in (0, 1)$.*

---

[4] Let $\{a_n\}_{n\ge1}$ be a sequence of real numbers such that the radius of convergence of the power series $\sum_{n=0}^{\infty} a_n z^n$ is 1. Suppose that the sum $\sum_{n=0}^{\infty} a_n$ is convergent. Then the power series $\sum_{n=0}^{\infty} a_n x^n$ is uniformly convergent in $[0, 1]$ and

$$\lim_{x\uparrow 1} \sum_{n=0}^{\infty} a_n x^n = \sum_{n=0}^{\infty} a_n ,$$

where $x \uparrow 1$ means that x tends to 1 strictly from below.

Proof. Just observe that for $x \in [0,1]$,

$$g'(x) = \sum_{n=1}^{\infty} nP(X = n)x^{n-1} \geq 0\,,$$

and therefore g is non-decreasing, and

$$g''(x) = \sum_{n=2}^{\infty} n(n-1)P(X - n)x^{n-2} \geq 0\,,$$

and therefore g is convex. For $g'(x)$ to be null for some $x \in (0,1)$, it is necessary to have $P(X = n) = 0$ for all $n \geq 1$, and therefore $P(X = 0) = 1$. For $g''(x)$ to be null for some $x \in (0,1)$, one must have $P(X = n) = 0$ for all $n \geq 2$, and therefore $P(X = 0) + P(X = 1) = 1$.

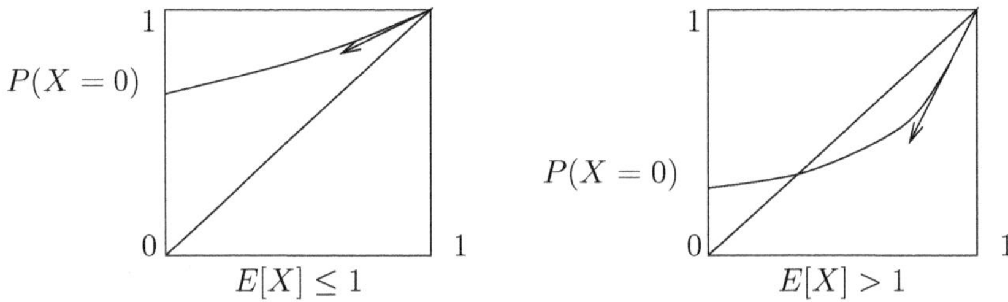

Two aspects of the generating function

The graph of $g : [0,1] \to \mathbb{R}$ has, in the strictly increasing strictly convex case $P(X = 0) + P(X = 1) < 1$, the general shape shown in the figure, where we distinguish two cases: $E[X] = g'(1) \leq 1$, and $E[X] = g'(1) > 1$. The rest of the proof is then easy. □

The next two examples are typical of the use of generating functions in combinatorics.

Example 2.3.10: The Lottery. We compute the probability that in a 6-digit lottery the sum of the first three digits equals the sum of the last three digits. (The digits from 0 to 9 are drawn equiprobably and independently and the result is presented in the order they appear.)

Let X_1, X_2, X_3, X_4, X_5, and X_6 be independent random variables uniformly distributed over $\{0, 1, \ldots, 9\}$. We first compute the generating function of

$Y = 27 + X_1 + X_2 + X_3 - X_4 - X_5 - X_6$. We have

$$E[z^{X_i}] = \frac{1}{10}(1 + z + \cdots + z^9) = \frac{1}{10}\frac{1 - z^{10}}{1 - z},$$

$$\begin{aligned} E[z^{-X_i}] &= \frac{1}{10}\left(1 + \frac{1}{z} + \cdots + \frac{1}{z^9}\right) \\ &= \frac{1}{10}\frac{1 - z^{-10}}{1 - z^{-1}} = \frac{1}{10}\frac{1}{z^9}\frac{1 - z^{10}}{1 - z}, \end{aligned}$$

and

$$\begin{aligned} E[z^Y] &= E[\left[z^{27 + \sum_{i=1}^{3} X_i - \sum_{i=4}^{6} X_i}\right] \\ &= E\left[z^{27}\prod_{i=1}^{3} z^{X_i}\prod_{i=4}^{6} z^{-X_i}\right] = z^{27}\prod_{i=1}^{3} E[z^{X_i}]\prod_{i=4}^{6} E[z^{-X_i}]. \end{aligned}$$

Therefore,

$$g_Y(z) = \frac{1}{10^6}\frac{(1 - z^{10})^6}{(1 - z)^6}.$$

But $P(X_1 + X_2 + X_3 = X_4 + X_5 + X_6) = P(Y = 27)$ is the factor of z^{27} in the power series expansion around the origin of g_Y. Since

$$(1 - z^{10})^6 = 1 - \binom{6}{1}z^{10} + \binom{6}{2}z^{20} + \cdots$$

and

$$(1 - z)^{-6} = 1 + \binom{6}{5}z + \binom{7}{5}z^2 + \binom{8}{5}z^3 + \cdots$$

(recall the *negative binomial formula* (1.18):

$$(1 - z)^{-p} = 1 + \binom{p}{p - 1}z + \binom{p + 1}{p - 1}z^2 + \binom{p + 2}{p - 1}z^3 + \cdots),$$

we find that

$$P(Y = 27) = \frac{1}{10^6}\left(\binom{32}{5} - \binom{6}{1}\binom{22}{5} + \binom{6}{2}\binom{12}{5}\right).$$

EXAMPLE 2.3.11: BIASED DICE. Does there exist two biased dice such that, when tossed independently, the sum of their values is uniformly distributed on $\{2,3,\dots,12\}$? The answer is no.

To see this, let us call X_1 and X_2 the values obtained by tossing the first and the second die respectively, and g_1 and g_2 the corresponding generating functions. The generating function of $X := X_1 + X_2$ is $g(z) = g_1(z) \times g_2(z)$ since the dice are supposed to be tossed independently. If the sum was uniformly distributed on $\{2,3,\dots,12\}$, then we would have

$$g_1(z) \times g_2(z) = \frac{1}{11}(z^2 + \cdots + z^{12}) = \frac{z^2}{11}\frac{z^{11}-1}{z-1}.$$

Equivalently,

$$P_1(z)P_2(z) = \frac{1}{11}(1 + z + \cdots + z^{10}), \tag{$\star$}$$

where the polynomials

$$P_i(z) := \frac{1}{z}g_i(z) = \sum_{k=0}^{5} P(X_i = k+1)z^k \quad (i = 1,2)$$

have common degree 5. Being of odd degree they each have at least one real root, whereas the right-hand side of ($\star$) has no real roots (its roots are the ten eleventh roots of unity not equal to 1). Hence a contradiction.

Random Sums

How to compute the distribution of random sums? Here again, generating functions help.

Theorem 2.3.12 *Let $\{Y_n\}_{n\geq 1}$ be an* IID *sequence of integer-valued random variables with the common generating function g_Y. Let T be another random variable, integer-valued, independent of the sequence $\{Y_n\}_{n\geq 1}$, and let g_T be its generating function. The generating function of*

$$X = \sum_{n=1}^{T} Y_n,$$

where by convention $\sum_{n=1}^{0} = 0$, is

$$g_X(z) = g_T(g_Y(z)). \tag{2.21}$$

Proof. Since $\{\{T = k\}\}_{k\geq 0}$ is a sequence of mutually exclusive and exhaustive subsets of Ω,

$$1 = \sum_{k=0}^{\infty} 1_{\{T=k\}},$$

and

$$z^X = z^{\sum_{n=1}^{T} Y_n} = \left(\sum_{k=0}^{\infty} 1_{\{T=k\}}\right) z^{\sum_{n=1}^{T} Y_n}$$
$$= \sum_{k=0}^{\infty} \left(z^{\sum_{n=1}^{T} Y_n}\right) 1_{\{T=k\}} = \sum_{k=0}^{\infty} \left(z^{\sum_{n=1}^{k} Y_n}\right) 1_{\{T=k\}}.$$

Therefore,

$$E[z^X] = \sum_{k=0}^{\infty} E\left[1_{\{T=k\}} \left(z^{\sum_{n=1}^{k} Y_n}\right)\right] = \sum_{k=0}^{\infty} E[1_{\{T=k\}}]E[z^{\sum_{n=1}^{k} Y_n}],$$

where we have used the assumption of independence of T and $\{Y_n\}_{n\geq 1}$. Now, $E[1_{\{T=k\}}] = P(T = k)$, and

$$E[z^{\sum_{n=1}^{k} Y_n}] = E[\prod_{n=1}^{k} z^{Y_n}] = \prod_{n=1}^{k} E[\prod_{n=1}^{k} z^{Y_n}] = g_Y(x)^k,$$

and therefore

$$E[z^X] = \sum_{k=0}^{\infty} P(T = k) g_Y(z)^k = g_T(g_Y(z)).$$

□

EXAMPLE 2.3.13: THINNING OF A POISSON RANDOM VARIABLE. Let $\{X_n\}_{n\geq 1}$ be a Bernoulli sequence of parameter p, and let T be a Poisson random variable with mean $\theta > 0$, independent of $\{X_n\}_{n\geq 1}$. We show that

$$S := X_1 + \cdots + X_T$$

is a Poisson random variable with mean $p\theta$. (In other words, if one "thins out" with thinning probability $1 - p$ a population sample of Poissonian size, the remaining sample also has a Poissonian size, with the obvious mean, that is, p times that of the original sample.) Indeed, in this case,

$$g_T(z) = \mathrm{e}^{\theta(z-1)}$$

and

$$g_Y(z) = pz + (1 - p),$$

so that

$$g_S(z) = g_T(g_Y(z)) = e^{p\theta(z-1)}.$$

Compare with a direct proof:

$$\begin{aligned}
P(S = k) &= P(X_1 + \cdots + X_T = k) \\
&= P\left(\cup_{n=k}^{\infty}\{X_1 + \cdots + X_n = k, T = n\}\right) \\
&= \sum_{n=k}^{\infty} P(X_1 + \cdots + X_n = k, T = n) \\
&= \sum_{n=k}^{\infty} P(X_1 + \cdots + X_n = k)P(T = n),
\end{aligned}$$

that is

$$\begin{aligned}
P(S = k) &= \sum_{n=k}^{\infty} \frac{n!}{k!(n-k)!} p^k q^{n-k} e^{-\theta} \frac{\theta^n}{n!} \\
&= e^{-\theta} \frac{(p\theta)^k}{k!} \sum_{n=k}^{\infty} \frac{(q\theta)^{n-k}}{(n-k)!} \\
&= e^{-\theta} \frac{(p\theta)^k}{k!} \sum_{i=0}^{\infty} \frac{(q\theta)^i}{i!} \\
&= e^{-\theta} \frac{(p\theta)^k}{k!} e^{q\theta} = e^{p\theta} \frac{(p\theta)^k}{k!}.
\end{aligned}$$

Branching Trees

Francis Galton, a cousin of Darwin, was interested in the survival probability of a given line of English peerage. He posed the problem in the *Educational Times* in 1873. In the same year and the same journal, Watson proposed the method of solution that has become a textbook classic, and thereby initiated an important domain of probability called *branching process theory.*

Here is the description of the Galton–Watson model, the statement of Galton's purpose and Watson's solution.

Let $Z_n = (Z_n^{(1)}, Z_n^{(2)}, \ldots)$, where the random variables $\{Z_n^{(j)}\}_{n\geq 1, j\geq 1}$ are IID and integer-valued. The recurrence equation

$$X_{n+1} = \sum_{k=1}^{X_n} Z_{n+1}^{(k)}\,, \tag{2.22}$$

with the convention $X_{n+1} = 0$ if $X_n = 0$, receives the following interpretation: X_n is the number of individuals in the nth generation of a given population (humans, particles, etc.). Individual number k of the nth generation gives birth to $Z_{n+1}^{(k)}$ descendants, and this accounts for Eqn. (2.22). The number X_0 of ancestors is assumed to be independent of $\{Z_n\}_{n\geq 1}$. The sequence of random variables $\{X_n\}_{n\geq 0}$ is called a branching process because of the genealogical tree that it generates (see Figure 2.1). The branching process is also known as the *Galton–Watson process.*

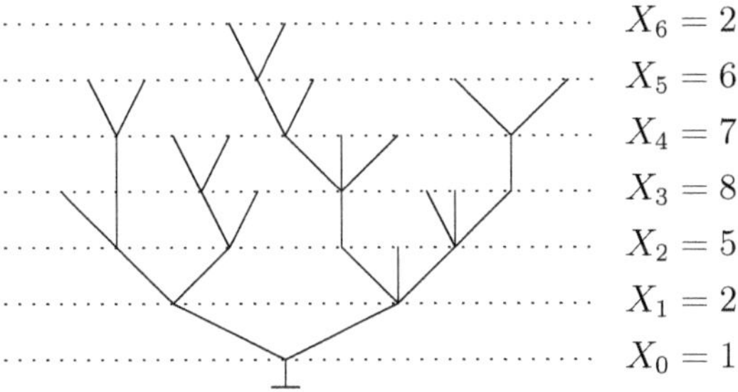

Figure 2.1: Sample tree of a branching process

With the purpose of obtaining the probability of extinction of the population, first observe that the event $\mathcal{E}$ = "an extinction occurs" is just "at least one generation is empty", that is,

$$\mathcal{E} = \cup_{n=1}^{\infty}\{X_n = 0\}\,,$$

In order to discard trivial cases, assume that $P(Z = 0) < 1$ and $P(Z \geq 2) > 0$.

Let g be the common generating function of the variables $Z_n^{(k)}$. Let

$$\psi_n(z) := E[z^{X_n}]$$

be the generating function of the number of individuals in the nth generation. We prove successively that

(a) $P(X_{n+1} = 0) = g(P(X_n = 0))$,

(b) $P(\mathcal{E}) = g(P(\mathcal{E}))$, and

(c) if $E[Z_1] < 1$ the probability of extinction is 1; and if $E[Z_1] > 1$, the probability of extinction is < 1 but nonzero.

Proof.

(a) In Equation (2.22), X_n is independent of the $Z_{n+1}^{(k)}$'s. Therefore, by Theorem 2.3.12,

$$\psi_{n+1}(z) = \psi_n(g(z)).$$

Iterating this equality, we obtain $\psi_{n+1}(z) = \psi_0(g^{(n+1)}(z))$, where $g^{(n)}$ is the nth iterate of g. If there is only *one ancestor*, then $\psi_0(z) = z$, and therefore $\psi_{n+1}(z) = g^{(n+1)}(z) = g(g^{(n)}(z))$, that is,

$$\psi_{n+1}(z) = g(\psi_n(z)).$$

In particular, since $\psi_n(0) = P(X_n = 0)$, we have (a).

(b) Since $X_n = 0$ implies $X_{n+1} = 0$, the family $\{X_n = 0\}$ is non-decreasing, and by monotone sequential continuity,

$$P(\mathcal{E}) = \lim_{n \uparrow \infty} P(X_n = 0).$$

The generating function g is continuous, and therefore from (a) and the last equation, the probability of extinction satisfies (b).

(c) Let Z be any of the random variables $Z_n^{(k)}$. Excluding the trivial cases where $P(Z = 0) = 1$ or $P(Z \geq 2) = 0$, we have by Theorem 2.3.9 that:

(α) if $E[Z] \leq 1$, the only solution of $x = g(x)$ in $[0, 1]$ is 1, and therefore $P(\mathcal{E}) = 1$. The branching process eventually becomes extinct, and

(β) if $E[Z] > 1$, there are two solutions of $x = g(x)$ in $[0, 1]$, 1 and x_0 such that $0 < x_0 < 1$. From the strict convexity of $g : [0, 1] \to [0, 1]$, it follows that the sequence $y_n = P(X_n = 0)$ that satisfies $y_0 = 0$ and $y_{n+1} = g(y_n)$ converges to x_0. Therefore, when the mean number of descendants $E[Z]$ is strictly larger than 1, $P(\mathcal{E}) \in (0, 1)$.

□

EXAMPLE 2.3.14: EXTINCTION PROBABILITY FOR A POISSON OFFSPRING. Take for the offspring distribution the Poisson distribution with mean $\lambda > 0$, whose generating function is $g(x) = e^{\lambda(x-1)}$. Suppose that $\lambda > 1$ (the supercritical case). The probability of extinction $P(\mathcal{E})$ is the unique solution in $(0, 1)$ of $x = e^{\lambda(x-1)}$.

EXAMPLE 2.3.15: EXTINCTION PROBABILITY FOR A BINOMIAL OFFSPRING. Take for the offspring distribution the binomial distribution $\mathcal{B}(N,p)$, with $0 < p < 1$. Its mean is $m = Np$ and its generating function is $g(x) = (px + (1-p))^N$. Suppose that $Np > 1$ (the supercritical case). The probability of extinction $P(\mathcal{E})$ is the unique solution in $(0,1)$ of $x = (px + (1-p))^N$.

EXAMPLE 2.3.16: POISSON BRANCHING AS THE LIMIT OF BINOMIAL BRANCHING. Suppose now that $p = \frac{\lambda}{N}$ with $\lambda > 1$ (therefore we are in the supercritical case) and the probability of extinction is given by the unique solution in $(0,1)$ of

$$x = \left(\frac{\lambda}{N}x + (1 - \frac{\lambda}{N})\right)^N = \left(1 - \frac{\lambda}{N}(1-x)\right)^N .$$

Letting $N \uparrow \infty$, we see that the right-hand side tends from below ($1 - x \le \mathrm{e}^{-x}$) to the generating function of a Poisson variable with mean λ. Using this fact and the concavity of the generating functions, it follows that the probability of extinction also tends to the probability of extinction relative to the Poisson distribution of the offspring.

Let T be the extinction time of the Galton–Watson branching process. The distribution of T is fully described by

$$P(T \le n) = P(X_n = 0) = \psi_n(0) \qquad (n \ge 0)$$

and $P(T = \infty) = 1 - P(\mathcal{E})$. In particular

$$\lim_{n\uparrow\infty} P(T \le n) = P(\mathcal{E}) . \tag{$\star$}$$

Theorem 2.3.17 *In the supercritical case ($m > 1$ and therefore $0 < P(\mathcal{E}) < 1$),*

$$P(\mathcal{E}) - P(T \le n) \le (g'(P(\mathcal{E})))^n . \tag{2.23}$$

Proof. The probability of extinction $P(\mathcal{E})$ is the limit of the sequence $x_n = P(X_n = 0)$ satisfying the recurrence equation $x_{n+1} = g(x_n)$ with initial value $x_0 = 0$. We have that

$$0 \le P(\mathcal{E}) - x_{n+1} = P(\mathcal{E}) - g(x_n) = g(P(\mathcal{E})) - g(x_n) ,$$

that is,

$$\frac{P(\mathcal{E}) - x_{n+1}}{P(\mathcal{E}) - x_n} = \frac{g(P(\mathcal{E})) - g(x_n)}{P(\mathcal{E}) - x_n} \leq g'(P(\mathcal{E})),$$

where we have taken into account the convexity of g and the inequality $x_n < P(\mathcal{E})$. The result follows from there by recurrence. □

EXAMPLE 2.3.18: CONVERGENCE RATE FOR THE POISSON OFFSPRING DISTRIBUTION. For a Poisson offspring with mean $m = \lambda > 1$, $g'(x) = \lambda g(x)$ and therefore $g'(P(\mathcal{E})) = \lambda P(\mathcal{E})$. Therefore

$$P(\mathcal{E}) - P(T \leq n) \leq (\lambda P(\mathcal{E}))^n .$$

EXAMPLE 2.3.19: CONVERGENCE RATE FOR THE BINOMIAL OFFSPRING DISTRIBUTION. For a $\mathcal{B}(N, p)$ offspring with mean $m = Np > 1$, $g'(x) = Np\frac{g(x)}{1-p(1-x)}$ and therefore

$$g'(P(\mathcal{E})) = Np\frac{P(\mathcal{E})}{1 - p(1 - P(\mathcal{E}))} .$$

Taking $p = \frac{\lambda}{N}$,

$$g'(P_N(\mathcal{E})) = \lambda\frac{P_N(\mathcal{E})}{1 - \frac{\lambda}{N}(1 - P_N(\mathcal{E}))},$$

where the notation stresses the dependence of the extinction probability on N.

2.4 Conditional Expectation I

Conditioning is the most important concept of probability theory after independence. We have already encountered this notion under the form of the conditional probability *of an event given an event* and the Bayes formulas. This book introduces the conditional expectation *of a random variable given a random variable* progressively, starting from the discrete case and then proceeding to the absolutely continuous case (Chapter 3), and finally giving the general theory in Chapter 5.

We start with the notion of conditional expectation *of a random variable given an event*. Let Z be a discrete random variable with values in E, and let $f : E \to \mathbb{R}$ be a non-negative function. Let A be some event of positive probability. The conditional expectation of $f(Z)$ given A, denoted by $E[f(Z) \mid A]$, is by definition the expectation when the distribution of Z is replaced by its conditional distribution given A:

$$E[f(Z) \mid A] := \sum_z f(z)P(Z = z \mid A).$$

Let $\{A_i\}_{i \in \mathbb{N}}$ be a partition of the sample space. The following formula is then a direct consequence of Bayes' formula of total causes:

$$E[f(Z)] = \sum_{i \in \mathbb{N}} E[f(Z) \mid A_i] P(A_i).$$

EXAMPLE 2.4.1: RANDOM QUICKSORT. We want to sort a sequence of numbers in increasing order, say $7, 6, 4, 2, 9, 3, 1, 8, 5$. The *quicksort algorithm* proposes to choose one of these numbers at random, say 4, called the *pivot*. It then scans the list from left to right, comparing each number to the pivot, placing the ones that are smaller than the pivot to the left, the others to the right. This creates three sets:

$$\{2, 3, 1\}, 4, \{7, 6, 9, 8, 5\}$$

It operates likewise on the two subsets of size > 1 of this list. For instance, starting with subset $\{2, 1, 3\}$, and choosing at random the pivot for this sublist, say 1, and then continuing with the subset $\{7, 6, 9, 8, 5\}$ with the pivot 7, we obtain:

$$1, \{2, 3\}, 4, \{6, 5\}, 7, \{9, 8\}.$$

Keep doing this until all the subsets have only one member. In this example just one more iteration is needed. The number of comparisons used in this specific example is $8 + (2 + 4) + (1 + 1 + 1) = 17$. One would like to know how well this algorithm performs (in terms of the number of comparisons) in the general case. The ideal situation would be if at each splitting the median number is chosen, resulting in a number of comparisons approximately equal to

$$n + 2\frac{n}{2} + 4\frac{n}{4} + \cdots$$

where there are approximately $\log_2 n$ terms in the sum.

In the random quicksort algorithm, pivots are chosen randomly uniformly among the existing possibilities. We will compare the average number of comparisons in the random quicksort to the ideal $n \log_2 n$.

Let C_n be the number of comparisons needed and let X be the rank of the initial pivot selected. We have, with $M_n = E[C_n]$,

$$M_n = \sum_{j=1}^{n} E[C_n | X = j] P(X = j)$$

$$= \sum_{j=1}^{n} (n - 1 + M_{j-1} + M_{n-j}) \times \frac{1}{n} = n - 1 + \frac{2}{n} \sum_{k=1}^{n-1} M_k ,$$

and therefore

$$nM_n = n(n-1) + 2 \sum_{k=1}^{n-1} M_k .$$

Subtracting the same expression with $n - 1$ instead of n, we have

$$nM_n = (n+1)M_{n-1} + 2(n-1) ,$$

or

$$\frac{M_n}{n+1} = \frac{M_{n-1}}{n} + \frac{2(n-1)}{n(n+1)} .$$

By iteration,

$$\frac{M_n}{n+1} = 2 \sum_{k=1}^{n} \frac{k-1}{k(k+1)} = 2 \sum_{k=1}^{n} \left(\frac{2}{k+1} - \frac{1}{k} \right)$$

and therefore, finally

$$M_n \sim 2n \ln n .$$

The conditional expectation of a discrete random variable Z given another discrete random variable Y is the expectation of Z using the probability measure modified by the observation of Y. For instance, if $Y = y$, instead of the original probability assigning the mass $P(A)$ to the event A, we use the conditional probability given $Y = y$ assigning the mass $P(A|Y = y)$ to this event.

Definition 2.4.2 *Let X and Y be two discrete random variables taking their values in the denumerable sets F and G, respectively, and let $g : F \times G \to \mathbb{R}_+$ be either non-negative, or such that $E[|g(X,Y)|] < \infty$. For $y \in G$ such that $P(Y = y) > 0$, let*

$$\psi(y) := \sum_{x \in F} g(x,y) P(X = x \mid Y = y) , \tag{2.24}$$

and otherwise, if $P(Y = y) = 0$, let $\psi(y) = 0$. This quantity is called the conditional expectation of $g(X,Y)$ given $Y = y$, *and is denoted by $E^{Y=y}[g(X,Y)]$, or $E[g(X,Y) \mid Y = y]$. The random variable $\psi(Y)$ is called the* conditional expectation of $g(X,Y)$ given Y, *and is denoted by $E^Y[g(X,Y)]$ or $E[g(X,Y) \mid Y]$.*

The sum in (2.24) is well defined (possibly infinite however) when g is non-negative. Note that in the non-negative case, we have that

$$\begin{aligned}\sum_{y\in G}\psi(y)P(Y=y) &= \sum_{y\in G}\sum_{x\in F} g(x,y)P(X=x\mid Y=y)P(Y=y)\\ &= \sum_{x}\sum_{y} g(x,y)P(X=x,Y=y) = E[g(X,Y)]\,.\end{aligned}$$

In particular, if $E[g(X,Y)] < \infty$, then

$$\sum_{y\in G}\psi(y)P(Y=y) < \infty\,,$$

which implies that $\psi(y) < \infty$ for all $y \in G$ such that $P(Y=y) > 0$. We observe (for reference in a few lines) that in this case, $\psi(Y) < \infty$ almost surely, that is to say $P(\psi(Y) < \infty) = 1$ (in fact, $P(\psi(Y) = \infty) = \sum_{y;\psi(y)=\infty} P(Y=y) = 0$).

Let now $g : F\times G \to \mathbb{R}$ be a function of arbitrary sign such that $E[|g(X,Y)|] < \infty$, and in particular $E[g^{\pm}(X,Y)] < \infty$. Denote by $\psi^{\pm}$ the functions associated to $g^{\pm}$ as in (2.24). As we just saw, for all $y \in G$, $\psi^{\pm}(y) < \infty$, and therefore $\psi(y) = \psi^{+}(y) - \psi^{-}(y)$ is well defined (not an indeterminate $\infty - \infty$ form). Thus, the conditional expectation is well defined also in the integrable case. From the observation made a few lines above, in this case, $|E^{Y}[g(X,Y)]| < \infty$.

EXAMPLE 2.4.3: BINOMIAL RANDOM VARIABLES AND CONDITIONING. Let X_1 and X_2 be independent binomial random variables of the same size N and same parameter p. We show that

$$E^{X_1+X_2}[X_1] = h(X_1+X_2) := \frac{X_1+X_2}{2}\,.$$

We have

$$\begin{aligned}P(X_1=k|X_1+X_2=n) &= \frac{P(X_1=k)P(X_2=n-k)}{P(X_1+X_2=n)}\\ &= \frac{\binom{N}{k}p^k(1-p)^{N-k}\binom{N}{n-k}p^{n-k}(1-p)^{N-n+k}}{\binom{2N}{n}p^n(1-p)^{N-n}} = \frac{\binom{N}{k}\binom{N}{n-k}}{\binom{2N}{n}}\,,\end{aligned}$$

where we have used the fact that the sum of two independent binomial random variables with size N and parameter p is a binomial random variable with size $2N$ and parameter p. The right-hand side of the last display is the probability of obtaining k black balls when a sample of n balls is randomly selected from an urn containing N black balls and N red balls. This is the *hypergeometric distribution.* The mean of such a distribution is (by symmetry) $\frac{n}{2}$, therefore

$$E^{X_1+X_2=n}[X_1] = \frac{n}{2} = h(n)$$

and this gives the announced result.

Exercise 5.7.14 will give a more elegant solution to the above example, and the reader will discover there that the result is more general.

EXAMPLE 2.4.4: POISSON VARIABLES AND CONDITIONING. Let X_1 and X_2 be two independent Poisson random variables with respective means $\theta_1 > 0$ and $\theta_2 > 0$. We compute $E^{X_1+X_2}[X_1]$, that is $E^Y[X]$, where $X := X_1$, $Y := X_1 + X_2$.

Following the instructions of Definition 2.4.2, we must first compute (only for $y \geq x$, why?)

$$\begin{aligned} P(X = x \mid Y = y) &= \frac{P(X = x, Y = y)}{P(Y = y)} = \frac{P(X_1 = x, X_1 + X_2 = y)}{P(X_1 + X_2 = y)} \\ &= \frac{P(X_1 = x, X_2 = y - x)}{P(X_1 + X_2 = y)} = \frac{P(X_1 = x)P(X_2 = y - x)}{P(X_1 + X_2 = y)} \\ &= \frac{\mathrm{e}^{-\theta_1}\frac{\theta_1^x}{x!}\mathrm{e}^{-\theta_2}\frac{\theta_2^{y-x}}{(y-x)!}}{\mathrm{e}^{-(\theta_1+\theta_2)}\frac{(\theta_1+\theta_2)^y}{y!}} = \binom{y}{x}\left(\frac{\theta_1}{\theta_1+\theta_2}\right)^x\left(\frac{\theta_2}{\theta_1+\theta_2}\right)^{y-x}. \end{aligned}$$

Therefore, with $\alpha := \frac{\theta_1}{\theta_1+\theta_2}$,

$$\psi(y) = E^{Y=y}[X] = \sum_{x=0}^{y} x\binom{y}{x}\alpha^x(1-\alpha)^{y-x} = \alpha y\,.$$

Finally, $E^Y[X] = \psi(Y) = \alpha Y$, that is,

$$E^{X_1+X_2}[X_1] = \frac{\theta_1}{\theta_1+\theta_2}(X_1 + X_2)\,.$$

We now give the main properties of conditional expectation:

The first one, *linearity*, is obvious from the definitions: For all $\lambda_1, \lambda_2 \in \mathbb{R}$,

$$E^Y[\lambda_1 g_1(X,Y) + \lambda_2 g_2(X,Y)] = \lambda_1 E^Y[g_1(X,Y)] + \lambda_2 E^Y[g_2(X,Y)]$$

whenever the conditional expectations thereof are well defined and do not produce $\infty - \infty$ forms. *Monotonicity* is equally obvious: if $g_1(x,y) \le g_2(x,y)$, then

$$E^Y[g_1(X,Y)] \le E^Y[g_2(X,Y)]\,.$$

Theorem 2.4.5 *If g is non-negative or such that $E[|g(X,Y)|] < \infty$, we have*

$$E[E^Y[g(X,Y)]] = E[g(X,Y)]\,.$$

Proof. We have

$$\begin{aligned} E[E^Y[g(X,Y)]] &= E[\psi(Y)]] = \sum_{y\in G} \psi(y)P(Y=y) \\ &= \sum_{y\in G}\sum_{x\in F} g(x,y)P(X=x \mid Y=y)P(Y=y) \\ &= \sum_x \sum_y g(x,y)P(X=x, Y=y) \\ &= E[g(X,Y)]\,. \end{aligned}$$

□

Theorem 2.4.6 *If w is non-negative or such that $E[|w(Y)|] < \infty$,*

$$E^Y[w(Y)] = w(Y)\,, \tag{2.25}$$

and more generally,

$$E^Y[w(Y)h(X,Y)] = w(Y)E^Y[h(X,Y)]\,, \tag{2.26}$$

assuming that the left-hand side of (2.26) is well defined.

Proof. We prove (2.26) ((2.25) follows by setting $h(x,y) \equiv 1$). We consider only the case where w and h are non-negative, since the general case follows easily from this special case. We have,

$$\begin{aligned} E^{Y=y}[w(Y)h(X,Y)] &= \sum_{x\in F} w(y)h(x,y)P(X=x \mid Y=y) \\ &= w(y)\sum_{x\in F} h(x,y)P(X=x \mid Y=y) \\ &= w(y)E^{Y=y}[h(X,Y)]\,. \end{aligned}$$

□

Theorem 2.4.7 *If X and Y are independent and if v is non-negative or such that $E[|v(X)|] < \infty$, then*

$$E^Y[v(X)] = E[v(X)] .$$

Proof. We have

$$\begin{aligned} E^{Y=y}[v(X)] &= \sum_{x \in F} v(x) P(X = x \mid Y = y) \\ &= \sum_{x \in F} v(x) P(X = x) = E[v(X)] . \end{aligned}$$

□

Theorem 2.4.8 *If X and Y are independent and if $g : F \times G \to \mathbb{R}$ is non-negative or such that $E[|g(X,Y)|] < \infty$, then, for all $y \in G$,*

$$E[g(X, Y \mid Y = y] = E[g(X, y)] .$$

Proof. Applying formula (2.24) with $P(X = x \mid Y = y) = P(X = x)$ (by independence), we obtain

$$\psi(y) = \sum_{x \in F} g(x, y) P(X = x) = E\left[g(X, y)\right] .$$

□

We now give the *successive conditioning rule*. Suppose that $Y = (Y_1, Y_2)$, where Y_1 and Y_2. In this situation, we use the more developed notation

$$E^Y[g(X, Y)] = E^{Y_1, Y_2}[g(X, Y_1, Y_2] .$$

Theorem 2.4.9 *Suppose that $Y = (Y_1, Y_2)$ as above. If g is non-negative or such that $E[|g(X,Y)|] < \infty$, then*

$$E^{Y_2}[E^{Y_1, Y_2}[g(X, Y_1, Y_2)]] = E^{Y_2}[g(X, Y_1, Y_2)] . \tag{2.27}$$

Proof. Let

$$\psi(Y_1, Y_2) := E^{Y_1, Y_2}[g(X, Y_1, Y_2)] .$$

We have to show that

$$E^{Y_2}[\psi(Y_1, Y_2)] = E^{Y_2}[g(X, Y_1, Y_2)] .$$

Here

$$\psi(y_1, y_2) = \sum_x g(x, y_1, y_2) P(X = x \mid Y_1 = y_1, Y_2 = y_2)$$

and

$$E^{Y_2=y_2}[\psi(Y_1, Y_2)] = \sum_{y_1} \psi(y_1, y_2) P(Y_1 = y_1 \mid Y_2 = y_2)\,,$$

that is,

$$\begin{aligned}
&E^{Y_2=y_2}[\psi(Y_1, Y_2)] \\
&= \sum_{y_1} \sum_{x} g(x, y_1, y_2) P(X = x \mid Y_1 = y_1, Y_2 = y_2) P(Y_1 = y_1 \mid Y_2 = y_2)\,.
\end{aligned}$$

But

$$\begin{aligned}
P(X = x \mid Y_1 = y_1, Y_2 = y_2)&P(Y_1 = y_1 \mid Y_2 = y_2) \\
&= \frac{P(X = x,\ Y_1 = y_1, Y_2 = y_2)}{P(Y_1 = y_1, Y_2 = y_2)} \frac{P(Y_1 = y_1, Y_2 = y_2)}{P(Y_2 = y_2)} \\
&= P(X = x, Y_1 = y_1 \mid Y_2 = y_2)\,.
\end{aligned}$$

Therefore

$$\begin{aligned}
E^{Y_2=y_2}[\psi(Y_1, Y_2)] &= \sum_{y_1} \sum_{x} g(x, y_1, y_2) P(X = x, Y_1 = y_1 \mid Y_2 = y_2) \\
&= E^{Y_2=y_2}[g(X, Y_1, Y_2)]\,.
\end{aligned}$$

□

We shall see later that the above rules are very general.

2.5 Exercises

Exercise 2.5.1. AN ALTERNATIVE PROOF OF POINCARÉ'S FORMULA
Let $A_1, \dots, A_n$ be events and let $X_1, \dots, X_n$ be their indicator functions. From the developed expression of $E\left[\Pi_{i=1}^n(1 - X_i)\right]$, deduce the formula:

$$P(\cup_{i=1}^n A_i) = \sum_i P(A_i) - \sum_{i<j} P(A_i \cap A_j) + \sum_{i<j<k} P(A_i \cap A_j \cap A_k) - \cdots + (-1)^{n+1} P(A_1 \cap A_2 \cap \cdots \cap A_n) .$$

Exercise 2.5.2. NON-ESSENTIAL SET
Let X be a discrete random variable taking its values in E, with probability distribution $(p(x), x \in E)$. Let $A := \{\omega; p(X(\omega)) = 0\}$. Show that $P(A) = 0$.

Exercise 2.5.3. THE MEAN IS THE CENTER OF INERTIA
Let X be a discrete random variable taking real values, with mean μ and finite variance σ^2. Show that, for all $a \in \mathbb{R}$, $a \neq \mu$,

$$E[(X-a)^2] > E[(X-\mu)^2] = \sigma^2 .$$

Exercise 2.5.4. NULL VARIANCE
Prove for an integer-valued random variable that a null variance implies that this random variable is almost surely constant.

Exercise 2.5.5. GIBBS'S INEQUALITY
Let $(p(x), x \in \mathcal{X})$ and $(q(x), x \in \mathcal{X})$ be two probability distributions on the finite space $\mathcal{X}$. Prove the *Gibbs inequality*

$$-\sum_{x \in \mathcal{X}} p(x) \log p(x) \leq -\sum_{x \in \mathcal{X}} p(x) \log q(x) , \tag{2.28}$$

with equality if and only if $p(x) = q(x)$ for all $x \in \mathcal{X}$.

Exercise 2.5.6. THE ARITHMETIC-GEOMETRIC INEQUALITY
Let x_i $(1 \leq i \leq n)$ be positive numbers, and let p_i $(1 \leq i \leq n)$ be non-negative numbers such that $\sum_{i=1}^n p_i = 1$. Prove that

$$p_1 x_1 + p_2 x_2 + \cdots + p_n x_n \geq x_1^{p_1} x_2^{p_2} \cdots x_n^{p_n} .$$

Exercise 2.5.7. THE GEOMETRIC DISTRIBUTION IS MEMORYLESS
Show that a geometric random variable T with parameter $p \in (0,1)$ is *memoryless* in the sense that for all integers $k, k_0 \geq 1$, $P(T = k + k_0 \mid T > k_0) = P(T = k)$.

Exercise 2.5.8. SUM OF INDEPENDENT GEOMETRIC VARIABLES
Let T_1 and T_2 be two independent geometric random variables with the same parameter $p \in (0,1)$. Give the probability distribution of the sum $X = T_1 + T_2$.

Exercise 2.5.9. FACTORIAL OF POISSON

1. Let X be a Poisson random variable with mean $\theta > 0$. Compute the mean of the random variable $X!$ (factorial, not exclamation mark).

2. Compute $E\left[\theta^X\right]$.

3. What is the probability that X is odd?

Exercise 2.5.10. PROFESSOR NEBULOUS
Professor Nebulous travels from Los Angeles to Paris with stopovers in New York and London. In each airport, his luggage is transferred to the departing plane. In each airport, with probability p, his luggage will not be placed in the right plane. Professor Nebulous finds that his suitcase has not reached Paris. What are the chances that the mishap took place in Los Angeles, New York and London respectively?

Exercise 2.5.11. THE RETURN OF THE COUPON COLLECTOR
In the coupon's collector problem of Example 2.2.6, prove that for all $c > 0$, $P\left(X > \lceil n \ln n + cn \rceil\right) \leq \mathrm{e}^{-c}$. Hint: you might find it useful to define A_i to be the event that a Type i coupon has not shown up during in first $\lceil n \ln n + cn \rceil$ tablets.

Exercise 2.5.12. MORE BERNOULLI
Let $X_1, \ldots, X_{2n}$ be independent random variables taking the values 0 or 1, and such that $P(X_i = 1) = p \in (0,1)$ $(1 \leq i \leq 2n)$. Let $Z := \sum_{i=1}^{n} X_i X_{n+i}$. Compute $P(Z = k)$ $(1 \leq k \leq n)$.

Exercise 2.5.13. STOCHASTICALLY LARGER
Let X and Y be two integer-valued random variables. Then X is said to be *stochastically larger* than Y if for all $n \geq 0$, $P(X \geq n) \geq P(Y \geq n)$. Show that in this case $E[u(X)] \geq E[u(Y)]$ whenever $u : \mathbb{N} \to \mathbb{R}$ is a non-negative non-decreasing function.

Exercise 2.5.14. THE MATCHBOX
A smoker has a matchbox containing N matches in each pocket. He reaches at random for one box or the other. What is the probability that, having eventually found an empty matchbox, there will be k matches left in the other box?

Exercise 2.5.15. THE ENTOMOLOGIST
Each individual of a specific breed of insects has, independently of the others, the probability θ of being a male.

(a) An entomologist seeks to collect exactly $M > 1$ males, and therefore stops hunting as soon as M males are captured. What is the distribution of X, the number of insects that must be caught in order to collect *exactly* M males?

(b) What is the distribution of X, the smallest number of insects that the entomologist must catch to collect *at least* M males and N females?

Exercise 2.5.16. THE ENTOMOLOGIST STRIKES AGAIN!
Recall the setting of Exercise 2.5.15. Each individual of a specific breed of insects has, independently of the others, the probability θ of being a male. An entomologist seeks to collect exactly $M > 1$ males, and therefore stops hunting as soon as she captures M males. She has to capture an insect in order to determine its gender. What is the expectation of X, the number of insects she must catch to collect *exactly* M males? (In Exercise 2.5.15, you computed the distribution of X, from which you can of course compute the mean. However you can find the solution more quickly, and this is what is required in the present exercise.)

Exercise 2.5.17. THE BLUE PINKO
The blue pinko, an extravagant and yet unregistered bird, lays T eggs, each egg blue or pink, with probability p for each given egg to be blue. The colors of the successive eggs are independent, and independent of the number of eggs laid. Example 2.3.13 showed that if the number of eggs is Poisson with mean θ, then the number of blue eggs is Poisson with mean θp and the number of pink eggs is Poisson with mean θq. Show that the number of blue eggs and the number of pink eggs are independent random variables.

Exercise 2.5.18. WALD'S EXPECTATION FORMULA
Let $\{Y_n\}_{n\geq 1}$ be a sequence of integer-valued integrable random variables such that $E[Y_n] = E[Y_1]$ for all $n \geq 1$. Let T be an integer-valued random variable such that for all $n \geq 1$, the event $\{T \geq n\}$ is independent of Y_n. Let $X := \sum_{n=1}^{T} Y_n$. Prove that

$$E[X] = E[Y_1]E[T].$$

Exercise 2.5.19. FAKE SYMMETRY!
Let $\{X_n\}_{n\geq 1}$ be an independent sequence of $\{H, T\}$-valued ($H=$ heads, $T=$ tails) random variables such that

$$P(X_n = H) = \frac{1}{2} \quad (n \geq 1).$$

Suppose "heads" first appears at the n-th toss. Is it true that the probability that n is even is equal to the probability that n is odd, and therefore equal to $\frac{1}{2}$, "by symmetry"?

Exercise 2.5.20. $\alpha g_1 + (1-\alpha)g_2$
Show that if g_1 and g_2 are the generating functions of some integer-valued random variables, then $\alpha g_1 + (1-\alpha)g_2$ is also the generating function of an integer-valued random variable. Which one?

Exercise 2.5.21. MEAN AND VARIANCE VIA GENERATING FUNCTIONS

(a) Compute the mean and variance of the binomial random variable of size n and parameter p from its generating function. Do the same for the Poisson random variable of mean θ.

(b) What is the generating function of the geometric random variable T with parameter $p \in (0,1)$. Compute its first two derivatives and deduce from the result the variance of T.

(c) What is the n-th factorial moment ($E[X(X-1)\cdots(X-n+1)]$) of a Poisson random variable X of mean $\theta > 0$?

Exercise 2.5.22. FROM THE GENERATING FUNCTION TO THE DISTRIBUTION
What is the probability distribution of the integer-valued random variable X with generating function $g(z) = \frac{1}{(2-z)^2}$ ($|z| < 2$)?

Exercise 2.5.23. THROW A DIE
You perform three independent tosses of an unbiased die. What is the probability that one of these tosses results in a number that is the sum of the two other numbers? (You are required to find a solution using generating functions.)

Exercise 2.5.24. RESIDUAL TIME
Let X be a random variable with values in $\mathbb{N}$ and with finite mean m. Show that $p_n := \frac{1}{m}P(X > n)$ ($n \in \mathbb{N}$) defines a probability distribution on $\mathbb{N}$. Compute its generating function G in terms of the generating function g and the mean m of X.

Exercise 2.5.25. A RECURRENCE EQUATION, TAKE 1
Recall the notation $a^+ = \max(a, 0)$. Consider the recurrence equation,

$$X_{n+1} = (X_n - 1)^+ + Z_{n+1} \qquad (n \geq 0)\,,$$

where X_0 is a random variable taking its values in $\mathbb{N}$, and $\{Z_n\}_{n\geq 1}$ is a sequence of independent random variables taking their values in $\mathbb{N}$, and independent of X_0. Express the generating function ψ_{n+1} of X_{n+1} in terms of the generating function φ of Z_1.

Exercise 2.5.26. POISSON AND MULTINOMIAL
Suppose we have N bins in which we place balls in such a manner that the number of balls in any given bin is a Poisson variable of mean $\frac{m}{N}$ and is independent of numbers in the other bins. In particular, the total number of balls $Y_1 + \cdots + Y_N$ is, as the sum of independent Poisson random variables, a Poisson random variable whose mean is the sum of the means, that is m.

For a given arbitrary integer k, compute the conditional probability that there are k_1 balls in bin 1, k_2 balls in bin 2, etc, given that the total number of balls is $k_1 + \cdots + k_N = k$.

Exercise 2.5.27. CONDITIONED POISSON
Let X_1 and X_2 be two independent Poisson random variables with respective means $\theta_1 > 0$ and $\theta_2 > 0$. Compute $E^{X_1+X_2}[X_1]$, that is $E^Y[X]$, where $X = X_1$, $Y = X_1 + X_2$.

Exercise 2.5.28. MULTINOMIAL DISTRIBUTION AND CONDITIONING
Let $(X_1, \ldots, X_k)$ be a multinomial random vector with size n and parameters $p_1, \ldots, p_k$. Compute $E^{X_1}[X_2]$.

Exercise 2.5.29. SEVERAL ANCESTORS
Give the survival probability of the branching process of Section 2.3 (subsection *Branching Trees*, page 58) with k ancestors, $k > 1$.

Exercise 2.5.30. VARIANCE OF THE BRANCHING PROCESS
Give the mean and variance of the size X_n of the branching process of Section 2.3 (subsection *Branching Trees*, page 58) with one ancestor, and then with k ancestors.

Exercise 2.5.31. BRANCHING WITH IMMIGRATION
The branching model with one ancestor is modified as follows. The n-th generation $(n \geq 1)$ is augmented by a random number of immigrants I_n. The sequence $\{I_n\}_{n\geq 1}$

is IID with common generating function g_I, and each I_n is independent of the state of the population before (<) time n. The immigrants and the other members of the population are indistinguishable. Show that the generating function Ψ_n of the number X_n of members of the total population satisfies the recurrence equation

$$\Psi_{n+1}(z) = \Psi_n(g(z))g_I(z),$$

where g is the common generating function corresponding to the progeny of the members of the population (indigenous or immigrants).

Exercise 2.5.32. EXTINCTION TIME
Consider the branching process with a single ancestor and typical progeny geometrically distributed ($P(Z = k) = qp^k$ ($k \geq 0$, $p \in (0,1)$, $q = 1 - p$). Find the distribution of the extinction time $T := \inf\{n\,;\, X_n = 0\}$. For what values of p is $E[T] < \infty$?

Exercise 2.5.33. CONDITIONAL INDEPENDENCE OF TWO VARIABLES GIVEN AN EVENT
Let A be some event of positive probability, and let P_A denote the probability P conditioned by A, that is,

$$P_A(\cdot) = P(\cdot \mid A).$$

The random variables X and Y are said to be conditionally independent given A if they are independent with respect to probability P_A. Prove that this is the case if and only if for all $u, v \in \mathbb{R}$,

$$P(A)E[\mathrm{e}^{iuX}\mathrm{e}^{ivY}1_A] = E[\mathrm{e}^{iuX}1_A]E[\mathrm{e}^{ivY}1_A].$$

Chapter 3

Continuous Random Vectors

Having studied discrete random variables, that is, random variables taking their values in a finite or countable set, we now introduce random variables taking real (possibly infinite) values, and random vectors with a probability density (the so-called "continuous" random vectors).

3.1 Random Variables with Real Values

We start with the definitions of a random variable and of its cumulative distribution function. Recall the notation $\overline{\mathbb{R}} := \mathbb{R} \cup \{-\infty + \infty\}$.

Definition 3.1.1 *A* random variable *is a function* $X : \Omega \to \overline{\mathbb{R}}$ *such that for all* $a \in \mathbb{R}$,

$$\{X \leq a\} \in \mathcal{F}\,.$$

This is a minimal requirement if one wants to assign a probability to $\{X \leq a\}$.

If X does not take infinite values, we say more precisely that X is a *real random variable*.

EXAMPLE 3.1.2: RANDOM POINT ON THE SQUARE, TAKE 2. (Example 1.2.4 ct'd) Here $\omega = (x, y)$, where $x, y \in [0, 1]$. Define the *coordinate* functions of Ω, X, and Y by

$$X(\omega) = x, Y(\omega) = y\,.$$

Since $\{\omega; X(\omega) \leq a\} = [0, a] \times [0, 1]$ is a set for which the area can be defined, X is a random variable. So is Y for similar reasons.

P. Brémaud, *An Introduction to Applied Probability*, Texts in Applied Mathematics 77,
https://doi.org/10.1007/978-3-031-49306-5_3

Definition 3.1.3 *¿From the probabilistic point of view, a random variable X is described by its* cumulative distribution function *(for short:* CDF*)*

$$F(x) = P(X \leq x) . \tag{3.1}$$

In particular, for all $a, b \in \mathbb{R}$ such that $a \leq b$,

$$P(a < X \leq b) = F(b) - F(a)$$

(watch the inequality signs). Indeed, $\{a < X \leq b\} + \{X \leq a\} = \{X \leq b\}$, and therefore $P(a < X \leq b) + P(X \leq a) = P(X \leq b)$, from which the announced identity follows.

Theorem 3.1.4 *The cumulative distribution function F has the following properties:*

(i) $F : \mathbb{R} \to [0, 1]$.

(ii) *F is non-decreasing.*

(iii) *F is right-continuous.*

(iv) *For each $x \in \mathbb{R}$ there exists $F(x-) := \lim_{h \downarrow 0} F(x - h)$.*

(v) $F(+\infty) := \lim_{a \uparrow \infty} F(a) = P(X < \infty) = 1 - P(X = +\infty)$.

(vi) $F(-\infty) := \lim_{a \downarrow -\infty} F(a) = P(X = -\infty)$.

(vii) *$P(X = a) = F(a) - F(a-)$ for all $a \in \mathbb{R}$.*

Proof. (i) is obvious; (ii) If $a \leq b$, then $\{X \leq a\} \subseteq \{X \leq b\}$, and therefore $P(X \leq a) \leq P(X \leq b)$; (iii) Let $B_n = \left\{X \leq a + \frac{1}{n}\right\}$. Since $\cap_{n \geq 1} \left\{X \leq a + \frac{1}{n}\right\} = \{X \leq a\}$ (see Exercise 4.5.2), we have, by sequential continuity,

$$\lim_{n \uparrow \infty} P\left(X \leq a + \frac{1}{n}\right) = P(X \leq a).$$

(iv) We know from Analysis that a non-decreasing function from $\mathbb{R}$ to $\mathbb{R}$ has at any point a limit to the left; (v) Let $A_n = \{X \leq n\}$ and observe that $\cup_{n=1}^{\infty} \{X \leq n\} = \{X < \infty\}$. The result again follows by sequential continuity; (vi) Apply (1.6) with $B_n = \{X \leq -n\}$ and observe that $\cap_{n=1}^{\infty} \{X \leq -n\} = \{X = -\infty\}$. The result follows by sequential continuity. (vii) The sequence $B_n = \left\{a - \frac{1}{n} < X \leq a\right\}$ is decreasing, and $\cap_{n=1}^{\infty} B_n = \{X = a\}$. Therefore, by sequential continuity,

$$P(X = a) = \lim_n P\left(a - \frac{1}{n} < X \leq a\right) = \lim_n \left(F(a) - F\left(a - \frac{1}{n}\right)\right),$$

that is to say, $P(X = a) = F(a) - F(a-)$. □

Remember in particular that

$$P(X = -\infty) = F(-\infty) \text{ and } P(X = +\infty) = 1 - F(+\infty).$$

From (vii), we see that the CDF is continuous at $a \in \mathbb{R}$ if and only if $P(X = a) = 0$.

Being a non-decreasing right-continuous function, F has *at most a countable set of discontinuity points* on $\mathbb{R}$, say $\{d_n, n \in D\}$, where $D \subseteq \mathbb{N}$. Define the discontinuous part F_d of F by

$$\begin{aligned} F_d(x) &:= F(-\infty) + \sum_{n \in D} (F(d_n) - F(d_n-))\, 1_{\{d_n \le x\}} + (1 - F(+\infty))1_{\{+\infty\}}(x) \\ &= P(X = -\infty) + \sum_{n \in D} P(X = d_n)1_{\{d_n \le x\}} + P(X = +\infty)1_{\{+\infty\}}(x). \end{aligned}$$

In particular, when a random variable takes its values in a denumerable subset (D to which one must possibly add $-\infty$ and $+\infty$), its CDF reduces to the discontinuous part F_d, and the sequence $p(d_n) = P(X = d_n)$ $(n \in D)$ together with the values $F(-\infty)$ and $F(+\infty)$ suffice to describe the probabilistic behavior of X.

An important special case is when X is a *real* random variable and

$$F(x) = \int_{-\infty}^{x} f(y)\, \mathrm{d}y \tag{3.2}$$

for some function $f \ge 0$ called the *probability density function* of X. The random variable and its CDF are then called *(absolutely) continuous.*

Note that, if X is real (no infinite values),

$$\int_{-\infty}^{\infty} f(y)\, \mathrm{d}y = 1.$$

Definition 3.1.5 *Two random variables X and Y are called* independent *if for all $a \in \mathbb{R}, b \in \mathbb{R}$,*

$$P(X \le a, Y \le b) = P(X \le a)P(Y \le b). \tag{3.3}$$

EXAMPLE 3.1.6: RANDOM POINT IN THE SQUARE, TAKE 3. Recall the model: $\Omega = [0, 1]^2$, $P(A) =$ area of A. Let X and Y be the coordinate random variables defined as follows; $\omega = (x, y)$, $X(\omega) = x$ and $Y(\omega) = y$. We are going to prove that these random variables are independent. Indeed,

$$\{(x, y) \in \mathbb{R}^2; x \le a, y \le b\} = \{X \le a\} \cap \{Y \le b\},$$

and therefore (with $0 \le a\,,\, b \le 1$)

$$\begin{aligned} P(\{X \le a\} \cap \{Y \le b\}) &= a \times b \\ &= P(X \le a)P(Y \le b). \end{aligned}$$

Expectation

For a function $g : \overline{\mathbb{R}} \to \overline{\mathbb{R}}$, the symbol

$$\int_{-\infty}^{+\infty} g(x)\,\mathrm{d}F(x) \tag{3.4}$$

denotes the *Stieltjes–Lebesgue integral* of the function g with respect to F.

The precise definition of this integral will be given in Chapter 4. For practical purposes, it suffices to mention that this integral is well defined for a large class of functions g, comprising the non-negative "measurable functions". The class of measurable functions is extremely large and one can say that for practical purposes, "all functions are measurable". The reader who does not feel comfortable with this provisional lack of precision is referred to Section 4.1 where these objects and the Stieltjes–Lebesgue integral are rigorously defined and where all the formal manipulations performed in the chapters preceding Chapter 4 will be shown to be licit. In this chapter, it will also be shown that a measurable function of a random variable is in turn a random variable, a result that we shall use a number of times.

In the special case of a *real* random variable for which the continuous component of the CDF is absolutely continuous, that is,

$$F_c(x) = \int_{-\infty}^{x} f_c(y)\,\mathrm{d}y, \tag{3.5}$$

the integral in Eqn. (3.4) is

$$\sum_{n\in D} g(d_n)(F(d_n) - F(d_{n-})) + \int_{-\infty}^{+\infty} g(x)f_c(x)\mathrm{d}x.$$

The most frequent cases arising are the purely discontinuous case where $F(t) = F_d(t)$, for which (in the case where X can take infinite values and when the values

$g(-\infty)$ and $g(+\infty)$ are defined)

$$\int_{-\infty}^{+\infty} g(x)\,\mathrm{d}F(x) =$$
$$F(-\infty)g(-\infty) + \sum_{n\in D} g(d_n)\{F(d_n) - F(d_{n-})\} + (1 - F(+\infty))g(+\infty),$$

and the absolutely continuous case, for which

$$\int_{-\infty}^{+\infty} g(x)\,\mathrm{d}F(x) = \int_{-\infty}^{+\infty} g(x)f(x)\,\mathrm{d}x. \tag{3.6}$$

Definition 3.1.7 *Let X be a random variable with the cumulative distribution function F and let the function $g : \mathbb{R} \to \overline{\mathbb{R}}$ be either non-negative, or such that $\int_{-\infty}^{+\infty} |g(x)|\mathrm{d}F(x) < \infty$ (one then says that $g(X)$ is* integrable*). The* expectation *of $g(X)$ is the quantity*

$$E[g(X)] := \int_{-\infty}^{+\infty} g(x)\,\mathrm{d}F(x)\,. \tag{3.7}$$

For a complex function $g = g_R + ig_I : \mathbb{R} \to \mathbb{C}$, we let

$$E[g(X)] := E[g_R(X)] + E[g_I(X)]$$

as long as $E[g_R(X)]$ and $E[g_I(X)]$ are finite quantities.

Theorem 3.1.8 *Let g, g_1, g_2 be (measurable) functions from $\overline{\mathbb{R}}$ to $\overline{\mathbb{R}}$. We have* (linearity)

$$E[\lambda_1 g_1(X) + \lambda_2 g_2(X)] = \lambda_1 E[g_1(X)] + \lambda_2 E[g_2(X)], \tag{3.8}$$

whenever

(a) *$\lambda_1, \lambda_2 \in \mathbb{R}_+$ and g_1 and g_2 are non-negative, or*

(b) *either $\lambda_1, \lambda_2 \in \mathbb{R}$, and g_1 and g_2 satisfy the integrability condition (3.5).*

Also (monotonicity),

$$E[g_1(X)] \le E[g_2(X)], \tag{3.9}$$

whenever both sides are well defined and $g_1 \le g_2$.

Also (triangle inequality),

$$|E[g(X)]| \le E[|g(X)|]\,. \tag{3.10}$$

Proof. These properties follow from the corresponding properties of the Stieltjes–Lebesgue integral and will be admitted for the time being until Chapter 4. □

Mean and Variance

Definition 3.1.9 *Let X be a real random variable such that $E[|X|] < \infty$. Then X is said to be* integrable, *and in this case (only in this case) we define the* mean *of X as the (finite) number*

$$m := E[X] .$$

From the inequality $|a| \leq 1+a^2$, true for all $a \in \overline{\mathbb{R}}$, we have that $|X| \leq 1+X^2$, and therefore, by the monotonicity and linearity properties, $E[|X|] \leq 1 + E[X^2]$ (we also used the fact that $E[1] = 1$). Therefore if $E[X^2] < \infty$ (in which case we say that X is *square-integrable*), then X is integrable. The following definition then makes sense.

Definition 3.1.10 *Let X be a square-integrable random variable. The* variance *σ^2 of X is the quantity*

$$\sigma^2 := E[(X-m)^2] .$$

The variance is also denoted by Var (X). From the linearity of expectation, it follows that $E[(X-m)^2] = E[X^2] - 2mE[X] + m^2$, that is,

$$\text{Var}\,(X) = E[X^2] - m^2 . \tag{3.11}$$

In a sense, "the mean is the center of inertia of X". By this unprecise remark, the following is meant:

Theorem 3.1.11 *For every square integrable random variable X with mean m,*

$$E[(X-c)^2] \geq E[(X-m)^2] \quad \textit{for all } c.$$

Proof.

$$\begin{aligned} E[(X-c)^2] &= E[(X-m)^2] + 2(m-c)E[X-m] + (m-c)^2 \\ &= E[(X-m)^2] + 0 + (m-c)^2 \geq E[(X-m)^2] . \end{aligned}$$

□

Theorem 3.1.12 *Let Z be a non-negative real random variable and let a be a positive number. Then* (Markov's inequality)*:*

$$P(Z \geq a) \leq \frac{E[Z]}{a} .$$

The proof is the same as in the discrete case (Theorem 2.1.23).

Taking $Z = (X-m)^2$ in Markov's inequality and $a = \varepsilon^2$, we obtain Chebyshev's inequality:

$$P(|X - m| \geq \epsilon) \leq \frac{\text{Var}(X)}{\varepsilon^2}.$$

Again, the proof is the same as the one in the discrete case.

The following analogue of Theorem 2.5.3 has the same proof.

Theorem 3.1.13 *Let X be a real random variable with mean m and variance σ^2. Then, for all $a \in \mathbb{R}$,*

$$\sigma^2 \leq E[(X - a)^2].$$

The following analogue of Theorem 2.1.25 has the same proof.

Theorem 3.1.14 *Let I be as above and let $\varphi : I \to \mathbb{R}$ be a convex function. Let X be an integrable real-valued random variable such that $P(X \in I) = 1$. Assume moreover that either φ is non-negative, or that $\varphi(X)$ is integrable. Then* (Jensen's inequality)

$$E[\varphi(X)] \geq \varphi(E[X]).$$

EXAMPLE 3.1.15: EXAMPLES. Let X be integrable. Then $E[X^2] \geq E[X]$ and $E\left[e^X\right] \geq e^{E[X]}$.

Remarkable Continuous Random Variables

Definition 3.1.16 *Let a and b be real numbers. A real random variable X with probability density function*

$$f(x) = \frac{1}{b-a} 1_{[a,b]}(x) \tag{3.12}$$

is called a uniform random variable on $[a, b]$. *This is denoted by $X \sim \mathcal{U}([a, b])$.*

Theorem 3.1.17 *The mean and the variance of a uniform random variable on $[a, b]$ are given by*

$$E[X] = \frac{a+b}{2}, \quad \text{Var}(X) = \frac{(b-a)^2}{12}. \tag{3.13}$$

Proof. Direct computation. □

Definition 3.1.18 *A real random variable X with probability density function*

$$f(x) = \frac{1}{\sigma\sqrt{2\pi}} e^{-\frac{1}{2}\frac{(x-m)^2}{\sigma^2}}, \tag{3.14}$$

where $m \in \mathbb{R}$ and $\sigma > 0$, is called a Gaussian random variable with mean m and variance σ^2. *This is denoted by $X \sim \mathcal{N}(m, \sigma^2)$.*

One can check that $E[X] = m$ and $\text{Var}\,(X) = \sigma^2$ (Exercise 3.6.13).

Definition 3.1.19 *A random variable X with probability density function*

$$f(x) = \lambda e^{-\lambda x} 1_{\{x \geq 0\}} \tag{3.15}$$

for some $\lambda > 0$ is called an exponential random variable *with parameter λ. This is denoted by $X \sim \mathcal{E}(\lambda)$.*

The CDF of the exponential random variable is

$$F(x) = \int_0^x \lambda e^{-\lambda u}\, du = (1 - e^{-\lambda x}) 1_{\{x \geq 0\}}\,.$$

Theorem 3.1.20 *The mean of an exponential random variable with parameter λ is*

$$E[X] = \lambda^{-1}\,. \tag{3.16}$$

Proof. Direct computation. Or, see the Gamma distribution below. □

The exponential distribution is *memoryless* in the following sense:

Theorem 3.1.21 *Let $X \sim \mathcal{E}(\lambda)$. For all $t, t_0 \in \mathbb{R}_+$, we have*

$$P(X \geq t_0 + t \mid X \geq t_0) = P(X \geq t).$$

Proof.

$$\begin{aligned} P(X \geq t_0 + t \mid X \geq t_0) &= \frac{P(X \geq t_0 + t\,,\; X \geq t_0)}{P(X \geq t_0)} \\ &= \frac{P(X \geq t_0 + t)}{P(X \geq t_0)} \\ &= \frac{e^{-\lambda(t_0+t)}}{e^{-\lambda(t_0)}} = e^{-\lambda t} = P(X \geq t). \end{aligned}$$

□

In preparation for the next definition, recall the definition of the *gamma function* Γ:

$$\Gamma(\alpha) := \int_0^\infty x^{\alpha-1}\, \mathrm{e}^{-x}\, \mathrm{d}x\,.$$

Integration by parts yields, for $\alpha > 0$,

$$\begin{aligned} 0 = u^\alpha \mathrm{e}^{-u}\Big|_0^\infty &= \int_0^\infty \alpha u^{\alpha-1}\mathrm{e}^{-u}\mathrm{d}u - \int_0^\infty \mathrm{e}^{-u}u^\alpha \mathrm{d}u \\ &= \alpha\Gamma(\alpha) - \Gamma(\alpha+1). \end{aligned}$$

Therefore

$$\Gamma(\alpha+1) = \alpha\,\Gamma(\alpha),$$

from which it follows in particular, since $\Gamma(1) = \int_0^\infty \mathrm{e}^{-\lambda x}\mathrm{d}x = 1$, that for all integers $n \geq 1$,

$$\Gamma(n) = (n-1)!$$

Definition 3.1.22 *Let α and β be two positive real numbers. A non-negative random variable X with the probability density function*

$$f(x) = \frac{\beta^\alpha}{\Gamma(\alpha)} x^{\alpha-1}\, \mathrm{e}^{-\beta x} 1_{\{x>0\}} \tag{3.17}$$

is called a Gamma random variable of parameters α and β. *This is denoted by $X \sim \gamma(\alpha, \beta)$.*

We must check that (3.17) defines a probability density of a *real* random variable (that is, the integral of f is 1).

Proof. Indeed:

$$\begin{aligned} \int_{-\infty}^{+\infty} f(x)\mathrm{d}x &= \frac{\beta^\alpha}{\Gamma(\alpha)} \int_0^\infty x^{\alpha-1}\mathrm{e}^{-\beta x}\mathrm{d}x \\ &= \frac{1}{\Gamma(\alpha)} \int_0^\infty y^{\alpha-1}\mathrm{e}^{-y}\mathrm{d}y \\ &= \frac{\Gamma(\alpha)}{\Gamma(\alpha)} = 1\,, \end{aligned}$$

where the second equality has been obtained with the change of variable $y = \beta x$. □

Theorem 3.1.23 *If $X \sim \gamma(\alpha, \beta)$, then*

$$E[X] = \frac{\alpha}{\beta} \text{ and } \mathrm{Var}\,(X) = \frac{\alpha}{\beta^2}\,. \tag{3.18}$$

Proof.

$$\begin{aligned} E[X] &= \int_0^\infty \frac{\beta^\alpha}{\Gamma(\alpha)} x\, x^{\alpha-1}\, \mathrm{e}^{-\beta x}\, \mathrm{d}x \\ &= \frac{\beta^\alpha}{\Gamma(\alpha)} \int_0^\infty x^\alpha\, \mathrm{e}^{-\beta x}\, \mathrm{d}x \\ &= \frac{\Gamma(\alpha+1)}{\Gamma(\alpha)} \frac{1}{\beta} = \frac{\alpha}{\beta}. \end{aligned}$$

Similarly,

$$E\left[X^2\right] = \frac{\Gamma(\alpha+2)}{\Gamma(\alpha)} \frac{1}{\beta^2} = \frac{\alpha(\alpha+1)}{\beta^2}.$$

Therefore

$$\begin{aligned} \mathrm{Var}\,(X) &= E\left[X^2\right] - E[X]^2 \\ &= \frac{\alpha(\alpha+1)}{\beta^2} - \left(\frac{\alpha}{\beta}\right)^2 = \frac{\alpha}{\beta^2}. \end{aligned}$$

□

The exponential distribution is a particular case of the Gamma distribution. In fact, $\gamma(1,\lambda) \equiv \mathcal{E}(\lambda)$.

Definition 3.1.24 *A* chi-square random variable with n degrees of freedom *is, by definition, a random variable X with the $\gamma(\frac{n}{2},\frac{1}{2})$ distribution. This is denoted by $X \sim \chi_n^2$.*

Its probability density function is therefore

$$f(x) = \frac{1}{2^{\frac{n}{2}}\Gamma(\frac{n}{2})} x^{\frac{n}{2}-1}\, \mathrm{e}^{-\frac{1}{2}x} 1_{\{x>0\}}\,. \tag{3.19}$$

This distribution plays an important role in Statistics.

Definition 3.1.25 *A random variable X with probability density function*

$$f(x) = \tfrac{1}{\pi(1+x^2)} \tag{3.20}$$

is called a Cauchy random variable.

It is important to observe that the mean of X is not defined since

$$\int_{\mathbb{R}} \frac{|x|}{\pi(1+x^2)}\, \mathrm{d}x = +\infty\,.$$

Of course, *a fortiori*, X does not have a variance.

Characteristic Functions

The notion of characteristic function brings Fourier analysis into the picture of probability theory. It provides a technique for manipulating probability distributions, just as the generating function does for integer-valued random variables, only this time for real random variables. It is also a fundamental tool for the study of convergence in distribution of a sequence of random variables, a notion that will be introduced in Chapter 7.

Definition 3.1.26 *The characteristic function (for short:* CF*) of a real random variable X is the function $\psi : \mathbb{R} \to \mathbb{C}$ given by*

$$\psi(u) := E[\mathrm{e}^{iuX}]. \tag{3.21}$$

Alternatively,

$$\psi(u) = \int_{\mathbb{R}} \mathrm{e}^{iux} \mathrm{d}F(x),$$

where F is the cumulative distribution function of X. In particular, if X is an absolutely continuous random variable with probability density function f,

$$\psi(u) = \int_{\mathbb{R}} \mathrm{e}^{iux} f(x)\, \mathrm{d}x,$$

that is, ψ is the Fourier transform of f.

In the case of integer-valued random variables the generating function g and the characteristic function ψ of such a variable X are linked by $\psi(u) = g(\mathrm{e}^{iu})$.

EXAMPLE 3.1.27: UNIFORM. Let X be a random variable uniformly distributed on $[a, b]$. Its CF is given by the integral $\frac{1}{b-a}\int_a^b \mathrm{e}^{iux}\, \mathrm{d}x$, and therefore

$$X \sim \mathcal{U}([a, b]) \,:\, \psi(u) = \tfrac{\mathrm{e}^{iub} - \mathrm{e}^{iua}}{iu(b-a)}.$$

In the frequent special case where X is uniformly distributed on $[-T, +T]$,

$$X \sim \mathcal{U}([-T, +T]) \,:\, \psi(u) = \tfrac{\sin(Tu)}{Tu}.$$

EXAMPLE 3.1.28: EXPONENTIAL, GAMMA AND CHI-SQUARE. One can check that the following table gives the characteristic function of the corresponding random variables:

(i) EXPONENTIAL

$$X \sim \mathcal{E}(\lambda) \, : \, \psi(u) = \frac{\lambda}{\lambda - iu}\,.$$

Indeed, integrating by parts:

$$\begin{aligned} 1 &= \left(-\mathrm{e}^{iux}\mathrm{e}^{-\lambda x}\right)_0^\infty \\ &= -\int_0^\infty iu\mathrm{e}^{iux}\mathrm{e}^{-\lambda x}\,\mathrm{d}x + \int_0^\infty \mathrm{e}^{iux}\lambda\mathrm{e}^{-\lambda x}\,\mathrm{d}x \\ &= -\frac{iu}{\lambda}\psi(u) + \psi(u)\,, \end{aligned}$$

from which the result follows.

(ii) GAMMA A standard computation gives

$$X \sim \gamma(\alpha, \beta) \, : \, \psi(u) = \left(1 - i\frac{u}{\beta}\right)^{-\alpha}\,.$$

In particular, with $\beta = \lambda$ and $\alpha = 1$, we recover the result for the exponential distribution. Also, with $\beta = \frac{1}{2}$ and $\alpha = \frac{n}{2}$:

(iii) CHI-SQUARE with n degrees of freedom

$$X \sim \chi_n^2 \, : \, \psi(u) = (1 - 2iu)^{-\frac{n}{2}}\,.$$

(This follows from (ii) since $\chi_n^2 = \gamma(\frac{n}{2}, \frac{1}{2})$.)

EXAMPLE 3.1.29: CAUCHY. An elementary computation gives

$$\psi(u) = \int_{-\infty}^{+\infty} \frac{1}{\pi}\frac{1}{1+x^2}\mathrm{e}^{iux}\mathrm{d}x = \mathrm{e}^{-|u|}\,.$$

Theorem 3.1.30 *The characteristic function of a real random variable characterizes its distribution.*

This means that if two random variables X and Y have the same characteristic function, then $P(X \le x) = P(Y \le x)$ for all $x \in \mathbb{R}$. The proof is omitted at this stage since a more general result is available in Section 5.2 (subsection *Characteristic Functions*, page 185).

Note that if f is continuous, and if ψ is integrable, a classical result of Fourier analysis tells us that

$$f(x) = \frac{1}{2\pi}\int_{\mathbb{R}} \mathrm{e}^{-iux}\psi(u)\,\mathrm{d}u\,.$$

This is a proof, in a particular case often encountered in practice, of the general result that the characteristic functions indeed characterize the distribution of a random variable.

Laplace Transforms

For non-negative random variables or random vectors with non-negative coordinates, one can also work with Laplace transforms rather than with characteristic functions.

Definition 3.1.31 *The* Laplace transform *of a non-negative random variable X (resp., of a cumulative probability distribution function F on $\mathbb{R}_+$) is the function $t \in \mathbb{R}_+ \mapsto E\left[\mathrm{e}^{-tX}\right]$ (resp. $t \in \mathbb{R}_+ \mapsto \int_{\mathbb{R}_+} \mathrm{e}^{-tx}\,\mathrm{d}F(x)$).*

The Laplace function characterizes the distribution of a non-negative random variable, in the sense that

Theorem 3.1.32 *Two non-negative random variables X and Y with the same Laplace transforms have the same distribution.*

The proof will be based on the following lemma of intrinsic interest.

Lemma 3.1.33 *Two bounded random variables X and Y such that $E[X^n] = E[Y^n]$ for all $n = 0, 1, \dots$ have the same distribution.*

Proof. Let $M < \infty$ be a common bound of these variables. The hypothesis implies that for any polynomial P, $E[P(X)] = E[P(Y)]$. By the Weierstrass approximation theorem[1] if $h : [0, M] \to \mathbb{R}$ is a continuous function, there exists for any $\varepsilon > 0$ a polynomial P_ε such that $\sup_{0 \le x \le M} |h(x) - P_\varepsilon(x)| \le \varepsilon$. Therefore

$$E[|h(X) - P_\varepsilon(X)|] \le \varepsilon \text{ and } E[|h(Y) - P_\varepsilon(Y)|] \le \varepsilon$$

and

$$\begin{aligned} &E[|h(X) - h(Y)|] \\ &\quad \le E[|h(X) - P_\varepsilon(X)|] + E[|P_\varepsilon(X) - P_\varepsilon(Y)|] + E[|h(Y) - P_\varepsilon(Y)|] \\ &\quad = E[|h(X) - P_\varepsilon(X)|] + E[|h(Y) - P_\varepsilon(Y)|] \le 2\varepsilon\,. \end{aligned}$$

Since ε can be chosen arbitrarily small, $E[h(X)] = E[h(Y)]$. By uniformly approximating the indicator function of any interval $(a, b] \subseteq [0, M]$ by a continuous function, we deduce that

$$P(X \in (a, b]) = E\left[1_{(a,b]}(X)\right] = E\left[1_{(a,b]}(Y)\right] = P(Y \in (a, b])\,,$$

[1] A refinement of this fundamental result was given in Example 2.1.24.

that is, $X \stackrel{\mathcal{D}}{=} Y$. □

We now prove Theorem 3.1.32.

Proof. The variables $U := \mathrm{e}^{-X}$ and $V := \mathrm{e}^{-Y}$ are bounded and such that for all $n = 0, 1, \ldots, E[U^n] = E[V^n]$, so that $U \stackrel{\mathcal{D}}{=} V$, and therefore $X = -\log U$ and $Y = -\log V$ have the same distribution. □

Random Vectors

We now consider random vectors, at first sight a notion not quite novel with respect to random variables. However it introduces dependency between two (or more) random variables.

Definition 3.1.34 *A* random vector *of dimension n is a collection of n real random variables*

$$X := (X_1, \ldots, X_n).$$

From a probabilistic point of view, each of the random variables $X_1, \ldots, X_n$ can be characterized by its cumulative distribution function. However, the CDF of each coordinate of a random vector does not completely describe the probabilistic behavior of the whole vector. See Exercise 3.6.16.

Throughout this book we shall use compact notations for multiple integrals, for instance

$$\int_{\mathbb{R}^n} g(x)\,\mathrm{d}x := \int_{-\infty}^{+\infty} \cdots \int_{-\infty}^{+\infty} g(x_1, \ldots, x_n)\,\mathrm{d}x_1 \cdots \mathrm{d}x_n.$$

.

A. *The absolutely continuous case.* Let $X = (X_1, \ldots, X_n)$ be a random vector taking its values in $\mathbb{R}^n$ and let $f : \mathbb{R}^n \to \mathbb{R}_+$ be a function such that

$$\int_{-\infty}^{+\infty} \cdots \int_{-\infty}^{+\infty} f(x_1, \ldots, x_n)\,\mathrm{d}x_1 \cdots \mathrm{d}x_n = 1. \tag{3.22}$$

Definition 3.1.35 *The random vector X taking its values in $\mathbb{R}^n$ is said to admit the* probability density $f : \mathbb{R}^n \to [0, 1]$ *if*

$$P(X \in C) = \int_C f(x)\,\mathrm{d}x \quad (C \in \mathcal{B}(\mathbb{R}^n)).$$

A random vector admitting a probability density is also called *absolutely continuous.*

B. *Discrete random vectors.* For convenience, we recall here a previously given definition with a slight change in the notation. Consider the random vector $X = (X_1, \dots, X_n)$ where all the random variables X_i take their values in the *same* (this restriction is not essential, but it simplifies the notation) denumerable space E. Let $f : E^n \to \mathbb{R}_+$ be a function such that

$$\sum_{x \in E^n} f(x) = 1 .$$

Definition 3.1.36 *The discrete random vector X above is said to admit the* probability distribution f *if for all sets $C \subseteq E^n$,*

$$P(X \in C) = \sum_{x \in C} f(x) .$$

In fact, as we already observed, there is nothing new here with respect to discrete random variables since X is a discrete random variable taking its values in the denumerable set $\mathcal{X} := E^n$.

C. *The mixed case.* Let $X = (X_1, \dots, X_n)$ be a random vector of the form $X = (Z, Y)$ where $Z = (X_1, \dots, X_k)$ $(k < n)$ is a discrete random variable taking its values in $\mathcal{Z} = E^k$ for some integer $k \geq 1$, where E is a denumerable set, and where $Y = (X_{k+1}, \dots, X_n)$ is a random vector with values in $\mathcal{Y} = \mathbb{R}^{n-k}$. Let $f : \mathcal{Z} \times \mathcal{Y} \to \mathbb{R}_+$ be a function such that

$$\sum_{z \in \mathcal{Z}} \int_{\mathbb{R}^{n-k}} f(z, y) \, \mathrm{d}y = 1 .$$

Definition 3.1.37 *The random vector X above is said to admit the* mixed density f *if*

$$P(X \in A, Y \in B) = \sum_{z \in A} \int_B f(z, y) \, \mathrm{d}y \quad (A \subseteq E^k, B \in \mathcal{B}(\mathbb{R}^n)) .$$

3.2 Continuous Random Vectors

The results below have obvious counterparts concerning discrete random vectors and random vectors with a mixed density, sums replacing integrals when necessary. The corresponding statements and proofs are left to the reader.

Theorem 3.2.1 *Let $X = (X_1, X_2) \in \mathbb{R}^2$ be a two-dimensional vector with probability density function f_{X_1,X_2}. The probability density function of X_1 is obtained by integrating out x_2:*

$$f_{X_1}(x_1) = \int_{-\infty}^{+\infty} f_{X_1,X_2}(x_1, x_2)\,\mathrm{d}x_2\,.$$

Proof. Indeed,

$$\begin{aligned} P(X_1 \leq a) &= P((X_1, X_2) \in (-\infty, a] \times \mathbb{R}) \\ &= \int_{-\infty}^{a} \int_{-\infty}^{+\infty} f_{X,X_2}(x_1, x_2)\,\mathrm{d}x_1\mathrm{d}x_2 \\ &= \int_{-\infty}^{a} \left(\int_{-\infty}^{+\infty} f_{X_1,X_2}(x_1, x_2)\,\mathrm{d}x_2 \right) \mathrm{d}x_1\,. \end{aligned}$$

□

Theorem 3.2.1 extends in an obvious way to the case where X_1 and X_2 are random *vectors*.

EXAMPLE 3.2.2: THE BUTTERFLY DISTRIBUTION. Let $X = (X_1, X_2)$ be an absolutely continuous two-dimensional random vector with probability density function

$$f_{X_1,X_2}(x_1, x_2) = 2 \times 1_C(x_1, x_2)$$

where $C := \left([0, \frac{1}{2}] \times [0, \frac{1}{2}]\right) \cup \left([\frac{1}{2}, 1] \times [\frac{1}{2}, 1]\right)$. (The support of this distribution has a "butterfly shape", hence the name.) Then $X_1 \sim \mathcal{U}[0, 1]$. Indeed, if $x_1 \in [0, \frac{1}{2}]$,

$$f_{X_1}(x_1) = \int_0^{\frac{1}{2}} 2\,\mathrm{d}x_2 = 1\,,$$

and if $x_1 \in [\frac{1}{2}, 1]$,

$$f_{X_1}(x_1) = \int_{\frac{1}{2}}^{1} 2\,\mathrm{d}x_2 = 1\,.$$

Similarly, $X_2 \sim \mathcal{U}[0, 1]$.

Definition 3.2.3 *For a function $g : \mathbb{R}^n \to \mathbb{R}$, the* expectation *of $g(X)$ when X admits a probability density f is, by definition, the quantity*

$$E[g(X)] := \int_{\mathbb{R}^n} g(x) f(x)\,\mathrm{d}x\,, \tag{3.23}$$

where it is required that either g be non-negative, or that

$$\int_{\mathbb{R}^n} |g(x)| f(x)\,\mathrm{d}x < \infty. \tag{3.24}$$

In the case that (3.24) holds, one says that $g(X)$ is *integrable*. Expectation so defined enjoys, *mutatis mutandis*, the properties mentioned for the scalar case: linearity (see (3.8)), monotonicity (see (3.9)), and the triangle inequality (see (3.10)).

In the following theorem the hypothesis of absolute continuity is crucial.

Theorem 3.2.4 *For any two independent absolutely continuous random variables X and Y, $P(X = Y) = 0$.*

More generally, if $(X_1, \ldots, X_n)$ is an absolutely continuous random vector, the probability of the event

$$C := \{\omega\,;\, X_i(\omega) = X_j(\omega) \text{ for some } i, j \text{ possibly depending on } \omega\}$$

is null.

Proof. We first do the proof for two random variables. Recall that for any event A, $E[1_A] = P(A)$. In particular,

$$P(X = Y) = E[1_{\{X=Y\}}] = \int_{-\infty}^{+\infty} \int_{-\infty}^{+\infty} g(x, y) \mathrm{d}x \, \mathrm{d}y,$$

where $g(x, y) = 1_{\{x=y\}} f_X(x) f_Y(y)$ is null outside the diagonal. Since the diagonal has a null area, the integral is null.

For the general case, we first observe that

$$C = \cup_{i \neq j} \{X_i = X_j\},$$

and therefore, by sub-σ-additivity,

$$P(C) \leq \sum_{i,j=1}^{n} P(X_i = X_j) = \sum_{i,j=1}^{n} 0 = 0.$$

□

Independence

For random vectors, we shall use the following type of abbreviation: $P(X \leq a)$ is short for $P(X_1 \leq a_1, \ldots, X_n \leq a_n)$, where $X = (X_1, \ldots, X_n)$ and $a = (a_1, \ldots, a_n)$

Definition 3.2.5 *Two random vectors X and Y of respective dimensions n and p are called* independent vectors *if for all $a \in \mathbb{R}^n, b \in \mathbb{R}^p$,*

$$P(X \leq a, Y \leq b) = P(X \leq a) P(Y \leq b). \tag{3.25}$$

Definition 3.2.6 *A sequence* $\{X_n\}_{n\in\mathbb{N}}$ *of real random vectors is called an* independent sequence *if for any* finite *collection of* distinct *random variables* $X_{i_1}, \ldots, X_{i_r}$ *from this sequence,*

$$\begin{aligned} &P(\{X_{i_1} \le a_1\} \cap \{X_{i_2} \le a_2\} \cap \cdots \cap \{X_{i_r} \le a_r\}) \\ &= P(X_{i_1} \le a_1) \times P(X_{i_2} \le a_2) \times \cdots \times P(X_{i_r} \le a_r) \end{aligned} \tag{3.26}$$

for all vectors (of appropriate dimensions) $a_1, \ldots, a_r$.

Definition 3.2.7 *The sequences of random vectors* $\{X_n\}_{n\in\mathbb{N}}$ *and* $\{Y_n\}_{n\in\mathbb{N}}$ *are called* independent sequences *if*

$$\begin{aligned} &P\left((\cap_{\ell=1}^{r}\{X_{i_\ell} \le a_\ell\}) \cap (\cap_{m=1}^{s}\{Y_{j_m} \le b_m\})\right) \\ &= P\left(\cap_{\ell=1}^{r}\{X_{i_\ell} \le a_\ell\}\right) P\left(\cap_{m=1}^{s}\{Y_{j_m} \le b_m\}\right) \end{aligned} \tag{3.27}$$

for all indices $i_1, \ldots, i_r$ *and* $j_1, \ldots, j_s \in \mathbb{N}$, *and all vectors (of appropriate dimensions)* $a_1, \ldots, a_r$ *and* $b_1, \ldots, b_s$.

Theorem 3.2.8 *A. If* $X_1, \ldots, X_n$ *are absolutely continuous random vectors with probability density functions* $f_1, \ldots, f_n$ *respectively, and if, moreover,* $X_1, \ldots, X_n$ *are independent, then the probability density function of the vector* $(X_1, \ldots, X_n)$ *is the product of the probability density functions of its components:*

$$f_X(x_1, \ldots, x_n) = f_1(x_1) \cdots f_n(x_n). \tag{3.28}$$

B. Conversely, if the vector X *has a probability density function factoring as in (3.28), where* $f_1, \ldots, f_n$ *are probability density functions, then* $X_1, \ldots, X_n$ *are independent random vectors with respective probability density functions* $f_1, \ldots, f_n$.

Proof. To simplify the writing we only consider the case $n = 2$ for random variables. A. If X_1, X_2 are absolutely continuous random variables with probability density functions f_1, f_2 respectively, and if, moreover, X_1, X_2 are independent, then

$$\begin{aligned} P(X_1 \le x_1, X_2 \le x_2) &= P(X_1 \le x_1)P(X_2 \le x_2) \\ &= \left(\int_{-\infty}^{x_1} f_1(y_1)\,dy_1\right)\left(\int_{-\infty}^{x_2} f_2(y)\,dy_2\right) \\ &= \int_{-\infty}^{x_1}\int_{-\infty}^{x_2} f_1(y_1)f_2(y_2)\,dy_1\,dy_2. \end{aligned}$$

B. We have

$$P(X_1 \le x_1, X_2 \le x_2) = \int_{-\infty}^{x_1}\int_{-\infty}^{x_2} f_1(y_1)f_2(y_2)\,dy_1\,dy_2,$$

that is, by Fubini's theorem,

$$P(X_1 \le x_1, X_2 \le x_2) = \left(\int_{-\infty}^{x_1} f_1(y_1)\,\mathrm{d}y_1\right) \times \left(\int_{-\infty}^{x_2} f_2(y_2)\,\mathrm{d}y_2\right).$$

Letting $x_2 = +\infty$ in the last identity yields

$$P(X_1 \le x_1) = \int_{-\infty}^{x_1} f_1(y_1)\,\mathrm{d}y_1,$$

which proves that X_1 has the probability density function f_1. Similarly, $P(X_2 \le x_2) = \int_{-\infty}^{x_2} f_2(y_2)\,\mathrm{d}y_2$, and therefore

$$P(X_1 \le x_1, X_2 \le x_2) = P(X_1 \le x_1)P(X_2 \le x_2),$$

which proves independence. □

EXAMPLE 3.2.9: THE UNIFORM DISTRIBUTION ON THE SQUARE. Let $X = (X_1, X_2)$ be an absolutely continuous two-dimensional random vector with probability density function

$$f_{X_1,X_2}(x_1, x_2) = 1_{[0,1]^2}(x_1, x_2).$$

Since this probability density function factors as the product $1_{[0,1]}(x_1) \times 1_{[0,1]}(x_2)$ of two probability density functions of uniform distributions on $[0, 1]$, $X_1 \sim \mathcal{U}[0, 1]$, $X_2 \sim \mathcal{U}[0, 1]$ and they are independent.

EXAMPLE 3.2.10: THE UNIFORM DISTRIBUTION ON THE DISK. Let $X = (X_1, X_2)$ be an absolutely continuous two-dimensional random vector uniformly distributed on the disk $D = \{(x_1, x_2);\ x_1^2 + x_2^2 \le 1\}$. Its probability density function is

$$f_{X_1,X_2}(x_1, x_2) = \frac{1}{\pi} 1_D(x_1, x_2).$$

Clearly, this probability density function does not factor as the product of two probability density functions and therefore X_1 X_2 are not independent.

Here again, a discrete and a mixed version of Theorem 3.2.8 are available.

Product Formula for Expectations

The following result was already given in particular cases in Theorems 2.1.27 and 2.3.6.

Theorem 3.2.11 *Let Y and Z be independent random vectors of dimension p and q respectively. If $g_1 : \mathbb{R}^p \to \mathbb{C}$ and $g_2 : \mathbb{R}^q \to \mathbb{C}$ are such that $g_1(Y)$ and $g_2(Z)$ are integrable, then the product $g_1(Y)g_2(Z)$ is integrable and*

$$E[g_1(Y)g_2(Z)] = E[g_1(Y)]E[g_2(Z)]. \tag{3.29}$$

Formula (3.29) holds true without condition if g_1 and g_2 are real non-negative functions.

Proof. We consider the case where Y and Z admit probability densities f_Y and f_Z. (The fully general result will be given in Theorem 5.4.4.) By Theorem 3.2.8, the vector $X = (Y, Z)$ admits the probability density

$$f_{Y,Z}(y, z) = f_Y(y)f_Z(z) \tag{3.30}$$

and the result follows by Fubini's theorem:

$$\begin{aligned} E[g_1(Y)g_2(Z)] &= \int\int g_1(y)g_2(z)f_Y(y)f_Z(z)\,\mathrm{d}y\,\mathrm{d}z \\ &= \int g_1(y)f_Y(y)\,\mathrm{d}y \times \int g_2(z)f_Z(z)\,\mathrm{d}z\,. \end{aligned}$$

□

Freeze and Integrate

The next result is convenient for computing expectations of functions of two independent vectors. It says that one may fix one of these vectors, compute the expectation with respect to the other vector, and take the expectation of the result with respect to the previously fixed vector. In other words: *freeze and integrate.*

Theorem 3.2.12 *Let X_1, X_2 be independent random vectors of dimensions n_1 and n_2, and with probability density functions f_1 and f_2 respectively. Then for any function $g : \mathbb{R}^{n_1} \times \mathbb{R}^{n_2} \to \overline{\mathbb{R}}$ that is either non-negative or such that $g(X_1, X_2)$ is integrable,*

$$E[g(X_1, X_2)] = \textstyle\int_{-\infty}^{+\infty} E[g(y, X_2)]f_1(y)\mathrm{d}y\,.$$

Proof. We have

$$\begin{aligned} E[g(X_1, X_2)] &= \int_{-\infty}^{+\infty}\int_{-\infty}^{+\infty} g(x_1, x_2)f_1(x_1)f_2(x_2)\mathrm{d}x_1\mathrm{d}x_2 \\ &= \int_{-\infty}^{+\infty} f_1(x_1)\left\{\int_{-\infty}^{+\infty} g(x_1, x_2)f_2(x_2)\mathrm{d}x_2\right\}\mathrm{d}x_1 \\ &= \int_{-\infty}^{+\infty} f_1(x)E[g(x, X_2)]\mathrm{d}x\,. \end{aligned}$$

□

EXAMPLE 3.2.13: THE COMPUTATION OF $P(X_1 > X_2)$. Let X_1, X_2 be as in Theorem 3.2.12. Then

$$P(X_1 > X_2) = \int_{-\infty}^{+\infty} P(y > X_2) f_1(y) \mathrm{d}y = \int_{-\infty}^{+\infty} (1 - F_2(y)) f_1(y) \mathrm{d}y .$$

To prove this, it suffices to apply Theorem 3.2.12 with $g(x_1, x_2) = 1_{\{x_1 > x_2\}}$ and to observe that $E[g(y, X_2)] = E[1_{\{y > X_2\}}] = P(y > X_2)$.

If for instance $X_1 \sim \mathcal{E}(\lambda_1)$ and $X_2 \sim \mathcal{E}(\lambda_2)$, we obtain by application of the last displayed formula,

$$\begin{aligned} P(X_1 > X_2) &= \int_{-\infty}^{+\infty} \mathrm{e}^{-\lambda_2 y} \lambda_1 \mathrm{e}^{-\lambda_1 y} \mathrm{d}y \\ &= \int_{-\infty}^{+\infty} \mathrm{e}^{-\lambda_2 y} \lambda_1 \mathrm{e}^{-(\lambda_1 + \lambda_2) y} \mathrm{d}y = \frac{\lambda_1}{\lambda_1 + \lambda_2} . \end{aligned}$$

We now give the *convolution formula* for the probability density function of a random vector that is the sum of two independent absolutely continuous random vectors.

Theorem 3.2.14 *The probability density function of the random vector $Z = X + Y$, where X and Y are independent random vectors with the same dimension n and with respective probability densities f_X and f_Y, is given by the convolution formula*

$$f_Z(z) = \int_{\mathbb{R}^n} f_Y(z - y) f_X(y) \mathrm{d}y . \tag{3.31}$$

Proof. ($n = 1$ for simplicity) The probability density function of the vector (X, Y) is $f_X(x) f_Y(y)$, and therefore, for all $a \in \mathbb{R}$,

$$\begin{aligned} P(Z \le a) &= P(X + Y \le a) = E[1_{\{X+Y \le a\}}] \\ &= \int_{-\infty}^{+\infty} \int_{-\infty}^{+\infty} 1_{\{x+y \le a\}} f_X(x) f_Y(y) \mathrm{d}x \mathrm{d}y . \end{aligned}$$

The latter integral can be written, by Fubini's theorem,

$$\begin{aligned} &\int_{-\infty}^{+\infty} \left\{ \int_{-\infty}^{+\infty} 1_{\{y \le a - x\}} f_Y(y) \mathrm{d}y \right\} f_X(x) \mathrm{d}x \\ &\qquad = \int_{-\infty}^{+\infty} \left\{ \int_{-\infty}^{a-x} f_Y(y) \mathrm{d}y \right\} f_X(x) \mathrm{d}x , \end{aligned}$$

that is, after an obvious change of variable,

$$P(Z \leq a) = \int_{-\infty}^{a} \left\{ \int_{-\infty}^{+\infty} f_Y(z-x) f_X(x) \mathrm{d}x \right\} \mathrm{d}z .$$

□

EXAMPLE 3.2.15: SUM OF INDEPENDENT UNIFORM RANDOM VARIABLES. Let $X \sim \mathcal{U}[0,1]$ and $Y \sim \mathcal{U}[0,1]$ be two independent random variables. Then $Z = X + Y$ admits the "triangular" probability density function

$$f_Z(z) = z\, 1_{[0,1]}(z) + (1-z) 1_{[1,2]}(z) .$$

This is an immediate application of (3.31) with $f_X(x) = 1_{[0,1]}(x)$ and $f_Y(y) = 1_{[0,1]}(y)$.

Characteristic Functions and Laplace Transforms of Random Vectors

The definition and properties of characteristic functions readily extend to the case of random vectors.

Definition 3.2.16 *The characteristic function of the real random vector $X = (X_1, \dots, X_n)$ is the function $\psi : \mathbb{R}^n \to \mathbb{C}$ defined by*

$$\psi(u) = E[\mathrm{e}^{iu^T X}] . \tag{3.32}$$

In the case where the X is an absolutely continuous random vector with continuous probability density f,

$$\psi(u) = \int_{\mathbb{R}^n} \mathrm{e}^{iu^T x} f(x)\, \mathrm{d}x .$$

If moreover,

$$\int_{\mathbb{R}^n} |\psi(u)|\, \mathrm{d}u < \infty ,$$

a theorem of analysis tells us that the probability density function of X is then given by the Fourier inversion formula

$$f(x) = \frac{1}{2\pi} \int_{\mathbb{R}^n} \psi(u)\, \mathrm{e}^{-iu^T x}\, \mathrm{d}u .$$

(The proof is given in Corollary 5.3.3.)

Theorem 3.2.17 *If two random vectors X and Y have the same characteristic function, they have the same distribution.*

The proof is postponed until Corollary 5.3.2.

Definition 3.2.18 *The* Laplace transform *of a vector of non-negative random variables* $(X_1, \dots, X_m)$ *is the function* $(t_1, \dots, t_m) \in \mathbb{R}_+^m \mapsto E\left[\mathrm{e}^{-(t_1 X_1 + t_m X_m)}\right] \in [0, 1]$.

The following result, which will be admitted,[2] generalizes Theorem 3.1.32.

Theorem 3.2.19 *Two non-negative random vectors*

$$(X_1, \dots, X_m) \text{ and } (Y_1, \dots, Y_m)$$

with the same Laplace transforms have the same distribution.

Characteristic Function Test for Independence

Characteristic functions give one of the most useful criteria for testing independence of random vectors.

Theorem 3.2.20 *Suppose that* Y *and* Z *are two random vectors of respective dimensions* p *and* q*, and that for all* $v \in \mathbb{R}^p$, $w \in \mathbb{R}^q$*, it holds that*

$$E[\mathrm{e}^{i(v^T Y + w^T Z)}] = \psi_1(v)\psi_2(w)\,, \tag{3.33}$$

where $\psi_1(v)$ *and* $\psi_2(w)$ *are the characteristic functions of some random vectors* $\tilde{Y}$ *and* $\tilde{Z}$ *of respective dimensions* p *and* q*. Then* Y *and* Z *are independent,* Y *has the same distribution as* $\tilde{Y}$*, and* Z *has the same distribution as* $\tilde{Z}$*.*

Proof. Define $X = (Y, Z)$ and $u = (v, w)$, so that (3.33) reads

$$E[\mathrm{e}^{iu^T X}] = \psi(u) = \psi_1(v)\psi_2(w)\,.$$

Consider two independent random vectors $\hat{Y}$ and $\hat{Z}$, where $\hat{Y}$ is distributed as Y, and $\hat{Z}$ is distributed as Z. Let $\hat{X} = (\hat{Y}, \hat{Z})$. Then, by the product formula for expectations,

$$\begin{aligned} E[\mathrm{e}^{iu^T \hat{X}}] &= E[\mathrm{e}^{iv^T \hat{Y}} \mathrm{e}^{iw^T \hat{Z}}] = E[\mathrm{e}^{iv^T \hat{Y}}]E[\mathrm{e}^{iw^T \hat{Z}}] \\ &= E[\mathrm{e}^{iv^T Y}]E[\mathrm{e}^{iw^T Z}] = \psi_1(v)\psi_2(w)\,. \end{aligned}$$

Therefore, (Y, Z) has the same distribution as $(\hat{Y}, \hat{Z})$ and in particular, Y and Z are independent. □

[2] See Chapter 6 of [11].

EXAMPLE 3.2.21: CONVOLUTION FORMULA VIA FOURIER. We give an alternative proof of Theorem 3.2.14. Define

$$(f_Y * f_Z)(x) := \int_{\mathbb{R}^n} f_Y(x-z)\, f_Z(z)\, \mathrm{d}z\,.$$

First observe that $f := f_Y * f_Z$ is a probability density function, that is, a non-negative function integrating to 1:

$$\begin{aligned}\int_{\mathbb{R}^n} f(x)\, \mathrm{d}x &= \int_{\mathbb{R}^n} \left(\int_{\mathbb{R}^n} f_Y(x-z)\, f_Z(z)\, \mathrm{d}z\right) \mathrm{d}x \\ &= \int_{\mathbb{R}^n} \left(\int_{\mathbb{R}^n} f_Y(x-z)\, \mathrm{d}x\right) f_Z(z)\, \mathrm{d}z \\ &= \int_{\mathbb{R}^n} 1 \times f_Z(z)\, \mathrm{d}z = 1\,.\end{aligned}$$

By a classical result of Fourier analysis, the Fourier transform of $f_Y * f_Z$ is the product of the Fourier transforms of f_Y and f_Z, that is

$$\int_{\mathbb{R}^n} \mathrm{e}^{iu^T x} f(x)\, \mathrm{d}x = E[\mathrm{e}^{iu^T Y}]E[\mathrm{e}^{iu^T Z}] = E[\mathrm{e}^{iu^T (Y+Z)}]\,,$$

where we have used the independence of Y and Z for the second equality. Thus $E[\mathrm{e}^{iu^T(Y+Z)}]$ is the characteristic function of $Y+Z$ and of a random vector with probability density function $f(x)$. Therefore f is the probability density function of $Y+Z$.

Random Sums and Wald's Identity

The next two results concerning random sums were already proved in the discrete case (Theorems 2.3.12 and 2.5.18).

Theorem 3.2.22 *Let $\{Y_n\}_{n\geq 1}$ be an* IID *sequence of random variables with the common characteristic function φ_Y. Let T be a random variable, integer-valued, independent of the sequence $\{Y_n\}_{n\geq 1}$, and let g_T be its generating function. The characteristic function of the random variable*

$$X := \sum_{n=1}^{T} Y_n\,,$$

where by convention $\sum_{n=1}^{0} = 0$, is

$$\varphi_X(u) = g_T\left(\varphi_Y(u)\right)\,. \tag{3.34}$$

Proof. We need only to adapt the proof of Theorem 2.3.12. We have

$$\begin{aligned} e^{iuX} &= e^{iu\sum_{n=1}^{T} Y_n} = \left(\sum_{k=0}^{\infty} 1_{\{T=k\}}\right) e^{iu\sum_{n=1}^{T} Y_n} \\ &= \sum_{k=0}^{\infty} \left\{ \left(e^{iu\sum_{n=1}^{T} Y_n}\right) 1_{\{T=k\}} \right\} = \sum_{k=0}^{\infty} \left(e^{iu\sum_{n=1}^{k} Y_n}\right) 1_{\{T=k\}} \,. \end{aligned}$$

Therefore,

$$E[e^{iuX}] = \sum_{k=0}^{\infty} E\left[1_{\{T=k\}} \left(e^{iu\sum_{n=1}^{k} Y_n}\right)\right] = \sum_{k=0}^{\infty} E[1_{\{T=k\}}]E[e^{iu\sum_{n=1}^{k} Y_n}]\,,$$

where we have used independence of T and $\{Y_n\}_{n\geq 1}$. Now, $E[1_{\{T=k\}}] = P(T = k)$, and $E[e^{iu\sum_{n=1}^{k} Y_n}] = \varphi_Y(u)^k$, and therefore

$$E[e^{iuX}] = \sum_{k=0}^{\infty} P(T = k)\varphi_Y(u)^k = g_T\left(\varphi_Y(u)\right) \,.$$

□

Theorem 3.2.23 *Let $\{Y_n\}_{n\geq 1}$ be a sequence of integrable random variables such that $E[Y_n] = E[Y_1]$ for all $n \geq 1$. Let T be an integer-valued random variable such that for all $n \geq 1$, the event $\{T \geq n\}$ is independent of Y_n. Let*

$$X := \textstyle\sum_{n=1}^{T} Y_n \,.$$

Then

$$E\left[X\right] = E[Y_1]E[T] \,. \tag{3.35}$$

Proof. Same as that of Theorem 2.5.18. □

Smooth Change of Variables

Let $X = (X_1, \ldots, X_n)$ be a random vector with the probability density function f_X, and define the random vector $Y = g(X)$, where $g : \mathbb{R}^n \to \mathbb{R}^n$. More explicitly,

$$\begin{cases} Y_1 = g_1(X_1, \ldots, X_n), \\ \vdots \\ Y_n = g_n(X_1, \ldots, X_n). \end{cases}$$

Under smoothness assumptions on g, the random vector Y is absolutely continuous, and its probability density function can be explicitly computed from g and the probability density function f_X. These assumptions are the following:

A_1: The function $g : U \to \mathbb{R}^n$, where U is an open subset of $\mathbb{R}^n$, is one-to-one (injective).

A_2: The coordinate functions g_i $(1 \le i \le n)$ are continuously differentiable.

A_2: Moreover, denoting the Jacobian matrix of the function g by

$$J_g(x_1, \dots, x_n) := \left\{ \frac{\partial g_i}{\partial x_j}(x_1, \dots, x_n) \right\}_{1 \le i,j \le n},$$

we assume that on U,

$$|\det J_g(x_1, \dots, x_n)| > 0\,.$$

A standard result of Analysis says that $V = g(U)$ is an open subset of $\mathbb{R}^n$, and that the invertible function $g : U \to V$ admits an inverse $g^{-1} : V \to U$ with the same properties as the direct function g. In particular, on V,

$$|\det J_{g-1}(y_1, \dots, y_n)| > 0\,.$$

Moreover,

$$J_{g-1}(y) = J_g(g^{-1}(y))^{-1}\,.$$

Also, under conditions $A_1 - A_3$, we have the basic rule of *change of variables* of calculus: For any function $u : \mathbb{R}^n \to \mathbb{R}^n$,

$$\int_U u(x)\mathrm{d}x = \int_{g(U)} u(g^{-1}(y))|\det J_{g^{-1}}(y)|\mathrm{d}y\,.$$

Theorem 3.2.24 *Under the conditions just stated for X, g, and U, and if moreover $P(X \in U) = 1$, then Y admits the probability density*

$$f_Y(y) = f_X(g^{-1}(y))|\det J_g(g^{-1}(y))|^{-1} 1_V(y)\,. \tag{3.36}$$

Proof. The proof consists in checking that for any bounded function $h : \mathbb{R} \to \mathbb{R}$,

$$E[h(Y)] = \int_{\mathbb{R}^n} h(y)\psi(y)\mathrm{d}y\,, \tag{3.37}$$

where ψ is the function on the right-hand side of (3.36). Indeed, taking $h(y) = 1_{y \le a} = 1_{y_1 \le a_1} \cdots 1_{y_n \le a_n}$, (3.37) reads

$$P(Y_1 \le a_1, \dots, Y_n \le a_n) = \int_{-\infty}^{a_1} \cdots \int_{-\infty}^{a_n} \psi(y_1, \dots, y_n)\mathrm{d}y_1 \cdots \mathrm{d}y_n\,.$$

To prove that (3.37) holds with the appropriate ψ, one just uses the basic rule of change of variables:

$$\begin{aligned} E[h(Y)] &= E[h(g(X))] = \int_U h(g(x)) f_X(x) \mathrm{d}x \\ &= \int_V h(y) f_X(g^{-1}(y)) |\det J_{g-1}(y)| \mathrm{d}y. \end{aligned}$$

□

Corollary 3.2.25 *Let X be an n-dimensional random vector with probability density f_X. Let A be an invertible $n \times n$ real matrix and b an n-dimensional real vector. Then, the random vector $Y = AX + B$ admits the density*

$$f_Y(y) = f_X(A^{-1}(y - b)) \frac{1}{|\det A|} . \tag{3.38}$$

Proof. Here $U = \mathbb{R}^n$, $g(x) = Ax + b$, and $|\det J_{g^{-1}}(y)| = \frac{1}{|\det A|}$. □

EXAMPLE 3.2.26: POLAR COORDINATES. Let (X_1, X_2) be a 2-dimensional random vector with probability density $f_{X_1,X_2}(x_1, x_2)$, and let (R, Θ) be its polar coordinates. The probability density of (R, Θ) is given by the formula

$$f_{R,\Theta}(r, \theta) = f_{X_1,X_2}(r \cos\theta,\ r \sin\theta)\, r\, .$$

Proof. Here g is the bijective function from the open set U consisting of $\mathbb{R}^2$ without the half-line $\{(x_1, 0)\,;\, x_1 \geq 0\}$ to the open set $V = (0, \infty) \times (0, 2\pi)$. The inverse function is

$$x = r\cos\theta \qquad y = r\sin\theta\, ,$$

with Jacobian

$$J_{g^{-1}}(r, \theta) = \begin{pmatrix} \cos\theta & -r\sin\theta \\ \sin\theta & r\cos\theta \end{pmatrix}$$

of determinant $\det J_{g^{-1}}(r, \theta) = r$. Applying formula (3.36), we obtain the announced result. □

There are cases of practical interest where the function g does not admit an inverse but where its domain U can be decomposed into disjoint open sets (say, 2): $U = U_1 + U_2$, such that the restrictions of g to U_1 and U_2, respectively

g_1 and g_2, satisfy the conditions of smoothness and of injectivity of the standard case. In this case the same method applies, but one must dissociate the integrals:

$$\int_U h(g(x))f_X(x)\,\mathrm{d}x = \int_{U_1} h(g_1(x))f_X(x)\,\mathrm{d}x + \int_{U_2} h(g_2(x))f_X(x)\,\mathrm{d}x$$

and apply the formula of smooth change of variables to each part separately. This gives

$$E\left[h(Y)\right] = \int_{g_1(U_1)} h(y)f_X(g_1^{-1}(y))\left|J_{g_1^{-1}}(y)\right|\,\mathrm{d}y + \int_{g_2(U_2)} h(y)f_X(g_2^{-1}(y))\left|J_{g_2^{-1}}(y)\right|\,\mathrm{d}y$$

and therefore

$$f_Y(y) = f_X(g_1^{-1}(y))\left|J_{g_1^{-1}}(y)\right| 1_{g_1(U_1)}(y) + f_X(g_2^{-1}(y))\left|J_{g_2^{-1}}(y)\right| 1_{g_2(U_1)}(y)\,.$$

Order Statistics

We now give a formula which allows us to compute the probability density function of the random vector obtained by reordering the coordinates of a given absolutely continuous random vector.

Let $X_1, \dots, X_n$ be independent random variables with the same probability density function f. We know (see Theorem 3.2.4) that the probability of two or more among $X_1, \dots, X_n$ taking the same value is null. Therefore one can define unambiguously the random variables $Z_1, \dots, Z_n$ obtained by arranging $X_1, \dots, X_n$ in increasing order:

$$\begin{cases} Z_i \in \{X_1, \dots, X_n\}, \\ Z_1 \le Z_2 \le \cdots \le Z_n. \end{cases}$$

In particular, $Z_1 = \min(X_1, \dots, X_n)$ and $Z_n = \max(X_1, \dots, X_n)$.

Theorem 3.2.27 *The probability density of the reordered vector $Z = (Z_1, \dots, Z_n)$ (defined above) is*

$$f_Z(z_1, \dots, z_n) = n!\left\{\prod_{j=1}^n f(z_j)\right\} 1_C(z_1, \dots, z_n)\,, \tag{3.39}$$

where $C := \{(z_1, \dots, z_n) \in \mathbb{R}^n\,;\; z_1 \le z_2 \le \cdots \le z_n\}$.

Proof. Let σ be the permutation of $\{1, \dots, n\}$ that orders $X_1, \dots, X_n$ in ascending order, that is,

$$X_{\sigma(i)} = Z_i$$

(note that σ is a *random permutation*). For any set $A \subseteq \mathbb{R}^n$,

$$P(Z \in A) = P(Z \in A \cap C)$$
$$= P(X_\sigma \in A \cap C) = \sum_{\sigma_o} P(X_{\sigma_o} \in A \cap C, \sigma = \sigma_o),$$

where the sum is over all permutations of $\{1, \ldots, n\}$. Observing that $X_{\sigma_o} \in A \cap C$ implies $\sigma = \sigma_o$,

$$P(X_{\sigma_o} \in A \cap C, \sigma = \sigma_o) = P(X_{\sigma_o} \in A \cap C)$$

and therefore since the probability distribution of X_{σ_o} does not depend upon a fixed permutation σ_o (here we invoke the independence and equidistribution assumption for the X_i's),

$$P(X_{\sigma_o} \in A \cap C) = P(X \in A \cap C).$$

Therefore,

$$\begin{aligned} P(Z \in A) &= \sum_{\sigma_o} P(X \in A \cap C) = n! P(X \in A \cap C) \\ &= n! \int_{A \cap C} f_X(x) \mathrm{d}x = \int_A n! f_X(x) 1_C(x) \mathrm{d}x. \end{aligned}$$

□

EXAMPLE 3.2.28: VOLUME OF A RIGHT-ANGLED PYRAMID. We shall apply the above result to prove the formula

$$\int_a^b \cdots \int_a^b 1_C(z_1, \ldots, z_n) \mathrm{d}z_1 \cdots \mathrm{d}z_n = \frac{(b-a)^n}{n!}. \tag{3.40}$$

Indeed, when the X_i's are uniformly distributed over $[a, b]$,

$$f_Z(z_1, \ldots, z_n) = \frac{n!}{(b-a)^n} 1_{[a,b]^n}(z_1, \ldots, z_n) 1_C(z_1, \ldots, z_n). \tag{3.41}$$

The result follows since $\int_{\mathbb{R}^n} f_Z(z)\, \mathrm{d}z = 1$.

EXAMPLE 3.2.29: THE i-TH SMALLEST UNIFORM. We seek the probability density function of the random variable Z_i, the ith smallest among $X_1, \ldots, X_n$, when the X_i's are independent random variables uniformly distributed on $[0, 1]$.

By Theorem 3.2.27, the distribution of $Z = (Z_1, \dots, Z_n)$ is

$$f_Z(z_1, \dots, z_n) = \frac{1}{n!} 1_{[0,1]^n \cap C}(z_1, \dots, z_n) ,$$

where $C = \{x_1 \le x_2 \le \dots \le x_n\}$. The density of Z_i is obtained by integrating f_Z with respect to $z_1, \dots, z_{i-1}, z_{i+1}, \dots, z_n$:

$$\begin{aligned} f_{Z_i}(z) &= n! \underbrace{\int_0^1 \dots \int_0^1}_{n-1} 1_{(z_1 \le \dots \le z_{i-1} \le z \le z_{i+1} \le \dots \le z_n)} \mathrm{d}z_1 \dots \mathrm{d}z_{i-1} \mathrm{d}z_{i+1} \dots \mathrm{d}z_n \\ &= n! \underbrace{\int_0^1 \dots \int_0^1}_{i-1} 1_{(z_1 \le \dots \le z_{i-1} \le z)} \mathrm{d}z_1 \dots \mathrm{d}z_{i-1} \times \cdots \\ &\qquad \cdots \underbrace{\int_0^1 \dots \int_0^1}_{n-i} 1_{(z \le z_{i+1} \le \dots \le z_n)} \mathrm{d}z_{i+1} \dots \mathrm{d}z_n , \end{aligned}$$

that is, in view of the result of Example 3.2.28,

$$f_{Z_i}(z) = n! \frac{z^{i-1}}{(i-1)!} \frac{(1-z)^{n-i}}{(n-i)!} = n \binom{n-1}{i-1} z^{i-1} (1-z)^{n-1} .$$

Sampling a Distribution

We now address a problem that arises in the context of simulation of stochastic systems. It consists in generating a random variable with prescribed CDF, or in other terms, *sampling* the said CDF. For this, one is allowed to use a random generator that produces a sequence $U_1, U_2, \dots$ of independent real random variables, uniformly distributed on $[0, 1]$. In practice, the numbers that such random generators produce are not quite random, but they look as if they are (they are called *pseudo-random generators*). The topic of how to devise a good pseudo-random generator is out of our scope, and we shall admit that we can trust our favorite computer to provide us with an IID sequence of random variables uniformly distributed on $[0, 1]$ (from now on we call them *random numbers*).

We now give two methods for constructing a random variable Z with CDF $F(z) = P(Z \le z)$.

In the case where Z is a discrete random variable with distribution $P(Z = a_i) = p_i$ $(0 \le i \le K)$, the basic principle of the sampling algorithm is the following

(α) Draw $U \sim \mathcal{U}([0,1])$.

(β) Set $Z = a_\ell$ if $p_0 + p_1 + \cdots + p_{\ell-1} \le U \le p_0 + p_1 + \cdots + p_\ell$.

This method is called the *method of the inverse.*

A crude generation algorithm would successively perform the tests $U \le p_0$?, $U \le p_0 + p_1$?, ..., until the answer is positive. The average number of iterations required would therefore be $\sum_{i\ge 0}(i+1)p_i = 1 + E[Z]$. This number may be too large, but there are ways of improving this, as Example 7.2.2 will show for the Poisson distribution.

The above method can be generalized to real random variables. Since for $u \in (0,1)$, the set $\{x\,;\, F(x) \ge u\}$ is an unbounded interval of $\mathbb{R}$, it admits a smallest element denoted by $F^{\leftarrow}(u)$:

$$\{x\,;\, F(x) \ge u\} = [F^{\leftarrow}(u), +\infty)\,.$$

The function $F^{\leftarrow}$ so defined on $(0,1)$ is non-decreasing. It is called the *pseudo-inverse* of F and coincides with the inverse function F^{-1} when F is continuous and strictly increasing.

Theorem 3.2.30 *If U is a uniform random variable on $(0,1)$, then $F^{\leftarrow}(U)$ has the same probability distribution as X.*

Proof. First note that for all $u \in (0,1)$, $F^{\leftarrow}(u) \le t$ implies $F(t) \ge u$. Indeed, in this case, for all $s > t$ there exists an $x < s$ such that $F(x) > u$ and therefore $F(s) > u$; and consequently, by right-continuity of F, $F(t) \ge u$. Conversely, $F(t) \ge u$ implies that $t \in \{x\,;\, F(x) \ge u\}$ and therefore $F^{\leftarrow}(u) \le t$. Taking all this into account,

$$\begin{aligned} F(t) = P(U < F(t)) &\le P(F^{\leftarrow}(U) \le t) \\ &\le P(F(t) \ge U) = F(t)\,. \end{aligned}$$

This forces $P(F^{\leftarrow}(U) \le t)$ to equal $F(t)$. □

EXAMPLE 3.2.31: EXPONENTIAL DISTRIBUTION. We want to sample from $\mathcal{E}(\lambda)$. The corresponding CDF is

$$F(z) = 1 - \mathrm{e}^{-\lambda z} \quad (z \ge 0)\,.$$

The solution of $y = 1 - \mathrm{e}^{-\lambda z}$ is $z = -\frac{1}{\lambda}\ln(1-y) = F^{-1}(y)$, and therefore, $Z = -\frac{1}{\lambda}\ln(1-U)$ will do, or since U and $1-U$ have the same distribution,

$$Z = -\tfrac{1}{\lambda}\ln U\,.$$

Here is an often useful trick.

EXAMPLE 3.2.32: SYMMETRIC EXPONENTIAL DISTRIBUTION. We want to sample from the symmetric exponential distribution with probability density function

$$f(x) = \frac{1}{2}\mathrm{e}^{-\lambda|x|}.$$

One way is to generate two independent random variables Y and Z where $Z \sim \mathcal{E}(\lambda)$ and $P(Y = +1) = P(Y = -1) = \frac{1}{2}$. Taking $X = YZ$ we have that

$$\begin{aligned} P(X \leq x) &= P(U = +1, Z \leq x) + P(U = -1, Z \geq -x) \\ &= \frac{1}{2}(F_Z(x) + 1 - F_Z(-x)), \end{aligned}$$

and therefore, taking derivatives,

$$f_X(x) = \frac{1}{2}(f_Z(x) + f_Z(-x)) = \frac{1}{2}f_Z(|x|).$$

The computation of the inverse of the cumulative distribution function of the random variable to be generated may be difficult. An alternative method is the *method of acceptance-rejection* below.

Let $\{Y_n\}_{n\geq 1}$ be a sequence of IID random variables with probability density g that satisfies the two requirements below:

(i) it is easy (or at least feasible) to sample it, and

(ii) for all $x \in \mathbb{R}$

$$\frac{f(x)}{g(x)} \leq c \tag{3.42}$$

for some finite constant c (necessarily larger or equal to 1).

Let $\{U_n\}_{n\geq 1}$ be a sequence of IID random variables uniformly distributed on $[0, 1]$.

Theorem 3.2.33 *Let τ be the first index $n \geq 1$ for which*

$$U_n \leq \frac{f(Y_n)}{cg(Y_n)}$$

and let $Z = Y_\tau$. Then

(a) *Z admits the probability density function f, and*

(b) *$E[\tau] = c$.*

Proof. We have

$$P(Z \leq x) = P(Y_\tau \leq x) = \sum_{n\geq 1} P(\tau = n, Y_n \leq x)\,.$$

Denote by A_k the event $\{U_k > \frac{f(Y_k)}{cg(Y_k)}\}$. Then

$$\begin{aligned} P(\tau = n, Y_n \leq x) &= P(A_1, \dots, A_{n-1}, \overline{A_n}, Y_n \leq x) \\ &= P(A_1) \cdots P(A_{n-1}) P(\overline{A_n}, Y_n \leq x)\,, \end{aligned}$$

$$\begin{aligned} P\left(\overline{A_k}\right) &= \int_{\mathbb{R}} P\left(U_k \leq \frac{f(y)}{cg(y)}\right) g(y)\, \mathrm{d}y \\ &= \int_{\mathbb{R}} \frac{f(y)}{cg(y)} g(y)\, \mathrm{d}y = \int_{\mathbb{R}} \frac{f(y)}{c} \mathrm{d}y = \frac{1}{c}\,, \end{aligned}$$

$$\begin{aligned} P\left(\overline{A_k}, Y_k \leq x\right) &= \int_{\mathbb{R}} P\left(U_k \leq \frac{f(y)}{cg(y)}\right) 1_{y\leq x} g(y)\, \mathrm{d}y \\ &= \int_{-\infty}^{x} \frac{f(y)}{cg(y)} g(y)\, \mathrm{d}y = \int_{-\infty}^{x} \frac{f(y)}{c}\, \mathrm{d}y = \frac{1}{c} \int_{-\infty}^{x} f(y)\, \mathrm{d}y\,. \end{aligned}$$

Therefore

$$P(Z \leq x) = \sum_{n\geq 1} \left(1 - \frac{1}{c}\right)^{n-1} \frac{1}{c} \int_{-\infty}^{x} f(y)\mathrm{d}y = \int_{-\infty}^{x} f(y)\mathrm{d}y\,.$$

Also, using the above calculations,

$$\begin{aligned} P(\tau = n) &= P\left(A_1, \dots, A_{n-1}, \overline{A_n}\right) \\ &= P(A_1) \cdots P(A_{n-1}) P\left(\overline{A_n}\right) = \left(1 - \frac{1}{c}\right)^{n-1} \frac{1}{c}\,, \end{aligned}$$

from which it follows that $E[\tau] = c$. □

We see that the method depends on our ability to easily generate random vectors with the probability density g. Also we have to select a probability density function satisfying the constraint (3.42), with c as small as possible.

3.3 Square-integrable Random Variables

Definition 3.3.1 *A complex random variable X is said to be* square-integrable *if* $E[|X|^2] < \infty$.

Theorem 3.3.2 *The set of complex square-integrable random variables, denoted $\mathcal{L}^2_{\mathbb{C}}(P)$, is a vector space with scalar field $\mathbb{C}$. Similarly, the set of real square-integrable random variables, denoted $\mathcal{L}^2_{\mathbb{R}}(P)$, is a vector space with scalar field $\mathbb{R}$.*

Proof. We show that if X and Y are complex square-integrable random variables and $\lambda \in \mathbb{C}$, then λX and $X+Y$ are square-integrable. The first assertion is obvious. For the last assertion use (for instance) the inequality $(a+b)^2 \le 2a^2 + 2b^2$, true for all $a, b \in \mathbb{R}$, to obtain

$$\begin{aligned} E[|X+Y|^2] &\le E[(|X|+|Y|)^2] \\ &\le E[(|X|+|Y|)^2] \le 2E[(|X|)^2] + 2E[(|Y|)^2] < \infty\,. \end{aligned}$$

□

Lemma 3.3.3 *A non-negative random variable Z such that $E[Z] = 0$ is almost surely equal to 0.*

Proof. By Markov's inequality, $P(Z \ge \frac{1}{n}) \le nE[Z] = 0$, and therefore, by the sequential continuity of probability

$$\begin{aligned} P(Z > 0) &= P\left(\bigcup_{n=1}^{\infty}\left\{Z \ge \frac{1}{n}\right\}\right) \\ &= \lim_{n\uparrow\infty} P\left(Z \ge \frac{1}{n}\right) = 0\,, \end{aligned}$$

that is, $P(Z = 0) = 1$. □

Inner Product and Schwarz's Inequality

Definition 3.3.4 *Let H be a vector space with scalar field $K = \mathbb{C}$ or $\mathbb{R}$, endowed with a mapping from $H \times H$ to K associating to the pair (x, y) of vectors of H the scalar $\langle x, y\rangle$, and such that for all $x, y, z \in H$ and all $\lambda \in K$,*

1. $\langle y, x\rangle = \langle x, y\rangle^*$,

2. $\langle \lambda y, x \rangle = \lambda \langle y, x \rangle$,
3. $\langle x, y + z \rangle = \langle x, y \rangle + \langle x, z \rangle$.

The map $(x, y) \mapsto \langle x, y \rangle$ *is called an* inner product, *and* $\langle x, y \rangle$ *is called the inner product of* x *and* y. *The vector space* H, *when endowed with such an inner product, is called a* pre-Hilbert space.

If we take

$$\langle X, Y \rangle := E[XY^*]$$

for inner product of $\mathcal{L}^2_{\mathbb{C}}(P)$, the three conditions above are obviously satisfied.

Two vectors x and y in H are called *orthogonal* if $\langle x, y \rangle = 0$.

For any $x \in H$, let

$$\|x\|^2 := \langle x, x \rangle .$$

Theorem 3.3.5 *For all* $x, y \in H$,

$$|\langle x, y \rangle| \leq \|x\| \times \|y\| ,$$

with equality if and only if x *and* y *are colinear.*

Proof. Say $K = \mathbb{C}$. If x and y are colinear, that is $x = \lambda y$ for some $\lambda \in \mathbb{C}$, the inequality is obviously an equality. If x and y are linearly independent, then for all $\lambda \in \mathbb{C}$, $x + \lambda y \neq 0$. Therefore

$$\begin{aligned} 0 < \|x + \lambda y\|^2 &= \|x\|^2 + |\lambda y|^2 \|\lambda y\|^2 + \lambda^* \langle x, y \rangle + \lambda \langle x, y \rangle^* \\ &= \|x\|^2 + |\lambda|^2 \|y\|^2 + 2\mathrm{Re}(\lambda^* \langle x, y \rangle) . \end{aligned}$$

Take $u \in \mathbb{C}$, $|u| = 1$, such that $u^* \langle x, y \rangle = |\langle x, y \rangle|$. For $t \in \mathbb{R}$, let $\lambda := tu$. Then

$$0 < \|x\|^2 + t^2 \|y\|^2 + 2t |\langle x, y \rangle| .$$

This is true for all $t \in \mathbb{R}$. Therefore the discriminant of this second degree equation in t must be strictly negative, that is, $4|\langle x, y \rangle|^2 - 4\|x\|^2 \times \|y\|^2 < 0$. □

The Correlation Coefficient

Schwarz's inequality for square-integrable random variables reads:

$$|E[XY]| \leq E[|XY|] \leq E[|Y|^2]^{\frac{1}{2}} \times E[|X|^2]^{\frac{1}{2}} . \tag{3.43}$$

In particular, with $Y = 1$,

$$E[|X|] \leq E[|X|^2]^{\frac{1}{2}} < \infty . \tag{3.44}$$

Definition 3.3.6 *Two complex square-integrable random variables are said to be* orthogonal *if* $E[XY^*] = 0$. *They are said to be* uncorrelated *if* $E[(X - m_X)(Y - m_Y)^*] = 0$.

Definition 3.3.7 *The* covariance *of the two complex square integrable variables* X *and* Y *is, by definition, the complex number* $E[(X - m_X)(Y - m_Y)^*]$. *It will be denoted by* σ_{XY}.

Definition 3.3.8 *Let* X *and* Y *be square-integrable real random variables with respective means* m_X *and* m_Y, *and respective variances* $\sigma_X^2 > 0$ *and* $\sigma_Y^2 > 0$. *Their* correlation coefficient *is the quantity*

$$\rho_{XY} := \frac{\sigma_{XY}}{\sigma_X \, \sigma_Y},$$

where σ_{XY} *is the covariance.*

By Schwarz's inequality, $|\sigma_{XY}| \leq \sigma_X \, \sigma_y$, and therefore

$$|\rho_{XY}| \leq 1,$$

with equality if and only if X and Y are colinear. Recall that when $\rho_{XY} = 0$ X and Y are said to be uncorrelated. If $\rho_{XY} > 0$, they are said to be *positively correlated*, whereas if $\rho_{XY} < 0$, they are said to be *negatively correlated.*

The next result provides an interesting interpretation of the correlation coefficient.

Theorem 3.3.9 *Let* X *be a square-integrable real random variable. Among all variables* $Z = aX + b$, *where* a *and* b *are real numbers, the one that minimizes the error* $E[(Z - Y)^2]$ *is*

$$\hat{Y} = m_Y + \frac{\sigma_{XY}}{\sigma_X^2}(X - m_X)$$

and the error is then

$$E[(\hat{Y} - Y)^2] = \sigma_Y^2(1 - \rho_{XY}^2).$$

This is a particular case of the forthcoming Theorem 3.3.16.

We see that if the variables are not correlated, then the best prediction is the trivial one $\hat{Y} = m_Y$ and the (maximal) error is then σ_Y^2. In imprecise but suggestive terms, high correlation implies high predictability.

Covariance Matrices

Recall the notation in use in this book for vectors and matrices: an asterisk superscript (*) denotes complex conjugates, a T superscript (^{T}) is for vector transposition, and the dagger superscript (†) is for conjugation-transposition. When x is a vector of $\mathbb{R}^n$, we shall always assume in the notation that it is a *column* vector, and therefore x^T will be the corresponding *line* vector.

Definition 3.3.10 *A random vector $X = (X_1, \dots, X_n)^T$ such that $X_1, \dots, X_n$ are square-integrable complex random variables is called a* square-integrable complex vector.

In particular, by (3.44),

$$E[|X_i|] < \infty \quad (1 \le i, j \le n)$$

and by Schwarz's inequality (3.43),

$$E[|X_i X_j|] < \infty \quad (1 \le i, j \le n)\,.$$

Therefore, the *mean*

$$m_X := E[X] = (E[X_1], \dots, E[X_n])^T$$

and the *covariance matrix* of X

$$\begin{aligned}\Gamma_X &:= E[(X - m_X)(X - m_X)^\dagger] \\ &= \{E\left[(X_i - m_{X_i})(X_j - m_{X_j})^*\right]\}_{1 \le i,j \le n} \\ &= \{\operatorname{cov}(X_i, X_j)\}_{1 \le i,j \le n}\end{aligned}$$

are well defined.

Theorem 3.3.11 *The matrix Γ_X is* symmetric Hermitian, *that is,*

$$\Gamma_X^\dagger = \Gamma_X\,, \tag{3.45}$$

and it is non-negative definite, *that is,*

$$\alpha^\dagger \Gamma_X \alpha \ge 0\,, \tag{3.46}$$

for all $\alpha \in \mathbb{C}^n$. This is denoted by $\Gamma_X \ge 0$.

Proof.

$$
\begin{aligned}
\alpha^\dagger \Gamma \alpha &= \alpha^T \Gamma \alpha^* \\
&= \sum_{i=1}^n \sum_{j=1}^n \alpha_i \alpha_j^* E[(X_i - E[X_i])(X_j - E[X_j])^*] \\
&= E\left[\sum_{i=1}^n \sum_{j=1}^n \alpha_i \alpha_j^* (X_i - E[X_i])(X_j - E[X_j])^*\right] \\
&= E\left[\left(\sum_{i=1}^n \alpha_i (X_i - E[X_i])\right)\left(\sum_{j=1}^n \alpha_j (X_j - E[X_j])\right)^*\right] \\
&= E[|\alpha^T (X - E[X])|^2] \geq 0 .
\end{aligned}
$$

□

Theorem 3.3.12 *Let X be a square-integrable real random vector of dimension $n \geq 2$ with a covariance matrix Γ_X which is* degenerate, *that is,*

$$\alpha^T \Gamma_X \alpha = 0 ,$$

for some $\alpha \in \mathbb{R}^n$, $\alpha \neq 0$. Then, X lies almost surely in a hyperplane of $\mathbb{R}^n$ of dimension strictly less than n, and cannot have a probability density.

Proof. For such α, $E[|\alpha^T(X - E[X])|^2] = \alpha^T \Gamma_X \alpha = 0$, and therefore, by Theorem 3.3.3,

$$\alpha^T (X - E[X]) = 0 ,$$

almost surely. Suppose the existence of such a probability density f. Then, denoting by Π the hyperplane in question,

$$P(X \in \Pi) = \int_\Pi f(x)\, \mathrm{d}x ,$$

a null quantity since the n-volume of an hyperplane of $\mathbb{R}^n$ is null. □

If Γ_X is non-degenerate, we write $\Gamma_X > 0$. A vector X with degenerate covariance matrix is also called *degenerate.*

We now examine the effects of an affine transformation of a random vector on its covariance matrix. Let X be a square-integrable n-dimensional complex random vector, with mean m_X and covariance matrix Γ_X. Let A be an $(n \times k)$-dimensional complex matrix, and b a k-dimensional real vector.

Theorem 3.3.13 *The k-dimensional complex vector $Z = AX + b$ has mean*

$$m_Z = Am_X + b$$

and covariance matrix

$$\Gamma_Z = A\Gamma_X A^\dagger.$$

Proof. The formula giving the mean is immediate. As for the other one, it suffices to observe that $(Z - m_Z) = A(X - m_X)$ and to write

$$\begin{aligned}\Gamma_Z &= E\left[(Z - m_Z)(Z - m_Z)^\dagger\right] \\ &= E\left[A(X - m_X)(A(X - m_X))^\dagger\right] \\ &= E\left[A(X - m_X)(X - m_X)^\dagger A^\dagger\right] \\ &= AE\left[(X - m_X)(X - m_X)^T\right] A^\dagger = A\Gamma_X A^\dagger.\end{aligned}$$

□

Let X and Y be square-integrable complex random vectors of respective dimensions n and q. We define the *covariance matrix* of X and Y—in this order—by

$$\Gamma_{XY} = E[(X - m_X)(Y - m_Y)^\dagger].$$

Note that

$$\Gamma_{YX} = \Gamma_{XY}^\dagger.$$

Also, for the $(n + q)$-dimensional vector

$$Z = (X_1, \dots, X_n, Y_1, \dots, Y_q)^T$$

the covariance matrix takes the block diagonal form

$$\Gamma_Z = \begin{pmatrix} \Gamma_X & \Gamma_{XY} \\ \Gamma_{YX} & \Gamma_Y \end{pmatrix}.$$

Linear Regression

Let $Y, X_1, \dots, X_N$ be square-integrable real random variables. We now consider the problem of the best *linear-quadratic approximation* of Y based on $X_1, \dots, X_N$. More precisely, we seek a real vector $a = (a_1, \dots, a_N)^T$ such that the linear combination $\hat{Y} := a^T X = \sum_{i=1}^N a_i X_i$ satisfies

$$E\left[||Y - \hat{Y}||^2\right] \le E\left[||Y - Z||^2\right]$$

for every linear combination $Z = \sum_{i=1}^N b_i X_i$, where $b = (b_1, \dots, b_N)^T$ is a real vector.

Definition 3.3.14 *The random variable $\hat{Y}$ achieving the minimum is called the* linear regression *of Y on $X_1, \dots, X_N$, or, again, the best* linear-quadratic approximation *of Y as a function of $X_1, \dots, X_N$. The vector a is called the* regression vector.

Letting

$$F(b) := E\left[\|Y - Z\|^2\right] = E\left[(Y - \sum_{i=1}^{N} b_i X_i)(Y - \sum_{i=1}^{N} b_i X_i)\right],$$

we have

$$\frac{\partial F}{\partial b_i}(b) = -2E\left[(Y - \sum_{i=1}^{N} b_i X_i) X_i\right] = -2E\left[(Y - Z)X_i\right].$$

On writing $\partial F / \partial b_i = 0$ $(1 \le i \le N)$, we see that a vector a realizing an extremum of F and the corresponding approximation $\hat{Y} = a^T X$ satisfy the system

$$E\left[(Y - \hat{Y})X_i\right] = 0 \quad (1 \le i \le N). \tag{3.47}$$

The N preceding equations may be written as a function of the unknowns $a_1, \dots, a_N$,

$$\sum_{i=1}^{N} a_j E\left[X_j X_i\right] = E\left[Y X_i\right] \quad (1 \le i \le N), \tag{3.48}$$

or, in matrix form:

$$\begin{pmatrix} E[X_1X_1] & E[X_1X_2] & \dots & E[X_1X_N] \\ E[X_2X_1] & E[X_2X_2] & \dots & E[X_2X_N] \\ \vdots & \vdots & \vdots & \vdots \\ E[X_NX_1] & E[X_NX_2] & \dots & E[X_NX_N] \end{pmatrix} \begin{pmatrix} a_1 \\ a_2 \\ \vdots \\ a_N \end{pmatrix} = \begin{pmatrix} E[YX_1] \\ E[YX_2] \\ \vdots \\ E[YX_N] \end{pmatrix}.$$

More compactly,

$$\Gamma_X a = \Gamma_{XY}. \tag{3.49}$$

In view of (3.47), we have $E\left[(Y - \hat{Y})\hat{Y}\right] = 0$, and therefore,

$$d^2 := E\left[(Y - \hat{Y})^2\right] = E\left[(Y - \hat{Y})Y\right]. \tag{3.50}$$

The covariance matrix Γ_X is non-singular if and only if $X_1, \dots, X_N$ are linearly independent vectors. In this case (3.48) admits a unique solution. Thus under the condition $\Gamma_X > 0$, we have a unique extremum, which we know to be a minimum because the coefficients of the squares of the quadratic form F are positive (at least if we assume, without loss of generality, that none of the X_i is the null vector).

In summary,

Theorem 3.3.15 *Let $Y, X_1, \ldots, X_N$ be real square-integrable centered random variables. A necessary and sufficient condition for $\hat{Y} = a_1X_1 + \ldots a_NX_N$ to be a best quadratic approximation of Y by a linear function of $X_1, \ldots, X_N$ is*

$$E\left[(Y - \hat{Y})X_i\right] = 0 \quad (1 \le i \le N).$$

The regression vector is given by (3.51)

$$\Gamma_X a = \Gamma_{XY} \tag{3.51}$$

and the minimum quadratic error $d^2 = E\left[|Y - \hat{Y}|^2\right]$ is given by $d^2 = \langle Y - \hat{Y}, Y\rangle$.

We now assume linear independence (in the algebraic sense) of $X_1, \ldots, X_N$, which is expressed by the condition

$$\Gamma_X > 0\,, \tag{3.52}$$

in which case Γ_X^{-1} exists and therefore there exists a unique regression vector $a = \Gamma_X^{-1}\Gamma_{XY}$, so that

$$\hat{Y} = \Gamma_{YX}\Gamma_X^{-1}X\,. \tag{3.53}$$

We now consider the case where the random variables Y and $X_1, \ldots, X_N$ are *no longer assumed to be centered* (but we keep the condition (3.52)). The problem is now to find the *affine* combination of $X_1, \ldots, X_N$ which best approximates Y in the least-squares sense. In other words, we seek to minimize

$$E[(Y - b_0 - b_1X_1 - \ldots - b_NX_N)^2]$$

with respect to the scalars $b_0, \ldots, b_N$. This problem can be reduced to the preceding one as follows. In fact, for every square integrable random variable U with mean m,

$$E[(U - c)^2] \ge E[(U - m)^2] \quad \text{for all } c.$$

(Exercise 3.1.11.) Therefore

$$E[Y - b^TX - b_0]^2 \ge E[(Y - b^TX - E[Y - b^TX])^2]\,,$$

where

$$b^T = (b_1, \ldots, b_N)\,.$$

This shows that b_0 is necessarily of the form $b_0 = m_Y - b^T m_X$. Therefore we have reduced the original problem to that of minimizing with respect to b the quantity $E[((Y - m_Y) - b^T(X - m_X))^2]$, and for this we can use the result obtained in the case of centred random variables.

Theorem 3.3.16 *If X is nondegenerate, the best linear-quadratic approximation of Y as an affine function of X is*

$$\hat{Y} = m_Y + \Gamma_{YX}\Gamma_X^{-1}(X - m_X). \tag{3.54}$$

The minimum quadratic error is then given by

$$E[(\hat{Y} - Y)^2] = \sigma_Y^2 - \Gamma_{YX}\Gamma_X^{-1}\Gamma_{XY}. \tag{3.55}$$

Proof. It remains to prove (3.55). From (3.54), we have

$$\begin{aligned} E[(\hat{Y} - Y)^2] &= E[(Y - m_Y - \Gamma_{YX}\Gamma_X^{-1}(X - m_X))^2] \\ &= E[(Y - m_Y)^2] - 2E[(\Gamma_{YX}\Gamma_X^{-1}(X - m_X))(Y - m_Y)] \\ &\qquad + E[(\Gamma_{YX}\Gamma_X^{-1}(X - m_X))^2]. \end{aligned}$$

But

$$\begin{aligned} E[(\Gamma_{YX}\Gamma_X^{-1}(X - m_X))(Y - m_Y)] &= \Gamma_{YX}\Gamma_X^{-1}E[((X - m_X))(Y - m_Y)] \\ &= \Gamma_{YX}\Gamma_X^{-1}\Gamma_{XY}. \end{aligned}$$

Also

$$\begin{aligned} E[(\Gamma_{YX}\Gamma_X^{-1}(X - m_X))^2] &= E[(\Gamma_{YX}\Gamma_X^{-1}(X - m_X))(\Gamma_{YX}\Gamma_X^{-1}(X - m_X))^T] \\ &= E[\Gamma_{YX}\Gamma_X^{-1}(X - m_X)(X - m_X)^T\Gamma_X^{-1}\Gamma_{YX}] \\ &= \Gamma_{YX}\Gamma_X^{-1}E[(X - m_X)(X - m_X)^T]\Gamma_X^{-1}\Gamma_{YX} \\ &= \Gamma_{YX}\Gamma_X^{-1}\Gamma_X\Gamma_X^{-1}\Gamma_{YX} = \Gamma_{YX}\Gamma_X^{-1}\Gamma_{XY}. \end{aligned}$$

□

When X is centered, we shall denote $P(Y|X)$ by $\hat{Y}$.

3.4 Gaussian Vectors

The importance of Gaussian vectors is due to their mathematical tractability, their stability with respect to linear transformations, and the fact that their distribution is entirely characterized by their mean vector and their covariance matrix.

We begin by slightly extending the definition of a Gaussian random variable. This extension will be useful in the definition of a Gaussian vector:

Definition 3.4.1 *An* extended Gaussian variable X *is any real random variable with a characteristic function of the form*

$$\phi_X(u) = \exp\{imu - \tfrac{1}{2}\sigma^2 u^2\}, \tag{3.56}$$

where $m \in \mathbb{R}$ *and* $\sigma^2 \in \mathbb{R}_+$.

The only difference with the standard definition is that a null variance σ^2 is allowed, in which case the random variable is (almost surely) a constant.

Definition 3.4.2 *A* standard Gaussian variable *is a Gaussian variable with mean* 0 *and variance* 1*:* $X \sim \mathcal{N}(0, 1)$.

Definition 3.4.3 *An n-dimensional real random vector X is called a* Gaussian random vector *if the random variable $\alpha^T X$ is an extended Gaussian random variable for all $\alpha \in \mathbb{R}^n$.*

Definition 3.4.4 *A* standard Gaussian vector *is a Gaussian vector with mean vector* 0 *and covariance matrix I (the identity matrix):* $X \sim \mathcal{N}(0, I)$.

The next result is an immediate consequence of the above definition and of Theorem 3.3.13.

Theorem 3.4.5 *Let X be an n-dimensional Gaussian vector with mean vector m_X and covariance matrix Γ_X. Let A be an $(n \times k)$-dimensional real matrix, and b a k-dimensional real vector. The k-dimensional vector $Z = AX + b$ is then a Gaussian vector with mean vector*

$$m_Z = Am_X + b,$$

and covariance matrix

$$\Gamma_Z = A\Gamma_X A^T.$$

We now make the connection with the classical definition of Gaussian vectors in terms of characteristic functions.

Theorem 3.4.6 *For a real n-dimensional random vector X to be a Gaussian vector it is necessary and sufficient that its characteristic function ϕ_X be of the following form:*

$$\phi_X(u) = \exp\{iu^T m_X - \tfrac{1}{2}u^T \Gamma_X u\}, \tag{3.57}$$

where $m_X \in \mathbb{R}^n$ and where Γ_X is a symmetric and non-negative definite $n \times n$ matrix. In this case the parameters m_X and Γ_X are respectively the mean vector and the covariance matrix of X.

Proof. Necessary condition. The characteristic function of a Gaussian vector as defined in Definition 3.4.3 is

$$E[e^{iu^T X}] = \varphi_Z(1)\,,$$

where φ_Z is the CF of $Z := u^T X$. The random variable Z being an extended Gaussian variable,

$$\phi_Z(1) = \exp\{im_Z - \tfrac{1}{2}\sigma_Z^2\}\,,$$

where

$$m_Z := E[Z] = u^T E[X] = u^T m_X$$

and

$$\begin{aligned}\sigma_Z^2 &:= E[(u^T(X - m_X))(u^T(X - m_X))^T] \\ &= u^T E[(X - m_X)(X - m_X)^T]u = u^T \Gamma_X u\,.\end{aligned}$$

Therefore, finally,

$$\phi_X(u) = \exp\{iu^T m_X - \tfrac{1}{2}u^T \Gamma_X u\}\,.$$

Sufficient condition. Let X be a random vector with characteristic function given by (3.57). Let $Z = \alpha^T X$, where $\alpha \in \mathbb{C}^n$. The characteristic function of the random variable Z is

$$\begin{aligned}\phi_Z(v) &= E[\exp\{ivZ\}] = E[\exp\{iv\alpha^T X\}] \\ &= \exp\{iv(\alpha^T m_X) - \tfrac{1}{2}v^2(\alpha^T \Gamma_X \alpha)\}\,.\end{aligned}$$

Therefore Z is an extended Gaussian random variable. □

Mixed Moments of Gaussian Vectors

We shall give two useful formulas concerning the moments of a centered (0-mean) n-dimensional Gaussian vector $X = (X_1, \dots, X_n)^T$ with the covariance matrix $\Gamma = \{\sigma_{ij}\}$.

First, we have

$$E[X_{i_1} X_{i_2}, \dots, X_{i_{2k}}] = \sum_{\substack{(j_1,\dots,j_{2k}) \\ j_1<j_2,\dots,j_{2k-1}<j_{2k}}} \sigma_{j_1 j_2}\sigma_{j_3 j_4}\dots\sigma_{j_{2k} j_{2k}}\,, \tag{3.58}$$

where the summation extends over all permutations $(j_1, \dots, j_{2k})$ of $\{i_1, \dots, i_{2k}\}$ such that $j_1 < j_2, \dots, j_{2k-1} < j_{2k}$. There are $1 \cdot 3 \cdot 5 \dots (2k-1)$ terms in the right-hand side of Eq. (3.58). The indices $i_1, \dots, i_{2k}$ are in $\{1, \dots, n\}$ and they may occur with repetitions. For instance

$$\begin{aligned} E[X_1X_2X_3X_4] &= \sigma_{12}\sigma_{34} + \sigma_{13}\sigma_{24} + \sigma_{14}\sigma_{23} \\ E[X_1^2X_2^2] &= \sigma_{11}\sigma_{22} + \sigma_{12}\sigma_{12} + \sigma_{12}\sigma_{12} = \sigma_1^2\sigma_2^2 - 2\sigma_{12}^2 \\ E[X_1^4] &= 3\sigma_{11}^2 = 3\sigma_1^4 \\ E[X_1^{2k}] &= 1 \cdot 3 \dots (2k-1)\sigma_1^{2k} . \end{aligned}$$

Also the odd moments of a centered gaussian vector are null, that is:

$$E[X_{i_1} \dots X_{i_{2k+1}}] = 0\,, \tag{3.59}$$

for all $(i_1, \dots, i_{2k+1}) \in \{1, 2, \dots, n\}^{2k+1}$.

The proof of the formulas above is required in Exercise 5.7.11.

Independence and Non-Correlation

In general, non-correlation does not imply independence. However, this is nearly (see Example 3.4.9 below) true in the case of Gaussian vectors. We start with a definition in view of correctly stating the announced result.

Definition 3.4.7 *Two random real vectors X and Y of respective dimensions n and q are said to be* jointly Gaussian *if the vector Z defined by*

$$Z^T = (X^T, Y^T) = (X_1, \dots, X_n, Y_1, \dots, Y_q)$$

is a Gaussian vector.

Theorem 3.4.8 *Two jointly Gaussian random vectors X and Y of respective dimensions n and q are independent if and only if they are uncorrelated (that is $\Gamma_{XY} = 0$).*

Proof. Necessity: If X and Y are independent then, by the product formula for expectations,

$$E[(X - m_X)(Y - m_Y)^T] = E[X - m_X]E[Y - m_Y]^T = 0\,.$$

Sufficiency: If X and Y are uncorrelated the vector Z has for covariance matrix

$$\Gamma_Z = \begin{pmatrix} \Gamma_X & 0 \\ 0 & \Gamma_Y \end{pmatrix},$$

and the mean

$$m_Z = \begin{pmatrix} m_X \\ m_Y \end{pmatrix}.$$

It is a Gaussian vector by hypothesis and therefore, with

$$w := (u_1, \ldots, u_n, v_1, \ldots, v_q)^T,$$

we have that

$$\begin{aligned} E[\exp\{i(u^T X + v^T Y)\}] &= E[\exp\{iw^T Z\}] \\ &= \exp\{iw^T m_Z - \frac{1}{2}w^T \Gamma_Z w\} \\ &= \exp\{i(u^T m_X + v^T m_Y) - \frac{1}{2}u^T \Gamma_X u - \frac{1}{2}v^T \Gamma_Y v\} \\ &= E[\exp\{iu^T X\}]E[\exp\{iv^T Y\}], \end{aligned}$$

and the conclusion follows from the factorization theorem of characteristic functions (Theorem 3.2.20). □

EXAMPLE 3.4.9: GAUSSIAN, UNCORRELATED, NOT JOINTLY GAUSSIAN. Let X and U be two independent random variables, where $X \sim \mathcal{N}(0,1)$ and $U \in \{-1, 1\}$, $P(U = \pm 1) = \frac{1}{2}$. We show that

$$Y = UX \sim \mathcal{N}(0,1)$$

and therefore X and Y are *separately* Gaussian. However, we also show that they are *not* jointly Gaussian, and that they are uncorrelated, and yet, not independent. The proof of the above statements is as follows:

$$\begin{aligned} P(Y \le x) &= P(UX \le x) = P(U = 1, X \le x) + P(U = -1, X \ge -x) \\ &= P(U = 1)P(X \le x) + P(U = -1)P(X \ge -x) \\ &= \frac{1}{2}P(X \le x) + \frac{1}{2}P(X \ge -x) = P(X \le x). \end{aligned}$$

Also, $E[YX] = E[UX^2] = E[U]E[X^2] = 0$, that is Y and Z are uncorrelated. We show that they are *not* independent. We have $P(X^2 = Y^2) = 1$. If X and Y were independent, since they are absolutely continuous, (X, Y) would admit a probability density, say, $f_{X,Y}(x, y)$. Then

$$P(X^2 = Y^2) = \int_{\mathbb{R}} \int_{\mathbb{R}} 1_{\{x^2 = y^2\}} f_{X,Y}(x, y)\, \mathrm{d}x\, \mathrm{d}y = 0,$$

since the set $\{(x, y); x^2 = y^2\}$ has a null area. Hence a contradiction.

The reason why Theorem 3.4.8 cannot be applied is that (X, Y) is not a Gaussian vector. If it were, then $X - Y$ would be an extended Gaussian random variable. Obviously $X - Y$ is not a constant. The only case remaining is that in which $X - Y$ has a probability distribution, and therefore $P(X - Y = 0) = 0$. But this is incompatible with $P(X - Y = 0) = P(U = -1) = \frac{1}{2}$.

Probability Density of a Non-degenerate Gaussian Vector

A Gaussian vector with a degenerate covariance matrix *cannot* have a probability density (Theorem 3.3.12). However:

Theorem 3.4.10 *Let X be an n-dimensional Gaussian vector with mean vector m and* non-degenerate *covariance matrix Γ_X (in particular, if $u^T\Gamma u = 0$, then $u = 0$). Then X admits the probability distribution function*

$$f_X(x) = \frac{1}{(2\pi)^{n/2}(\det \Gamma_X)^{1/2}} \exp\{-\tfrac{1}{2}(x - m)^T\Gamma_X^{-1}(x - m)\}. \tag{3.60}$$

Proof. Since $\Gamma_X > 0$, there exists a non-singular matrix A of the same dimension such that $\Gamma_X = AA^T$. Let $Z := A^{-1}(X - m)$. By Definition (3.4.3), it is a Gaussian vector with mean 0 and covariance matrix

$$\Gamma_Z = A^{-1}\Gamma_X A^{-T} = A^{-1}AA^TA^{-T} = I.$$

Therefore its characteristic function is

$$E[\exp\{iu^TZ\}] = \exp\left\{-\frac{1}{2}\sum_{i=1}^{n} u_i^2\right\}.$$

This is the characteristic function of a centered Gaussian vector having independent coordinates, and therefore $Z_1, \ldots, Z_n$ are independent standard Gaussian random variables. In particular, the probability density of Z has the form of a product:

$$f_Z(z) = \prod_{i=1}^{n} \frac{1}{\sqrt{2\pi}} \mathrm{e}^{-\frac{1}{2}z_i^2/2} = \frac{1}{(2\pi)^{n/2}} \exp\{-\frac{1}{2}\|z\|^2\}.$$

Now, $X = AZ + m$ and therefore, by the formula for a smooth change of variables,

$$\begin{aligned} f_X(x) &= \frac{1}{|\det A|} f_Y(A^{-1}(x - m)) \\ &= \frac{1}{(\det \Gamma_X)^{1/2}} \frac{1}{(2\pi)^{n/2}} \exp\{-\frac{1}{2}\|A^{-1}(x - m)\|^2\}, \end{aligned}$$

and this is precisely (3.60) since

$$\begin{aligned}\|A^{-1}(x-m)\|^2 &= \left(A^{-1}(x-m)\right)^T \left(A^{-1}(x-m)\right) \\ &= (x-m)^T A^{-T} A^{-1}(x-m) \\ &= (x-m)^T \Gamma_X^{-1}(x-m)\,.\end{aligned}$$

□

Empirical Mean and Variance of the Gaussian Distribution

A *Gaussian sample* of size n is, by definition, a random vector $X = (X_1, \dots, X_n)$ of IID $\mathcal{N}(m, \sigma^2)$ Gaussian variables. Any random variable of the form $f(X_1, \dots, X_n)$ is called a *statistic* of this sample. The two main statistics are the *empirical mean*

$$\overline{X} := \frac{X_1 + \ldots + X_n}{n}$$

and the *empirical variance*

$$S^2 = \frac{1}{n-1} \sum_{i=1}^{n} (X_i - \overline{X})^2\,.$$

The perhaps surprising factor $\frac{1}{n-1}$ (instead of $\frac{1}{n}$) is motivated by the result of Exercise 3.6.36.

Theorem 3.4.11 *The empirical means and the empirical variance of the above Gaussian sample are independent and $[(n-1)/\sigma^2]S^2$ has a chi-square distribution with $n-1$ degrees of freedom.*

Proof. We first treat the case where $m = 0$ and $\sigma^2 = 1$. For this, we rely on the next lemma (Cochran's lemma).

Recall that a *unitary* square complex matrix is one for which the conjugate transpose is its inverse.

Lemma 3.4.12 *There exists an $n \times n$ unitary matrix C such that if the n-vectors x and y are related by $y = Cx$, then (with the obvious notation)*

$$y_n = \sqrt{n}\,\overline{x} \text{ and } y_1^2 + \cdots + y_{n-1}^2 = (n-1)s^2\,, \tag{3.61}$$

where

$$\overline{x} := \frac{x_1 + \ldots + x_n}{n} \text{ and } s^2 = \frac{1}{n-1} \sum_{i=1}^{n} (x_i - \overline{x})^2\,. \tag{3.62}$$

The random vector $Y = CX$ is a Gaussian vector, and a standard one since

$$\Gamma_Y = C\Gamma_X C^T = CIC^T = CC^T = I$$

(the transpose of a unitary matrix is its inverse). According to (3.61) and (3.62)

$$\overline{X} = \frac{Y_n}{\sqrt{n}} \text{ and } S^2 = \frac{1}{n-1}(Y_1^2 + \cdots Y_{n-1}^2)\,.$$

The independence of $\overline{X}$ and S^2 then follows from the independence of Y_n and $(Y_1, \dots, Y_{n-1})$.

For the general case, apply the above result to the variables $X_i' := \frac{X_i - m}{\sigma}$ $(1 \le i \le n)$ and observe that $\overline{X'} = \frac{\overline{X}-m}{\sigma}$ and $(S')^2 = \frac{S^2}{\sigma^2}$. □

3.5 Conditional Expectation II

The difference with the discrete case is that for all y, $P(Y = y) = 0$, and this calls for a new definition, that of conditional probability density. Otherwise, this case is completely similar to the discrete case, with integrals replacing sums.

Definition 3.5.1 *Let X and Y be the random vectors of dimensions p and n respectively, with joint probability density $f_{X,Y}$, and let f_Y be the probability density function of Y. Let $y \in \mathbb{R}^n$ be fixed. The function $f_X^{Y=y} : \mathbb{R}^P \to \mathbb{R}$ defined by*

$$f_X^{Y=y}(x) = \frac{f_{X,Y}(x,y)}{f_Y(y)}\,,$$

with the convention $f_X^{Y=y}(x) = 0$ (or any other arbitrary value) when $f_Y(y) = 0$, is called the conditional probability density *of X given $Y = y$.*

Note that when $f_Y(y) > 0$, $f_{X,Y}(x,y) = f_X^{Y=y}(x) f_Y(y)$.

EXAMPLE 3.5.2: CORRELATED GAUSSIAN VARIABLES, TAKE 1. Let X_1 and X_2 be two random variables with the joint probability density

$$f_{X_1,X_2}(x_1,x_2) = \frac{1}{2\pi\sigma_1\sigma_2\sqrt{1-\rho^2}} \exp\left\{-\frac{1}{2(1-\rho^2)}\left(\frac{x_1^2}{\sigma_1^2} - 2\rho\frac{x_1x_2}{\sigma_1\sigma_2} + \frac{x_2^2}{\sigma_2^2}\right)\right\}.$$

The random variable X_2 is Gaussian random with mean 0 and variance σ_2^2, that is

$$f_{X_2}(x_2) = \frac{1}{\sqrt{2\pi}\sigma_2} \exp\left\{-\frac{1}{2}\frac{x_2^2}{\sigma_2^2}\right\}.$$

We then find that

$$f_{X_1}^{X_2=x_2}(x_1) = \frac{1}{\sqrt{2\pi}\sigma_1\sqrt{1-\rho^2}} \exp\left\{-\frac{1}{2\sigma_1^2(1-\rho^2)}(x_1 - \rho\frac{\sigma_1}{\sigma_2}x_2)\right\}.$$

Note that this is the probability density (in x_1) of a Gaussian random variable with mean $\rho\frac{\sigma_1}{\sigma_2}x_2$ and variance $\sigma_1^2(1-\rho^2)$.

Definition 3.5.3 *Let X and Y be two random vectors of dimensions p and n respectively, with joint probability density $f_{X,Y}$, and let $g : \mathbb{R}^p \times \mathbb{R}^n \to \mathbb{R}_+$ be either non-negative or such that $E[|g(X,Y)|] < \infty$. One defines the function $\psi : \mathbb{R}^n \to \mathbb{R}$ by*

$$\psi(y) = \int_{\mathbb{R}^p} g(x,y) f_X^{Y=y}(x)\mathrm{d}x \tag{3.63}$$

on the set $C = \{y \in \mathbb{R}^n\,;\, f_Y(y) > 0\}$, 0 otherwise. For each $y \in \mathbb{R}^n$, $\psi(y)$ is called the conditional expectation of $g(X,Y)$ given $Y = y$, and is denoted by $E^{Y=y}[g(X,Y)]$, or $E[g(X,Y) \mid Y = y]$:

$$E^{Y=y}[g(X,Y)] = \psi(y)\,. \tag{3.64}$$

The random variable $\psi(Y)$ is called the conditional expectation of $g(X,Y)$ given Y, and is denoted by $E^Y[g(X,Y)]$ or $E[g(X,Y) \mid Y]$.

The integral in (3.63) is well defined (possibly infinite however) when g is non-negative. It remains to check that it is also well defined when g is of arbitrary sign and satisfies the integrability condition $E[|g(X,Y)|] < \infty$. For this we proceed just as in the discrete case. First we note that in the non-negative case, we have that

$$\begin{aligned}
\int_{\mathbb{R}^n} \psi(y) f_Y(y)\,\mathrm{d}y &= \int_{\mathbb{R}^n}\int_{y\in\mathbb{R}^p} g(x,y) f_X^{Y=y}(x) f_Y(y) 1_C(y)\,\mathrm{d}x\,\mathrm{d}y \\
&= \int_{\mathbb{R}^n}\int_{\mathbb{R}^p} g(x,y) f_{X,Y}(x,y) 1_C(y)\,\mathrm{d}x\,\mathrm{d}y \\
&\leq \int_{\mathbb{R}^n}\int_{\mathbb{R}^p} g(x,y) f_{X,Y}(x,y)\,\mathrm{d}x\,\mathrm{d}y = E[g(X,Y)]\,.
\end{aligned}$$

Therefore, if $E[g(X,Y)] < \infty$, then

$$\int_{\mathbb{R}^n} \psi(y) f_Y(y)\,\mathrm{d}y < \infty\,,$$

which implies that $\psi(y) < \infty$ for all $y \in \mathbb{R}^n$ such that $f_Y(y) > 0$. In particular $\psi(Y) < \infty$ almost surely (that is, $P(\psi(Y) < \infty) = 1$). Indeed, $P(\psi(Y) = \infty) = \int_{\{y;\psi(y)=\infty\}} f_Y(y)\,\mathrm{d}y = 0$.

Let now $g : \mathbb{R}^p \times \mathbb{R}^n \to \mathbb{R}_+$ be a function of arbitrary sign such that $E[|g(X,Y)|] < \infty$, and in particular $E[g^{\pm}(X,Y)] < \infty$. Denote by $\psi^{\pm}$ the functions associated to $g^{\pm}$ as in (3.63). As we just saw, for all $y \in C$, $\psi^{\pm}(y) < \infty$, and therefore $\psi(y) = \psi^+(y) - \psi^-(y)$ is not an indeterminate form $\infty - \infty$. Thus the conditional expectation is well defined in the integrable case, and moreover $|E^Y[g(X,Y)]| < \infty$.

Properties of the Conditional Expectation

The properties will be given without proofs since they are easy adaptations of the proofs given in the discrete case, integrals replacing sums.

The first property of conditional expectation, *linearity*, is obvious from the definitions: For all $\lambda_1, \lambda_2 \in \mathbb{R}$,

$$E^Y[\lambda_1 g_1(X,Y) + \lambda_2 g_2(X,Y)] = \lambda_1 E^Y[g_1(X,Y)] + \lambda_2 E^Y[g_2(X,Y)]$$

whenever the conditional expectations thereof are well defined and do not produce $\infty - \infty$ forms. *Monotonicity* is equally obvious: if $g_1(x,y) \leq g_2(x,y)$, then

$$E^Y[g_1(X,Y)] \leq E^Y[g_2(X,Y)]\,.$$

Theorem 3.5.4 *If g is non-negative or such that $E[|g(X,Y)|] < \infty$, we have*

$$E[E^Y[g(X,Y)]] = E[g(X,Y)]\,.$$

Proof. Same as in Theorem 2.4.5. □

Theorem 3.5.5 *If w is non-negative or such that $E[|w(Y)|] < \infty$,*

$$E^Y[w(Y)] = w(Y)\,, \tag{3.65}$$

and more generally,

$$E^Y[w(Y)h(X,Y)] = w(Y)E^Y[h(X,Y)]\,, \tag{3.66}$$

assuming that the left-hand side of (3.66) is well defined.

Proof. Same as in Theorem 2.4.6. □

Theorem 3.5.6 *If X and Y are independent and if v is non-negative or such that $E[|v(X)|] < \infty$, then*

$$E^Y[v(X)] = E[v(X)] .$$

Proof. Same as in Theorem 2.4.7. □

Theorem 3.5.7 *If X and Y are independent and if $g : F \times G \to \mathbb{R}$ is non-negative or such that $E[|g(X,Y)|] < \infty$, then, for all $y \in G$,*

$$E[g(X,Y \mid Y = y] = E[g(X,y)] .$$

Proof. Same as in Theorem 2.4.8. □

We now give the *successive conditioning rule*. Suppose that $Y = (Y_1, Y_2)$, where Y_1 and Y_2. In this situation, we use the more developed notation

$$E^Y[g(X,Y)] = E^{Y_1,Y_2}[g(X,Y_1,Y_2] .$$

Theorem 3.5.8 *Suppose that $Y = (Y_1, Y_2)$ as above. If g is non-negative or such that $E[|g(X,Y)|] < \infty$, then*

$$E^{Y_2}[E^{Y_1,Y_2}[g(X,Y_1,Y_2)]] = E^{Y_2}[g(X,Y_1,Y_2)] . \tag{3.67}$$

Proof. Same as in Theorem 2.4.9. □

EXAMPLE 3.5.9: CORRELATED GAUSSIAN VARIABLES, TAKE 2. In the situation of Example 3.5.2, we have

$$E^{X_2}[X_1] = \rho \frac{\sigma_1}{\sigma_2} X_2 .$$

This follows from the remark at the end of the Example 3.5.2, because

$$E^{X_2}[X_1] = \psi(X_2)$$

where

$$\psi(x_2) = \int_{\mathbb{R}^p} x_1 f_{X_1}^{X_2=x_2}(x_1) \mathrm{d}x_1 = \rho \frac{\sigma_1}{\sigma_2} x_2 .$$

Similarly,

$$E^{X_2}[X_1^2] = \sigma_1^2(1-\rho^2) + \rho^2 \frac{\sigma_1^2}{\sigma_2^2} X_2^2 .$$

Indeed

$$E^{X_2}[X_1^2] = \gamma(X_2)$$

where

$$\gamma(x_2) = \int_{\mathbb{R}^P} x_1^2 f_{X_1}^{X_2=x_2}(x_1)\mathrm{d}x_1 .$$

This is the second moment of a Gaussian random variable of mean $\rho\frac{\sigma_1}{\sigma_2}x_2$ and variance $\sigma_1^2(1-p^2)$, and therefore

$$\gamma(x_2) = \sigma_1^2(1-\rho^2) + \left(\rho\frac{\sigma_1}{\sigma_2}x_2\right)^2 .$$

Bayesian Tests of Hypotheses

Let Θ be a discrete random variable with values in $\{1, 2, ..., K\}$ and let X be a random vector with values in $\mathbb{R}^m$. The joint distribution of Θ and X is specified as follows:

$$P(\Theta = i) = \pi(i), \qquad P(X \in C | \Theta = i) = \int_C f_i(x)\mathrm{d}x \quad (1 \le i \le K),$$

where the f_i's are probability densities on $\mathbb{R}^m$.

The interpretation in terms of *tests of hypotheses* is the following. The random variable Θ represents *the state of Nature*, and X — called *the observation* — is the (random) result of an experiment that depends on the actual state of Nature. If Nature happens to be in state i, then X admits a distribution with probability density f_i.

In view of the observation X, we wish to infer the actual value of Θ. For this, we design a guess strategy, that is a function $g : \mathbb{R}^m \to \{1, 2, ..., K\}$ with the interpretation that $\widehat{\Theta} := g(X)$ is our guess (based only on the observation X) of the (not directly observed) state Θ of Nature. An equivalent description of the strategy g is the partition $\mathcal{A} = \{A_1, \ldots, A_K\}$ of $\mathbb{R}^m$ given by $A_i := \{x \in \mathbb{R}^m;\ g(x) = i\}$.

The decision rule is then

$$X \in A_i \Rightarrow \widehat{\Theta} = i .$$

The probability of error associated with this strategy is, by the Bayes rule of total causes,

$$\begin{aligned} P_E(\mathcal{A}) = P\left(\Theta \neq \hat{\Theta}\right) &= \sum_{i=1}^{K} P\left(\hat{\Theta} \neq i | \Theta = i\right)\pi(i) \\ &= \sum_{i=1}^{K} P(X \notin A_i | \Theta = i)\,\pi(i) . \end{aligned}$$

Equivalently, the probability of *correct decision* is

$$1 - P_E(\mathcal{A}) = \sum_{i=1}^{K} P(X \in A_i | \Theta = i)\, \pi(i)$$
$$= \int_{\mathbb{R}^n} \left(\sum_{i=1}^{K} \pi(i) 1_{A_i} f_i(x) \right) \mathrm{d}x \,.$$

The following result is then obvious in view of the above expression for the probability of *correct decision*:

Theorem 3.5.10 *Any partition $\mathcal{A}^*$ such that*

$$x \in A_i^* \Rightarrow \pi(i) f_i(x) = \max_k (\pi(k) f_k(x))$$

minimizes the probability of error.

EXAMPLE 3.5.11: TWO GAUSSIAN HYPOTHESES, TAKE 1. In this example, the hypotheses are Gaussian. More specifically, there are two equiprobable Gaussian hypotheses: Nature chooses its state Θ equiprobably in $\{1, 2\}$, and the observation X is a Gaussian random variable, and $X \sim \mathcal{N}(m_i, \sigma^2)$ $(i = 1, 2)$. The two hypotheses differ only by the mean of the observation. Since the hypotheses are equiprobable, an optimal strategy is

$$f_1(X) > f_2(X) \Rightarrow \widehat{\Theta} = 1, \qquad f_1(X) \le f_2(X) \Rightarrow \widehat{\Theta} = 2.$$

Since

$$f_i(x) = \frac{1}{\sqrt{2\pi\sigma^2}} \mathrm{e}^{-\frac{1}{2}\frac{(x-m_i)^2}{\sigma^2}} \qquad (i = 1, 2)\,,$$

the optimal rule is

$$(X - m_1)^2 < (X - m_2)^2 \Rightarrow \widehat{\Theta} = 1\,.$$

Equivalently, supposing that $m_1 < m_2$,

$$X < \frac{m_1 + m_2}{2} \Rightarrow \widehat{\Theta} = 1\,.$$

The probability of error can be expressed as

$$P_E(\mathcal{A}) = \sum_{i=1}^{K} P\left(X \in \overline{A_i} | \Theta = i\right) \pi(i) = \sum_{i=1}^{K} P_{E_i}(\mathcal{A})\,,$$

where $P_{E_i}(\mathcal{A})$ is the probability of making a wrong decision when Nature is in state i.

EXAMPLE 3.5.12: TWO GAUSSIAN HYPOTHESES, TAKE 2. This is the continuation of Example 3.5.11. The probability of error P_E is given by

$$P_E = Q\left(\frac{|m_2 - m_1|}{2\sigma}\right),$$

where the function Q is the *tail of the standard normal distribution*:

$$Q(x) := \frac{1}{\sqrt{2\pi}} \int_x^{\infty} \mathrm{e}^{-\frac{1}{2}y^2} \mathrm{d}y .$$

Proof. We evaluate P_{E_1}, the probability of error when $X \sim \mathcal{N}(m_1, \sigma^2)$ supposing that $m_1 < m_2$:

$$P_{E_1} = \frac{1}{2} \int_{\frac{m_1+m_2}{2}}^{\infty} f_1(x)\mathrm{d}x .$$

By symmetry, $P_{E_1} = P_{E_2} = P_E$. Therefore, with $X_1 \sim \mathcal{N}(m_1, \sigma^2)$, and observing that X_1 then has the same distribution as the variable $\sigma Z + m_1$, where $Z \sim \mathcal{N}(0,1)$,

$$\begin{aligned} P_E &= P\left(X_1 \geq \frac{m_1 + m_2}{2}\right) = P\left(\sigma Z + m_1 \geq \frac{m_1 + m_2}{2}\right) \\ &= P\left(Z \geq \frac{m_2 - m_1}{2\sigma}\right) = Q\left(\frac{|m_2 - m_1|}{2\sigma}\right). \end{aligned}$$

□

Let now the observation be a discrete random variable taking its values in some finite set E, and suppose that

$$f_i(x) = \Pr(X = x | \Theta = i) \quad (i = 1, 2) .$$

The result of the continuous observations case applies *mutatis mutandis*. Any partition $\mathcal{A}^*$ of E such that:

$$x \in A_i^* \Rightarrow \pi(i) f_i(x) = \max_k \left(\pi(k) f_k(x)\right)$$

minimizes the probability of error.

EXAMPLE 3.5.13: THE BINARY CHANNEL WITH FLIP NOISE. In this example $E = \{0,1\}^n$.The addition $\oplus$ defined on E being the componentwise addition modulo 2, the observation is $X = m_\Theta \oplus Z$ where

$$m_i = (m_i(1), ..., m_i(n)) \in \{0,1\}^n , \quad Z = (Z_1, ..., Z_n) ,$$

where Z and Θ are independent, the Z_i's ($1 \le i \le n$) are independent and identically distributed with $\Pr(Z_i = 1) = p$. A possible interpretation is in terms of digital communications, when one wishes to transmit the information Θ chosen among a finite set of "messages" which are binary strings of length n: $m_1, \ldots, m_K$. The vector Z is the "noise" inherent to all digital communications channels: if $Z_k = 1$ the k-th bit of the message Θ is flipped. In the simplest model, this error occurs with probability p, independently for all the bits of the message, and the hypotheses are equiprobable. One may suppose without loss of generality that $p < \frac{1}{2}$. We have:

$$P(X = x|\Theta = i) = P(Z \oplus m_i = x) = P(Z = m_i \oplus x) .$$

Denoting by $h(y)$ the *Hamming weight* of $y \in \{0,1\}^n$ (equal to the number of components of y that are equal to 1), and by

$$d(x,y) := \sum_{i=1}^{n} 1_{\{x_i \neq y_1\}} = \sum_{i=1}^{n} x_i \oplus y_i = h(x \oplus y)$$

the *Hamming distance* between x and y in E^n, we have

$$\begin{aligned} P(Z = y) &= p^{\sum_{i=1}^n y_i} (1-p)^{n - \sum_{i=1}^n} \\ &= \left(\frac{p}{1-p}\right)^{\sum_{i=1}^n y_i} (1-p)^n = (1-p)^n \left(\frac{p}{1-p}\right)^{h(y)} . \end{aligned}$$

Therefore

$$f_i(x) = (1-p)^n \left(\frac{p}{1-p}\right)^{h(m_i \oplus y)} = (1-p)^n \left(\frac{p}{1-p}\right)^{d(m_i,y)} .$$

Therefore the optimal strategy consists in choosing the hypothesis corresponding to the message closest to the observation in terms of the Hamming distance.

3.6 Exercises

Exercise 3.6.1. SUM OF IID UNIFORM VARIABLES
A point inside the unit square $[0,1]^2 = [0,1]\times[0,1]$ is chosen at random according to the following model: $\Omega = [0,1]^2$, $P(A) =$ area of A. In other words, $\omega = (x,y) \in \Omega$ is a point uniformly distributed on the unit square. Let $X(\omega) := x$ and $Y(\omega) = x$.

1. Compute the probability density function of $Z = X + Y$.

2. Compute $E[Z^2]$.

Exercise 3.6.2. UNIFORM DISTRIBUTION ON A DISK
Consider the following probability model: $\Omega = \{(x,y) \in \mathbb{R}^2, x^2+y^2 \leq 1\}$, $P(A) = \frac{1}{\pi}\times$ (area of A). In other words, $\omega = (x,y) \in \Omega$ is a point uniformly distributed on the unit disk. Letting $X(\omega) := x$ and $Y(\omega) = x$, show that X and Y are *not* independent random variables.

Exercise 3.6.3. SQUARE ROOT OF A RANDOM VARIABLE
Let X be a non-negative real random variable with probability density function f_X. What is the probability density function of Z, the non-negative square root of X?

Exercise 3.6.4. QUANTIZATION NOISE
In the digital world, measurements are not recorded in continuous form, but in quantized form. For instance, a random variable X taking its values in the range $[0,+A]$ will be recorded as $Y = i\Delta$ if $X \in [i\Delta, (i+1)\Delta)$, where $\Delta = \frac{A}{2^n}$. Therefore there are 2^n possible values for Y, and it is then said that X has been quantized on n bits. In the applied literature, the error $X - Y \in [0, \Delta)$ is often assumed to be uniformly distributed on this interval. Compute its variance under this (generally wrong) assumption.

Exercise 3.6.5. CAUCHY DISTRIBUTION
(a) Show that the characteristic function of a Cauchy random variable (that is, with the probability density function $f(x) := \frac{1}{\pi}\frac{1}{1+x^2}$) is $\psi_X(u) = \mathrm{e}^{-|u|}$.

(b) Let $\{X_n\}_{n\geq 1}$ be a sequence of independent Cauchy random variables. Let T be a positive integer-valued random variable, independent of this sequence. Define $Y = \sum_{n=1}^{T} X_n$. What is the probability distribution of $Z = \frac{Y}{T}$?

Exercise 3.6.6. CONTINUOUS $\times$ DISCRETE
1) Let X be a real-valued random variable with the probability density function f_X. Let Y be a positive integer-valued random variables ($Y \in \{1,2,...\}$) with the

distribution $P(Y=k) = p_k$, $k \geq 1$. Suppose that X and Y are independent. Show that the random variable $Z = XY$ is absolutely continuous and give its probability density function f_Z.
2) Consider the same setting as in 1) except that Y may take the value 0, with positive probability p_0. What is the cumulative distribution function of Z?

Exercise 3.6.7. CONTINUOUS + DISCRETE
Let X be a real-valued random variable with probability density function $f_X(x)$ and let Y be an integer-valued random variable with distribution $P(Y=k) = p_k$, $k \geq 0$. Suppose that X and Y are independent. Show that the sum $Z = X+Y$ is an absolutely continuous random variable, and give its probability density function.

Exercise 3.6.8. HAZARD RATE, I
The hazard rate function $\lambda : \mathbb{N} \to [0,1]$ of an integer-valued function X is defined by $\lambda(n) = P(X = n | X \geq n)$.

(i) Compute $P(X \geq n)$ and $P(X = n)$ in terms of $\lambda(0), \cdots, \lambda(n)$.

(ii) Let $\{U_n\}_{n\geq 0}$ be a sequence of IID random variables uniformly distributed on $[0,1]$. Show that the random variable $Z := \min\{n \geq 0 : U_n \leq \lambda(n)\}$ has the same distribution as X.

Exercise 3.6.9. HAZARD RATE, II
Let F be the CDF of a non-negative random variable with a probability density function f. Let $I := [-\infty, t_0)$ (t_0 possibly infinite) be the set of $t \in \mathbb{R}_+$ such that $F(t) < 1$. Define, for $t \in I$, the hazard rate

$$\lambda(t) := \frac{f(t)}{1 - F(t)} .$$

1. Show that for $t \in I$,

$$f(t) = \lambda(t)\, \mathrm{e}^{-\int_0^t \lambda(s) ds} .$$

2. Compute the hazard rate of the exponential variable.

Exercise 3.6.10. MORE HAZARD RATES
Alle 1. Let T_1 and T_2 be two non-negative random variables admitting a probability density function and with respective hazard rates (see Exercise 3.6.9) $\lambda_1(t)$ and $\lambda_2(t)$. What is the hazard rate of $T = \min(T_1, T_2)$?

2. Show that the property $P(T_2 > t) = P(T_1 > t)^{\alpha}$ (for some $\alpha > 0$) is equivalent to: $\lambda_2(t) = \alpha \lambda_1(t)$.

Exercise 3.6.11. $\cos(\Phi)$
Let Φ be a random variable uniformly distributed on the interval $[0, 2\pi]$ and define $X = \cos(\Phi)$. Compute the mean and the variance of X.

Exercise 3.6.12. MAXIMUM OF IID VARIABLES
Let $X_1, X_2, \ldots, X_n$ be independent random variables uniformly distributed on $[0, 1]$, that is to say, with the probability density $f(x) = 1_{[0,1]}(x)$. Compute the expectation of $Z = \max(X_1, \ldots, X_n)$.

Exercise 3.6.13. GAUSSIAN MEAN AND VARIANCE
Let σ, $m \in \mathbb{R}$, $\sigma > 0$.
i) Prove that $\int_{\mathbb{R}} \mathrm{e}^{-\frac{1}{2}x^2} \mathrm{d}x = 2\pi$, and deduce that $f(x) = \frac{1}{\sqrt{2\pi\sigma^2}} \mathrm{e}^{-\frac{x^2}{2\sigma^2}}$ is a probability density function on $\mathbb{R}$.
ii) Prove that $\frac{1}{\sqrt{2\pi\sigma^2}} \int_{\mathbb{R}} x \mathrm{e}^{-\frac{1}{2}\left(\frac{x-m}{\sigma}\right)^2} \mathrm{d}x = m$.
iii) Prove that $\frac{1}{\sqrt{2\pi\sigma^2}} \int_{\mathbb{R}} (x-m)^2 \mathrm{e}^{-\frac{1}{2}\left(\frac{x-m}{\sigma}\right)^2} \mathrm{d}x = \sigma^2$.

Exercise 3.6.14. SQUARE OF A GAUSSIAN VARIABLE
Let X be a real random variable with the probability density function $f_X(x) = \frac{1}{(2\pi\sigma^2)^{\frac{1}{2}}} \mathrm{e}^{-\frac{x^2}{2\sigma^2}}$. Compute the probability density function f_Y of $Y = X^2$.

Exercise 3.6.15. TWO RANDOM NUMBERS IN $[0, 1]$
Two numbers are drawn independently and completely at random on $[0, 1]$. The smaller is larger than $\frac{1}{3}$. Given this information, what is the probability that the larger number exceeds $\frac{3}{4}$.

Exercise 3.6.16. COUNTEREXAMPLE
Give a simple example showing that the cumulative distributions of each coordinate of a random vector does not completely describe the probabilistic behavior of the whole vector.

Exercise 3.6.17. POLAR COORDINATES
Let (X, Y) be a random vector uniformly distributed on $D \backslash \{0\}$, the closed unit disk of $\mathbb{R}^2$ centered at the origin without the origin. Let (Z, Θ) be its polar coordinates ($Z \in (0, 1], \Theta \in (0, 2\pi]$). Show that Z and Θ are independent.

Exercise 3.6.18. QUOTIENT OF UNIFORM RANDOM VARIABLES
Let X_1 and X_2 be two independent random variables uniformly distributed over $(0, 1]$. Find the probability density function of X_1/X_2.

Exercise 3.6.19. PRODUCT OF UNIFORM VARIABLES

Let U and V be two independent random variables uniformly distributed on $[0, 1]$. Show that the variable $Z = UV$ has a probability density and compute it.

Exercise 3.6.20. RANDOM ROOTS

The numbers A and B are selected independently and uniformly on the segment $[-1, +1]$. Find the probability that the roots of the equation $x^2 + 2Ax + B$ are real.

Exercise 3.6.21. ISN'T THIS PUZZLING?

Some guy uses his wild imagination to preselect two different numbers, and he does not tell you which ones. Then he chooses one of the two preselected numbers at random (probability $\frac{1}{2}, \frac{1}{2}$). He shows this number to you and asks you to guess if it is the largest of the 2 numbers he preselected. Are you interested in playing (meaning: do you think that you have a better guess than a random guess (yes-no probability $\frac{1}{2}, \frac{1}{2}$)? Hint: you might fix for yourself a "reference number", and compare it with the showned number.

Exercise 3.6.22. INFIMUM OF INDEPENDENT EXPONENTIALS

Let $X_1, \ldots, X_n$ be independent exponential random variables with the respective parameters λ_i, $i \in [0, n]$. Define $Z = \inf(X_1, \ldots, X_n)$ and let J be the (random) index such that $X_J = Z$ (J is for almost all $\omega \in \Omega$ unambiguously defined because, P-almost surely, $X_1, \ldots, X_n$ take different values). Show that Z and J are independent, and give their respective distributions.

Exercise 3.6.23. RANDOM SUM OF EXPONENTIAL VARIABLES

Let $\{X_n\}_{n \geq 1}$ be a sequence of IID exponential random variables with common mean $\lambda^{-1} > 0$, and let T be a geometric random variable with mean $p^{-1} > 0$, and independent of the above sequence. Show that $Z := X_1 + \cdots + X_T$ admits a probability density function. Which one?

Exercise 3.6.24. SUM OF IID EXPONENTIALS

Let $\{X_n\}_{n \geq 1}$ be an IID sequence of exponential random variables with mean $1/\theta$, where $\theta \in (0, \infty)$. What is the distribution of $Z = X_1 + \cdots + X_n$?

Exercise 3.6.25. CHARACTERISTIC FUNCTION OF $Y = AX + b$

Let $\psi_X(u)$ be the characteristic function of the random vector X. What is the characteristic function of the random vector $Y = AX + b$, where A is a matrix and b is a vector of appropriate dimensions?

Exercise 3.6.26. CHARACTERISTIC FUNCTION OF THE MULTINOMIAL RANDOM VECTOR
Let $(X_1, \ldots, X_k)$ be a multinomial random vector of size k and parameters $p_1, \ldots, p_k$ ($p_i > 0$, $p_1 + \cdots + p_k = 1$). Compute the characteristic function of $(X_1, \ldots, X_{k-1})$.

Exercise 3.6.27. PRODUCT OF UNIFORM VARIABLES
Let $U_1, \ldots, U_n$ be independent uniform random variables on $[0, 1]$. Give the cdf of the random variable $U_1 \times U_2 \times \cdots \times U_n$. (Hint: logarithms and Exercise 3.6.24.)

Exercise 3.6.28. QUOTIENT OF EXPONENTIAL RANDOM VARIABLES
Let X_1 and X_2 be two independent random variables with a common exponential distribution of mean θ^{-1}. Give the probability density function of the variable X_1/X_2.

Exercise 3.6.29. CORRELATION COEFFICIENT
Let X and Y be square-integrable random variables. Let a, b, c, d be real numbers, $a \neq 0$, $d \neq 0$. Give the correlation coefficient of $aX + b$ and $cY + d$ in terms of the correlation coefficient ρ_{XY} of X and Y.

Exercise 3.6.30. SIGNAL PLUS NOISE ON TWO CHANNELS
Let Y, Z_1, Z_2 be square-integrable *centered* real random variables, and suppose that Y is independent of Z_1 and Z_2. A useful interpretation is that Y represents an informative "signal" that is observed via two channels, one producing the observation $Y + Z_1$ and the other the observation $Y + Z_2$, where Z_1 and Z_2 are considered as "noises". The following questions are then natural.

What is the best linear-quadratic estimate of Y in terms of $(Y + Z_1, Y + Z_2)$? Give the minimum quadratic error. Write this error in the following particular cases: (a): Z_1 and Z_2 are uncorrelated, and (b): Z_1 and Z_2 have the same variance.

Exercise 3.6.31. COVARIANCE MATRIX OF THE MULTINOMIAL VECTOR
Compute the covariance matrix of a multinomial random vector of size k and with parameters $p_1 \ldots, p_k$.

Exercise 3.6.32. AUTOREGRESSIVE GAUSSIAN MODEL, TAKE 1
Consider the stochastic sequence $\{X_n\}_{n \geq 0}$ defined by

$$X_{n+1} = aX_n + \epsilon_{n+1} \quad (n \geq 0)\,,$$

where X_0 is a Gaussian random variable of mean 0 and variance c^2, and $\{\epsilon_n\}_{n \geq 0}$ is a sequence of IID Gaussian variables of mean 0 and variance σ^2, and independent of X_0.

1. Show that for all $n \geq 1$, the vector $(X_0, \dots, X_n)$ is a Gaussian vector.

2. Express X_n in terms of $X_0, \epsilon_1, \dots, \epsilon_n$ (and a). Give the mean and variance of X_n.

Exercise 3.6.33. PROBABILITY OF THE QUADRANT
Let (X, Y) be a 2-dimensional Gaussian vector with probability density

$$f(x, y) = \frac{1}{2\pi(1-\rho^2)^{1/2}} \exp\left\{-\frac{1}{2(1-\rho^2)}\left(x^2 - 2\rho xy + y^2\right)\right\},$$

where $|\rho| < 1$. Show that X and $(Y - \rho X) / (1 - \rho^2)^{1/2}$ are independent Gaussian random variables with mean 0 and variance 1. Deduce from this that

$$P(X > 0, Y > 0) = \tfrac{1}{4} + \tfrac{1}{2\pi} \sin^{-1}(\rho).$$

Exercise 3.6.34. $\frac{X+Y}{\sqrt{2}}$ AND $\frac{X-Y}{\sqrt{2}}$
Let X and Y be two independent Gaussian random variables with mean 0 and variance 1. Show that the random variables $\frac{X+Y}{\sqrt{2}}$ and $\frac{X-Y}{\sqrt{2}}$ are independent Gaussian random variables, and give their means and variances.

Exercise 3.6.35. QUOTIENT OF χ^2 DISTRIBUTIONS
Let X and Y be two independent random variables such that

$$X \sim \chi^2_n \text{ and } Y \sim \chi^2_m .$$

Compute the probability density function of (Z, Y) where $Z := \frac{X}{Y}$ and deduce from the result the probability density function of Z.

Exercise 3.6.36. UNBIASEDNESS OF THE EMPIRICAL VARIANCE
Let $\{X_n\}_{n\geq 1}$ be an IID sequence of square integrable random variables with mean θ and variance σ^2. Show that the variance estimate

$$\hat{\sigma}^2_N := \frac{\sum_{i=1}^N (X_i - \hat{\theta}_N)^2}{n-1},$$

where $\hat{\theta}_N := \frac{1}{N}\sum_{i=1}^N X_i$, is unbiased, that is $E\left[\sigma^2_N\right] = \sigma^2$.

Exercise 3.6.37. $X_1 - X_2$
Let X_1 and X_2 be two independent random variables admitting the probability density functions f_1 and f_2 respectively. What is the probability density function of $X_1 - X_2$?

Exercise 3.6.38. CUMULATIVE DISTRIBTION FUNCTIONS
Let X be a real-valued random variable with CDF F, and let g be a strictly increasing function. Find the CDFs of the random variables X^2, $\sqrt{X}$ (X assumed non-negative), $F(X)$, $g^{-1}(X)$ and $g^{-1}(F(X))$.

Exercise 3.6.39. SUM OF IID EXPONENTIALS
Let $X_1, \dots, X_n$ be IID exponential random variables with mean λ^{-1}. Give the characteristic function of $X_1 + \cdots + X_n$, and deduce from the result its probability density function.

Exercise 3.6.40. $(S, T - S)$
Let $\{X_n\}_{n \geq 1}$ be independent random variables taking the values 0 and 1 with probability $q = 1 - p$ and p, respectively, where $p \in (0, 1)$. Let T be a Poisson random variable with mean $\theta > 0$, independent of $\{X_n\}_{n \geq 1}$. Define

$$S = X_1 + \cdots + X_T.$$

Compute the characteristic function of the vector $(S, T - S)$. Deduce from this that S and $T - S$ are independent Poisson random variable with respective means $p\theta$ and $q\theta$.

Exercise 3.6.41. PROBABILITY DENSITY FUNCTION OF $X_1 - X_2$
Let X_1 and X_2 be two independent random variables admitting the probability density functions f_1 and f_2 respectively. What is the probability density function of $X_1 - X_2$?

Exercise 3.6.42. THE FIRST BOX
Let $X = (X_1, \cdots, X_K)$ be a multinomial vector of size (n, K) and parameters $p_1, \dots, p_K$. Show that X_1 is a binomial random variable of size n and parameter p_1.

Exercise 3.6.43. SUM OF MULTINOMIALS
Let $X = (X_1, \dots, X_k)$ and $Y = (Y_1, \dots, Y_k)$ be two independent multinomial random vectors of sizes (n, K) and (m, K), respectively, and with the same parameters $p_1, \dots, p_K$. What is the distribution of $Z = X + Y$?

Exercise 3.6.44. POISSON COVARIANCE MATRIX
Let $Z_1, Z_2, \dots, Z_n$ be independent Poisson random variables with respective means $\theta_1, \theta_2, \dots, \theta_n$. Let

$$X_i := Z_1 + \cdots + Z_i \quad (1 \leq i \leq n).$$

Give the covariance matrix of $X = (X_1, \dots, X_n)^T$.

Exercise 3.6.45. UNCORRELATED, YET DEPENDENT
Let X and Y be IID random variables with the equiprobable values 0 or 1. Show that $X+Y$ and $|X-Y|$ are uncorrelated, yet dependent.

Exercise 3.6.46. USELESS INFORMATION
In the theory of Bayesian tests of hypotheses of Section 3.5 (page 129), suppose that the observation is of the form $X = (Y, Z) \in \mathbb{R}^m = \mathbb{R}^{n+p}$ where $Y \in \mathbb{R}^n$ and $Z \in \mathbb{R}^p$, and that under each hypothesis $\Theta = i$, the probability density of X admits the factorization

$$f_i(y, z) = g_i(y)h(z)\,,$$

where g_i and h are probability densities on $\mathbb{R}^n$ and $\mathbb{R}^p$ respectively. Show that the optimal test based on X does not use the information on Z, and is the same as the optimal test based on Y alone.

Exercise 3.6.47. TWO GAUSSIAN HYPOTHESES, TAKE 3
We consider a Bayesian test of hypotheses with two equiprobable hypotheses. The observation $X \in \mathbb{R}^m$ is a random vector. For $i = 1, 2$, $X \sim \mathcal{N}(m_i, \Gamma)$ where Γ is an invertible covariance matrix (the two hypotheses differ only by the mean of X). Describe the optimal Bayesian test of hypotheses in this situation. Give details for the case where the coordinate of the observation vector X are independent and identically distributed. In the latter case, compute the probability of error. Compare with Examples 3.5.11 and 3.5.12.

Exercise 3.6.48. BAYESIAN TEST AND VARIATION DISTANCE
(a) Let X and Y be two absolutely continuous random variables with the respective probability densities f and g. Define their distance in variation by

$$d_V(X, Y) := \sup_{A \in \mathbb{R}} (P(X \in A) - P(Y \in A))\,.$$

Show that

$$d_V(X, Y) = \frac{1}{2} \int_{\mathbb{R}} |f(x) - g(x)|\, \mathrm{d}x\,.$$

(b) In the Bayesian test with an observation $X \in \mathbb{R}^n$ and two equiprobable hypotheses for the probability density function of the observation: f_1 and f_2, compute the probability of error of the optimal test.

Chapter 4

The Lebesgue Integral

The previous chapters concerned what one may call the basic "calculus of probability", that is, the acquisition of the skills that suffice to deal with elementary stochastic models involving discrete random variables and absolutely continuous random vectors. This chapter will considerably increase the expertise of the reader at the expense of a reasonable amount of abstraction. It contains a short summary of the abstract Lebesgue integral that will then be interpreted in probabilistic terms in the next chapter.

4.1 Measurable Functions and Measures

σ-fields

Denote by $\mathcal{P}(X)$ the collection of all subsets of an arbitrary set X. Recall the definition of a σ-field:

Definition 4.1.1 *A family $\mathcal{X} \subseteq \mathcal{P}(X)$ of subsets of X is called a* σ-field *on X if:*

(α) $X \in \mathcal{X}$;

(β) $A \in \mathcal{X} \implies \bar{A} \in \mathcal{X}$;

(γ) $A_n \in \mathcal{X}$ *for all* $n \in \mathbb{N} \implies \cup_{n=0}^{\infty} A_n \in \mathcal{X}$.

One then says that $(X, \mathcal{X})$ is a measurable space.

Also recall:

Definition 4.1.2 *The σ-field* generated *by a non-empty collection of subsets $\mathcal{C} \subseteq \mathcal{P}(X)$ is, by definition, the smallest σ-field on X containing all the sets in $\mathcal{C}$. It is denoted by $\sigma(\mathcal{C})$.*

P. Brémaud, *An Introduction to Applied Probability*, Texts in Applied Mathematics 77,
https://doi.org/10.1007/978-3-031-49306-5_4

The Borel σ-field on $\mathbb{R}^n$, $\mathcal{B}(\mathbb{R}^n)$, already briefly introduced in the first chapter, receives a convenient definition in terms of the Euclidean topology.

First recall that a set $O \subseteq \mathbb{R}^n$ is called *open* if for any $x \in O$, one can find a non-empty open ball centered on x and contained in O.

Definition 4.1.3 *The Borel σ-field $\mathcal{B}(\mathbb{R}^n)$ on $\mathbb{R}^n$ is, by definition, the σ-field generated by the open sets of $\mathbb{R}^n$.*

The next result gives a more convenient way of defining the Borel σ-field, of the type given in the first chapter.

Theorem 4.1.4 *The σ-field $\mathcal{B}(\mathbb{R}^n)$ is also generated by the collection $\mathcal{C}$ of all rectangles of the type $\prod_{i=1}^{n}(-\infty, a_i]$, where $a_i \in \mathbb{Q}$ (the rationals) for all $i \in \{1, \dots, n\}$.*

Proof. Exercise 4.5.4. □

Definition 4.1.5 *$\mathcal{B}(\overline{\mathbb{R}})$ is, by definition, the σ-field on $\overline{\mathbb{R}} := \mathbb{R}^n \cup \{+\infty, -\infty\}$ generated by the intervals of type $(-\infty, a]$ $(a \in \mathbb{R})$.*

It can be readily checked that it consists of the collection of sets of the form

$$A, A \cup \{+\infty\}, A \cup \{-\infty\}, A \cup \{+\infty, -\infty\} \quad (A \in \mathcal{B}(\mathbb{R}))$$

Measurable Functions

This is the first fundamental notion of Lebesgue's integration theory.

Definition 4.1.6 *Let $(X, \mathcal{X})$ and $(E, \mathcal{E})$ be two measurable spaces. A function $f : X \to E$ is called a* measurable function *with respect to $\mathcal{X}$ and $\mathcal{E}$ if*

$$f^{-1}(C) := \{x \in X \,;\, f(x) \in C\} \in \mathcal{X} \text{ for all } C \in \mathcal{E}\,.$$

This is denoted by

$$f : (X, \mathcal{X}) \to (E, \mathcal{E}) \quad \text{or} \quad f \in \mathcal{E}/\mathcal{X}\,.$$

A function $f : (X, \mathcal{X}) \to (\overline{\mathbb{R}}, \mathcal{B}(\overline{\mathbb{R}}))$, where $(X, \mathcal{X})$ is an arbitrary measurable space, is called an *extended* measurable function. Functions $f : (X, \mathcal{X}) \to (\mathbb{R}, \mathcal{B}(\mathbb{R}))$ are called *real* measurable functions.

Definition 4.1.7 *A measurable function* $f : (X, \mathcal{X}) \to (\mathbb{R}, \mathcal{B}(\mathbb{R}))$ *of the type*

$$f(x) = \sum_{i=1}^{k} a_i 1_{A_i}(x), \tag{4.1}$$

where $k \in \mathbb{N}_+$, $a_1, \ldots, a_k \in \mathbb{R}$, $A_1, \ldots, A_k \in \mathcal{X}$, *is called a* simple measurable function *(defined on* X*).*

It seems difficult to prove measurability since most σ-fields are not defined explicitly (see the definition of $\mathcal{B}(\mathbb{R}^n)$ for instance). However, the following result renders the task feasible.

Theorem 4.1.8 *Let* $(X, \mathcal{X})$ *and* $(E, \mathcal{E})$ *be two measurable spaces, where* $\mathcal{E} = \sigma(\mathcal{C})$ *for some collection* $\mathcal{C}$ *of subsets of* E*. Then* $f : (X, \mathcal{X}) \to (E, \mathcal{E})$ *if and only if* $f^{-1}(C) \in \mathcal{X}$ *for all* $C \in \mathcal{C}$*.*

Proof. We shall first make two obvious preliminary observations. Let X and E be arbitrary sets, $f : X \to E$ an arbitrary function from X to E, $\mathcal{G}$ an arbitrary σ-field on E, and let $\mathcal{C}, \mathcal{C}_1, \mathcal{C}_2$ be arbitrary non-empty collections of subsets of E. Then

(i) $\sigma(\mathcal{G}) = \mathcal{G}$,

(ii) $\mathcal{C}_1 \subseteq \mathcal{C}_2 \Rightarrow \sigma(\mathcal{C}_1) \subseteq \sigma(\mathcal{C}_2)$.

Now, the collection $\mathcal{G} := \{C \subseteq E; f^{-1}(C) \in \mathcal{X}\}$ is a σ-field and, by hypothesis, $\mathcal{C} \subseteq \mathcal{G}$. Therefore, by (ii) and (i), $\mathcal{E} = \sigma(\mathcal{C}) \subseteq \sigma(\mathcal{G}) = \mathcal{G}$. □

An immediate application of this result and Theorem 4.1.4 is:

Corollary 4.1.9 *Let* $(X, \mathcal{X})$ *be a measurable space and let* $n \geq 1$ *be an integer. Then* $f = (f_1, \ldots, f_n) : (X, \mathcal{X}) \to (\mathbb{R}^n, \mathcal{B}(\mathbb{R}^n))$ *if and only if for all* i $(1 \leq i \leq n)$, $\{f_i \leq a_i\} \in \mathcal{X}$ *for all* $a_i \in \mathbb{Q}$.

A function $f : \mathbb{R}^k \to \mathbb{R}^m$ is said to be continuous if the inverse image of an open set is open,[1] that is, for all open sets $O_m \subset \mathbb{R}^m$, the set $\{x \in \mathbb{R}^k ; f(x) \in O_m\}$ is an open set of $\mathbb{R}^k$. The following result is then a direct application of Theorem 4.1.8 in view of the Definition 4.1.3 of $\mathcal{B}(\mathbb{R}^n)$.

It follows from this definition of continuity and Theorem 4.1.8 that

[1] This definition is equivalent to the usual $\varepsilon - \delta$ definition of continuity, which we shall admit here.

Corollary 4.1.10 *Any continuous function $f : \mathbb{R}^k \to \mathbb{R}^m$ is measurable with respect to $\mathcal{B}(\mathbb{R}^k)$ and $\mathcal{B}(\mathbb{R}^m)$.*

Another nice feature of the notion of measurability is its stability under composition.

Theorem 4.1.11 *Let $(X, \mathcal{X})$, $(Y, \mathcal{Y})$ and $(E, \mathcal{E})$ be three measurable spaces, and let $\phi : (X, \mathcal{X}) \to (Y, \mathcal{Y})$, $g : (Y, \mathcal{Y}) \to (E, \mathcal{E})$. Then $g \circ \phi : (X, \mathcal{X}) \to (E, \mathcal{E})$.*

Proof. Let $f = g \circ \phi$ (meaning: $f(x) = g(\phi(x))$ for all $x \in X$). For all $C \in \mathcal{E}$,

$$f^{-1}(C) = \phi^{-1}(g^{-1}(C)) = \phi^{-1}(D) \in \mathcal{X},$$

because $D = g^{-1}(C)$ is a set in $\mathcal{Y}$ since $g \in \mathcal{E}/\mathcal{Y}$, and therefore $\phi^{-1}(D) \in \mathcal{X}$ since $\phi \in \mathcal{Y}/\mathcal{X}$. □

Corollary 4.1.12 *Let $\varphi = (\varphi_1, \ldots, \varphi_n)$ be a measurable function from $(X, \mathcal{X})$ to $(\mathbb{R}^n, \mathcal{B}(\mathbb{R}^n))$, and let $g : \mathbb{R}^n \to \mathbb{R}$ be a continuous function. Then $g \circ \phi : (X, \mathcal{X}) \to (\mathbb{R}, \mathcal{B}(\mathbb{R}))$.*

Proof. Follows directly from Theorem 4.1.11 and Corollary 4.1.10. □

This corollary in turn allows us to show that the elementary operations (addition, multiplication and quotient) preserve measurability.

Corollary 4.1.13 *Let $\varphi_1, \varphi_2 : (X, \mathcal{X}) \to (\mathbb{R}, \mathcal{B}(\mathbb{R}))$, and let $\lambda \in \mathbb{R}$. Then $\varphi_1 \times \varphi_2$, $\varphi_1 + \varphi_2$, $\lambda\varphi_1$, $(\varphi_1/\varphi_2)1_{\varphi_2 \neq 0}$ are real measurable functions. Moreover, the set $\{\varphi_1 = \varphi_2\}$ is a measurable set.*

Proof. For the first three functions, take in the previous corollary $g(x_1, x_2) = x_1 \times x_2$, $= x_1 + x_2$, $= \lambda x_1$ successively.

For $(\varphi_1/\varphi_2)1_{\varphi_2 \neq 0}$, define $\psi_2 = \frac{1_{\varphi_2 \neq 0}}{\varphi_2}$, check that the latter function is measurable, and use the just proven fact that the product $\varphi_1 \psi_2$ is then measurable.

Finally, $\{\varphi_1 = \varphi_2\} = \{\varphi_1 - \varphi_2 = 0\} = (\varphi_1 - \varphi_2)^{-1}(\{0\})$ is a measurable set since $\varphi_1 - \varphi_2$ is a measurable function and $\{0\}$ is a measurable set. □

Finally, and most importantly, taking limits preserves measurability. By contrast, it is far from being true that limits of continuous functions are continuous functions.

Theorem 4.1.14 *Let* $f_n : (X, \mathcal{X}) \to (\overline{\mathbb{R}}, \mathcal{B}(\overline{\mathbb{R}}))$, $n \in \mathbb{N}$. *Then* $\liminf_{n\uparrow\infty} f_n$ *and* $\limsup_{n\uparrow\infty} f_n$ *are measurable functions, and the set*

$$\{\limsup_{n\uparrow\infty} f_n = \liminf_{n\uparrow\infty} f_n\} = \{\exists \lim_{n\uparrow\infty} f_n\}$$

belongs to $\mathcal{X}$. *In particular, if* $\{\exists \lim_{n\uparrow\infty} f_n\} = X$, *the function* $\lim_{n\uparrow\infty} f_n$ *is a measurable function.*

Proof. We first prove the result in the particular case when the sequence of functions is non-decreasing. Denote by f the limit of this sequence. By Theorem 4.1.8 it suffices to show that $\{f \le a\} \in \mathcal{X}$ for all $a \in \mathbb{R}$. But since the sequence $\{f_n\}_{n\ge1}$ is non-decreasing, we have that $\{f \le a\} = \cap_{n=1}^{\infty} \{f_n \le a\}$, which is indeed in $\mathcal{X}$, being a countable intersection of sets in $\mathcal{X}$.

Now recall that, by definition,

$$\liminf_{n\uparrow\infty} f_n = \lim_{n\uparrow\infty} g_n,$$

where $g_n = \inf_{k\ge n} f_k$. The function g_n is measurable since for all $a \in \mathbb{R}$, $\{\inf_{k\ge n} f_k \le a\}$ is a measurable set, being the complement of $\{\inf_{k\ge n} f_k > a\} = \cap_{k\ge n}\{f_k > a\}$, a measurable set, being the countable intersection of measurable sets. Since the sequence $\{g_n\}_{n\ge1}$ is non-decreasing, the measurability of $\liminf_{n\uparrow\infty} f_n$ follows from the particular case of non-decreasing functions.

Similarly, $\limsup_{n\uparrow\infty} f_n = -\liminf_{n\uparrow\infty}(-f_n)$ is measurable.

The set $\{\limsup_{n\uparrow\infty} f_n = \liminf_{n\uparrow\infty} f_n\}$ is the set on which two measurable functions are equal, and therefore, by the last assertion of Corollary 4.1.13, it is a measurable set.

Finally, if $\lim_{n\uparrow\infty} f_n$ exists, it is equal to $\limsup_{n\uparrow\infty} f_n$, which is, as we just proved, a measurable function. □

The results above give substance to the assertion that "basically all functions are measurable". However, beware! One can prove (not in this book) that there exist functions from $\mathbb{R}$ to $\mathbb{R}$ that are not measurable with respect to $\mathcal{B}(\mathbb{R})$. Hence all the fuss. In fact, there are subsets of $\mathbb{R}$ that are not in $\mathcal{B}(\mathbb{R})$.

The basis of the construction of the Lebesgue integral is the following *fundamental approximation theorem.*

Theorem 4.1.15 *Let* $f : (X, \mathcal{X}) \to (\overline{\mathbb{R}}, \mathcal{B}(\overline{\mathbb{R}})$ *be a non-negative measurable function. There exists a non-decreasing sequence* $\{f_n\}_{n\ge1}$ *of non-negative simple measurable functions that converges pointwise to* f.

Proof. Take

$$f_n(x) = \sum_{k=0}^{n2^{-n}-1} k2^{-n} 1_{A_{k,n}}(x) + n1_{A_n}(x) ,$$

where

$$A_{k,n} = \{x \in X : k2^{-n} < f(x) \leq (k+1)2^{-n}\}, \; A_n = \{x \in X : f(x) > n\} .$$

This sequence of functions has the announced properties. In fact, for any $x \in X$ such that $f(x) < \infty$, and n large enough,

$$|f(x) - f_n(x)| \leq 2^{-n} ,$$

and for any $x \in X$ such that $f(x) = \infty$, $f_n(x) = n$ indeed converges to $f(x) = +\infty$. □

Measure

Definition 4.1.16 *Let $(X, \mathcal{X})$ be a measurable space and let $\mu : \mathcal{X} \to [0, \infty]$ be a set function such that $\mu(\varnothing) = 0$ and such that for any* countable *family $\{A_n\}_{n\geq 1}$ of* mutually disjoint *sets in $\mathcal{X}$,*

$$\mu\left(\cup_{n=1}^{\infty} A_n\right) = \sum_{n=1}^{\infty} \mu(A_n) . \tag{4.2}$$

The set function μ is called a measure *on $(X, \mathcal{X})$, and $(X, \mathcal{X}, \mu)$ is called a* measure(d) space.

Property (4.2) is the *sigma-additivity* property.

EXAMPLE 4.1.17: THE DIRAC MEASURE. Let $a \in X$ and let $\mathcal{X}$ be an arbitrary σ-field on X. The measure ε_a defined on $(X, \mathcal{X})$ by $\varepsilon_a(C) = 1_C(a)$ is called the *Dirac measure* at $a \in X$. The set function $\mu : \mathcal{X} \to [0, \infty]$ defined by

$$\mu(C) := \sum_{i=0}^{\infty} \alpha_i 1_{a_i}(C) ,$$

where $\alpha_i \in \mathbb{R}_+$ for all $i \in \mathbb{N}$, is a measure on $(X, \mathcal{X})$ denoted by $\sum_{i=0}^{\infty} \alpha_i \varepsilon_{a_i}$.

EXAMPLE 4.1.18: WEIGHTED COUNTING MEASURE. Let $\{\alpha_n\}_{n\geq 1}$ be a sequence of non-negative numbers. The set function $\mu : \mathcal{P}(\mathbb{Z}) \to [0, \infty]$ defined by $\mu(C) :=$

$\sum_{n\in C} \alpha_n$ is a measure on $(\mathbb{Z}, \mathcal{P}(\mathbb{Z}))$. When $\alpha_n \equiv 1$, it is called the *counting measure* on $\mathbb{Z}$.

EXAMPLE 4.1.19: THE LEBESGUE MEASURE. The measure ℓ on $(\mathbb{R}, \mathcal{B}(\mathbb{R}))$ such that

$$\ell((a,b]) = b - a$$

is called the *Lebesgue measure* on $\mathbb{R}$. Measure theory tells us that there exists one and only one such measure. (See the more general result below, Theorem 4.1.27.)

The proofs of existence and uniqueness of measures are in general not given. They are usually very technical and tedious, and their omission has no bearing in the rest of the book. See Theorem 4.1.27 and the comment following it.

Definition 4.1.20 *Let μ be a measure on $(X, \mathcal{X})$. If $\mu(X) < \infty$ the measure μ is called a* finite *measure. If $\mu(X) = 1$ the measure μ is called a* probability *measure. If there exists a sequence $\{K_n\}_{n\geq 1}$ of $\mathcal{X}$ such that $\mu(K_n) < \infty$ for all $n \geq 1$, and $\cup_{n=1}^{\infty} K_n = X$, the measure μ is called a* sigma-finite *measure. A measure μ on $(\mathbb{R}^n, \mathcal{B}(\mathbb{R}^n))$ such that $\mu(C) < \infty$ for all bounded sets in $\mathcal{B}(\mathbb{R}^n)$ is called a* locally finite *measure.*

For instance, the Dirac measure ε_a is a probability measure, the counting measure ν on $\mathbb{Z}$ is a sigma-finite measure, the Lebesgue measure is a locally finite measure, and any locally finite measure on $(\mathbb{R}^n, \mathcal{B}(\mathbb{R}^n))$ is sigma-finite.

The following result is the *sequential continuity theorem* for measures.

Theorem 4.1.21 *Let $(X, \mathcal{X}, \mu)$ be a measure space. Let $\{A_n\}_{n\geq 1}$ be a sequence of $\mathcal{X}$, non-decreasing (that is, $A_n \subseteq A_{n+1}$ for all $n \geq 1$). Then*

$$\mu(\cup_{n=1}^{\infty} A_n) = \lim_{n\uparrow\infty} \mu(A_n)\,. \tag{4.3}$$

Proof. The proof is the same as that of Theorem 1.2.8. □

μ-negligible sets

The notion of a negligible set of Definition 1.2.10 will be repeated in a more general setting.

Definition 4.1.22 *Let $(X, \mathcal{X}, \mu)$ be a measure space. A μ-*negligible set *is a set contained in a measurable set $N \in \mathcal{X}$ such that $\mu(N) = 0$. One says that some* *property $\mathcal{P}$ relative to the elements $x \in X$ holds μ-almost everywhere* (μ-a.e.) *if the set* $\{x \in X : x$ does not satisfy $\mathcal{P}\}$ *is a μ-negligible set.*

For instance, if f and g are two measurable functions defined on X, the expression

$$f \leq g \quad \mu\text{-a.e.}$$

means that $\mu(\{x : f(x) > g(x)\}) = 0$.

Theorem 4.1.23 *A countable union of μ-negligible sets is a μ-negligible set.*

Proof. Same proof as for Theorem 1.2.11. □

EXAMPLE 4.1.24: CONTINUOUS FUNCTIONS. We show that two continuous functions $f, g : \mathbb{R} \to \mathbb{R}$ that are ℓ-a.e. equal, are in fact *everywhere* equal.

Proof. Let $t \in \mathbb{R}$ be such that $f(t) \neq g(t)$. For any $c > 0$, there exists an $s \in [t - c, t + c]$ such that $f(s) = g(s)$ (Otherwise, the set $\{t; f(t) \neq g(t)\}$ would contain the whole interval $[t-c, t+c]$, and therefore could not be of null Lebesgue measure. Therefore, one can construct a sequence $\{t_n\}_{n\geq 1}$ converging to t and such that $f(t_n) = g(t_n)$ for all $n \geq 1$. Letting n tend to ∞ yields $f(t) = g(t)$, a contradiction. □

Cumulative Distribution Function

Definition 4.1.25 *A function $F : \mathbb{R} \to \mathbb{R}$ is called a* cumulative distribution function (CDF) *if the following properties are satisfied:*

1. *F is non-decreasing;*
2. *F is right-continuous;*
3. *F admits a left-hand limit, denoted by $F(x-)$, at all $x \in \mathbb{R}$.*

EXAMPLE 4.1.26: THE CDF OF A MEASURE. Let μ be a locally finite measure on $(\mathbb{R}, \mathcal{B}(\mathbb{R}))$ and define

$$F_\mu(t) := \begin{cases} +\mu((0,t]) & \text{if } t \geq 0, \\ -\mu((t,0]) & \text{if } t < 0. \end{cases}$$

This is a cumulative distribution function (CDF), and moreover,

$$F_\mu(b) - F_\mu(a) = \mu((a, b]),$$
$$F_\mu(a) - F_\mu(a-) = \mu(\{a\}).$$

The proof that this function is indeed a CDF follows the same lines as the proof of Theorem 3.1.4. The function F_μ is called the CDF of μ.

Theorem 4.1.27 *Let $F : \mathbb{R} \to \mathbb{R}$ be a* CDF. *There exists a unique locally finite measure μ on $(\mathbb{R}, \mathcal{B}(\mathbb{R}))$ such that $F_\mu = F$.*

The last result is easily stated, but it is not trivial, even in the case of the Lebesgue measure (Example 4.1.19). It is typical of the existence and uniqueness results which answer the following type of question:

Let $\mathcal{C}$ be a collection of subsets of X with $\mathcal{C} \subseteq \mathcal{X}$, where $\mathcal{X}$ is a σ-field on X. Given a set function $u : \mathcal{C} \to [0, \infty]$, does there exist a measure μ on $(X, \mathcal{X})$ such that $\mu(C) = u(C)$ for all $C \in \mathcal{C}$, and is it unique? As mentioned in the introduction, this issue will not be treated in this book.[2]

However, we shall now quote a fundamental result that we shall need in the chapter on martingales (Chapter 8).

Caratheodory's Theorem

Definition 4.1.28 *Let X be a set. The collection $\mathcal{A} \subseteq \mathcal{P}(X)$ is called an* algebra *if*

(α) $X \in \mathcal{A}$;

(β) $A, B \in \mathcal{A} \implies A \cup B \in \mathcal{A}$;

(γ) $A \in \mathcal{A} \implies \overline{A} \in \mathcal{A}$.

The only difference with a σ-field is that we require it to be closed under *finite* (instead countable) unions. (This is why a σ-field is also called a *σ-algebra.*) Note that, similarly to the σ-field case, $\varnothing \in \mathcal{A}$ and $\mathcal{A}$ is closed under finite intersections.

EXAMPLE 4.1.29: FINITE UNIONS OF DISJOINT INTERVALS. On $\mathbb{R}$, the collection of finite sums of disjoint intervals is an algebra. (By interval, we mean any type of interval: open, closed, semi-open, semi-closed, infinite, etc., in other words a connected subset of $\mathbb{R}$.[3])

[2] See [1], [3], or [11].

[3] A subset C of $\mathbb{R}$ is called connected if for all $a, b \in C$, the segment $[a, b] \subseteq C$.

Definition 4.1.30 *Let X be a set. The collection $\mathcal{C} \subseteq \mathcal{P}(X)$ is called a* semi-algebra *if*

(α) $X \in \mathcal{C}$,

(β) $A, B \in \mathcal{C} \implies A \cup B \in \mathcal{C}$, *and*

(γ) *when $A \in \mathcal{C}$, $\overline{A}$ can be expressed as a* finite union of disjoint sets *of $\mathcal{C}$.*

EXAMPLE 4.1.31: THE COLLECTION OF INTERVALS. On $\mathbb{R}$, the collection of intervals is a semi-algebra.

Theorem 4.1.32 *Let $\mathcal{C}$ be either an algebra or a semi-algebra defined on X. Let μ be a σ-finite measure on $(X, \mathcal{C})$. Then there exists a unique extension of μ to $(X, \sigma(\mathcal{C}))$ that is a measure.*

The proof is omitted[4].

4.2 The Integral

We are now in a position to define (when it exists) the Lebesgue integral of a measurable function $f : (X, \mathcal{X}) \to (\overline{\mathbb{R}}, \mathcal{B}(\overline{\mathbb{R}}))$ with respect to a measure μ. This integral will be denoted by

$$\int_X f \, d\mu, \quad \text{or} \quad \int_X f(x) \, \mu(dx), \quad \text{or} \quad \mu(f).$$

Let $\mathcal{S}^+(X)$ (or $\mathcal{S}^+$ if the context is clear) be the set of *non-negative* simple functions $f : (X, \mathcal{X}) \to (\mathbb{R}, \mathcal{B}(\mathbb{R}))$, and by $\mathcal{M}^+(X)$ (or $\mathcal{M}^+$) the set of *non-negative* functions $f : (X, \mathcal{X}) \to (\overline{\mathbb{R}}, \mathcal{B}(\overline{\mathbb{R}}))$.

The integral is defined in three steps. Firstly for simple functions, where the definition imposes itself. Secondly for non-negative measurable functions, by a natural limiting procedure involving the approximation theorem (Theorem 4.1.15), and finally for (some) functions of arbitrary sign by considering their negative and positive parts.

STEP 1. One first defines the integral for integrands in $\mathcal{S}^+$. Let $f : X \to \mathbb{R}$ be a non-negative simple Borel function as in Definition 4.1.7. The integral of f with respect to μ is defined by

$$\int_X f \, d\mu := \sum_{i=1}^{k} a_i \, \mu(A_i). \tag{4.4}$$

[4] See for instance [12], Theorem 1.41

In order to check that this definition does not depend on the representation of f, one must show that if f admits another representation

$$f(x) = \sum_{j=1}^{m} b_j \, 1_{B_j}(x)\,,$$

where $m \in \mathbb{N}_+$, $b_1, \dots, b_m \in \mathbb{R}$, and $B_1, \dots, B_m$ are sets in $\mathcal{X}$, then

$$\sum_{j=1}^{m} b_j \, \mu(B_j) = \sum_{i=1}^{k} a_i \, \mu(A_i)\,. \tag{4.5}$$

This verification is easy and left for the reader.

The next lemma collects a few intermediary results.

Lemma 4.2.1 *Let $f, f_1, f_2, \dots$ be in $\mathcal{S}^+$. Then*

(a) *for all $\lambda \geq 0$, $\lambda f \in \mathcal{S}^+$ and $\int_X (\lambda f)\,\mathrm{d}\mu = \lambda \int_X f\,\mathrm{d}\mu$,*

(b) *$f_1 + f_2 \in \mathcal{S}^+$ and $\int_X (f_1 + f_2)\,\mathrm{d}\mu = \int_X f_1\,\mathrm{d}\mu + \int_X f_2\,\mathrm{d}\mu$,*

(c) *$f_1 \leq f_2$ implies $\int_X f_1\,\mathrm{d}\mu \leq \int_X f_2\,\mathrm{d}\mu$,*

(d) *$f_1 \wedge f_2$ and $f_1 \vee f_2$ are in $\mathcal{S}^+$, and*

(e) *if $f_n \leq f_{n+1} \leq f$ for all $n \geq 1$ and $\lim_{n\uparrow\infty} f_n = f$, then $\lim_{n\uparrow\infty} \int_X f_n\,\mathrm{d}\mu = \int_X f\,\mathrm{d}\mu$.*

Proof. Properties (a)–(d) are immediate. For (e), first consider the case $f = 1_A$. Fix $m \geq 1$. For all $n \geq 1$, define $A_{n,m} = \{x : f_n(x) \geq 1 - \frac{1}{m}\}$. Since $\{f_n\}_{n\geq 1}$ is non-decreasing, we have that $A_{n,m} \subseteq A_{n+1,m}$; and since $f_n \uparrow 1_A$, we have that $\cup_{n=1}^{\infty} A_{n,m} = A$. Note that

$$\left(1 - \frac{1}{m}\right) 1_{A_{n,m}} \leq f_n \leq 1_A\,,$$

and therefore

$$\left(1 - \frac{1}{m}\right) \mu(A_{n,m}) \leq \int_X f_n\,\mathrm{d}\mu \leq \mu(A)\,,$$

from which the announced result follows in this particular case by first letting $n \uparrow +\infty$ and then $m \uparrow +\infty$.

Consider now the general case where $f = \sum_{i=1}^{k} a_i 1_{A_i}$. We may suppose that $\{A_i\}_{i=1}^{k}$ is a partition of X, so that $f_n = \sum_{i=1}^{k} f_n 1_{A_i}$ and therefore, by (a),

$$\int_X f_n \, d\mu = \sum_{i=1}^{k} \int_X f_n 1_{A_i} \, d\mu = \sum_{i=1}^{k} a_i \int_X \frac{f_n}{a_i} 1_{A_i} \, d\mu \, .$$

Passing to the limit $n \uparrow \infty$ yields the desired result, since $\frac{f_n}{a_i} 1_{A_i} \uparrow 1_{A_i}$. □

STEP 2. The integral will now be defined for integrands $f \in \mathcal{M}^+$. For such f, let

$$\mu(f) := \sup \left\{ \int_X \varphi \, d\mu \, ; \, \varphi \leq f, \, \varphi \in \mathcal{S}^+ \right\} .$$

The function f is called μ-integrable if $\mu(f) < \infty$.

We first check that if $f \in \mathcal{S}^+$, $\mu(f) = \int_X f \, d\mu$. For this, let

$$A_f := \left\{ \int_X \varphi \, d\mu \, ; \, \varphi \leq f, \, \varphi \in \mathcal{S}^+ \right\} .$$

Since $f \in \mathcal{S}^+$, $\int_X f \, d\mu \in A_f$ and therefore $\mu(f) \geq \int_X f \, d\mu$. On the other hand, for all $\varphi \in \mathcal{S}^+$ such that $\varphi \leq f$, $\int_X \varphi \, d\mu \leq \int_X f \, d\mu$ and therefore $\mu(f) \leq \int_X f \, d\mu$. Therefore $\mu(f) = \int_X f \, d\mu$.

Having checked this point, it is now safe to call $\mu(f)$ the integral of f with respect to μ, and denote it also by $\int_X f \, d\mu$. Indeed the two ways of defining $\int_X f \, d\mu$ for $f \in \mathcal{S}^+$ (as in Step 1 and Step 2) give the same result.

The next result, due to Beppo Levi, is the *monotone convergence theorem.*

Theorem 4.2.2 *Let $\{f_n\}_{n \geq 1}$ be a non-decreasing sequence of non-negative measurable functions from X to $\overline{\mathbb{R}}$. Then*

$$\lim_{n \uparrow \infty} \int_X f_n \, d\mu = \int_X (\lim_{n \uparrow \infty} f_n) \, d\mu \, .$$

Proof. We shall use the following monotonicity property: if f_1, f_2 in $\mathcal{M}^+$ are such that $f_1 \leq f_2$, then $\int_X f_1 \, d\mu \leq \int_X f_2 \, d\mu$. In fact, $A_{f_1} \subseteq A_{f_2}$, and therefore

$$\mu(f_1) = \sup A_{f_1} \leq \sup A_{f_2} = \mu(f_2) \, .$$

We now turn to the proof of the theorem. Denote $\lim_{n \uparrow \infty} f_n$ by f. By the just proved monotonicity property of the integral of functions in $\mathcal{M}^+$,

$$\int_X f_n \, d\mu \leq \int_X f_{n+1} \, d\mu \leq \int_X f \, d\mu \, .$$

Being a non-decreasing sequence, $\{\int_X f_n \,\mathrm{d}\mu\}_{n\geq 1}$ has a limit, and by the previous inequality

$$\lim_{n\uparrow\infty} \int_X f_n \,\mathrm{d}\mu \leq \int_X f \,\mathrm{d}\mu \,.$$

It remains to prove the converse inequality. For $\varphi \in \mathcal{S}^+$ such that $\varphi \leq f$, $\lambda \in (0,1)$ and $n \geq 1$, define the (measurable) set $E_n = \{f_n \geq \lambda\varphi\}$. We have that $E_n \subseteq E_{n+1}$. Moreover $\cup_{n\geq 1} E_n = X$. Since $\lambda\varphi 1_{E_n} \leq f_n$,

$$\int_X \lambda\varphi 1_{E_n} \,\mathrm{d}\mu \leq \int_X f_n \,\mathrm{d}\mu \leq \lim_{k\uparrow\infty} \int_X f_k \,\mathrm{d}\mu \,.$$

On the other hand, since $E_n \subseteq E_{n+1}$ and $\cup_{n\geq 1} E_n = X$, we have that $1_{E_n} \uparrow 1$ and in particular $1_{E_n}\varphi \uparrow \varphi$. Therefore by (e) of Lemma 4.2.1, $\lim_{n\uparrow\infty} \int_X \lambda\varphi 1_{E_n} \,\mathrm{d}\mu = \int_X \lambda\varphi \,\mathrm{d}\mu$. Passing to the limit $n \uparrow \infty$ in the last displayed inequalities, we have that for all $\lambda \in (0,1)$,

$$\lambda \int_X \varphi \,\mathrm{d}\mu \leq \lim_{n\uparrow\infty} \int_X f_n \,\mathrm{d}\mu \,.$$

This equality remains true at the limit $\lambda = 1$. This being true of all $\varphi \in \mathcal{S}^+$ such that $\varphi \leq f$, we have

$$\int_X f \,\mathrm{d}\mu \leq \lim_{n\uparrow\infty} \int_X f_n \,\mathrm{d}\mu \,.$$

□

Here is another collection of intermediary results that we group in a lemma for later reference:

Lemma 4.2.3 *Let f, f_1, f_2 be in $\mathcal{M}^+$. Then*

(i) *for all $\lambda \geq 0$, $\int_X (\lambda f) \,\mathrm{d}\mu = \lambda \int_X f \,\mathrm{d}\mu$,*

(ii) *$\int_X (f_1 + f_2) \,\mathrm{d}\mu = \int_X f_1 \,\mathrm{d}\mu + \int_X f_2 \,\mathrm{d}\mu$, and*

(iii) *if $f_1 \leq f_2$, then $\int_X f_1 \,\mathrm{d}\mu \leq \int_X f_2 \,\mathrm{d}\mu$.*

Proof. (iii) was obtained in the proof of Theorem 4.2.2. Properties (i) and (ii) are satisfied for functions in $\mathcal{S}^+$ (Lemma 4.2.1). Using non-decreasing sequences of functions in $\mathcal{S}^+$, $\{f_{1,n}\}_{n\geq 1}$ and $\{f_{2,n}\}_{n\geq 1}$, converging respectively to f_1 and f_2, we have that for all $n \geq 1$,

$$\int_X (\lambda f_{1,n}) \,\mathrm{d}\mu = \lambda \int_X f_{1,n} \,\mathrm{d}\mu$$

and

$$\int_X (f_{1,n} + f_{2,n})\, \mathrm{d}\mu = \int_X f_{1,n}\, \mathrm{d}\mu + \int_X f_{2,n}\, \mathrm{d}\mu\,.$$

Letting $n \uparrow \infty$, the monotone convergence theorem 4.2.2 yields (i) and (ii). □

The next result is a fundamental technical tool, called *Fatou's lemma*:

Theorem 4.2.4 *Let $\{f_n\}_{n\geq 1}$ be a sequence of non-negative measurable functions from X to $\overline{\mathbb{R}}_+$. Then*

$$\int_X (\liminf_{n\uparrow\infty} f_n)\, \mathrm{d}\mu \leq \liminf_{n\uparrow\infty} \int_X f_n\, \mathrm{d}\mu\,.$$

Proof. Define $f := \liminf_{n\uparrow\infty} f_n := \lim_{n\uparrow\infty} (\inf_{k\geq n} f_k)$. By the monotone convergence theorem (Theorem 4.2.2) for the second equality, we obtain

$$\int_X f\, \mathrm{d}\mu = \int_X (\lim_{n\uparrow\infty} \inf_{k\geq n} f_k)\, \mathrm{d}\mu = \lim_{n\uparrow\infty} \int_X (\inf_{k\geq n} f_k)\, \mathrm{d}\mu\,.$$

On the other hand, since for all $i \geq n$, $\int_X (\inf_{k\geq n} f_k)\, \mathrm{d}\mu \leq \int_X f_i\, \mathrm{d}\mu$, we have that $\int_X (\inf_{k\geq n} f_k)\, \mathrm{d}\mu \leq \inf_{i\geq n}(\int_X f_i\, \mathrm{d}\mu)$. Therefore

$$\int_X f\, \mathrm{d}\mu \leq \liminf_{n\uparrow\infty\, i\geq n} \left(\int_X f_i\, \mathrm{d}\mu\right) = \liminf_{n\uparrow\infty} \int_X f_n\, \mathrm{d}\mu\,.$$

□

STEP 3. Integrals of functions of arbitrary sign.

Definition 4.2.5 *A measurable function $f : (X, \mathcal{X}) \to (\overline{\mathbb{R}}, \mathcal{B}(\overline{\mathbb{R}}))$ satisfying*

$$\int_X |f|\, \mathrm{d}\mu < \infty$$

is called a μ-integrable function.

Define $f^+ := \max(f, 0)$ and $f^- := \max(-f, 0)$. In particular, $f = f^+ - f^-$ and $f^\pm \leq |f|$. Therefore, by monotonicity (Property (iii) of Lemma 4.2.3),

$$\int_X f^\pm\, \mathrm{d}\mu \leq \int_X |f|\, \mathrm{d}\mu\,.$$

Thus, if f is integrable, the right-hand side of

$$\int_X f\, \mathrm{d}\mu := \int_X f^+\, \mathrm{d}\mu - \int_X f^-\, \mathrm{d}\mu \tag{4.6}$$

is meaningful (no $-\infty + \infty$ form) and defines the integral of the left-hand side. Moreover, the integral of f with respect to μ defined in this way is finite.

The integral can be defined for some *non-integrable* measurable functions, for instance, as we have seen, for all measurable non-negative functions. More generally, if $f : (X, \mathcal{X}) \to (\overline{\mathbb{R}}, \mathcal{B}(\overline{\mathbb{R}}))$ is such that at least one of the integrals $\int_X f^+ \, \mathrm{d}\mu$ or $\int_X f^- \, \mathrm{d}\mu$ is finite, one defines the integral as in (4.6). This leads to one of the forms "finite minus finite", "finite minus infinite", and "infinite minus finite". The case which is rigorously excluded is that in which $\mu(f^+) = \mu(f^-) = +\infty$.

Let $A \in \mathcal{X}$. The following equality is a definition of the left-hand side provided the right-hand side is well defined:

$$\int_A f(x) \, \mu(\mathrm{d}x) := \int_X 1_A(x) \, f(x) \, \mu(\mathrm{d}x) \, .$$

For a complex Borel function $f : X \to \mathbb{C}$ (that is, $f = f_1 + i f_2$, where $f_1, f_2 : (X, \mathcal{X}) \to (\mathbb{R}, \mathcal{B}(\mathbb{R}))$) such that $\mu(|f|) < \infty$, let

$$\int_X f \, \mathrm{d}\mu := \int_X f_1 \, \mathrm{d}\mu + i \int_X f_2 \, \mathrm{d}\mu \, .$$

EXAMPLE 4.2.6: INTEGRAL WITH RESPECT TO THE DIRAC MEASURE. Let $(X, \mathcal{X})$ be an arbitrary measurable space and let ε_a be the Dirac measure at a point $a \in X$. Let $f : (X, \mathcal{X}) \to (\mathbb{R}, \mathcal{B}(\mathbb{R}))$. We shall prove formally that it is ε_a-integrable and that

$$\varepsilon_a(f) = f(a) \, .$$

For a simple function f as in (4.1), we have

$$\varepsilon_a(f) = \sum_{i=1}^{k} a_i \, \varepsilon_a(A_i) = \sum_{i=1}^{k} a_i \, 1_{A_i}(a) = f(a) \, .$$

For a non-negative function f, and any non-decreasing sequence of simple non-negative measurable functions $\{f_n\}_{n \geq 1}$ converging to f, we have

$$\varepsilon_a(f) = \lim_{n \uparrow \infty} \varepsilon_a(f_n) = \lim_{n \uparrow \infty} f_n(a) = f(a) \, .$$

Finally, for any $f : (X, \mathcal{X}) \to (\mathbb{R}, \mathcal{B}(\mathbb{R}))$

$$\varepsilon_a(f) = \varepsilon_a(f^+) - \varepsilon_a(f^-) = f^+(a) - f^-(a) = f(a)$$

is a well defined quantity.

What we have finally obtained is the formula

$$\int_X f(x)\,\varepsilon_a(\mathrm{d}x) = f(a)\,.$$

The *aficionados* of the so-called *Dirac "function"* δ like to write the left-hand side

$$\int_X f(x)\,\delta_a(x)\,\mathrm{d}x \text{ or } \int_X f(x)\,\delta(x-a)\,\mathrm{d}x\,.$$

EXAMPLE 4.2.7: LEBESGUE-INTEGRABLE BUT NOT RIEMANN-INTEGRABLE. The function f defined by $f := 1_{\mathbb{Q}}$ ($\mathbb{Q}$ is the set of rational numbers) is a Borel function and it is Lebesgue integrable with its integral equal to zero because $\{f \neq 0\}$ is the set of rational numbers, which has null Lebesgue measure. However, f is not Riemann integrable.

We finally define the *Stieljes–Lebesgue integral.*

Definition 4.2.8 *Let F be cumulative distribution function on $(\mathbb{R}, \mathcal{B}(\mathbb{R}))$ and let μ_F be the associated locally finite measure on $(\mathbb{R}, \mathcal{B}(\mathbb{R})))$ (see Example 4.1.26). By definition, the Stieltjes–Lebesgue integral of the measurable function $g : (\mathbb{R}, \mathcal{B}(\mathbb{R})) \to (\mathbb{R}, \mathcal{B}(\mathbb{R}))$ with respect to F is the integral of g with respect to μ_F. It is denoted by $\int_{\mathbb{R}} g(x)\,\mathrm{d}F(x)$. Therefore*

$$\int_{\mathbb{R}} g(x)\,\mathrm{d}F(x) := \int_{\mathbb{R}} g(x)\,\mu_F(\mathrm{d}x)\,.$$

4.3 Basic Properties of the Integral

We first state and prove one of the most important results of integration theory, the *Lebesgue theorem,* also called the *dominated convergence theorem.*

Theorem 4.3.1 *Let $\{f_n\}_{n\geq 1}$ be a sequence of measurable functions from $(X, \mathcal{X})$ to $(\mathbb{R}, \mathcal{B}(\mathbb{R}))$ that converges to a (necessarily) measurable function f. Suppose moreover that for all $n \geq 1$, $|f_n| \leq g$, where g is integrable. Then*

$$\lim_{n\uparrow\infty} \int_X f_n\,\mathrm{d}\mu = \int_X (\lim_{n\uparrow\infty} f_n)\,\mathrm{d}\mu\,.$$

Proof. By Fatou's lemma applied to the sequence of non-negative functions $\{g + f_n\}_{n\geq 1}$,

$$\int_X (g+f)\,d\mu = \int_X \lim_{n\uparrow\infty}(g+f_n)\,d\mu$$
$$\leq \liminf_{n\uparrow\infty}\int_X (g+f_n)\,d\mu = \int_X g\,d\mu + \liminf_{n\uparrow\infty}\int_X f_n\,d\mu\,.$$

Therefore,

$$\int_X f\,d\mu \leq \liminf_{n\uparrow\infty}\int_X f_n\,d\mu\,.$$

Similarly, replacing f and f_n by $-f$ and $-f_n$ respectively,

$$\int_X f\,d\mu \geq \limsup_{n\uparrow\infty}\int_X f_n\,d\mu\,.$$

In particular, $\lim_{n\uparrow\infty}\int_X f_n\,d\mu$ exists and is equal to $\int_X f\,d\mu$. □

Recall that for all $A \in \mathcal{X}$,

$$\int_X 1_A\,d\mu = \mu(A) \tag{4.7}$$

by definition and that the notation $\int_A f\,d\mu$ stands for $\int_X 1_A f\,d\mu$.

Theorem 4.3.2 *Let $f, g : (X, \mathcal{X}) \to (\overline{\mathbb{R}}, \mathcal{B}(\overline{\mathbb{R}}))$ be μ-integrable functions, and let $a, b \in \mathbb{R}$. Then*

(a) *$af + bg$ is μ-integrable and $\mu(af + bg) = a\mu(f) + b\mu(g)$,*

(b) *if $f = 0$ μ-a.e., then $\mu(f) = 0$; if $f = g$ μ-a.e., then $\mu(f) = \mu(g)$,*

(c) *if $f \leq g$ μ-a.e., then $\mu(f) \leq \mu(g)$,*

(d) *$|\mu(f)| \leq \mu(|f|)$,*

(e) *if $f \geq 0$ μ-a.e. and $\mu(f) = 0$, then $f = 0$ μ-a.e.,*

(f) *if $\mu(1_A f) = 0$ for all $A \in \mathcal{X}$, then $f = 0$ μ-a.e., and*

(g) *if f is μ-integrable, then $|f| < \infty$ μ-a.e.*

Proof. The (easy) proofs of (a)–(c) are omitted.

(d) $\mu(f) = \mu(f_+) - \mu(f_-)$. Therefore $|\mu(f)| \leq \mu(f_+) + \mu(f_-) = \mu(f_+ + f_-) = \mu(|f|)$.

(e) Define $A_n = \{f \geq \frac{1}{n}\}$. Since f is non-negative, $f \geq \frac{1}{n} 1_{A_n}$, and therefore,

$$\mu(f) \geq \frac{1}{n} \mu(A_n) ,$$

from which it follows that, since $\mu(f) = 0$, $\mu(A_n) = 0$, and $\lim_{n\uparrow\infty} \mu(A_n) = 0$. But the sequence of sets $\{A_n\}_{n\geq 1}$ increases to $\{f > 0\}$ and therefore, by sequential continuity, $\mu(\{f > 0\}) = 0$, that is, $f \leq 0, \mu$-a.e. On the other hand, by hypothesis, $f \geq 0, \mu$-a.e. Therefore $f = 0, \mu$-a.e.

(f) With $A = \{f > 0\}$, $1_A f$ is a non-negative measurable function. By (e), $1_A f = 0, \mu$-a.e. This implies that $1_A = 0, \mu$-a.e., that is to say $f \leq 0, \mu$-a.e. Similarly, $f \geq 0, \mu$-a.e. Therefore, $f = 0, \mu$-a.e.

(g) It is enough to consider the case $f \geq 0$. Since $f \geq n 1_{\{f=\infty\}}$ for all $n \geq 1$, we have

$$\infty > \mu(f) \geq n\mu(\{f = \infty\}) ,$$

and therefore $n\mu(\{f = \infty\}) < \infty$. This cannot be true for all $n \geq 1$ unless $\mu(\{f = \infty\}) = 0$. □

The extension to complex Borel functions of the properties (a), (b), (d) and (f) is immediate.

Beppo Levi, Fatou and Lebesgue

The following versions of the theorems of Beppo Levi, Fatou and Lebesgue differ from the previous ones by the introduction of "μ-almost everywhere" in the statements of the conditions. No other proofs are needed since integrals of almost everywhere equal functions are equal and countable unions of negligible sets are negligible. Only a convention must be stated: if the limit of a sequence of real measurable functions exists μ-almost everywhere, that is, outside a μ-negligible set, then the limit is typically assigned some arbitrary value on this μ-negligible set.

Remember that we are looking for conditions guaranteeing that

$$\int_X \lim_{n\uparrow\infty} f_n \, \mathrm{d}\mu = \lim_{n\uparrow\infty} \int_X f_n \, \mathrm{d}\mu . \tag{4.8}$$

We start by restating the *monotone convergence* theorem.

Theorem 4.3.3 *Let $f_n : (X, \mathcal{X}) \to (\overline{\mathbb{R}}, \mathcal{B}(\overline{\mathbb{R}}))$ $(n \geq 1)$ be such that*

(i) $f_n \geq 0$ *μ-a.e., and*

(ii) $f_{n+1} \geq f_n$ *μ-a.e.*

Then, there exists a non-negative function $f : (X, \mathcal{X}) \to (\overline{\mathbb{R}}, \mathcal{B}(\overline{\mathbb{R}}))$ such that

$$\lim_{n \uparrow \infty} f_n = f \quad \mu\text{-a.e.},$$

and (4.8) holds true.

Next, we restate *Fatou's lemma*.

Theorem 4.3.4 *Let $f_n : (X, \mathcal{X}) \to (\overline{\mathbb{R}}, \mathcal{B}(\overline{\mathbb{R}}))$ $(n \geq 1)$ be such that $f_n \geq 0$ μ-a.e. $(n \geq 1)$. Then*

$$\int_X (\liminf_{n \uparrow \infty} f_n) \, \mathrm{d}\mu \leq \liminf_{n \uparrow \infty} \left(\int_X f_n \, \mathrm{d}\mu \right). \tag{4.9}$$

Finally, we restate the *Lebesgue* or *dominated convergence* theorem.

Theorem 4.3.5 *Let $f_n : (X, \mathcal{X}) \to (\overline{\mathbb{R}}, \mathcal{B}(\overline{\mathbb{R}}))$ $(n \geq 1)$ be such that, for some function $f : (X, \mathcal{X}) \to (\overline{\mathbb{R}}, \mathcal{B}(\overline{\mathbb{R}}))$ and some μ-integrable function $g : (X, \mathcal{X}) \to (\overline{\mathbb{R}}, \mathcal{B}(\overline{\mathbb{R}}))$:*

(i) $\lim_{n \uparrow \infty} f_n = f$, *$\mu$-a.e., and*

(ii) $|f_n| \leq |g|$ *μ-a.e. for all $n \geq 1$.*

Then, (4.8) holds true.

EXAMPLE 4.3.6: THE CLASSICAL COUNTEREXAMPLE.
Let $(X, \mathcal{X}, \mu) = (\mathbb{R}, \mathcal{B}(\mathbb{R}), \ell)$, and let

$$f_n(x) := n \, 1_{(0, \frac{1}{n}]}(x).$$

One has $\lim_{n \uparrow \infty} f_n = 0$. Therefore $\mu(\lim_{n \uparrow \infty} f_n) = 0$. However, $\mu(f_n) = 1$ for all $n \geq 1$.

Differentiation under the Integral Sign

Let $(X, \mathcal{X}, \mu)$ be a measure space and let $(a, b) \subseteq \mathbb{R}$. Let $f : (a, b) \times X \to \mathbb{R}$ and for all $t \in (a, b)$, define $f_t : X \to \mathbb{R}$ by $f_t(x) := f(t, x)$. Suppose that for all $t \in (a,b)$, f_t is measurable with respect to $\mathcal{X}$, and define, when possible, the function $I : (a,b) \to \mathbb{R}$ by the formula

$$I(t) = \int_X f(t,x)\, \mu(\mathrm{d}x)\,. \tag{4.10}$$

Theorem 4.3.7 *Assume that for μ-almost all x the function $t \mapsto f(t,x)$ is continuous at $t_0 \in (a,b)$ and that there exists a μ-integrable function $g : (X,\mathcal{X}) \to (\overline{\mathbb{R}}, \mathcal{B}(\overline{\mathbb{R}}))$ such that $|f(t,x)| \leq |g(x)|$ μ-a.e. for all t in a neighborhood V of t_0. Then:*

A. *I is well defined and is continuous at t_0.*

B. *We now assume in addition that*

   (α) *$t \to f(t,x)$ is continuously differentiable on V for μ-almost all x, and*

   (β) *for some μ-integrable function $h : (X,\mathcal{X}) \to (\overline{\mathbb{R}}, \mathcal{B}(\overline{\mathbb{R}}))$ and all $t \in V$,*

$$|(\mathrm{d}f/\mathrm{d}t)\,(t,x)| \leq |h(x)| \quad \mu\text{-a.e.}$$

Then I is differentiable at t_0 and

$$I'(t_0) = \int_X (\mathrm{d}f/\mathrm{d}t)\,(t_0,x)\, \mu(\mathrm{d}x)\,. \tag{4.11}$$

Proof. A. Let $\{t_n\}_{n\geq 1}$ be a sequence in $V \setminus \{t_0\}$ such that $\lim_{n\uparrow\infty} t_n = t_0$, and define $f_n(x) = f(t_n,x)$, $f(x) = f(t_0,x)$. By dominated convergence,

$$\lim_{n\uparrow\infty} I(t_n) = \lim_{n\uparrow\infty} \mu(f_n) = \mu(f) = I(t_0)\,.$$

B. Let $\{t_n\}_{n\geq 1}$ be a sequence in $V \setminus \{t_0\}$ such that $\lim_{n\uparrow\infty} t_n = t_0$, and define $f_n(x) = f(t_n,x)$, $f(x) = f(t_0,x)$. By dominated convergence,

$$\lim_{n\uparrow\infty} I(t_n) = \lim_{n\uparrow\infty} \mu(f_n) = \mu(f) = I(t_0)\,.$$

Also

$$\frac{I(t_n) - I(t_0)}{t_n - t_0} = \int_X \frac{f(t_n,x) - f(t_0,x)}{t_n - t_0}\, \mu(\mathrm{d}x)\,,$$

and for some $\theta \in (0,1)$, possibly depending upon n,

$$\left| \frac{f(t_n,x) - f(t_0,x)}{t_n - t_0} \right| \leq |(\mathrm{d}f/\mathrm{d}t)\,(t_0 + \theta(t_n - t_0), x)|\,.$$

The latter quantity is bounded by $|h(x)|$. Therefore, by dominated convergence,

$$\begin{aligned}\lim_{n\uparrow\infty}\frac{I(t_n)-I(t_0)}{t_n-t_0} &= \int_X\left(\lim_{n\uparrow\infty}\frac{f(t_n,x)-f(t_0)}{t_n-t_0}\right)\mu(\mathrm{d}x)\\ &= \int_X(\mathrm{d}f/\mathrm{d}t)\,(t_0,x)\,\mu(\mathrm{d}x)\,.\end{aligned}$$

□

4.4 The Big Theorems

The Image Measure Theorem

Definition 4.4.1 *Let $(X,\mathcal{X})$ and $(E,\mathcal{E})$ be two measurable spaces, let $h:(X,\mathcal{X})\to(E,\mathcal{E})$ be a measurable function, and let μ be a measure on $(X,\mathcal{X})$. The measure $\mu\circ h^{-1}$ on $(E,\mathcal{E})$, called the* image of μ by h, *is defined by*

$$(\mu\circ h^{-1})(C)=\mu(h^{-1}(C)),\quad C\in\mathcal{E}.$$

(One easily checks that it is indeed a measure.)

In the proof of the following theorem, the combination of the approximation theorem for measurable functions and of the monotone convergence theorem is typical.

Theorem 4.4.2 *For $f:(X,\mathcal{X})\to(\overline{\mathbb{R}},\mathcal{B}(\overline{\mathbb{R}})$ an arbitrary non-negative measurable function*

$$\int_X(f\circ h)(x)\,\mu(\mathrm{d}x)=\int_E f(y)(\mu\circ h^{-1})\,(\mathrm{d}y)\,. \tag{4.12}$$

For functions $f:(X,\mathcal{X})\to(\overline{\mathbb{R}},\mathcal{B}(\overline{\mathbb{R}})$ of arbitrary sign either one of the conditions

(a) *$f\circ h$ is μ-integrable, or*

(b) *f is $\mu\circ h^{-1}$-integrable,*

implies the other, and equality (4.12) then holds.

Proof. The equality (4.12) is readily verified when f is a non-negative simple measurable function. In the general case one approximates f by a non-decreasing sequence of non-negative simple measurable functions $\{f_n\}_{n\geq 1}$ and (4.12) then follows from the same equality written with $f=f_n$, by letting $n\uparrow\infty$ and using the monotone convergence theorem. For the case of functions of arbitrary sign, apply (4.12) with f^+ and f^-. □

The Radon–Nikodým Theorem

Definition 4.4.3 *Let $(X, \mathcal{X}, \mu)$ be a measure space and let $h : (X, \mathcal{X}) \to (\overline{\mathbb{R}}, \mathcal{B}(\overline{\mathbb{R}}))$ be a non-negative measurable function. Define the set function $\nu : \mathcal{X} \to [0, \infty]$ by*

$$\nu(C) = \int_C h(x)\,\mu(\mathrm{d}x)\,.$$

Then ν is a measure on $(X, \mathcal{X})$ called the product of μ by the function h. *This is denoted by $\mathrm{d}\nu = h\,\mathrm{d}\mu$.*

That ν is a measure is easily checked. First of all, it is obvious that $\nu(\varnothing) = 0$. As for the σ-additivity property, write for any sequence of mutually disjoint measurable sets $\{A_n\}_{n\geq 1}$,

$$\begin{aligned}
\nu(\cup_{n\geq 1} A_n) &= \int_{\cup_{n\geq 1} A_n} h\,\mathrm{d}\mu = \int_X 1_{\cup_{n\geq 1} A_n} h\,\mathrm{d}\mu \\
&= \int_X \left(\sum_{n\geq 1} 1_{A_n}\right) h\,\mathrm{d}\mu = \int_X \left(\lim_{k\uparrow\infty} \sum_{n=1}^{k} 1_{A_n}\right) h\,\mathrm{d}\mu \\
&= \lim_{k\uparrow\infty} \int_X \left(\sum_{n=1}^{k} 1_{A_n}\right) h\,\mathrm{d}\mu = \lim_{k\uparrow\infty} \sum_{n=1}^{k} \int_X 1_{A_n} h\,\mathrm{d}\mu \\
&= \lim_{k\uparrow\infty} \sum_{n=1}^{k} \nu(A_n) = \sum_{n\geq 1} \nu(A_n)\,,
\end{aligned}$$

where the fifth equality is by monotone convergence.

Theorem 4.4.4 *Let μ, h and ν be as in Definition 4.4.3.*

(i) *For non-negative $f : (X, \mathcal{X}) \to (\overline{\mathbb{R}}, \mathcal{B}(\overline{\mathbb{R}}))$,*

$$\int_X f(x)\, \nu(\mathrm{d}x) = \int_X f(x) h(x)\, \mu(\mathrm{d}x)\,. \tag{4.13}$$

(ii) *If $f : (X, \mathcal{X}) \to (\overline{\mathbb{R}}, \mathcal{B}(\overline{\mathbb{R}}))$ has arbitrary sign, then either one of the following conditions*

(a) *f is ν-integrable,*

(b) *fh is μ-integrable,*

implies the other, and the equality (4.13) then holds.

Proof. Verify (4.13) for elementary non-negative functions and, approximating f by a non-decreasing sequence of such functions, use the monotone convergence theorem as in the proof of (4.12). For the case of functions of arbitrary sign, apply (4.13) with $f = f^+$ and $f = f^-$. □

Observe that in the situation of Theorem 4.4.4,

$$\mu(C) = 0 \implies \nu(C) = 0 \quad (C \in \mathcal{X})\,. \tag{4.14}$$

Definition 4.4.5 *Let μ and ν be two measures on $(X, \mathcal{X})$. If (4.14) holds, ν is said to be* absolutely continuous *with respect to μ. This is denoted by $\nu \ll \mu$.*

The proof of the next theorem is omitted [5]

Theorem 4.4.6 *Let μ and ν be two σ-finite measures on $(X, \mathcal{X})$ such that $\nu \ll \mu$. Then there exists a non-negative function $h : (X, \mathcal{X}) \to (\overline{\mathbb{R}}, \mathcal{B}(\overline{\mathbb{R}}))$ such that*

$$\nu(\mathrm{d}x) = h(x)\, \mu(\mathrm{d}x)\,.$$

The function h is called the *Radon–Nikodým derivative* of ν with respect to μ and is denoted $\mathrm{d}\nu/\mathrm{d}\mu$. With such a notation, we have that

$$\int_X f(x)\, \nu(\mathrm{d}x) = \int_X f(x)\, \tfrac{\mathrm{d}\nu}{\mathrm{d}\mu}(x)\, \mu(\mathrm{d}x)$$

for all non-negative $f = (X, \mathcal{X}) \to (\overline{\mathbb{R}}, \mathcal{B}(\overline{\mathbb{R}}))$.

[5] See Subsection 2.3.1 of [5].

The Fubini–Tonelli Theorem

Let $(X_1, \mathcal{X}_1, \mu_1)$ and $(X_2, \mathcal{X}_2, \mu_2)$ be two measure spaces where μ_1 and μ_2 are sigma-finite. Define the product set $X = X_1 \times X_2$ and the *product σ-field* $\mathcal{X} = \mathcal{X}_1 \otimes \mathcal{X}_2$, where by definition the latter is the smallest σ-field on X containing all sets of the form $A_1 \times A_2$, where $A_1 \in \mathcal{X}_1$, $A_2 \in \mathcal{X}_2$.

Theorem 4.4.7 *There exists a unique measure μ on $(X_1 \times X_2, \mathcal{X}_1 \otimes \mathcal{X}_2)$ such that*

$$\mu(A_1 \times A_2) = \mu_1(A_1)\mu_2(A_2) \tag{4.15}$$

for all $A_1 \in \mathcal{X}_1$, $A_2 \in \mathcal{X}_2$.

The proofs of this theorem and of the next one are omitted[6].

The measure μ is the *product measure* of μ_1 and μ_2, and is denoted $\mu_1 \times \mu_2$.

The above result and the following ones are stated for products of two sigma-finite measures, but extend in an obvious manner to a finite number of sigma-finite measures.

The typical example of a product measure is the Lebesgue measure on the space $(\mathbb{R}^n, \mathcal{B}(\mathbb{R}^n))$: It is the unique measure ℓ^n on that space that is such that $\ell^n(\Pi_{i=1}^n A_i) = \Pi_{i=1}^n \ell(A_i)$ for all $A_1, \dots, A_n \in \mathcal{B}$.

Theorem 4.4.8 *Let $(X_1, \mathcal{X}_1, \mu_1)$ and $(X_1, \mathcal{X}_2, \mu_2)$ be two measure spaces in which μ_1 and μ_2 are sigma-finite. Let $(X, \mathcal{X}, \mu) := (X_1 \times X_2, X_1 \otimes \mathcal{X}_2, \mu_1 \times \mu_2)$.*

(A) Tonelli. *If f is non-negative, then, for μ_1-almost all x_1, the function $x_2 \mapsto f(x_1, x_2)$ is measurable with respect to $\mathcal{X}_2$, and*

$$x_1 \mapsto \int_{X_2} f(x_1, x_2)\, \mu_2(\mathrm{d}x_2)$$

is a measurable function with respect to $\mathcal{X}_1$. Furthermore,

$$\int_X f\, \mathrm{d}\mu = \int_{X_1} \left[\int_{X_2} f(x_1, x_2)\, \mu_2(\mathrm{d}x_2) \right] \mu_1(\mathrm{d}x_1). \tag{4.17}$$

(B) Fubini. *If f is μ-integrable, then, for μ_1-almost all x_1, the function $x_2 \mapsto f(x_1, x_2)$ is μ_2-integrable and $x_1 \mapsto \int_{X_2} f(x_1, x_2)\, \mu_2(\mathrm{d}x_2)$ is μ_2-integrable, and (4.17) is true.*

We shall refer to the global result as the *Fubini–Tonelli* Theorem. Part (A)

[6] See Subsection 2.3.2 of [5].

says that one can integrate a non-negative measurable function in any order of its variables. Part (B) says that the same is true of an arbitrary measurable function if that function is μ-integrable. In general, in order to apply Part (B) one must use Part (A) in order to ascertain whether or not f is μ-integrable. The next example should convince the reader of the necessity of checking this integrability condition.

EXAMPLE 4.4.9: WHEN FUBINI IS NOT APPLICABLE. Consider the function f defined on $X_1 \times X_2 = (1, \infty) \times (0, 1)$ by the formula

$$f(x_1, x_2) = \mathrm{e}^{-x_1 x_2} - 2\mathrm{e}^{-2x_1 x_2}.$$

We have

$$\begin{aligned}\int_{(1,\infty)} f(x_1, x_2)\, \mathrm{d}x_1 &= \frac{\mathrm{e}^{-x_2} - \mathrm{e}^{-2x_2}}{x_2} = h(x_2) \geq 0, \\ \int_{(0,1)} f(x_1, x_2)\, \mathrm{d}x_2 &= -\frac{\mathrm{e}^{-x_1} - \mathrm{e}^{-2x_1}}{x_1} = -h(x_1).\end{aligned}$$

However,

$$\int_0^1 h(x_2)\, \mathrm{d}x_2 \neq \int_1^\infty (-h(x_1))\, \mathrm{d}x_1,$$

since $h \geq 0$ ℓ-a.e. on $(0, \infty)$. We therefore see that successive integrations yields different results according to the order in which they are performed. As a matter of fact, f is *not* integrable on $(0, 1) \times (1, \infty)$.

The Formula of Integration by Parts

Theorem 4.4.10 *Let μ_1 and μ_2 be two σ-finite measures on $(\mathbb{R}, \mathcal{B}(\mathbb{R}))$. For any interval $(a, b) \subseteq \mathbb{R}$*

$$\mu_1((a, b])\mu_2((a, b]) = \int_{(a,b]} \mu_1((a, t])\, \mu_2(\mathrm{d}t) + \int_{(a,b]} \mu_2((a, t))\, \mu_1(\mathrm{d}t). \tag{4.18}$$

Observe that the first integral features the interval $(a, t]$ (closed on the right), whereas in the second integral, the interval is of the type (a, t) (open on the right).

In terms of Lebesgue–Stieltjes integrals,

$$F_1(b)F_2(b) - F_1(a)F_1(a) = \int_{(a,b]} F_1(x)\, \mathrm{d}F_2(x) + \int_{(a,b]} F_2(x-)\, \mathrm{d}F_1(x),$$

where F_1 and F_2 are CDFs on $\mathbb{R}$. This is the Lebesgue–Stieltjes version of the integration by parts formula of calculus.

Proof. The proof consists in computing the $\mu_1 \times \mu_2$-measure of the square $D := (a,b] \times (a,b]$ in two ways. The first one is obvious and gives the left-hand side of (4.18). The second one consists in observing that $\mu(D) = \mu(D_1) + \mu(D_2)$, where $D_1 = \{(x,y); a < y \le b, a < x \le y\}$ and $D_2 = \{(a,b] \times (a,b]\} \cap \overline{D_1}$. Then $\mu(D_1)$ and $\mu(D_2)$ are computed using Tonelli's theorem. For instance,

$$\mu(D_1) = \int_{\mathbb{R}} \left(\int_{\mathbb{R}} 1_{D_1}(x,y)\mu_1(\mathrm{d}x) \right) \mu_2(\mathrm{d}y)$$
$$\int_{\mathbb{R}} \left(\int_{\mathbb{R}} 1_{\{a<x\le y\}}\mu_1(\mathrm{d}x) \right) \mu_2(\mathrm{d}y) = \int_{\mathbb{R}} \mu_1((a,y])\, \mu_2(\mathrm{d}y)\,.$$

□

L^p-spaces and the Riesz–Fischer Theorem

For a given integer $p \ge 1$, $L^p_{\mathbb{C}}(\mu)$ is, roughly speaking (see the details below), the collection of complex-valued measurable functions f defined on X such that $\int_X |f|^p \,\mathrm{d}\mu < \infty$. We shall see that it is a complete normed vector space over $\mathbb{C}$, that is, a Banach space.

Let $(X, \mathcal{X}, \mu)$ be a measure space and let f, g be two complex-valued measurable functions defined on X. The relation $\mathcal{R}$ defined by

$$f\mathcal{R}g \text{ if and only if } f = g \quad \mu\text{-a.e.}$$

is an equivalence relation. Denote the equivalence class of f by $\{f\}$. Note that for any $p > 0$ (using property (b) of Theorem 4.3.2),

$$f\mathcal{R}g \implies \int_X |f|^p \,\mathrm{d}\mu = \int_X |g|^p \,\mathrm{d}\mu\,.$$

The operations $+$, $\times$, * and multiplication by a scalar $\alpha \in \mathbb{C}$ are defined on the equivalence class by

$$\{f\} + \{g\} = \{f+g\}\,,\ \{f\}\{g\} = \{fg\}\,,\ \{f\}^* = \{f^*\}\,,\ \alpha\{f\} = \{\alpha f\}\,.$$

The first equality means that $\{f\} + \{g\}$ is, by definition, the equivalence class consisting of the functions $f + g$, where f and g are members of $\{f\}$ and $\{g\}$, respectively. Similar interpretations hold for the other equalities.

By definition, for a given $p \ge 1$, $L^p_{\mathbb{C}}(\mu)$ is the collection of equivalence classes $\{f\}$ such that $\int_X |f|^p \,\mathrm{d}\mu < \infty$. Clearly, it is a vector space over $\mathbb{C}$ (for the proof recall that

$$\left(\frac{|f|+|g|}{2}\right)^p \le \tfrac{1}{2}|f|^p + \tfrac{1}{2}|g|^p$$

since $t \to t^p$ is a convex function when $p \geq 1$). In order to avoid cumbersome notation, in this section and in general whenever we consider L^p-spaces, we shall write f for $\{f\}$. This abuse of notation is harmless since two members of the same equivalence class have the same integral if that integral is defined. Therefore, using this loose notation, we may write

$$L^p_{\mathbb{C}}(\mu) = \{f : \int_X |f|^p \, d\mu < \infty\} \,. \tag{4.19}$$

When the measure is the counting measure on the set $\mathbb{Z}$ of relative integers, the traditional notation is $\ell^p_{\mathbb{C}}(\mathbb{Z})$. This is the space of random complex sequences $\{x_n\}_{n\in\mathbb{Z}}$ such that

$$\sum_{n\in\mathbb{Z}} |x_n|^p < \infty.$$

The following is a simple and often used observation.

Theorem 4.4.11 *Let p and q be positive real numbers such that $p > q$. If the measure μ on $(X, \mathcal{X}, \mu)$ is* finite, *then $L^p_{\mathbb{C}}(\mu) \subseteq L^q_{\mathbb{C}}(\mu)$. In particular, $L^2_{\mathbb{C}}(\mu) \subseteq L^1_{\mathbb{C}}(\mu)$.*

Proof. From the inequality $|a|^q \leq 1 + |a|^p$, true for all $a \in \mathbb{C}$, it follows that $\mu(|f|^q) \leq \mu(1) + \mu(|f|^p)$. Since $\mu(1) = \mu(\mathbb{R}) < \infty$, $\mu(|f|^q) < \infty$ whenever $\mu(|f|^p) < \infty$. □

This inclusion is not true in general if μ is not a finite measure, for instance consider the Lebesgue measure ℓ on $\mathbb{R}$: there exist functions in $L^1_{\mathbb{C}}(\ell)$ that are not in $L^2_{\mathbb{C}}(\ell)$ and vice versa.

In the case of the counting measure on $\mathbb{Z}$, the order of inclusion is the reverse of the one concerning finite measures:

Theorem 4.4.12 *$\ell^p_{\mathbb{C}}$ inclusions. If $p > q$, $\ell^q_{\mathbb{C}}(\mathbb{Z}) \subset \ell^p_{\mathbb{C}}(\mathbb{Z})$. In particular, $\ell^1_{\mathbb{C}}(\mathbb{Z}) \subset \ell^2_{\mathbb{C}}(\mathbb{Z})$.*

Proof. Exercise 4.5.19. □

Theorem 4.4.13 *Let p and q be positive real numbers in $(0, 1)$ such that*

$$\frac{1}{p} + \frac{1}{q} = 1$$

(p and q are then said to be conjugate) and let $f, g : (X, \mathcal{X}) \mapsto (\mathbb{R}, \mathcal{B}(\mathbb{R}))$ be non-negative real functions. Then,

$$\int_X fg \, \mathrm{d}\mu \le \left[\int_X f^p \, \mathrm{d}\mu\right]^{1/p} \left[\int_X g^q \, \mathrm{d}\mu\right]^{1/q} . \tag{4.20}$$

In particular, if $f, g \in L^2_{\mathbb{C}}(\mathbb{R})$, then $fg \in L^1_{\mathbb{C}}(\mathbb{R})$.

Proof. Let

$$A = \left(\int_X f^p \, \mathrm{d}\mu\right)^{1/p}, \; B = \left(\int_X g^q \, \mathrm{d}\mu\right)^{1/q}.$$

It may be assumed that $0 < A, B < \infty$, because otherwise Hölder's inequality is trivially satisfied. Let $F := f/A$, $G := g/B$, so that

$$\int_X F^p \, \mathrm{d}\mu = \int_X G^q \, \mathrm{d}\mu = 1.$$

Suppose that we have been able to prove that

$$F(x)G(x) \le \frac{1}{p} F(x)^p + \frac{1}{q} G(x)^q . \tag{4.21}$$

Integrating this inequality yields

$$\int_X (FG) \, \mathrm{d}\mu \le \frac{1}{p} + \frac{1}{q} = 1,$$

and this is just (4.20).

Inequality (4.21) is trivially satisfied if x is such that $F \equiv 0$ or $G \equiv 0$. It is also satisfied in the case when F and G are not μ-almost everywhere null. Indeed, letting

$$s(x) := p \ln(F(x)), \qquad t(x) := q \ln(G(x)) ,$$

from the convexity of the exponential function and the assumption that $1/p+1/q = 1$,

$$\mathrm{e}^{s(x)/p+t(x)/q} \le \frac{1}{p} \mathrm{e}^{s(x)} + \frac{1}{q} \mathrm{e}^{t(x)},$$

and this is precisely inequality (4.21).

For the last assertion of the theorem, take $p = q = 2$. □

Theorem 4.4.14 *Let $p \geq 1$ and let $f, g : (X, \mathcal{X}) \mapsto (\mathbb{R}, \mathcal{B}(\mathbb{R}))$ be non-negative functions in $L^p_{\mathbb{C}}(\mu)$. Then,*

$$\left[\int_X (f+g)^p\right]^{1/p} \leq \left[\int_X f^p \,\mathrm{d}\mu\right]^{1/p} + \left[\int_X g^p \,\mathrm{d}\mu\right]^{1/p}. \tag{4.22}$$

Proof. For $p = 1$ the inequality (in fact an equality) is obvious. Therefore, assume $p > 1$. From Hölder's inequality

$$\int_X f(f+g)^{p-1} \,\mathrm{d}\mu \leq \left[\int_X f^p \,\mathrm{d}\mu\right]^{1/p} \left[\int_X (f+g)^{(p-1)q}\right]^{1/q}$$

and

$$\int_X g(f+g)^{p-1} \,\mathrm{d}\mu \leq \left[\int_X g^p \,\mathrm{d}\mu\right]^{1/p} \left[\int_X (f+g)^{(p-1)q}\right]^{1/q}.$$

Adding up the above two inequalities and observing that $(p-1)q = p$, we obtain

$$\int_X (f+g)^p \,\mathrm{d}\mu \leq \left(\left[\int_X f^p \,\mathrm{d}\mu\right]^{1/p} + \left[\int_X g^p \,\mathrm{d}\mu\right]^{1/p}\right) \left[\int_x (f+g)^p\right]^{1/q}.$$

One may assume that the right-hand side of (4.22) is finite and that the left-hand side is positive (otherwise the inequality is trivial). Therefore $\int_X (f+g)^p \,\mathrm{d}\mu \in (0, \infty)$ and we may therefore divide both sides of the last display by $\left[\int_X (f+g)^p \,\mathrm{d}\mu\right]^{1/q}$. Observing that $1 - 1/q = 1/p$ yields the announced inequality (4.22). □

Theorem 4.4.15 *Let $p \geq 1$. The mapping $\nu_p : L^p_{\mathbb{C}}(\mu) \mapsto [0, \infty)$ defined by*

$$\nu_p(f) := \left(\int_X |f|^p \,d\mu\right)^{1/p} \tag{4.23}$$

is a norm on $L^p_{\mathbb{C}}(\mu)$.

Proof. Clearly, $\nu_p(\alpha f) = |\alpha|\nu_p(f)$ for all $\alpha \in \mathbb{C}$, $f \in L^p_{\mathbb{C}}(\mu)$. Also, $\nu_p(f) = 0$ if and only if $\left(\int_X |f|^p \,\mathrm{d}\mu\right)^{1/p} = 0$, which in turn is equivalent to $f = 0$, μ-a.e. Finally, $\nu_p(f+g) \leq \nu_p(f) + \nu_p(g)$ for all $f, g \in L^p_{\mathbb{C}}(\mu)$, by Minkowski's inequality. □

Denoting $\nu_p(f)$ by $\|f\|_p$, $L^p_{\mathbb{C}}(\mu)$ is a normed vector space over $\mathbb{C}$, with the norm $\|\cdot\|_p$ and the induced metric $d_p(f,g) := \|f - g\|_p$.

Theorem 4.4.16 *Let $p \geq 1$. The metric d_p makes of $L^p_{\mathbb{C}}(\mu)$ a complete normed vector space.*

In other words, $L^p_{\mathbb{C}}(\mu)$ is a *Banach space* for the norm $\|\cdot\|_p$.

Proof. To show completeness one must prove that for any sequence $\{f_n\}_{n\geq 1}$ of $L^p_{\mathbb{C}}(\mu)$ that is a Cauchy sequence (that is, such that $\lim_{m,n\uparrow\infty} d_p(f_n, f_m) = 0$), there exists an $f \in L^p_{\mathbb{C}}(\mu)$ such that $\lim_{n\uparrow\infty} d_p(f_n, f) = 0$.

Since $\{f_n\}_{n\geq 1}$ is a Cauchy sequence, one can select a subsequence $\{f_{n_i}\}_{i\geq 1}$ such that

$$d_p(f_{n_{i+1}} - f_{n_i}) \leq 2^{-i}. \tag{4.24}$$

Let

$$g_k = \sum_{i=1}^{k} |f_{n_{i+1}} - f_{n_i}|,\ g = \sum_{i=1}^{\infty} |f_{n_{i+1}} - f_{n_i}|.$$

By (4.24) and Minkowski's inequality, $\|g_k\|_p \leq 1$. Fatou's lemma applied to the sequence $\{g_k^p\}_{k\geq 1}$ gives $\|g\|_p \leq 1$. In particular, any member of the equivalence class of g is μ-almost everywhere finite and therefore

$$f_{n_1}(x) + \sum_{i=1}^{\infty} (f_{n_{i+1}}(x) - f_{n_i}(x))$$

converges absolutely for μ-almost all x. Call the corresponding limit $f(x)$ (set $f(x) = 0$ when this limit does not exist). Since

$$f_{n_1} + \sum_{i=1}^{k-1} (f_{n_{i+1}} - f_{n_i}) = f_{n_k}$$

we see that

$$f = \lim_{k\uparrow\infty} f_{n_k} \quad \mu\text{-a.e.}$$

One must show that f is the limit in $L^p_{\mathbb{C}}(\mu)$ of $\{f_{n_k}\}_{k\geq 1}$. Let $\epsilon > 0$. There exists an integer $N = N(\epsilon)$ such that $\|f_n - f_m\|_p \leq \epsilon$ whenever $m, n \geq N$. For all $m > N$, by Fatou's lemma we have

$$\int_X |f - f_m|^p \,\mathrm{d}\mu \leq \liminf_{i\to\infty} \int_x |f_{n_i} - f_m|^p \,\mathrm{d}\mu \leq \epsilon^p.$$

Therefore $f - f_m \in L^p_{\mathbb{C}}(\mu)$ and consequently $f \in L^p_{\mathbb{C}}(\mu)$. It also follows from the last inequality that

$$\lim_{m\to\infty} \|f - f_m\|_p = 0.$$

□

The next result is a by-product of the proofs of Theorems 4.4.16.

Theorem 4.4.17 *Let $p \geq 1$ and let $\{f_n\}_{n\geq 1}$ be a convergent sequence in $L^p_{\mathbb{C}}(\mu)$. Let f be the corresponding limit in $L^p_{\mathbb{C}}(\mu)$. Then, there exists a subsequence $\{f_{n_i}\}_{i\geq 1}$ such that*

$$\lim\nolimits_{i\uparrow\infty} f_{n_i} = f \quad \mu\text{-a.e.} \tag{4.25}$$

Note that the statement in (4.25) is about functions and not about equivalence classes. The functions thereof are *any* members of the corresponding equivalence class. In particular, when a given sequence of functions converges μ-a.e. to two functions, these two functions are necessarily equal μ-a.e. Therefore,

Theorem 4.4.18 *If $\{f_n\}_{n\geq 1}$ converges both to f in $L^p_{\mathbb{C}}(\mu)$ and to g μ-a.e., then $f = g$ μ-a.e.*

Of special interest for applications is the space $L^2_{\mathbb{C}}(\mu)$ of complex measurable functions $f : X \to \mathbb{R}$ such that

$$\int_X |f(x)|^2 \, \mu(\mathrm{d}x) < \infty,$$

where two functions f and f' such that $f(x) = f'(x)$, μ-a.e. are not distinguished. We have by the *Riesz–Fischer theorem*:

Theorem 4.4.19 *$L^2_{\mathbb{C}}(\mu)$ is a vector space with scalar field $\mathbb{C}$, and when endowed with the inner product*

$$\langle f, g \rangle := \int_X f(x) g(x)^* \, \mu(\mathrm{d}x) \,, \tag{4.26}$$

it is a Hilbert space.

The norm of a function $f \in L^2_{\mathbb{C}}(\mu)$ is

$$\|f\| = \left(\int_X |f(x)|^2 \, \mu(\mathrm{d}x)\right)^{\frac{1}{2}}$$

and the distance between two functions f and g in $L^2_{\mathbb{C}}(\mu)$ is

$$d(f, g) = \left(\int_X |f(x) - g(x)|^2 \, \mu(\mathrm{d}x)\right)^{\frac{1}{2}} .$$

The completeness property of $L^2_{\mathbb{C}}(\mu)$ reads in this case as follows. If $\{f_n\}_{n\geq 1}$ is a sequence of functions in $L^2_{\mathbb{C}}(\mu)$ such that

$$\lim_{m,n\uparrow\infty} \int_X |f_n(x) - f_m(x)|^2 \, \mu(\mathrm{d}x) = 0,$$

then, there exists a function $f \in L^2_{\mathbb{C}}(\mu)$ such that

$$\lim_{n\uparrow\infty} \int_X |f_n(x) - f(x)|^2 \, \mu(\mathrm{d}x) = 0.$$

In $L^2_{\mathbb{C}}(\mu)$, Schwarz's inequality reads as follows:

$$\left|\int_X f(x) g(x)^* \, \mu(\mathrm{d}x)\right| \leq \left(\int_X |f(x)|^2 \, \mu(\mathrm{d}x)\right)^{\frac{1}{2}} \left(\int_X |g(x)|^2 \, \mu(\mathrm{d}x)\right)^{\frac{1}{2}} .$$

EXAMPLE 4.4.20: COMPLEX SEQUENCES. The set of complex sequences $a = \{a_n\}_{n\in\mathbb{Z}}$ such that

$$\sum_{n\in\mathbb{Z}} |a_n|^2 < \infty$$

is, when endowed with the inner product

$$\langle a, b\rangle = \sum_{n\in\mathbb{Z}} a_n b_n^* ,$$

a Hilbert space, denoted by $\ell^2_{\mathbb{C}}(\mathbb{Z})$. This is indeed a particular case of a Hilbert space $L^2_{\mathbb{C}}(\mu)$, where $X = \mathbb{Z}$ and μ is the counting measure. In this example, Schwarz's inequality takes the form

$$\left|\sum_{n\in\mathbb{Z}} a_n b_n^*\right| \le \left(\sum_{n\in\mathbb{Z}} |a_n|^2\right)^{\frac{1}{2}} \times \left(\sum_{n\in\mathbb{Z}} |b_n|^2\right)^{\frac{1}{2}} .$$

4.5 Exercises

Exercise 4.5.1. SET INVERSE FUNCTION
Let U and E be arbitrary sets and let f be some function from U to E. For any subset $A \subseteq E$, let

$$f^{-1}(A) := \{u \in U\,;\, f(u) \in A\} .$$

(i) Show that for all $u \in U$, $1_A(f(u)) = 1_{f^{-1}(A)}(u)$.

(ii) Prove that if $\mathcal{E}$ is a σ-field on E, then the collection of subsets $f^{-1}(\mathcal{E}) := \{f^{-1}(A)\,;\, A \in \mathcal{E}\}$ is a σ-field on U.

Exercise 4.5.2. NO TITLE
Let f be a function from $\mathbb{R}$ to $\mathbb{R}$. Prove that for any $a \in \mathbb{R}$,

$$\cap_{n\ge 1}\{x\,;\, f(x) \le a + 1/n\} = \{x\,;\, f(x) \le a\} .$$

Exercise 4.5.3. σ-FIELD GENERATED BY A COLLECTION OF SETS
Let I be an arbitrary non-empty index set.

(1) Let $\{\mathcal{F}_i\}_{i\in I}$ be an arbitrary non-empty family of σ-fields on some set Ω. Show that the family $\mathcal{F} := \cap_{i\in I}\mathcal{F}_i$ ($A \in \mathcal{F}$ if and only if $A \in \mathcal{F}_i$ for all $i \in I$) is a σ-field.

(2) Let $\mathcal{C}$ be an arbitrary family of subsets of some set Ω. Prove the existence of a smallest σ-field $\mathcal{F}$ containing $\mathcal{C}$. (This means, by definition, that $\mathcal{F}$ is a σ-field on Ω containing $\mathcal{C}$, such that if $\mathcal{F}'$ is a σ-field on Ω containing $\mathcal{C}$, then $\mathcal{F} \subseteq \mathcal{F}'$.)

Exercise 4.5.4. $\mathcal{B}(\mathbb{R}^n)$
Prove that $\mathcal{B}(\mathbb{R}^n)$ is generated by the collection $\mathcal{C}$ of all rectangles of the type $\prod_{i=1}^{n}(-\infty, a_i]$, where $a_i \in \mathbb{Q}$ $(i \in \{1, \ldots, n\})$. ($\mathbb{Q}$ is the set of rational numbers.)

Exercise 4.5.5. GROSS SIGMA-FIELD
Show that a function $f : X \to \mathbb{R}$ that is measurable with respect to the gross sigma-field on X and the Borel sigma-field on $\mathbb{R}$ is a constant (takes only one value).

Exercise 4.5.6. $|f|$ MEASURABLE, f NOT MEASURABLE
Let $(X, \mathcal{X})$ be a measurable space such that $\mathcal{X} \not\equiv \mathcal{P}(X)$ (for instance if $(X, \mathcal{X}) = (\mathbb{R}, \mathcal{B}(\mathbb{R}))$, a fact that we shall admit here). Let $f : X \to E$ be a function. Is it true that if $|f|$ is measurable with respect to $\mathcal{X}$ and $\mathcal{E}$, then so is f itself?

Exercise 4.5.7. SEQUENTIAL CONTINUITY
In Theorem 4.1.21, show by means of a counterexample the necessity of the condition $\mu(B_{n_0}) < \infty$ for some n_0.

Exercise 4.5.8. THE RATIONALS ARE LEBESGUE-NEGLIGIBLE
Prove that any singleton $\{a\}$ $(a \in \mathbb{R})$ is a Borel set of null Lebesgue measure and that the set of rationals $\mathbb{Q}$ is a Borel set of null Lebesgue measure.

Exercise 4.5.9. INTEGRAL OF A SIMPLE FUNCTION
Prove (4.5).

Exercise 4.5.10. INTEGRAL WITH RESPECT TO THE WEIGHTED COUNTING MEASURE
Any function $f : \mathbb{Z} \to \mathbb{R}$ is measurable with respect to $\mathcal{P}(\mathbb{Z})$ and $\mathcal{B}(\mathbb{R})$. With the measure μ defined in Example 4.1.18, and with $f \geq 0$ for instance, show that

$$\mu(f) = \sum_{n=1}^{\infty} \alpha_n f(n)$$

by following exactly the steps of the general construction.

Exercise 4.5.11. FOURIER TRANSFORM
The *Fourier transform* of a function $f : \mathbb{R} \to \mathbb{R}$ that is integrable with respect to Lebesgue measure is the function $\hat{f} : \mathbb{R} \to \mathbb{C}$ defined by:

$$\hat{f}(\nu) := \int_{\mathbb{R}} f(t)\, \mathrm{e}^{-2i\pi\nu t}\, \mathrm{d}t\,.$$

This is denoted by $\mathcal{F} : f \to \hat{f}$.

Prove that $\hat{f}$ is bounded and uniformly continuous.

Exercise 4.5.12. CONVOLUTION OF INTEGRABLE FUNCTIONS
Let $h, f : \mathbb{R} \to \mathbb{R}$ be functions that are integrable with respect to Lebesgue measure. Prove that the right-hand side of

$$g(t) := \int_{\mathbb{R}} h(t-s) f(s)\, \mathrm{d}s$$

is well defined almost everywhere (for the Lebesgue measure), and defines an integrable function. (The function g is the *convolution* of h with f, and is denoted by $g = h * f$.)

Exercise 4.5.13. THE FOURIER CONVOLUTION–MULTIPLICATION RULE.
(Continuation of Exercise 4.5.12) Prove that

$$\mathcal{F} : h * f \to \hat{h}\hat{f}\,.$$

Exercise 4.5.14. IMAGE MEASURE
What is the measure on $(\mathbb{R}, \mathcal{B}(\mathbb{R}))$ that is the image of the Lebesgue measure ℓ on $(\mathbb{R}, \mathcal{B}(\mathbb{R}))$ by the map $x \mapsto |x|$?

Exercise 4.5.15. SCHEFFÉ'S LEMMA
Let f and f_n $(n \geq 1)$ be μ-integrable *non-negative* functions such that $\lim_{n\uparrow\infty} f_n = f$ μ-a.e. and $\lim_{n\uparrow\infty} \int_X f_n \,\mathrm{d}\mu = \int_X f \,\mathrm{d}\mu$. Show that $\lim_{n\uparrow\infty} \int_X |f_n - f| \,\mathrm{d}\mu = 0$. (Hint: $|a - b| = a + b - \inf(a, b)$.)

Exercise 4.5.16. FUBINI TILES
Consider any bounded closed rectangle of $\mathbb{R}^2$. We say that it has Property (A) if *at least one* of its sides "is an integer" (meaning: its length is an integer). Now you are given a bounded closed rectangle Δ that is the union of a finite number of disjoint closed rectangles with Property (A). Show that Δ itself must have Property (A).

Exercise 4.5.17. Integrals and sums
Prove that for all $a, b \in \mathbb{R}$,

$$\int_{\mathbb{R}_+} \frac{t\,\mathrm{e}^{-at}}{1-\mathrm{e}^{-bt}}\,\mathrm{d}t = \sum_{n=0}^{+\infty} \frac{1}{(a+nb)^2}\,.$$

Exercise 4.5.18. Fubini again
Define $f : [0,1]^2 \to \mathbb{R}$ by

$$f(x,y) = \frac{x^2-y^2}{(x^2+y^2)^2}\,1_{\{(x,y)\neq(0,0)\}}\,.$$

Compute $\int_{[0,1]}\left(\int_{[0,1])f(x,y)\,\mathrm{d}x}\right)\,\mathrm{d}y$ and $\int_{[0,1]}\left(\int_{[0,1])f(x,y)\,\mathrm{d}y}\right)\,\mathrm{d}x$. Is f Lebesgue integrable on $[0,1]^2$?

Exercise 4.5.19. $\ell^p_{\mathbb{C}}(\mathbb{Z})$
Prove that $\ell^1_{\mathbb{C}}(\mathbb{Z}) \subset \ell^2_{\mathbb{C}}(\mathbb{Z})$

4.6 Solutions

Solution (Exercise 4.5.1).
(i) $1_A(f(u)) = 1 \iff f(u) \in A \iff u \in f^{-1}(A) \iff 1_{f^{-1}(A)}(u) = 1$.
(ii) is a direct consequence of the definition of a σ-field and of the following set identities: for any subsets $A, A_1, A_2, \dots$ of E,

$$f^{-1}(\bar{A}) = \overline{f^{-1}(A)}\,,$$

$$f^{-1}\left(\bigcap_{n=1}^{\infty} A_n\right) = \bigcap_{n=1}^{\infty} f^{-1}(A_n)\,,$$

$$f^{-1}\left(\bigcup_{n=1}^{\infty} A_n\right) = \bigcup_{n=1}^{\infty} f^{-1}(A_n)\,.$$

Solution (Exercise 4.5.2).
If $f(x) \le a + \frac{1}{n}$ for all $n \ge 1$, one cannot have $f(x) > a$, because if it were the case, there would certainly exist a sufficiently large n such that $f(x) > a + \frac{1}{n}$. Therefore

$$\cap_{n\ge 1}\{x\,;\, f(x) \le a + \frac{1}{n}\} \subseteq \{x\,;\, f(x) \le a\}\,.$$

If $f(x) \le a$, then obviously $f(x) \le a + 1/n$ for all $n \ge 1$. therefore

$$\{x\,;\, f(x) \le a\} \subseteq \cap_{n\ge 1}\{x\,;\, f(x) \le a + 1/n\}\,.$$

Solution (Exercise 4.5.3).
Obvious.

Solution (Exercise 4.5.4).
It suffices to show that $\mathcal{B}(\mathbb{R}^n)$ is generated by the collection $\mathcal{C}'$ of all rectangles $\prod_{i=1}^n (a_i, b_i)$ with rational endpoints. Note that $\mathcal{C}'$ is a countable collection and that all its elements are open sets for the Euclidean topology (the latter we denote by $\mathcal{O}$). It follows that $\mathcal{C}' \subseteq \mathcal{O}$ and therefore $\sigma(\mathcal{C}') \subseteq \sigma(\mathcal{O}) = \mathcal{B}(\mathbb{R}^n)$.

It remains to show that $\mathcal{O} \subseteq \sigma(\mathcal{C}')$, since this implies that $\sigma(\mathcal{O}) \subseteq \sigma(\mathcal{C}')$. For this it suffices to show that any set $O \in \mathcal{O}$ is a countable union of elements in $\mathcal{C}'$. Take $x \in O$. By definition of the Euclidean topology, there exists a non-empty open ball $B(x, r)$ centered at x and contained in O. Now we can always choose a rational rectangle $R_x \in \mathcal{C}'$ that contains x and that is contained in $B(x, r)$. Clearly $\cup_{x\in O} R_x = O$. Since the R_x are chosen in a countable family of sets, the union $\cup_{x\in O} R_x$ is in fact countable. As a countable union of sets in $\mathcal{C}'$ it is in $\sigma(\mathcal{C}')$. Therefore $O \in \sigma(\mathcal{C}')$.

Solution (Exercise 4.5.5).
Suppose it takes two distinct values a and b. Then $\{f = a\} := \{x \in X\,;\, f(x) = a\}$ and $\{f = b\}$ are two distinct members of the gross sigma-field on X. One of them must therefore be X itself, say $\{f = a\} = \Omega$. Therefore f is the constant function equal to a.

Solution (Exercise 4.5.6).
No. Take $f = 1_A - 1_{\overline{A}}$ where A is a non-measurable set. This function is clearly non-measurale (for instance $\{f = 1\} = A \notin \mathcal{X}$), but $|f| \equiv 1$ is measurable.

Solution (Exercise 4.5.7).
Let ν be the counting measure on $\mathbb{Z}$ and let $B_n := \{i \in \mathbb{Z} : |i| \ge n\}$ $(n \ge 1)$. Then $\nu(B_n) = +\infty$ for all $n \ge 1$, whereas

$$\nu\left(\bigcap_{n=1}^{\infty} B_n\right) = \nu(\varnothing) = 0.$$

SOLUTION (Exercise 4.5.8).
The Borel σ-field $\mathcal{B}(\mathbb{R})$ is generated by the intervals $I_a = (-\infty, a]$, $a \in \mathbb{R}$ (Theorem 4.1.4), and therefore $\{a\} = \cap_{n\geq 1}(I_a - I_{a-1/n})$ is also in $\mathcal{B}(\mathbb{R})$. Denoting by ℓ the Lebesgue measure, $\ell(I_a - I_{a-1/n}) = 1/n$, and therefore $\ell(\{a\}) = \lim_{n\geq 1} \ell(I_a - I_{a-1/n}) = 0$. $\mathbb{Q}$ is a countable union of sets in $\mathcal{B}(\mathbb{R})$ (singletons) and is therefore in $\mathcal{B}(\mathbb{R})$. It has Lebesgue measure 0 as a countable union of sets of Lebesgue measure 0.

SOLUTION (Exercise 4.5.9).
This follows from the following chain of equalities, where it is noted that if $A_i \cap B_j \neq \varnothing$, then $a_i = b_j$:

$$\begin{aligned}
\sum_{j=1}^{m} b_j\, \mu(B_j) &= \sum_{j=1}^{m} b_j \left(\sum_{i=1}^{k} \mu(B_j \cap A_i) \right) \\
&= \sum_{i=1}^{k} \sum_{j=1}^{m} b_j\, \mu(B_j \cap A_i) \\
&= \sum_{i=1}^{k} \sum_{j=1}^{m} a_i\, \mu(B_j \cap A_i) \\
&= \sum_{i=1}^{k} a_i \left(\sum_{j=1}^{m} \mu(B_j \cap A_i) \right) = \sum_{i=1}^{k} a_i\, \mu(A_i).
\end{aligned}$$

SOLUTION (Exercise 4.5.10).
It suffices to consider the approximating sequence of simple functions

$$f_n(k) = \sum_{j=-n}^{+n} f(j) 1_{\{j\}}(k)$$

whose integral is

$$\nu(f_n) = \sum_{j=-n}^{+n} f(j)\mu(\{j\}) = \sum_{j=-n}^{+n} f(j)\alpha_j$$

and to let n tend to ∞. When $\alpha_n \equiv 1$, the integral reduces to the sum of a series:

$$\nu(f) = \sum_{n\in\mathbb{Z}} f(n)\,.$$

In this case, integrability means that the series is absolutely convergent.

SOLUTION (Exercise 4.5.11).
From the definition, we have that

$$|\hat{f}(\nu)| \le \int_{\mathbb{R}} |f(t)\, \mathrm{e}^{-2i\pi\nu t}|\, \mathrm{d}t = \int_{\mathbb{R}} |f(t)|\, \mathrm{d}t,$$

where the last term does not depend on ν and is finite. Also, for all $h \in \mathbb{R}$,

$$\begin{aligned} |\hat{f}(\nu+h) - \hat{f}(\nu)| &\le \int_{\mathbb{R}} |f(t)|\, |\mathrm{e}^{-2i\pi(\nu+h)t} - \mathrm{e}^{-2i\pi\nu t}|\, \mathrm{d}t \\ &= \int_{\mathbb{R}} |f(t)|\, |\mathrm{e}^{-2i\pi h t} - 1|\, \mathrm{d}t\,. \end{aligned}$$

The last term is independent of ν and tends to 0 as $h \to 0$ by dominated convergence.

SOLUTION (Exercise 4.5.12).

By Tonelli's theorem and the integrability assumptions

$$\int_{\mathbb{R}}\int_{\mathbb{R}} |h(t-s)|\, |f(s)|\, \mathrm{d}t\, \mathrm{d}s = \left(\int_{\mathbb{R}} |h(t)|\, \mathrm{d}t\right)\left(\int_{\mathbb{R}} |f(t)|\, \mathrm{d}t\right) < \infty\,.$$

This implies that, for ℓ-almost all t,

$$\int_{\mathbb{R}} |h(t-s) f(s)|\, \mathrm{d}s < \infty\,.$$

The integral $\int_{\mathbb{R}} h(t-s) f(s)\, \mathrm{d}s$ is therefore well defined for ℓ-almost all t. Also

$$\begin{aligned} \int_{\mathbb{R}} |g(t)|\, \mathrm{d}t &= \int_{\mathbb{R}} \left| \int_{\mathbb{R}} h(t-s) f(s)\, \mathrm{d}s \right| \mathrm{d}t \\ &\le \int_{\mathbb{R}}\int_{\mathbb{R}} |h(t-s) f(s)|\, \mathrm{d}t\, \mathrm{d}s < \infty\,, \end{aligned}$$

that is, g is integrable.

SOLUTION (Exercise 4.5.13).
We have

$$\begin{aligned} &\int_{\mathbb{R}} \left(\int_{\mathbb{R}} h(t-s) f(s)\, \mathrm{d}s \right) \mathrm{e}^{-2i\pi\nu t}\, \mathrm{d}t \\ &\quad = \int_{\mathbb{R}}\int_{\mathbb{R}} h(t-s) \mathrm{e}^{-2i\pi\nu(t-s)} f(s) \mathrm{e}^{-2i\pi\nu s}\, \mathrm{d}s\, \mathrm{d}t \\ &\quad = \int_{\mathbb{R}} f(s) \mathrm{e}^{-2i\pi\nu s} \left(\int_{\mathbb{R}} h(t-s) \mathrm{e}^{-2i\pi\nu(t-s)} \mathrm{d}t \right) \mathrm{d}s = \hat{h}(\nu) \hat{f}(\nu)\,, \end{aligned}$$

by Fubini's theorem, which is applicable here because the function

$$(t, s) \mapsto \left|h(t-s)f(s)\mathrm{e}^{-2i\pi\nu t}\right| = |h(t-s)f(s)|$$

is integrable with respect to the product measure $\mathrm{d}t \times \mathrm{d}s$ (Exercise 4.5.12).

Solution (Exercise 4.5.14).
$2 1_{x\geq 0}\ell(dx)$.

Solution (Exercise 4.5.15).
The function $\inf(f_n, f)$ is bounded by the (μ-integrable) function f (this is where the non-negativeness assumption is used). Moreover, it converges to f. Therefore, by dominated convergence, $\lim_{n\uparrow\infty} \int_X \inf(f_n, f)\, d\mu = \int_X f\, d\mu$. The rest of the proof follows from

$$\int_X |f_n - f|\, d\mu = \int_X f_n\, d\mu + \int_X f\, d\mu - \int_X \inf(f_n, f)\, d\mu\,.$$

Solution (Exercise 4.5.16).
Let I be a finite interval of $\mathbb{R}$. Observe that $\int_I e^{2i\pi x} dx = 0$ if and only if the length of I is an integer. Let now $I \times J$ be a finite rectangle. It has Property (A) if and only if $\int\int_{I\times J} e^{2i\pi(x+y)} dx\, dy = \int_I e^{2i\pi x} dx \times \int_J e^{2i\pi y} dy = 0$. (This is where we use Fubini.) Now

$$\begin{aligned}\int\int_\Delta e^{2i\pi(x+y)} dx\, dy &= \int\int_{\cup_{n=1}^K \Delta_n} e^{2i\pi(x+y)} dx\, dy \\ &= \sum_{n=1}^{K} \int\int_{\Delta_n} e^{2i\pi(x+y)} dx\, dy = 0\,,\end{aligned}$$

since the Δ_n's form a partition of Δ and all have Property (A).

Solution (Exercise 4.5.17).

$$\begin{aligned}\int_{\mathbb{R}_+} \frac{t\, e^{-at}}{1-e^{-bt}}\, dt &= \int_{\mathbb{R}_+} \left(\sum_{n=0}^{+\infty} t\, e^{-(a+nb)}\right) dt \\ &= \sum_{n=0}^{+\infty} \int_{\mathbb{R}_+} t\, e^{-(a+nb)}\, dt = \sum_{n=0}^{+\infty} \frac{1}{(a+nb)^2}\end{aligned}$$

where the second equality is justified by Tonelli's theorem applied to the product of the Lebesgue measure by the counting measure.

SOLUTION (Exercise 4.5.18).
If $y \neq 0$, the function $x \mapsto f(x,y)$ is continuous on $[0,1]$ and therefore Lebesgue integrable, being bounded. We have

$$\int_{[0,1]} f(x,y)\,dx = \left(\frac{x}{x^2+y^2}\right)_0^1 = \frac{1}{1+y^2}$$

For $y = 0$, $\int_{[0,1]} f(x,0)\,dx = \int_{[0,1]} \frac{1}{x^2}\,dx = +\infty$. Therefore,

$$\int_{[0,1]} f(x,y)\,dx = \frac{1}{1+y^2},\ l-\text{a.e},$$

and

$$\int_{[0,1]}\left(\int_{[0,1]} f(x,y)\,dx\right)dy = \int_{[0,1]} \frac{1}{1+y^2}\,dy = \frac{\pi}{4}.$$

Observing that $f(x,y) = -f(y,x)$ we obtain

$$\int_{[0,1]}\left(\int_{[0,1]} f(x,y)\,dy\right)dx = -\frac{\pi}{4}.$$

f cannot be integrable on $[0,1]^2$, otherwise the two integrals would be the same, by Fubini's theorem.

SOLUTION (Exercise 4.5.19).
If $\sum_n |a_n| < \infty$, there exists n_0 such that if $n \geq n_0$, $|a_n| < 1$. In particular, for $n \geq n_0$, $|a_n|^2 < |a_n|$. Therefore $\sum_{n \geq n_0} |a_n|^2 < \sum_{n \geq n_0} |a_n| < \infty$, from which it follows that $\sum_n |a_n|^2 < \infty$.

Chapter 5

From Integral to Expectation

Probability theory is from a formal point of view, a particular chapter of measure and integration theory. Since the terminologies of the two theories are different, we shall first proceed to the "translation" of the theory of measure and integration into the theory of probability and expectation.

5.1 Translation

Recall the probabilistic trinity, the triple $(\Omega, \mathcal{F}, P)$, where P (the probability) is a measure on the measurable space $(\Omega, \mathcal{F})$ with total mass $P(\Omega) = 1$.

Most of the results of the present section follow from those of the previous chapter by a mere change of notation: $X \rightsquigarrow \Omega$, $\mathcal{X} \rightsquigarrow \mathcal{F}$, $\mu \rightsquigarrow P$ and $f \rightsquigarrow X$, so that for instance

$$\int_X f(x)\,\mu(\mathrm{d}x) \rightsquigarrow \int_\Omega X(\omega)\,P(\mathrm{d}\omega)\,.$$

(Of course, the reader is aware that the "X's" in both sides are of a different nature. But this notational collision will not happen any more in the sequel.)

Definition 5.1.1 *A measurable function X from $(\Omega, \mathcal{F})$ to a measurable space $(E, \mathcal{E})$ is called a* random element *with values in $(E, \mathcal{E})$ (or in E, for short, when the context is unambiguous).*

When $(E, \mathcal{E}) = (\mathbb{R}, \mathcal{B}(\mathbb{R}))$ or $(\overline{\mathbb{R}}, \mathcal{B}(\overline{\mathbb{R}}))$, X is also called a *random variable* (r.v.) (*real* r.v. if $E = \mathbb{R}$, *extended* r.v. if $E = \overline{\mathbb{R}}$). If $(E, \mathcal{E}) = (\mathbb{R}^n, \mathcal{B}(\mathbb{R}^n))$, X is called a *random vector* (of dimension n), and then $X = (X_1, \dots, X_n)$ where the X_i are random variables. A *complex random variable* is a function $X : \Omega \to \mathbb{C}$ of the form $X = X_R + iX_I$ where X_R and X_I are real random variables.

P. Brémaud, *An Introduction to Applied Probability*, Texts in Applied Mathematics 77,
https://doi.org/10.1007/978-3-031-49306-5_5

If X is a random element with values in $(E, \mathcal{E})$ and if g is a measurable function from $(E, \mathcal{E})$ to $(\overline{\mathbb{R}}, \mathcal{B}(\overline{\mathbb{R}}))$, then $g(X)$ is, by the composition theorem for measurable functions (Theorem 4.1.11), a random variable.

Since a random variable X is a measurable function, we can define, under rather general circumstances, its integral with respect to the probability measure P, called the *expectation* of X. Therefore

$$E[X] := \int_\Omega X(\omega) P(\mathrm{d}\omega).$$

Recall the construction of the integral given in Section 4.2 in the special case of a probability. First, if $A \in \mathcal{F}$,

$$E[1_A] := P(A)$$

and, more generally, if X is a simple random variable, that is, $X(\omega) = \sum_{i=1}^N \alpha_i 1_{A_i}(\omega)$ where $N \in \mathbb{N}_+$, $\alpha_i \in \mathbb{R}$ and $A_i \in \mathcal{F}$ $(1 \le i \le N)$, then

$$E[X] := \sum_{i=1}^N \alpha_i P(A_i).$$

For a non-negative random variable X, the expectation is defined by

$$E[X] := \lim_{n \uparrow \infty} E[X_n],$$

where $\{X_n\}_{n \ge 1}$ is any non-decreasing sequence of non-negative simple random variables that converges to X.

This definition is consistent, that is, it does not depend on the approximating non-decreasing sequence of non-negative simple random variables admitting X for limit. When X is of arbitrary sign, the expectation is defined by $E[X] := E[X^+] - E[X^-]$ if $E[X^+]$ and $E[X^-]$ are not both infinite. If $E[X^+]$ *and* $E[X^-]$ are infinite, the expectation is not defined. If $E[|X|] < \infty$, X is said to be *integrable*, and then $E[X]$ is a finite number.

The basic properties of expectation follow from the general case of the integral with respect to an arbitrary measure. These are *linearity* and *monotonicity*: If X_1 and X_2 are random variables with expectations, then for all λ_1, $\lambda_2 \in \mathbb{R}$,

$$E[\lambda_1 X_1 + \lambda_2 X_2] = \lambda_1 E[X_1] + \lambda_2 E[X_2],$$

whenever the right-hand side has meaning (i.e., is not an $\infty - \infty$ form). Also, if $X_1 \le X_2$, P-a.s., then

$$E[X_1] \le E[X_2].$$

It follows from this that if $E[X]$ is well defined, then

$$|E[X]| \le E[|X|] .$$

Given a sequence $\{X_n\}_{n\ge1}$ of random variables, one seeks conditions guaranteeing that, provided the limits thereafter exist,

$$\lim_{n\uparrow\infty} E\left[X_n\right] = E\left[\lim_{n\uparrow\infty} X_n\right] . \tag{5.1}$$

The next theorem (*monotone convergence theorem*) is, again, nothing but a rephrasing of the general result of the previous chapter (Theorem 4.3.3) in terms of expectations.

Theorem 5.1.2 *Let $\{X_n\}_{n\ge1}$ and X be real random variables such that*

(i) $P(\lim_{n\uparrow\infty} X_n = X) = 1$, *and*

(ii) $P(X_n \le X_{n+1} = 1)$ *for all* $n \ge 1$.

Then (5.1) holds true.

Finally, we have the *dominated convergence theorem*, which is a rephrasing of Theorem 4.3.5 in terms of expectations:

Theorem 5.1.3 *Let $\{X_n\}_{n\ge1}$ and X be real random variables such that*

(i) $P(\lim_{n\uparrow\infty} X_n = X) = 1$, *and*

(ii) *there exists a non-negative real random variable Z with finite expectation such that $P(|X_n| \le Z) = 1$ for all $n \ge 1$.*

Then (5.1) holds true.

5.2 The Distribution of a Random Element

Definition 5.2.1 *Let X be a random element with values in $(E, \mathcal{E})$. Its (probability)* distribution *is, by definition, the probability measure Q_X on $(E, \mathcal{E})$, the image of the probability measure P by the mapping X from $(\Omega, \mathcal{F})$ to $(E, \mathcal{E})$, that is,*

$$Q_X(C) = P(X \in C) \quad (C \in \mathcal{E}) .$$

EXAMPLE 5.2.2: DISTRIBUTION OF $X + a$. Let X be a random vector with values in $\mathbb{R}^m$ and distribution Q_X, and let $a \in \mathbb{R}^m$. The distribution Q_{X+a} of the random vector $X + a$ is given by

$$Q_{X+a}(C) := P(X + a \in C) = P(X \in C - a) = Q_X(C - a) \quad (C \in \mathcal{B}(\mathbb{R}^m)).$$

In particular, for all measurable non-negative functions $f : E \to \mathbb{R}$,

$$E[f(X + a)] = \int_E f(x) \, Q_x(\mathrm{d}x - a).$$

As a special case of Theorem 4.4.2, we have:

Theorem 5.2.3 *If g is a measurable function from $(E, \mathcal{E})$ to $(\overline{\mathbb{R}}, \mathcal{B}(\overline{\mathbb{R}}))$, then*

$$E[g(X)] = \int_E g(x) Q_X(\mathrm{d}x),$$

this formula requiring that one of the sides of the equality be well defined, in which case the other is also well defined.

In the particular case where $(E, \mathcal{E}) = (\mathbb{R}, \mathcal{B}(\mathbb{R}))$, taking $C = (-\infty, x]$,

$$Q_X((-\infty, x]) = P(X \leq x) = F_X(x),$$

where F_X is the *cumulative distribution function* (*c.d.f.*) of X, and

$$E[g(X)] = \int_{\mathbb{R}} g(x) \, \mathrm{d}F_X(x),$$

by definition of the Stieltjes–Lebesgue integral.

In the particular case where $(E, \mathcal{E}) = (\mathbb{R}^n, \mathcal{B}(\mathbb{R}^n))$ and the random vector X admits a probability density f_X (that is, if Q_X is the product of the Lebesgue measure on $(\mathbb{R}^n, \mathcal{B}(\mathbb{R}^n))$ with the function f_X), Theorem 4.4.4 tells us that

$$E[g(X)] = \int_{\mathbb{R}^n} g(x) f_X(x) \, \mathrm{d}x.$$

The following result is an example of the efficiency of Tonelli's theorem.

Theorem 5.2.4 (The telescope formula) *For any* non-negative *random variable X, we have the so-called* telescope formula

$$E[X] = \int_0^\infty [1 - F(x)] \mathrm{d}x.$$

Proof. This follows from Tonelli's theorem applied to the product measure $\ell \times P$. Indeed,

$$E[X] = E\left[\int_0^{\infty} 1_{\{X>x\}}\, \mathrm{d}x\right] = \int_0^{\infty} E\left[1_{\{X>x\}}\right]\, \mathrm{d}x = \int_0^{\infty} [1 - F(x)]\mathrm{d}x\,.$$

□

5.3 Characteristic Functions

Recall that the characteristic function $\varphi : \mathbb{R}^d \to \mathbb{C}$ of a real random vector $X \in \mathbb{R}^d$ is defined by

$$\varphi(u) := E\left[\mathrm{e}^{iu^T X}\right] \quad (u \in \mathbb{R}^d)\,.$$

Theorem 5.3.1 *Let $X \in \mathbb{R}^d$ be a random vector with characteristic function φ. Then* (Paul Lévy's formula) *for all $1 \le j \le d$, all $a_j, b_j \in \mathbb{R}^d$ such that $a_j < b_j$,*

$$\lim_{c\uparrow+\infty} \frac{1}{(2\pi)^d} \int_{-c}^{+c} \cdots \int_{-c}^{+c} \left(\prod_{j=1}^{d} \frac{\mathrm{e}^{-iu_j a_j} - \mathrm{e}^{-iu_j b_j}}{iu_j}\right) \varphi(u_1, \dots, u_d)\, du_1 \cdots du_d$$
$$= E\left[\prod_{j=1}^{d} \left(\frac{1}{2} 1_{\{X_j = a_j \text{ or } b_j\}} + 1_{\{a_j < X_j < b_j\}}\right)\right] .$$

Proof. We prove this result in the univariate case. The multivariate case is a straightforward adaptation of it. Let X be a real-valued random variable with cumulative distribution function F and characteristic function φ. We show that for any pair of points a, b $(a < b)$,

$$\lim_{c\uparrow+\infty} \frac{1}{2\pi} \int_{-c}^{+c} \frac{\mathrm{e}^{-iua} - \mathrm{e}^{-iub}}{iu} \varphi(u)\, \mathrm{d}u = E\left[\left(\frac{1}{2} 1_{\{X=a \text{ or } b\}} + 1_{\{a<X<b\}}\right)\right] . \qquad (\star)$$

For this, write

$$\begin{aligned}
\Phi_c :&= \frac{1}{2\pi} \int_{-c}^{+c} \frac{\mathrm{e}^{-iua} - \mathrm{e}^{-iub}}{iu} \varphi(u)\, \mathrm{d}u \\
&= \frac{1}{2\pi} \int_{-c}^{+c} \frac{\mathrm{e}^{-iua} - \mathrm{e}^{-iub}}{iu} \left(\int_{-\infty}^{+\infty} \mathrm{e}^{iux}\, \mathrm{d}F(x)\right) \mathrm{d}u \\
&= \frac{1}{2\pi} \int_{-\infty}^{+\infty} \left(\int_{-c}^{+c} \frac{\mathrm{e}^{-iua} - \mathrm{e}^{-iub}}{iu} \mathrm{e}^{iux}\, \mathrm{d}u\right) \mathrm{d}F(x) = \int_{-\infty}^{+\infty} \Psi_c(x)\, \mathrm{d}F(x)\,,
\end{aligned}$$

where

$$\Psi_c(x) := \frac{1}{2\pi}\int_{-c}^{+c} \frac{\mathrm{e}^{-iua} - \mathrm{e}^{-iub}}{iu}\mathrm{e}^{+iux}\,\mathrm{d}u\,.$$

The above computations are justified by Fubini's theorem. The conditions of this theorem are satisfied since, observing that

$$\left|\frac{\mathrm{e}^{-iua} - \mathrm{e}^{-iub}}{iu}\right| = \left|\int_a^b \mathrm{e}^{-iux}\mathrm{d}x\right| \leq (b-a)\,,$$

we have

$$\begin{aligned}
&\int_{-c}^{+c}\int_{-\infty}^{+\infty} \left|\frac{\mathrm{e}^{-iua} - \mathrm{e}^{-iub}}{iu}\mathrm{e}^{iux}\right| \mathrm{d}F(x)\,\mathrm{d}u \\
&\qquad = \int_{-c}^{+c}\int_{-\infty}^{+\infty} \left|\frac{\mathrm{e}^{-iua} - \mathrm{e}^{-iub}}{iu}\right| \mathrm{d}F(x)\,\mathrm{d}u \\
&\qquad \leq \int_{-c}^{+c}\int_{-\infty}^{+\infty} (b-a)\,\mathrm{d}F(x)\,\mathrm{d}u = 2c(b-a) < \infty\,.
\end{aligned}$$

Since the function $u \to \frac{\cos(au)}{u}$ is antisymmetric, $\int_{-c}^{+c}\frac{\cos(au)}{u}\,\mathrm{d}u = 0$, and therefore

$$\begin{aligned}
\Psi_c(x) &= \frac{1}{2\pi}\int_{-c}^{+c} \frac{\sin u(x-a) - \sin u(x-b)}{u}\,\mathrm{d}u \\
&= \frac{1}{2\pi}\int_{-c(x-a)}^{+c(x-a)} \frac{\sin u}{u}\,\mathrm{d}u - \frac{1}{2\pi}\int_{-c(x-b)}^{+c(x-b)} \frac{\sin u}{u}\,\mathrm{d}u\,.
\end{aligned}$$

The function $c \mapsto \int_0^c \frac{\sin u}{u}\,\mathrm{d}u = \int_{-c}^0 \frac{\sin u}{u}$ is uniformly continuous in c and tends to $\int_0^{+\infty}\frac{\sin u}{u}\,\mathrm{d}u = \frac{1}{2}\pi$ as $c \uparrow +\infty$. Therefore the function $(c, x) \to \Psi_c(x)$ is uniformly bounded. Moreover, in view of the above expression for Ψ_c,

$$\lim_{c\uparrow\infty}\Psi_c(x) := \Psi(x) = \begin{cases} 0 & \text{if} \quad x < a \text{ or } x > b \\ \frac{1}{2} & \text{if} \quad x = a \text{ or } x = b \\ 1 & \text{if} \quad a < x < b\,. \end{cases}$$

Therefore, by dominated convergence,

$$\begin{aligned}
\lim_{c\uparrow\infty}\Phi_c &= \int_{-\infty}^{+\infty}\lim_{c\uparrow\infty}\Psi_c(x)\,\mathrm{d}F(x) \\
&= \int_{-\infty}^{+\infty}\Psi(x)\,\mathrm{d}F(x) = E\left[\left(\frac{1}{2}1_{\{X=a \text{ or } b\}} + 1_{\{a<X<b\}}\right)\right].
\end{aligned}$$

□

Note that, in the univariate case, denoting by F the cumulative distribution function of the random variable X,

$$E\left[\left(\frac{1}{2}1_{\{X=a \text{ or } b\}} + 1_{\{a<X<b\}}\right)\right] = \frac{F(b)+F(b-)}{2} - \frac{F(a)+F(a-)}{2},$$

so that formula $(\star)$ takes the perhaps more familiar form

$$\frac{F(b)+F(b-)}{2} - \frac{F(a)+F(a-)}{2} = \lim_{c\uparrow+\infty} \frac{1}{2\pi} \int_{-c}^{+c} \frac{\mathrm{e}^{-iua} - \mathrm{e}^{-iub}}{iu} \varphi(u)\, \mathrm{d}u\,.$$

Corollary 5.3.2 *The distribution of a random vector of* $\mathbb{R}^d$ *is uniquely determined by its characteristic function.*

Corollary 5.3.3 *If the random variable* X *admits a probability density* f *and if moreover its characteristic function* φ *is integrable, then*

$$f(x) = \tfrac{1}{2\pi} \int_{-\infty}^{+\infty} \varphi(u)\mathrm{e}^{-iux}\, \mathrm{d}u\,. \tag{5.2}$$

Proof. With f defined as in (5.2), we have, by Fubini,

$$\begin{aligned}
\int_a^b f(x)\,\mathrm{d}x &= \int_a^b \frac{1}{2\pi} \int_{-\infty}^{+\infty} \varphi(u)\mathrm{e}^{-iux}\,\mathrm{d}u\,\mathrm{d}x \\
&= \frac{1}{2\pi} \int_{-\infty}^{+\infty} \varphi(u) \left(\int_a^b \mathrm{e}^{-iux}\,\mathrm{d}u\right) \mathrm{d}x \\
&= \lim_{c\uparrow\infty} \frac{1}{2\pi} \int_{-c}^{+c} \varphi(u) \left(\int_a^b \mathrm{e}^{-iux}\,\mathrm{d}u\right) \mathrm{d}x \\
&= \lim_{c\uparrow\infty} \frac{1}{2\pi} \int_{-c}^{+c} \varphi(u) \frac{\mathrm{e}^{-iua} - \mathrm{e}^{-iub}}{iu}\,\mathrm{d}x = F(b) - F(a)\,,
\end{aligned}$$

by Paul Lévy's inversion formula. This proves that f is a probability density of X. □

The next result says in particular how under certain conditions of integrability, the moments of a random variable can be extracted from its characteristic function.

Theorem 5.3.4 *Let X be a real random variable with characteristic function ψ, and suppose that $E[|X|^n] < \infty$ for some integer $n \geq 1$. Then for all integers $r \leq n$, the r-th derivative $\psi^{(r)}$ of ψ exists and is given by*

$$\psi^{(r)}(u) = i^r E\left[X^r e^{iuX}\right], \tag{5.3}$$

and in particular $E[X^r] = \frac{\psi^{(r)}(0)}{i^r}$. Moreover,

$$\psi(u) = \sum_{r=0}^{n} \frac{(iu)^r}{r!} E[X^r] + \frac{(iu)^n}{n!} \varepsilon_n(u), \tag{5.4}$$

where $\lim_{n\uparrow\infty} \varepsilon_n(u) = 0$ and $|\varepsilon_n(u)| \leq 3E[|X|^n]$.

Proof. First we observe that for any non-negative real number a, and all integers $r \leq n$, $a^r \leq 1 + a^n$ (indeed, if $a \leq 1$, then $a^r \leq 1$, and if $a \geq 1$, then $a^r \leq a^n$). In particular,

$$E[|X|^r] \leq E[1 + |X|^n] = 1 + E[|X|^n] < \infty.$$

Suppose that for some $r < n$,

$$\psi^{(r)}(u) = i^r E\left[X^r e^{iuX}\right].$$

In

$$\begin{aligned}\frac{\psi^{(r)}(u+h) - \psi^{(r)}(u)}{h} &= i^r E\left[X^r \frac{e^{i(u+h)X} - e^{iuX}}{h}\right] \\ &= i^r E\left[X^r e^{iuX} \frac{e^{ihX} - 1}{h}\right],\end{aligned}$$

the quantity under the expectation sign tends to $X^{r+1}e^{iuX}$ as $h \to 0$, and moreover, it is bounded in absolute value by an integrable function since

$$\left|X^r e^{iuX} \frac{e^{ihX} - 1}{h}\right| \leq \left|X^r \frac{e^{ihX} - 1}{h}\right| \leq |X|^{r+1}.$$

(For the last inequality, use the fact that $|e^{ia} - 1|^2 = 2(1 - \cos a) \leq a^2$.) Therefore, by dominated convergence,

$$\begin{aligned}\psi^{(r+1)}(u) &= \lim_{h\to 0} \frac{\psi(u+h) - \psi(u)}{h} \\ &= i^r E\left[\lim_{h\to 0} X^r e^{iuX} \frac{e^{ihX} - 1}{h}\right] = i^r E\left[X^{r+1} e^{iuX}\right].\end{aligned}$$

Equality (5.3) follows since the induction hypothesis is trivially true for $r = 0$.

We now prove (5.4). By Taylor's formula, for $y \in \mathbb{R}$,

$$e^{iy} = \cos y + i \sin y = \sum_{k=0}^{n-1} \frac{(iy)^k}{k!} + \frac{(iy)^n}{n!} (\cos(\theta_1 y) + i \sin(\theta_2 y))$$

for some $\theta_1, \theta_2 \in [-1, +1]$. Therefore

$$e^{iuX} = \sum_{k=0}^{n-1} \frac{(iuX)^k}{k!} + \frac{(iuX)^n}{n!} (\cos(\theta_1 uX) + i \sin(\theta_2 uX)) ,$$

where $\theta_1 = \theta_1(\omega), \theta_2 = \theta_2(\omega) \in [-1, +1]$, and

$$E\left[e^{iuX}\right] = \sum_{k=0}^{n-1} \frac{(iu)^k}{k!} E[X^k] + \frac{(iu)^n}{n!} (E[X^n] + \varepsilon_n(u)) ,$$

where

$$\varepsilon_n(u) = E\left[X^n (\cos \theta_1 uX + i \sin \theta_2 uX - 1)\right] .$$

Clearly $|\varepsilon_n(u)| \leq 3E[|X|^n]$. Also, since the random variable

$$X^n (\cos \theta_1 uX + i \sin \theta_2 uX - 1)$$

is bounded in absolute value by the integrable random variable $3|X|^n$ and tends to 0 as $u \to 0$, we have by dominated convergence $\lim_{u \to 0} \varepsilon_n(u) = 0$. □

Theorem 5.3.4 can be extended to random vectors, with a proof similar to that of the univariate case. We just quote the formula giving the *mixed moments* of a random vector in terms of its characteristic function:

Let $X = (X_1, \cdots, X_d) \in \mathbb{R}^d$ be a random vector with characteristic function

$$\varphi(u) := E\left[e^{u^T X}\right] \quad (u = (u_1, \cdots, u_d)) .$$

Theorem 5.3.5 *Suppose that $E[|X_i|^n] < \infty$ $(1 \leq i \leq d)$ for some $n \geq 1$. Then for all $\nu = (\nu_1, \cdots, \nu^d)$ such that $\nu_1 + \cdots + \nu^d \leq n$, the partial derivative*

$$\frac{\partial^{\nu_1 + \cdots + \nu^d}}{\partial u_1^{\nu_1} \cdots \partial u_1^{\nu_d}} \varphi(u_1, \cdots, u_d)$$

exists and is continuous, and

$$E\left[X_1^{\nu_1} \cdots X_d^{\nu_d}\right] = \frac{\partial^{\nu_1 + \cdots + \nu^d}}{\partial u_1^{\nu_1} \cdots \partial u_1^{\nu_d}} \varphi(0, \cdots, 0) . \tag{5.5}$$

The proof is required in Exercise 5.7.11.

5.4 Independence

This section revisits the notion of independence an rigorous and more general way than was done in the first three chapters.

Recall that two events A and B are said to be *independent* if

$$P(A \cap B) = P(A)P(B).$$

More generally, a family $\{A_i\}_{i \in I}$ of events, where I is an arbitrary index, is called an *independent family* if, for every *finite* subset $J \in I$,

$$P\left(\bigcap_{j \in J} A_j\right) = \prod_{j \in J} P(A_j).$$

Two random elements $X : (\Omega, \mathcal{F}) \to (E, \mathcal{E})$ and $Y : (\Omega, \mathcal{F}) \to (G, \mathcal{G})$ are called independent if for all $C \in \mathcal{E}$, $D \in \mathcal{G}$

$$P(\{X \in C\} \cap \{Y \in D\}) = P(X \in C)P(Y \in D).$$

More generally, a family $\{X_i\}_{i \in I}$ (where I is an arbitrary index) of random elements $X_i : (\Omega, \mathcal{F}) \to (E_i, \mathcal{E}_i)$ $(i \in I)$ is said to be independent if, for every finite subset $J \in I$,

$$P\left(\bigcap_{j \in J} \{X_j \in C_j\}\right) = \prod_{j \in J} P(X_j \in C_j)$$

for all $C_j \in \mathcal{E}_j$ $(j \in J)$.

The σ-fields $\mathcal{F}_1$ and $\mathcal{F}_2$ on Ω are called independent if for any $k_1, k_2 \geq 1$, any $A_1, \dots, A_{k_1} \in \mathcal{F}_1$, and any $B_1, \dots, B_{k_2} \in \mathcal{F}_2$,

$$P(A_1, \dots, A_{k_1}, B_1, \dots, B_{k_2}) = P(A_1, \dots, A_{k_1})P(B_1, \dots, B_{k_2}).$$

This definition extends in an obvious way to the independence of a finite number of σ-fields.

By definition, the σ-field generated by a random element X with values in the measurable space $(E, \mathcal{E})$ is the σ-field $\sigma(X)$ generated by the collection of events $\{X \in C\}$ $(C \in \mathcal{E})$.

Therefore a family $\{X_i\}_{i \in I}$, where I is an arbitrary index, of random elements $X_i : (\Omega, \mathcal{F}) \to (E_i, \mathcal{E}_i)$ $(i \in I)$, is an independent family of random elements if for every finite subset $J \in I$, the family of σ-fields $\{\sigma(X_i)\}_{i \in I}$ is independent.

The next result says that the independence property is preserved when taking functions of the random elements and is an immediate consequence of the definition of independent random elements. This "natural" result was used without further justification in the first three chapters.

Theorem 5.4.1 *If the random elements X and Y, taking their values in $(E, \mathcal{E})$ and $(G, \mathcal{G})$ respectively, are independent, then so are the random elements $\varphi(X)$ and $\psi(Y)$, where $\varphi : (E, \mathcal{E}) \to (E', \mathcal{E}')$, $\psi : (G, \mathcal{G}) \to (G', \mathcal{G}')$.*

Proof. For all $C' \in \mathcal{E}'$, $D' \in \mathcal{G}'$, the sets $C = \varphi^{-1}(C')$ and $D = \psi^{-1}(D')$ are in $\mathcal{E}$ and $\mathcal{G}$ respectively, since φ and ψ are measurable. We have

$$\begin{aligned} P(\varphi(X) \in C', \psi(Y) \in D') &= P(X \in C, Y \in D) \\ &= P(X \in C)\, P(Y \in D) \\ &= P(\varphi(X) \in C')\, P(\psi(Y) \in D'). \end{aligned}$$

□

The above result is stated for two random elements for simplicity, and it extends in the obvious way to a finite number of independent random elements.

In order to prove that two σ-fields are independent, it suffices to prove that certain subclasses of these σ-fields are independent.

More precisely:

Theorem 5.4.2 *Let $(\Omega, \mathcal{F}, P)$ be a probability space, and let $\mathcal{S}_1$ and $\mathcal{S}_2$ be two collections of events that are stable under finite intersections. If $\mathcal{S}_1$ and $\mathcal{S}_2$ are independent, then so are $\sigma(\mathcal{S}_1)$ and $\sigma(\mathcal{S}_2)$.*

This is not proved in this book.[1]

The next corollary brings us back to the elementary definition of independence of two random variables.

Corollary 5.4.3 *Let $(\Omega, \mathcal{F}, P)$ be a probability space on which are given two real random variables X and Y. For these two random variables to be independent, it is necessary and sufficient that for all $a, b \in \mathbb{R}$, $P(X \leq a, Y \leq b) = P(X \leq a)P(Y \leq b)$.*

Proof. This follows from Theorem 5.4.2, since the collection $\{(-\infty, a];\ a \in \mathbb{R}\}$ is stable under finite intersection and generates $\mathcal{B}(\mathbb{R})$. □

Similar arguments lead to a similar result for several random variables.

[1] See for instance Theorem 3.1.39 of [7].

The Product Formula

The independence of two random variables X and Y is equivalent to the factorisation of their joint distribution:

$$Q_{(X,Y)} = Q_X \times Q_Y \,,$$

where $Q_{(X,Y)}$, Q_X, and Q_Y are the distributions of (X,Y), X, and Y, respectively. Indeed, for all sets of the form $C \times D$, where $C \in \mathcal{E}$, $D \in \mathcal{G}$,

$$\begin{aligned} Q_{(X,Y)}(C \times D) &= P((X,Y) \in C \times D) = P(X \in C, Y \in D) \\ &= P(X \in C)P(Y \in D) = Q_X(C)Q_Y(D) \,. \end{aligned}$$

In particular, by the Fubini–Tonelli theorem,

Theorem 5.4.4 *Let X and Y be independent random elements taking their values in $(E,\mathcal{E})$ and $(G,\mathcal{G})$ respectively. Then for all $g : (E,\mathcal{E}) \to (\mathbb{R},\mathcal{B}(\mathbb{R}))$, $h : (G,\mathcal{G}) \to (\mathbb{R},\mathcal{B}(\mathbb{R}))$ such that $E[|g(X)|] < \infty$ and $E[|h(Y)|] < \infty$, or $g \geq 0$ and $h \geq 0$, we have the* product formula for expectations

$$E[g(X)h(Y)] = E[g(X)]\,E[h(Y)] \,.$$

EXAMPLE 5.4.5: CONVOLUTION PRODUCT. Let X and Y be two independent random vectors with values in $\mathbb{R}^m$ and with respective distributions Q_X and Q_Y. We compute the distribution of the random vector $Z := X + Y$:

$$\begin{aligned} P(Z \in C) &= P(X + Y \in C) = E[1_C(X+Y)] \\ &= \int_{\mathbb{R}^m}\int_{\mathbb{R}^m} 1_C(x+y)Q_X(\mathrm{d}y)Q_Y(\mathrm{d}y) \\ &= \int_{\mathbb{R}^m}\left(\int_{\mathbb{R}^m} 1_C(x+y)Q_X(\mathrm{d}x)\right)Q_Y(\mathrm{d}y) \\ &= \int_{\mathbb{R}^m}\left(\int_{\mathbb{R}^m} 1_{C-y}(x)Q_X(\mathrm{d}x)\right)Q_Y(\mathrm{d}y) \\ &= \int_{\mathbb{R}^m} Q_X(C-y)Q_Y(\mathrm{d}y) \,, \end{aligned}$$

that is,

$$Q_Z(C) = \int_{\mathbb{R}^m} Q_X(C-y)Q_Y(\mathrm{d}y) \,.$$

This probability distribution is called the *convolution product* of Q_X and Q_Y.

In the scalar case $m = 1$, and with $C := (-\infty, z]$, we have the following version of the convolution product formula in terms of cumulative distribution functions and Stieltjes–Lebesgue integrals:

$$F_Z(z) = \int_{\mathbb{R}} F_X(z-y)\, F_Y(\mathrm{d}y)\,.$$

The next result generalizes Theorem 3.2.20, with a similar proof.

Theorem 5.4.6 *For the random vectors $X_1, \dots, X_d$ to be independent, a necessary and sufficient condition is that the characteristic function φ_X of $X = (X_1, \dots, X_d)$ factorizes as*

$$\varphi_X(u_1, \dots, u_d) = \prod_{j=1}^{d} \varphi_j(u_j)\,,$$

where for all $1 \le j \le d$, φ_j is a characteristic function. In this case, for all $1 \le j \le d$, $\varphi_j = \varphi_{X_j}$, the characteristic function of X_j.

Proof. Exercise 5.7.7. □

5.5 Conditional Expectation III

In Chapters 2 and 3, the theory of conditional expectation was developed in the discrete and the absolutely continuous cases respectively. This chapter now gives the theory of conditional expectation of a random variable with respect to another random variable when the joint distribution is arbitrary.

We start with a preliminary observation. Let X and Y be two random vectors of dimensions p and n respectively, with the joint probability density $f_{X,Y}(x,y)$. Let the function $g : \mathbb{R}^p \times \mathbb{R}^n \to \mathbb{R}_+$ be either non-negative or such that $g(X,Y)$ is integrable. For any non-negative bounded function $\varphi : \mathbb{R}^n \to \mathbb{R}$, we have

$$E\left[E^Y\left[g(X,Y)\right]\varphi(Y)\right] = E\left[g(X,Y)\varphi(Y)\right]. \tag{5.6}$$

Indeed,

$$\begin{aligned}
E\left[E^Y\left[g(X,Y)\right]\varphi(Y)\right] &= E\left[\psi(Y)\varphi(Y)\right] = \textstyle\int_{\mathbb{R}^n} \psi(y)\varphi(y) f_Y(y)\,\mathrm{d}y \\
&= \textstyle\int_{\mathbb{R}^n} \left(\int_{\mathbb{R}^p} g(x,y)\frac{f_{X,Y}(x,y)}{f_Y(y)}\,\mathrm{d}x\right)\varphi(y) f_Y(y)\,\mathrm{d}y \\
&= \textstyle\int_{\mathbb{R}^n}\int_{\mathbb{R}^p} g(x,y)\varphi(y) f_{X,Y}(x,y)\,\mathrm{d}x\,\mathrm{d}y \\
&= E\left[g(X,Y)\varphi(Y)\right].
\end{aligned}$$

In the discrete case, a similar computation yields a similar result. This suggests to adopt, in the general case (not necessarily discrete or absolutely continuous), (5.6) as a definition of conditional expectation, where X and Y are random elements taking their values in spaces E and F respectively that can be either discrete or some $\mathbb{R}^k$. This would include mixed cases of the type discrete/absolutely continuous, but also more complex situations, such as the following: $E = \mathbb{R}^p$ and $F = \mathbb{R}^n$, X and Y admit a probability density function, but the couple (X, Y) does not admit a probability density function (for instance, in the univariate case, if $Y = X^2$).

We shall now give more generality to the study (but only in appearance) and take for the conditioned variable a real random variable Z (previously, we chose $Z = g(X, Y)$).

Definition 5.5.1 *A Y-measurable random variable is a random variable U of the form $U = \varphi(Y)$ where $\varphi : \mathbb{R}^n \to \mathbb{R}$ is a measurable function.*

Definition 5.5.2 *Let Z and Y be as above, and suppose that Z is either non-negative or integrable. The conditional expectation $E^Y[Z]$ is by definition the "essentially unique" variable of the form $\psi(Y)$, where ψ is measurable, such that equality*

$$E[\psi(Y)U] = E[ZU] \tag{5.7}$$

holds for any non-negative bounded Y-measurable real random variable $U = \varphi(Y)$.

By "essentially unique" the following is meant: If there are two functions ψ_1 and ψ_2 that meet the requirement, then $\psi_1(Y) = \psi_2(Y)$ almost surely, that is, $P(\psi_1(Y) = \psi_2(Y)) = 1$. We then say that $\psi_1(Y)$ and $\psi_2(Y)$ are two "versions" of $E^Y[Z]$.

Theorem 5.5.3 *In the situation described in the above definition, the conditional expectation exists and is essentially unique.*

Proof. The proof of existence is omitted at this point (see Section 5.6). In practice, one is usually able to find "a" function ψ by construction, as the examples and the exercises will show. The uniqueness part that we now agree to prove will guarantee that it is "the" function ψ.

Indeed, suppose that ψ_1 and ψ_2 meet the requirement. In particular,

$$E[\psi_1(Y)\varphi(Y)] = E[\psi_2(Y)\varphi(Y)]\,(= E[Z\varphi(Y)])$$

and therefore

$$E[(\psi_1(Y) - \psi_2(Y))\varphi(Y)] = 0$$

for a non-negative bounded measurable functions $\varphi : \mathbb{R}^n \to \mathbb{R}$. Choosing $\varphi(Y) = 1_{\{\psi_1(Y)-\psi_2(Y)>0\}}$, we obtain

$$E\left[(\psi_1(Y) - \psi_2(Y))1_{\{\psi_1(Y)-\psi_2(Y)>0\}}\right] = 0\,.$$

Since the random variable $(\psi_1(Y)-\psi_2(Y))1_{\{\psi_1(Y)-\psi_2(Y)>0\}}$ is non-negative and has a null expectation, it must be almost surely null (Lemma 3.3.3). In other terms $\psi_1(Y)-\psi_2(Y) \le 0$ almost surely. Exchanging the roles of ψ_1 and ψ_2, we have that $\psi_1(Y) - \psi_2(Y) \ge 0$ almost surely. Therefore $\psi_1(Y) - \psi_2(Y) = 0$ almost surely. □

EXAMPLE 5.5.4: THE DISCRETE CASE REVISITED. If Y is a positive integer-valued random variable, then

$$E^Y[Z] = \sum_{n=1}^{\infty} \frac{E[Z1_{\{Y=n\}}]}{P(Y=n)} 1_{\{Y=n\}}\,,$$

where, by convention, $\frac{E[Z1_{\{Y=n\}}]}{P(Y=n)} = 0$ when $P(Y=n) = 0$.

Proof. We must verify (5.7) for all bounded measurable $\varphi : \mathbb{R} \to \mathbb{R}$. The right-hand side is equal to

$$\begin{aligned}
&E\left[\left(\sum_{n\geq 1} \frac{E[Z1_{\{Y=n\}}]}{P(Y=n)} 1_{\{Y=n\}}\right)\left(\sum_{k\geq 1} \varphi(k)1_{\{Y=k\}}\right)\right] \\
&= E\left[\sum_{n\geq 1} \frac{E[Z1_{\{Y=n\}}]}{P(Y=n)} \varphi(n)1_{\{Y=n\}}\right] = \sum_{n\geq 1} \frac{E[Z1_{\{Y=n\}}]}{P(Y=n)} \varphi(n)E[1_{\{Y=n\}}] \\
&= \sum_{n\geq 1} \frac{E[Z1_{\{Y=n\}}]}{P(Y=n)} \varphi(n)P(Y=n) = \sum_{n\geq 1} E[Z1_{\{Y=n\}}]\varphi(n) = E[Z\varphi(Y)]\,.
\end{aligned}$$

□

EXAMPLE 5.5.5: THE ABSOLUTELY CONTINUOUS CASE REVISITED. Let X and Y be random vectors of dimensions p and n respectively, admitting the joint probability density $f_{X,Y}(x,y)$. Let $g : \mathbb{R}^{p+n} \to \mathbb{R}$ be a measurable function, and suppose that the random variable $Z = g(X,Y)$ is integrable. The conditional expectation of Z given Y is the random variable $\psi(Y)$, where

$$\psi(y) = \int_{\mathbb{R}^p} g(x,y) f_X^{Y=y}(x)\mathrm{d}x\,.$$

Proof. We first verify that $\psi(Y)$ is integrable. We have

$$|\psi(y)| \leq \int_{\mathbb{R}^P} |g(x,y)| f_Z^{Y=y}(x)\mathrm{d}x$$

and therefore

$$\begin{aligned}
E[|\psi(Y)|] &= \int_{\mathbb{R}^n} |\psi(y)| f_Y(y)\mathrm{d}y \leq \int_{\mathbb{R}^n} \left(\int_{\mathbb{R}^P} |g(x,y)| f_Z^{Y=y}(x)\mathrm{d}x \right) f_Y(y)\mathrm{d}y \\
&= \int_{\mathbb{R}^P}\int_{\mathbb{R}^n} |g(x,y)| f_X^{Y=y}(x) f_Y(y)\mathrm{d}x\mathrm{d}y \\
&= \int_{\mathbb{R}^P}\int_{\mathbb{R}^n} |g(x,y)| f_{X,Y}(x,y)\,\mathrm{d}x\,\mathrm{d}y = E[|g(X,Y)|] = E[|Z|] < \infty\,.
\end{aligned}$$

We check that (5.7) is true, with $U = \varphi(Y)$ bounded. The right-hand side is

$$\begin{aligned}
E[\psi(Y)\varphi(Y)] &= \int_{\mathbb{R}^n} \psi(y)\varphi(y) f_Y(y)\mathrm{d}y \\
&= \int_{\mathbb{R}^n} \left(\int_{\mathbb{R}^P} g(x,y) f_X^{Y=y}(x)\mathrm{d}x \right) \varphi(y) f_Y(y)\mathrm{d}y \\
&= \int_{\mathbb{R}^P}\int_{\mathbb{R}^n} g(x,y)\varphi(y) f_X^{Y=y}(x) f_Z(y)\,\mathrm{d}x\,\mathrm{d}y \\
&= \int_{\mathbb{R}^P}\int_{\mathbb{R}^n} g(x,y)\varphi(y) f_{X,Y}(x,y)\,\mathrm{d}x\,\mathrm{d}y \\
&= E[g(X,Y)\varphi(Y)] = E[Z\varphi(Y)]\,.
\end{aligned}$$

□

EXAMPLE 5.5.6: MIXED CASE, I. We shall consider the situation, often encountered in practice, where X is a random vector of dimension p and where Y takes its values in $\mathbb{N}_+$. We denote $P(Y = k)$ by $\pi(k)$. We suppose that for all $k \geq 1$, there is a probability density function f_k such that

$$P(X \in A | Y = k) = \int_A f_k(x)\,\mathrm{d}x \quad (A \in \mathcal{B}(\mathbb{R}^p))\,.$$

Then, for any function $g : \mathbb{R}^P \times \mathbb{N}_+ \to \mathbb{R}$ that is non-negative or such that $g(X,Y)$ is integrable, we have

$$E^X[g(X,Y)] = \psi(Y)\,,$$

where

$$\psi(k) := \int_{\mathbb{R}^p} g(x,k) f_k(y)\mathrm{d}y\,.$$

The proof is similar to the proof when (X,Y) has a joint probability distribution and is left to the reader.

EXAMPLE 5.5.7: MIXED CASE, II. We now treat the second type of mixed case, where the conditioning variable Y is a random vector of dimension n, X is a $\mathbb{N}_+$-valued random variable, with the joint distribution of (X,Y) given by

$$P(X=k)=\pi(k)\quad (k\geq 1)$$

and

$$P(Y\in A|X=k)=\int_A f_k(y)\mathrm{d}y\quad (k\geq 1, A\in\mathcal{B}(\mathbb{R}^n))\,.$$

For all $k\geq 1, y\in\mathbb{R}^n$, let

$$\pi_{X|Y}(k|y):=\frac{\pi(k)f_k(y)}{f_Y(y)}$$

if $f_Y(y)=\sum_{k\geq 1}\pi(k)f_k(y)>0$, and $\pi(k|y)=0$ otherwise. We let the reader verify that for all $g:\mathbb{N}\times\mathbb{R}^n\to\mathbb{R}$ such that $E[|g(X,Y)|<\infty$,

$$E^X[g(X,Y)]=\psi(Y)\,,$$

where

$$\psi(y)=\sum_{k\geq 1}g(k,y)\pi_{X|Y}(k|y)\,.$$

We now list, in the more general setting, the main rules that are useful in computing conditional expectations.

Let Y be a random variable, and let Z, Z_1, Z_2 be integrable (resp. non-negative finite) random variables, $\lambda_1,\lambda_2\in\mathbb{R}$ (resp. $\in\mathbb{R}_+$).

Theorem 5.5.8 Rule 1 *(linearity)*

$$E^Y[\lambda_1 Z_1+\lambda_2 Z_2]=\lambda_1 E^Y[Z_1]+\lambda_2 E^Y[Z_2]\,.$$

Proof. We consider only the integrable case. The non-negative case follows, *mutatis mutandis*. We must check that $\lambda_1 E^Y[Z_1]+\lambda_2 E^Y[Z_2]$ is Y-measurable (which is part of the definition of a conditional expectation with respect to Y) and that for all bounded Y-measurable random variables U

$$E[(\lambda_1 E^Y[X_1]+\lambda_2 E^Y[X_1])U]=E[(\lambda_1 Z_1+\lambda_2 Z_2)U]\,.$$

This follows immediately from the definition of $E^Y[Z_i]$, which says that $E[E^Y[Z_i]U] = E[Z_iU]$ $(i = 1, 2)$. □

Theorem 5.5.9 Rule 2. *If Z is independent of Y, then*

$$E^Y[Z] = E[Z]\,.$$

Proof. (Non-negative case.) The constant $E[Z]$ (as any constant) is Y-measurable. Moreover, for all bounded Y-measurable random variable U, $E[E[Z]U] = E[ZU]$. In fact, Z and U are independent and therefore $E[ZU] = E[Z]E[U]$. □

Theorem 5.5.10 Rule 3. *If Z is Y-measurable, then*

$$E^Y[Z] = Z\,.$$

Proof. (Non-negative case.) In fact, Z is Y-measurable by hypothesis and $E[ZU] = E[ZU]$! □

Theorem 5.5.11 Rule 4. *If $Z_1 \le Z_2$ P-a.s. Then*

$$E^Y[Z_1] \le E^Y[Z_2], P\text{-}a.s.$$

In particular, if Z is a non-negative random variable $E^Y[Z] \ge 0, P$- a.s.

Proof. We consider only the integrable case. For any bounded Y-measurable random variable U,

$$E[E^Y[Z_1]U] = E[Z_1U] \le E[Z_2U] = E[E^Y[Z_2]U]\,.$$

Therefore

$$E[(E^Y[Z_2] - E^Y[Z_1])U] \ge 0\,.$$

In particular,

$$E[(E^Y[Z_2] - E^Y[Z_1])1_{\{E^Y[Z_2]<E^Y[Z_1]\}}] \ge 0\,.$$

Since the left-hand side is non-positive, it follows that $P(E^Y[Z_2] < E^Y[Z_1])$. □

Theorem 5.5.12 Rule 5. *Let Y_1 and Y_2 be random variables, and let Z be either integrable or non-negative. Then*

$$E^{Y_2}[E^{Y_1,Y_2}[Z]] = E^{Y_2}[Z]\,.$$

Proof. We just have to check that $E^{Y_2}[E^{Y_1,Y_2}[Z]]$ is a version of $E^{Y_2}[Z]$. Since it is a Y_2-measurable variable it remains to show that it satisfies

$$E[E^{Y_2}[E^{Y_1,Y_2}[Z]]U] = E[ZU]\,,$$

for any bounded (resp. bounded non-negative) Y_2-measurable variable U. Since such a variable is *a fortiori* (Y_1, Y_2)-measurable,

$$E[[E^{Y_1,Y_2}[Z]U] = E[ZU]\,.$$

Moreover

$$E[E^{Y_2}[E^{Y_1,Y_2}[Z]]U] = E[[E^{Y_1,Y_2}[Z]U]\,,$$

by definition of $E^{Y_2}[E^{Y_1,Y_2}[Z]]$. □

Theorem 5.5.13 *Let Y be a random vector and let Z be of the form $Z = VZ'$, where V is a Y-measurable bounded (resp. non-negative finite) random variable, and Z' is an integrable (resp. non-negative finite) random variable. Then*

$$E^Y[VZ'] = VE^Y[Z']\,.$$

Proof. We consider only the integrable case. We observe that $VE^Y[Z]$ is Y-measurable, and it remains to prove that for all bounded Y-measurable random variables U,

$$E[VZ'U] = E[VE^Y[Z']U].$$

But, since VU is bounded, by definition of $E^Y[Z']$,

$$E[VE^Y[Z']U] = E[VZ'U]\,.$$

□

The theorems allowing interversion of limit and integral (monotone convergence theorem and dominated convergence theorem) have conditional versions.

We start with the monotone convergence theorem:

Theorem 5.5.14 *Let X be some random vector and let $\{Y_n\}_{n\geq 1}$ be a P-a.s. non-decreasing sequence of non-negative random variables converging P-a.s. to the random variable Y. Then $\{E^X[Y_n]\}_{n\geq 1}$ is a P-a.s. non-decreasing sequence of random variables converging P-a.s. to $E^X[Y]$.*

Proof. By monotonicity of conditional expectation, $\{E^X[Y_n]\}_{n\geq 1}$ is a P-a.s. non-decreasing sequence of X-measurable random variables. In particular, there exists a P-a.s. limit W, X-measurable, of this sequence. By monotone convergence, for any bounded non-negative X-measurable random variable U,

$$\lim_{n\uparrow\infty} E[Y_nU] = E[YU],$$

and

$$\lim_{n\uparrow\infty} E[E^X[Y_n]U] = E[WU].$$

Therefore, since $E[Y_nU] = E[E^X[Y_n]U]$ for all $n \geq 1$, $E[YU] = E[WU]$. This being true for all bounded X-measurable random variables U, $W = E[Y \mid X]$. $\square$

We now turn to the conditioned version of the dominated convergence theorem:

Theorem 5.5.15 *Let X be some random vector, and let $\{Y_n\}_{n\geq 1}$ be a sequence of random variables converging P-a.s. to the random variable Y, and such that $|Y_n| \leq Z$ for some integrable random variable Z. Then $\{E^X[Y_n]\}_{n\geq 1}$ converges P-a.s. to $E^X[Y]$.*

Proof. Let $W_n := \sup_{m\geq n} |Y_m - Y|$. The sequence $\{W_n\}_{n\geq 1}$ decreases P-a.s. to 0. We have

$$\begin{aligned} |E^X[Y_n] - E^X[Y]| &= |E^X[Y_n - Y]| \\ &\leq E^X[|Y_n - Y|] \leq E^X[W_n]. \end{aligned}$$

The non-negative sequence $\{E^X[W_n]\}_{n\geq 1}$ decreases P-a.s. (rule 4). Let $H \geq 0$ be its limit. Then

$$0 \leq |E[H]| \leq E\left[E^X[W_n]\right] = E[W_n],$$

where the latter quantity tends to 0 by dominated convergence (because $0 \leq W_n \leq 2Z$). Therefore $E[H] = 0$, which implies that $P(H = 0) = 1$ since H is P-a.s non-negative. $\square$

5.6 General Theory of Conditional Expectation

We shall need later a more general and abstract theory of conditional expectation. Previously, the conditioning was with respect to random variables or vectors. We now condition with respect to σ-fields.

Definition 5.6.1 *Let Y be an integrable (resp. finite non-negative) random variable, and let $\mathcal{G}$ be a sub-σ-field of $\mathcal{F}$. A version of the conditional expectation of Y given $\mathcal{G}$ is any integrable (resp. finite non-negative) $\mathcal{G}$-measurable random variable Z such that*

$$E[YU] = E[ZU] \tag{5.8}$$

for all bounded (resp. bounded non-negative) $\mathcal{G}$-measurable random variables U.

Theorem 5.6.2 *Let Y and $\mathcal{G}$ be as above. There exists at least one version of the conditional expectation of Y given $\mathcal{G}$, and it is essentially unique, that is, if Z' is another version of the conditional expectation of Y given $\mathcal{G}$, then $Z = Z'$, P-a.s.*

There will be no problem in representing two versions of this conditional expectation by the same symbol, since, as we just saw, they are P-almost surely equal. We choose the symbol $E[Y|\mathcal{G}]$ or $E^{\mathcal{G}}[Y]$ indifferently. From now on we say: $E^{\mathcal{G}}[Y]$ (or $E[Y|\mathcal{G}]$) is the conditional expectation of Y given $\mathcal{G}$. The defining equality (5.8) reads

$$E[YU] = E[E^{\mathcal{G}}[Y]U]$$

for all bounded (resp. bounded non-negative) $\mathcal{G}$-measurable random variables U.

Proof. Uniqueness. The integrable case will be treated, the other case being similar. First observe that

$$0 = E[ZU] - E[Z'U] = E[(Z - Z')U]$$

for all bounded $\mathcal{G}$-measurable random variable U. In particular, with $U = 1_{\{Z>Z'\}}$,

$$E[(Z - Z')1_{\{Z>Z'\}}] = 0\,.$$

Since the random variable in the expectation is non-negative, it can have a null expectation only if it is P-a.s. null, that is if P-a.s., $Z \le Z'$. By symmetry, P-a.s., $Z \ge Z'$, and therefore, as announced, $Z = Z'$, P-a.s.

Existence. We do this for the non-negative integrable case, the general case following easily from this special case. Consider the measure ν on $(\Omega, \mathcal{G})$ defined by

$$\nu(A) = \int_A Y \,\mathrm{d}P \quad (A \in \mathcal{G})\,.$$

It is finite (resp. σ-finite) since Y is assumed integrable (resp. finite non-negative). Moreover, if $P(A) = 0$ then $\nu(A) = 0$. Therefore the measure μ on $(\Omega, \mathcal{G})$ that is the restriction of P to $(\Omega, \mathcal{G})$ is absolutely continuous with respect to ν, so that, by the Radon–Nikodým theorem (Theorem 4.4.6), there exists an integrable (resp. finite non-negative) random variable of $(\Omega, \mathcal{G})$, that is, an integrable (resp. finite non-negative)
$\mathcal{G}$-measurable random variable Z of $(\Omega, \mathcal{F})$, such that

$$\nu(A) = \int_A Z \mathrm{d}P \quad (A \in \mathcal{G})\,.$$

In particular,

$$\int_\Omega U\,Y \,\mathrm{d}P = \int_\Omega U\,Z \,\mathrm{d}P$$

for all bounded (resp. non-negative bounded) $\mathcal{G}$-measurable random variables U.

□

A Special Case

Let

$$\mathcal{G} := \sigma(X)\,, \tag{5.9}$$

where $X = (X_1, \dots, X_N)$ is an arbitrary random vector defined on $(\Omega, \mathcal{F})$ and $\sigma(X)$ is, by definition, the smallest σ-field that contains all the sets of the form $\{X \in C\}$ where $C \in \mathcal{B}(\mathbb{R}^N)$. In this situation, we adopt the notation $E^X[Y]$ for $E^{\mathcal{G}}[Y]$ (or $E[Y|X]$ for $E[Y|\mathcal{G}]$), and we call this (equivalence class of) random variable(s) the conditional expectation of Y given X.

Theorem 5.6.3 *Let X be a random vector with values in the measurable space $(\mathbb{R}^k, \mathcal{B}(\mathbb{R}^k))$. A random variable $Z : (\Omega, \mathcal{F}) \to (\overline{\mathbb{R}}, \mathcal{B}(\overline{\mathbb{R}}))$ is $\sigma(X)$-measurable if and only if there exists a measurable function $g : (\mathbb{R}^k, \mathcal{B}(\mathbb{R}^k)) \to (\overline{\mathbb{R}}, \mathcal{B}(\overline{\mathbb{R}}))$ such that $Z = g(X)$.*

Proof. The "if" part is just the stability of measurability under composition (Theorem 4.1.11). For the necessity, first observe that this is true of simple random variables. It therefore remains to show that it is true for a non-negative random variable Z (from which the general case straightforwardly follows). Such a random variable is the limit of a non-decreasing sequence $\{Z_n\}_{n\geq 1}$ of non-negative simple random variables of the form $g_n(X)$ for some measurable function $g_n : (\mathbb{R}^k, \mathcal{B}(\mathbb{R}^k)) \to (\overline{\mathbb{R}}, \mathcal{B}(\overline{\mathbb{R}}))$. Let M be the (measurable) set on which the sequence $\{g_n\}_{n\geq 1}$ admits a limit. Define $g(x) = \lim g_n(x) 1_M(x)$ (a measurable function). For each ω, $Z(\omega) = \lim g_n(X(\omega))$, which implies that $Z(\omega) \in M$ and that $Z(\omega) = \lim g_n(X(\omega)) = g(X(\omega))$. □

Properties of the Conditional Expectation

The main rules that are useful in computing conditional expectations will be given once more, but this time in the general abstract framework.

Let $\mathcal{G}$ be a sub-σ-field of $\mathcal{F}$, and let Y, Y_1, Y_2 be integrable (resp. non-negative finite) random variables, $\lambda_1, \lambda_2 \in \mathbb{R}$ (resp. $\in \mathbb{R}_+$).

Theorem 5.6.4 Rule 1. *(linearity)*

$$E^{\mathcal{G}}[\lambda_1 Y_1 + \lambda_2 Y_2] = \lambda_1 E^{\mathcal{G}}[Y_1] + \lambda_2 E^{\mathcal{G}}[Y_2]\,.$$

Proof. We consider the integrable case. We must check that $\lambda_1 E\mathcal{G}[X_1] + \lambda_2 E\mathcal{G}[X_2]$ is $\mathcal{G}$-measurable (which is part of the definition of a conditional expectation with respect to $\mathcal{G}$) and that for all bounded $\mathcal{G}$-measurable random variables U

$$E[(\lambda_1 E\mathcal{G}[X_1] + \lambda_2 E\mathcal{G}[X_1])U] = E[(\lambda_1 Y_1 + \lambda_2 Y_2)U].$$

This follows immediately from the definition of $E\mathcal{G}[X_i]$, which says that $E[E\mathcal{G}[X_i]U] = E[Y_iU]$ $(i = 1, 2)$. □

Theorem 5.6.5 Rule 2. *If Y is independent of $\mathcal{G}$, then*

$$E^{\mathcal{G}}[Y] = E[Y].$$

Proof. We consider the integrable case. First recall that the constant $E[Y]$ is $\mathcal{G}$-measurable. It remains to prove that for all bounded $\mathcal{G}$-measurable random variables U, $E[E[Y]U] = E[YU]$. This is the case since Y and U are independent and therefore $E[YU] = E[Y]E[U]$. □

Theorem 5.6.6 Rule 3. *If Y is $\mathcal{G}$-measurable,*

$$E^{\mathcal{G}}[Y] = Y.$$

Proof. We consider the integrable case. We must check that Y is $\mathcal{G}$-measurable and that $E[YU] = E[YU]$. □

Theorem 5.6.7 Rule 4. *If $Y_1 \leq Y_2$ P-a.s., then*

$$E^{\mathcal{G}}[Y_1] \leq E^{\mathcal{G}}[Y_2] \quad P\text{-a.s.} \tag{5.10}$$

In particular, if Y is a non-negative random variable, then $E^{\mathcal{G}}[Y] \geq 0$, P-a.s.

Proof. We consider the integrable case. The non-negative case follows *mutatis mutandis*. For any bounded $\mathcal{G}$-measurable random variable $U \geq 0$,

$$E[E^{\mathcal{G}}[Y_1]U] = E[Y_1U] \leq E[Y_2U] = E[E^{\mathcal{G}}[Y_2]U].$$

Therefore

$$E[(E^{\mathcal{G}}[Y_2] - E^{\mathcal{G}}[Y_1])U] \geq 0.$$

Taking $U = 1_{\{E^{\mathcal{G}}[Y_2] < E^{\mathcal{G}}[Y_2]\}}$, we obtain (5.10). □

Theorem 5.6.8 Rule 5. (successive conditioning). *Let $\mathcal{H}$ be a sub-σ-field of $\mathcal{F}$ such that $\mathcal{H} \subseteq \mathcal{G}$. Then*

$$E^{\mathcal{H}}[E^{\mathcal{G}}[Y]] = E^{\mathcal{H}}[Y].$$

Proof. We just have to check that $E^{\mathcal{H}}[E^{\mathcal{G}}[Y]]$ is a version of $E^{\mathcal{H}}[Y]$. Since it is an $\mathcal{H}$-measurable variable, it remains to show that it satisfies the equality

$$E[E^{\mathcal{H}}[E^{\mathcal{G}}[Y]]U] = E[YU],$$

for any bounded (resp. bounded non-negative) $\mathcal{H}$-measurable variable U. Since such a variable is *a fortiori* $\mathcal{G}$-measurable,

$$E[[E^{\mathcal{G}}[Y]]U] = E[YU].$$

Moreover,

$$E[E^{\mathcal{H}}[E^{\mathcal{G}}[Y]]U] = E[[E^{\mathcal{G}}[Y]]U],$$

by definition of $E^{\mathcal{H}}[E^{\mathcal{G}}[Y]]$. □

Theorem 5.6.9 *Let Y be of the form $Y = VZ$, where V is a $\mathcal{G}$-measurable bounded (resp. non-negative finite) random variable, and Z is an integrable (resp. non-negative finite) random variable. Then*

$$E^{\mathcal{G}}[VZ] = VE^{\mathcal{G}}[Z].$$

Proof. We consider the integrable case. We observe that $VE^{\mathcal{G}}[Z]$ is $\mathcal{G}$-measurable, and it remains to prove that for all bounded $\mathcal{G}$-measurable random variables U,

$$E[VZU] = E[VE^{\mathcal{G}}[Z]U].$$

But, since VU is bounded, by definition of $E^{\mathcal{G}}[Z]$,

$$E[VE^{\mathcal{G}}[Z]U] = E[VZU].$$

□

The theorems allowing interversion of limit and integral (monotone convergence theorem and dominated convergence theorem) have conditional versions.

We start with the monotone convergence theorem:

Theorem 5.6.10 *Let $\mathcal{G}$ be a sub-σ-field of $\mathcal{F}$, and let $\{Y_n\}_{n\geq 1}$ be a P-a.s. non-decreasing sequence of non-negative random variables converging P-a.s. to the random variable Y. Then $\{E^{\mathcal{G}}[Y_n]\}_{n\geq 1}$ is a P-a.s. non-decreasing sequence of random variables converging P-a.s. to $E^{\mathcal{G}}[Y]$.*

Proof. By monotonicity of conditional expectation, $\{E^{\mathcal{G}}[Y_n]\}_{n\geq 1}$ is a P-a.s. non-decreasing sequence of $\mathcal{G}$-measurable random variables. In particular, there exists a P-a.s. limit W, $\mathcal{G}$-measurable, of this sequence. By monotone convergence, for any bounded non-negative $\mathcal{G}$-measurable random variable U,

$$\lim_{n\uparrow\infty} E[Y_nU] = E[YU],$$

and

$$\lim_{n\uparrow\infty} E[E^{\mathcal{G}}[Y_n]U] = E[WU].$$

Therefore, since $E[Y_nU] = E[E^{\mathcal{G}}[Y_n]U]$ for all $n \geq 1$, $E[YU] = E[WU]$. This being true for all bounded X-measurable random variables U, $W = E[Y \mid \mathcal{G}]$. □

We now turn to the conditioned version of the dominated convergence theorem:

Theorem 5.6.11 *Let $\mathcal{G}$ be a sub-σ-field of $\mathcal{F}$, and let $\{Y_n\}_{n\geq 1}$ be a sequence of random variables converging P-a.s. to the random variable Y, and such that $|Y_n| \leq Z$ for some integrable random variable Z. Then $\{E^{\mathcal{G}}[Y_n]\}_{n\geq 1}$ converges P-a.s. to $E^{\mathcal{G}}[Y]$.*

Proof. Let $W_n := \sup_{m\geq n} |Y_m - Y|$. The sequence $\{W_n\}_{n\geq 1}$ decreases P-a.s. to 0. We have

$$|E^{\mathcal{G}}[Y_n] - E^{\mathcal{G}}[Y]| = |E^{\mathcal{G}}[Y_n - Y]| \\ \leq E^{\mathcal{G}}[|Y_n - Y|] \leq E^{\mathcal{G}}[W_n] .$$

The non-negative sequence $\{E^{\mathcal{G}}[W_n]\}_{n\geq 1}$ decreases P-a.s. (rule 4). Let $H \geq 0$ be its limit. Then

$$0 \leq |E[H]| \leq E\left[E^{\mathcal{G}}[W_n]\right] = E[W_n] ,$$

where the latter quantity tends to 0 by dominated convergence (because $0 \leq W_n \leq 2Z$). Therefore $E[H] = 0$, which implies that $P(H = 0) = 1$ since H is P-a.s non-negative. □

The L^2-theory of Conditional Expectation

This paragraph gives another approach to conditional expectation that avoids the use of Radon–Nikodým's theorem (that was admitted in this book).

Conditional expectation will be first defined for square-integrable random variables in terms of projection from a Hilbert space onto a Hilbert subspace of the latter[2]. More precisely, let $(\Omega, \mathcal{F}, P)$ be a probability space and let $\mathcal{G}$ be a sub-σ-field of $\mathcal{F}$. Denote by $L^2_{\mathbb{R}}(\mathcal{F}, P)$ and $L^2_{\mathbb{R}}(\mathcal{G}, P)$ the Hilbert spaces of $\mathcal{F}$-measurable (resp. $\mathcal{G}$-measurable) square-integrable real random variables. Clearly, $L^2_{\mathbb{R}}(\mathcal{G}, P)$ is a Hilbert subspace of $L^2_{\mathbb{R}}(\mathcal{F}, P)$, and therefore, one can define the projection of an $\mathcal{F}$-measurable square-integrable variable X on $L^2_{\mathbb{R}}(\mathcal{G}, P)$, denoted by $P^{\mathcal{G}}(X)$. From the general theory of projection (see Theorems A.0.12 and A.0.11), this random variable is the unique (in the L^2-sense) variable Y such that

(i) $Y \in L^2_{\mathbb{R}}(\mathcal{G}, P)$, and

(ii) $\langle U, Y\rangle_{L^2_{\mathbb{R}}(\mathcal{F},P)} = \langle U, X\rangle_{L^2_{\mathbb{R}}(\mathcal{F},P)}$ for all $U \in L^2_{\mathbb{R}}(\mathcal{G}, P)$.

In other terms, $P^{\mathcal{G}}(X)$ is the unique (in the L^2-sense) square-integrable random variable Y such that

[2] See Appendix A for a review of Hilbert spaces.

(a) Y is $\mathcal{G}$-measurable, and

(b) $E[UY] = E[UX]$ for all square-integrable $\mathcal{G}$-measurable variables U.

This shows that $P^{\mathcal{G}}(X) = E[X \mid \mathcal{G}]$.

Starting from there, a proof of Theorem 5.6.2 is easy and left as an exercise.

Nonlinear Regression

We have previously obtained in Section 3.3 the best linear least-squares estimator of the square-integrable random variable Y in terms of the second-order random vector $X = (X_1, \dots, X_N)$. We shall now obtain the best non-linear least-square estimator of Y in terms of X, that is to say the square integrable random variable of the form $g(X)$, where $g : \mathbb{R}^N \to \mathbb{R}$ is measurable, which minimizes the quadratic risk $E[|Y - g(X)|^2]$. Whereas the best nonlinear estimator is chosen among all square integrable variables $g(X)$, $g : \mathbb{R}^N \to \mathbb{R}$ measurable, the best linear estimator is chosen among the variables $g(X)$ where $g(x) = a_0 + \sum_{j=1}^N a_j X_j$. In particular if $\hat{g}(X)$ is the best nonlinear estimator, and $\hat{Y}$ is the best linear estimator $E[|Y - \hat{g}(X)|^2] \le E[|Y - \hat{Y}|]^2$. It is therefore theoretically advantageous to use a nonlinear estimator. However, as we have seen, the construction of $\hat{Y}$ only requires the knowledge of the covariance structure of the vector (Y, X), whereas the construction of $\hat{g}(X)$ requires the knowledge of the joint distribution of (Y, X).

As we shall now see, the best nonlinear estimator is

$$\hat{g}(X) = E^X[Y].$$

Since Y is square-integrable, and in particular integrable, the conditional expectation of Y given X is well is square-integrable.

Theorem 5.6.12 *Let X be a random vector and let Y be a square integrable random variable. Then, for all measurable $g : \mathbb{R}^n \to \mathbb{R}$ such that $g(X)$ is square integrable,*

$$E[|Y - E^X[Y]|^2] \le E[|Y - g(X)|^2].$$

Proof. Developing both sides of this inequality, we have to show that,

$$E[E^X[Y]^2] - 2E[YE^X[Y]] \le E[g(X)^2] - 2E[Yg(X)].$$

Since $E^X[Y]$ is a square-integrable function of X,

$$E[E^X[Y]^2] = E[E^X[Y]E^X[Y]] = E[YE^X[Y]].$$

The left-hand side of the last inequality therefore equals $-E[E^X[Y]^2]$. But $E[Yg(X)] = E[E^X[Y]g(X)]$ therefore have to show that

$$-E[E^X[Y]^2] \leq E[g(X)^2] - 2E[E^X[Y]g(X)]\,.$$

But this is just

$$E[(g(X) - E^X[Y])^2] \geq 0\,.$$

□

If (Y, X) is jointly Gaussian, then the best linear estimator and the best non-linear estimator, of Y given X, coincide.

Theorem 5.6.13 *Let Y be a random variable and let X be an n-dimensional random vector. Suppose that (Y, X) is jointly Gaussian, and that X is non-degenerate (its covariance matrix is strictly positive). Then*

$$E^X[Y] = m_Y + \Gamma_{YX}\Gamma_X^{-1}(X - m_X)\,.$$

Proof. Consider the random variable

$$U = Y - (m_Y + \Gamma_{YX}\Gamma_X^{-1}(X - m_X))\,.$$

We have $E[U] = 0$ and

$$\begin{aligned} &E[U(X - m_X)^T] \\ &= E[(Y - m_Y)(X - m_X)^T]\Gamma_{YX}\Gamma_X^{-1}E[(X - m_X)(X - m_X)^T] \\ &= \Gamma_{YX} + \Gamma_{YX}\Gamma_X^{-1}\Gamma_X = \Gamma_{YX} - \Gamma_{YX} = 0\,. \end{aligned}$$

Therefore U and X are uncorrelated. Since (U, X) is jointly Gaussian, this implies that U and X are independent. In particular (Theorem 5.5.13),

$$E^X[U] = E[U] = 0\,.$$

Also, by linearity,

$$E^X[U] = E^X[Y] - E^X[m_Y + \Gamma_{YX}\Gamma_X^{-1}(X - m_X)]\,.$$

By (Theorem 5.5.10),

$$E^X[m_Y + \Gamma_{YX}\Gamma_X^{-1}(X - m_X)] = m_Y + \Gamma_{YX}\Gamma_X^{-1}(X - m_X)\,.$$

□

5.7 Exercises

Exercise 5.7.1. $P(f(X)=0)=0$
Let X be a random vector of $\mathbb{R}^d$ admitting the probability density function f. Show that $P(f(X)=0)=0$.

Exercise 5.7.2. EXTENSION OF THE TELESCOPE FORMULA
Let X be a non-negative random variable and let $G:\mathbb{R}_+\to\mathbb{C}$ be the primitive function of $g:\mathbb{R}_+\to\mathbb{C}$, that is, for all $x\geq 0$,

$$G(x)=G(0)+\int_0^x g(u)\,\mathrm{d}u\,.$$

Let X be a non-negative random variable with finite mean μ and such that $E\left[G(X)\right]<\infty$. Show that

$$E\left[G(X)\right]=G(0)+\int_0^\infty g(x)P(X\geq x)\,\mathrm{d}x\,.$$

Exercise 5.7.3. A FORMULA FOR MOMENTS
Let X be a non-negative random variable with the probability density function f. Let $r>0$ be such that $E\left[|X|^r\right]<\infty$. Prove that

$$E\left[X^r\right]=\int_0^\infty rx^{r-1}P(X>x)\,\mathrm{d}x\,.$$

Exercise 5.7.4. INFINITE SUMS AND EXPECTATIONS

In the first chapters, we have sometimes surreptitiously taken for granted that the expectation of an infinite sum of random variables is equal to the sum of their expectations. The result (to be proved) that justifies this, when it is true, is the following:
(a) Let $\{S_n\}_{n\geq 1}$ be a sequence of non-negative random variables. Then:

$$E\left[\sum_{n=1}^\infty S_n\right]=\sum_{n=1}^\infty E[S_n]\,.$$

(b) Let $\{S_n\}_{n\geq 1}$ be a sequence of real random variables such that $\sum_{n\geq 1}E[|S_n|]<\infty$. Then:

$$E\left[\sum_{n=1}^\infty S_n\right]=\sum_{n=1}^\infty E[S_n]\,.$$

Exercise 5.7.5. LAPLACE TRANSFORM
Let X be a non-negative random variable. Prove that

$$\lim_{0<\theta\uparrow\infty} E\left[e^{-\theta X}\right] = P(X=0).$$

Exercise 5.7.6. LADDER RANDOM VARIABLES
A real random variable X is called a *ladder random variable* if there exist a and h in $\mathbb{R}$ such that

$$\sum_{n\in\mathbb{Z}} P(X = a + nh) = 1.$$

Let φ be the characteristic function of a real random variable X. Prove that if $|\varphi(t_0)| = 1$ for some $t_0 \in \mathbb{R}$, $t_0 \neq 1$, then X is a ladder random variable.

Exercise 5.7.7. CHARACTERISTIC FUNCTIONS AND INDEPENDENCE
Prove Theorem 5.4.6.

Exercise 5.7.8. RADON–NIKODÝM
Let $\{P_n\}_{n\geq 1}$ be a sequence of probability measures on $(\Omega, \mathcal{F})$. Show the existence of a probability measure P on $(\Omega, \mathcal{F})$ and of a sequence $\{f_n\}_{n\geq 1}$ of P-integrable non-negative measurable functions such that for all $A \in \mathcal{F}$ and all $n \geq 1$,

$$P_n(A) = \int_A f_n(\omega)\, P(\mathrm{d}\omega).$$

Exercise 5.7.9. CONDITIONAL INDEPENDENCE
Let A be some event of positive probability, and let P_A denote the probability P conditioned by A, that is,

$$P_A(\cdot) = P(\cdot \mid A).$$

The random variables X and Y are said to be conditionally independent given A if they are independent with respect to probability P_A. Prove that this is the case if and only if for all $u, v \in \mathbb{R}$,

$$P(A)E[e^{iuX}e^{ivY}1_A] = E[e^{iuX}1_A]E[e^{ivY}1_A].$$

Exercise 5.7.10. $E[X] - E[Y]$
Let X and Y be real integrable random variables. Prove the following:

$$E[X] - E[Y] = \int_{\mathbb{R}} \left(P(X < t \leq Y) - P(Y < t \leq X)\right) \mathrm{d}t.$$

Exercise 5.7.11. MOMENTS OF GAUSSIAN VECTORS
A. Give the proof of Theorem 5.3.5.

B. Let $X = (X_1, \dots, X_n)^T$ be a centered (0-mean) n-dimensional Gaussian vector with the covariance matrix $\Gamma = \{\sigma_{ij}\}$. Show that

$$E[X_{i_1}X_{i_2}, \dots, X_{i_{2k}}] = \sum_{\substack{(j_1,\dots,j_{2k}) \\ j_1<j_2,\dots,j_{2k-1}<j_{2k}}} \sigma_{j_1j_2}\sigma_{j_3j_4}\dots\sigma_{j_{2k}j_{2k}}, \tag{5.11}$$

where the summation extends over all permutations $(j_1, \dots, j_{2k})$ of $\{i_1, \dots, i_{2k}\}$ such that $j_1 < j_2, \dots, j_{2k-1} < j_{2k}$. There are $1 \cdot 3 \cdot 5 \dots (2k-1)$ terms in the right-hand side of Eq. (5.11). The indices $i_1, \dots, i_{2k}$ are in $\{1, \dots, n\}$ and they may occur with repetitions. Show that the odd moments of X are null, that is:

$$E[X_{i_1} \dots X_{i_{2k+1}}] = 0,$$

for all $(i_1, \dots, i_{2k+1}) \in \{1, 2, \dots, n\}^{2k+1}$.

Exercise 5.7.12. CONDITIONING BY THE SQUARE.
Let X be a real random variable with probability density f_X. Let $h : \mathbb{R} \to \mathbb{R}$ be a function such that $h(X)$ is integrable. We prove that

$$E[h(X)|X^2] = h(\sqrt{X^2})\frac{f_X(\sqrt{X^2})}{f_X(\sqrt{X^2})+f_X(-\sqrt{X^2})} + h(-\sqrt{X^2})\frac{f_X(-\sqrt{X^2})}{f_X(\sqrt{X^2})+f_X(-\sqrt{X^2})}.$$

(Some people may find this result intuitive. Others will need a formal proof, which is given below.)

Exercise 5.7.13. MIXED CASE: A SPECIFIC EXAMPLE
Let Y be an $\mathbb{N}_+$-valued random variable, and let X be a real random variable of the form

$$X = Y + \xi,$$

where ξ is a random variable admitting a probability density f_ξ and independent of Y. Let $h : \mathbb{R} \to \mathbb{R}$ be a measurable function such that $E[|h(X)|] < \infty$. Give the function ψ such that $\psi(Y) = E^Y[h(X)]$.

Exercise 5.7.14. CONDITIONING BY THE SUM
Let X_1 and X_2 be two integrable independent identically distributed random variables. Show that

$$E^{X_1+X_2}[X_1] = \frac{X_1 + X_2}{2}.$$

Exercise 5.7.15. MIN CONDITIONED BY MAX

Let X_1 and X_2 be two independent random variables uniformly distributed on the interval $[0,1]$. Compute $E^{\max(X_1,X_2)}[\min(X_1,X_2]$.

Exercise 5.7.16. EXPONENTIAL DISTRIBUTIONS
Let $\{S_n\}_{n\geq 1}$ be a sequence of IID non-negative real random variables with exponential distribution of parameter λ. Define $T_1 = S_1$, $T_2 = T_1 + S_2$, ..., $T_{n+1} = T_n + S_{n+1}$, etc. Let $f : \mathbb{R}_+ \to \mathbb{R}_+$ be a non-negative function. Give for $m \geq 2$ the conditional distribution of $(T_1, \dots, T_m)$ given $T_m = s$. Show that it is the same as the distribution of the vector obtained by reordering $(sU_1, \dots, sU_m)$, where $(U_1, \dots, U_m)$ is a vector of m independent variables uniformly distributed on $[0,1]$). Compute $E\left[\mathrm{e}^{-\sum_{i=1}^{\infty} f(T_i)}\right]$.

Exercise 5.7.17. WILL THE SUN RISE NEXT DAY?
At the beginning of time, God chose (a probabilist once claimed) a number p at random in the interval $[0,1]$ and devised a biased coin with probability p for heads. Since then, He tosses the same coin once every morning and decides to let the sun rise this day if and only if the result is heads. The common belief, which will be taken to be true in this exercise, is that the sun has never failed to rise in the n days separating us from the beginning of time. What then is the probability that the sun will rise the next day ($n+1$-th)?

Exercise 5.7.18.
Let X and Y be two real random variables, and let $h : \mathbb{R} \to \mathbb{R}$ be one-to-one and onto. Show that for all $v : \mathbb{R} \to \mathbb{R}$ such that $E[|v(X)|] < \infty$, $E^Y[v(X)] = E^Z[v(X)]$, where $Z = h(Y)$.

Exercise 5.7.19. THE CONDITIONAL VARIANCE FORMULA
Prove the following formula

$$\operatorname{Var}(X) = E\left[\operatorname{Var}(X|Y)\right] + \operatorname{Var}\left(E\left[X|Y\right]\right),$$

where X is a square-integrable random variable, and

$$\operatorname{Var}(X|Y) := E\left[\left(X - E\left[X|Y\right]\right)^2 |Y\right]$$

is the so-called *conditional variance* of X given Y.

Exercise 5.7.20. CONDITIONAL JENSEN'S INEQUALITY
Let I be a general interval of $\mathbb{R}$ (closed, open, semi-closed, infinite, etc.) and let (a,b) be its interior, assumed non-empty. Let $\varphi : I \to \mathbb{R}$ be a convex function. Let X be an integrable real-valued random variable such that $P(X \in I) = 1$. Assume

moreover that either φ is non-negative, or that $\varphi(X)$ is integrable. Prove that for any sub-σ-field $\mathcal{G} \subseteq \mathcal{F}$

$$E[\varphi(X) \mid \mathcal{G}] \geq \varphi(E[X \mid \mathcal{G}]).$$

5.8 Solutions

SOLUTION (Exercise 5.7.1).

$$P(f(X) = 0) = E\left[1_{\{f(X)=0\}}\right] = \int_{\mathbb{R}^d} 1_{\{f(x)=0\}} f(x)\, dx = \int_{\mathbb{R}^d} 0\, dx = 0.$$

SOLUTION (Exercise 5.7.2).

$$\begin{aligned} E[G(X)] &= G(0) + E\left[\int_0^X g(u)\, du\right] = E\left[\int_0^\infty g(u) 1_{\{u \leq X\}}\, du\right] \\ &= G(0) + \int_0^\infty g(u) E\left[1_{\{u \leq X\}}\right] du = G(0) + \int_0^\infty g(u) P(X \geq u)\, du, \end{aligned}$$

where the third equality is due to Tonelli's theorem applied to the product measure $P \times \ell$.

SOLUTION (Exercise 5.7.3).
A direct consequence of Exercise 5.7.2 with $G(x) = x^r$.

SOLUTION (Exercise 5.7.4).
(a) Apply the monotone convergence theorem (Theorem 5.1.2) with $X_n = \sum_{k=1}^n S_k$ and $X = \sum_{n=1}^\infty S_n$.

(b) Apply the dominated convergence theorem (Theorem 5.1.3 with $X_n = \sum_{k=1}^n S_k$, $X = \sum_{n=1}^\infty S_n$ and $Z = \sum_{k=1}^\infty |S_k|$. (By (a), $E[Z] = \sum_{k=1}^\infty E[|S_k|] < \infty$.)

SOLUTION (Exercise 5.7.5).

$$E\left[e^{-\theta X}\right] = E\left[1_{\{X>0\}} e^{-\theta X}\right] + E\left[1_{\{X=0\}}\right]$$

But since $\lim_{0<\theta\uparrow\infty} 1_{\{X>0\}}e^{-\theta X} = 0$, $\lim_{0<\theta\uparrow\infty} E\left[1_{\{X>0\}}e^{-\theta X}\right] = 0$ by dominated convergence ($1_{\{X>0\}}e^{-\theta X} \leq 1$, an integrable random variable). Therefore the limit in question is $E\left[1_{\{X=0\}}\right] = P(f(X) = 0)$

SOLUTION (Exercise 5.7.6).
The hypothesis implies that there exists an $a \in \mathbb{R}$ such that $e^{iat_0} = E\left[e^{it_0X}\right]$. In particular (considering the real parts),

$$1 - E\left[\cos(t_0(X-a))\right] = E\left[1 - \cos(t_0(X-a))\right] = 0\,.$$

Since $1 - \cos(t_0(X-a)) \geq 0$, this implies that, P-a.s., $1 = \cos(t_0(X-a))$, which in turn implies the announced result.

SOLUTION (Exercise 5.7.7).
Necessity. Write

$$\begin{aligned}\varphi_X(u) &= E\left[e^{i\sum_{j=1}^{d} u_jX_j}\right] \\ &= E\left[\prod_{j=1}^{d} e^{iu_jX_j}\right] = \prod_{j=1}^{d} E\left[e^{iu_jX_j}\right] = \prod_{j=1}^{d} \varphi_{X_j}(u_j)\,,\end{aligned}$$

by the product formula for expectations.

Sufficiency. Let $X' := (X'_1, \dots, X'_d) \in \mathbb{R}^d$ be a random vector whose *independent* coordinate random variables $X'_1, \dots, X'_d$ have the respective characteristic functions $\varphi_1, \dots, \varphi_d$. The characteristic function of X' is $\prod_{j=1}^{d} \varphi_j(u_j)$ and therefore X and X' have the same distribution. In particular, $X_1, \dots, X_d$ are independent random variables with respective characteristic functions $\varphi_1, \dots, \varphi_d$.

SOLUTION (Exercise 5.7.8).
Take $P := \sum_{n\geq 1} 2^{-n}P_n$, check that it is a probability measure and that $P_n \ll P$ ($n \geq 1$). Then apply the Radon-Nikodým theorem.

SOLUTION (Exercise 5.7.9).
By Theorem 3.2.20, a necessary and sufficient condition for this is that for all $u, v \in \mathbb{R}$,

$$E_A[e^{iuX}e^{ivY}] = E_A[e^{iuX}]E_A[e^{ivY}],$$

where E_A denotes expectation with respect to P_A. Then, observe that for an integrable or non-negative random variable Z,

$$P(A)E_A[Z] = E[Z1_A].$$

SOLUTION (Exercise 5.7.10).
Observe that, by Fubini,

$$\int_{\mathbb{R}} P(X < t \leq Y)\mathrm{d}t = E\left[\int_{\mathbb{R}} 1_{X<t\leq Y}\mathrm{d}t\right]$$

and

$$\int_{\mathbb{R}} P(Y < t \leq X)\mathrm{d}t = E\left[\int_{\mathbb{R}} 1_{Y<t\leq X}\mathrm{d}t\right], .$$

SOLUTION (Exercise 5.7.11).
A. Apply Theorem 4.3.7.
B. Apply A with φ the characteristic function of this Gaussian vector.

SOLUTION (Exercise 5.7.12).
The right-hand side is a function of X^2. It remains to show that for all bounded measurable functions φ,

$$\begin{aligned}&E\left[h(X)\varphi(X^2)\right]\\ &= E\left[\left(h(\sqrt{X^2})\tfrac{f_X(\sqrt{X^2})}{f_X(\sqrt{X^2})+f_X(-\sqrt{X^2})} + h(-\sqrt{X^2})\tfrac{f_X(-\sqrt{X^2})}{f_X(\sqrt{X^2})+f_X(-\sqrt{X^2})}\right)\varphi(X^2)\right].\end{aligned}$$

It suffices to show that

$$E\left[h(X)1_{\{X>0\}}\varphi(X^2)\right] = E\left[\left(h(\sqrt{X^2})\tfrac{f_X(\sqrt{X^2})}{f_X(\sqrt{X^2})+f_X(-\sqrt{X^2})}\right)\varphi(X^2)\right]$$

and

$$E\left[h(X)1_{\{X<0\}}\varphi(X^2)\right] = E\left[\left(h(-\sqrt{X^2})\tfrac{f_X(-\sqrt{X^2})}{f_X(\sqrt{X^2})+f_X(-\sqrt{X^2})}\right)\varphi(X^2)\right].$$

We prove the first of these two equalities. Its right-hand side equals

$$\int_{-\infty}^{+\infty}\left(h(\sqrt{x^2})\frac{f_X(\sqrt{x^2})}{f_X(\sqrt{x^2})+f_X(-\sqrt{x^2})}\right)\varphi(x^2)f_X(x)\mathrm{d}x$$

or (splitting the domain of integration)

$$\begin{aligned}&\int_{0}^{+\infty}\left(h(\sqrt{x^2})\frac{f_X(\sqrt{x^2})}{f_X(\sqrt{x^2})+f_X(-\sqrt{x^2})}\right)\varphi(x^2)f_X(x)\mathrm{d}x\\ &\qquad+\int_{-\infty}^{0}\left(h(\sqrt{x^2})\frac{f_X(\sqrt{x^2})}{f_X(\sqrt{x^2})+f_X(-\sqrt{x^2})}\right)\varphi(x^2)f_X(x)\,\mathrm{d}x,\end{aligned}$$

that is, by the change of variable $x \mapsto -x$ in the second term,

$$\int_0^{+\infty} \left(h(\sqrt{x^2}) \frac{f_X(\sqrt{x^2}) + f_X(-\sqrt{x^2})}{f_X(\sqrt{x^2}) + f_X(-\sqrt{x^2})} \right) \varphi(x^2) f_X(x) \, \mathrm{d}x$$
$$= \int_0^{+\infty} h(\sqrt{x^2}) \varphi(x^2) f_X(x) \, \mathrm{d}x = E\left[h(X) 1_{X>0} \varphi(X^2) \right] .$$

SOLUTION (Exercise 5.7.13).
$E^Y[h(X)] = \psi(Y)$, where

$$\psi(k) = \int_{\mathbb{R}} h(x) f_k(x) dx$$

and where f_k is defined by

$$\int_A f_k(x) \, dx := P(X \in A | Y = k) \quad (A \in \mathcal{B}(\mathbb{R})) ,$$

that is,

$$\begin{aligned} \int_A f_k(x) dx &= \frac{P(X \in A, Y = k)}{P(Y = k)} \\ &= \frac{P(k + \xi \in A, Y = k)}{P(Y = k)} = \frac{P(k + \xi \in A) P(Y = k)}{P(Y = k)} \\ &= P(k + \xi \in A) = P(\xi \in A - k) = \int_A f_\xi(x + k) \, dx . \end{aligned}$$

Therefore

$$f_k(x) = f_\xi(x + k)$$

and

$$\psi(k) = \int_{\mathbb{R}} h(x) f_\xi(x + k) \, dx = \int_{\mathbb{R}} h(x + k) f_\xi(x) \, dx ,$$

that is,

$$\psi(k) = E[h(\xi + k)] .$$

Finally:

$$E^Y[h(X)] = \int_{\mathbb{R}} E[h(x + Y)] f_\xi(x) \, dx .$$

SOLUTION (Exercise 5.7.14).
$E^{X_1+X_2}[X_1]$ is of the form $h(X_1+X_2)$. By symmetry, $E^{X_1+X_2}[X_2] = h(X_1+X_2)$. But

$$E^{X_1+X_2}[X_1] + E^{X_1+X_2}[X_2] = E^{X_1+X_2}[X_1+X_2] = X_1+X_2 .$$

Therefore $2h(X_1+X_2) = X_1+X_2$.

SOLUTION (Exercise 5.7.15).
We must find a measurable function $h : [0,1]$ such that for all bounded measurable functions $\varphi : [0,1]$

$$E\left[\min(X_1,X_2)\varphi(\max(X_1,X_2))\right] = E\left[h(\max(X_1,X_2))\varphi(\max(X_1,X_2))\right] ,$$

in which case $E^{\max(X_1,X_2)}[\min(X_1,X_2] = h(\max(X_1,X_2))$. Now

$$\begin{aligned} &E\left[\min(X_1,X_2)\varphi(\max(X_1,X_2))\right] \\ &= \int_0^1\int_0^1 1_{\{x_1<x_2\}}\, x_1\, \varphi(x_2)\, dx_1\, dx_2 \\ &\qquad + \int_0^1\int_0^1 1_{\{x_2\le x_1\}}\, x_2\, \varphi(x_1)\, dx_1\, dx_2 \\ &= 2\int_0^1\left(\int_0^{x_2} x_1\, dx_1\right)\varphi(x_2)\, dx_2 = \int_0^1 x^2\varphi(x)\, dx . \end{aligned}$$

On the other hand, for $x \in [0,1]$,

$$P(\max(X_1,X_2)\le x) = P(X_1\le x,\, X_2\le x) = P(X_1\le x)P(X_2\le x) = x^2$$

and therefore, $\max(X_1,X_2)$ admits the probability density function $2x\, 1_{[0,1]}(x)$ and

$$E\left[h(\max(X_1,X_2))\varphi(\max(X_1,X_2))\right] = \int_0^1 h(x)\varphi(x)2x\, dx .$$

so that $2xh(x) = x^2$ and finally $E^{\max(X_1,X_2)}[\min(X_1,X_2] = \frac{1}{2}\max(X_1,X_2)$.

SOLUTION (Exercise 5.7.16).
Let $S := (S_1,\dots,S_m)$ and $T := (T_1,\dots,T_m)$. The probability density function of S is

$$f_S(s_1,\dots,s_m) = \lambda^n\left(e^{-\lambda\sum_{i=1}^n s_i}\right)1_{\{s_1>0,\dots,s_m>0\}}\} .$$

Since $S_1 = T_1$, $\quad S_2 = T_2 - T_1, \dots, \quad S_n = T_n - T_{n-1}$, the formula of smooth change of variables gives

$$f_T(t_1,\dots,t_m) = f_S(t_1, t_2-t_1,\dots,t_m-t_{m-1}) = \lambda^m e^{-\lambda t_m}1_C(t_1,\dots,t_m),$$

where $C := \{(t_1, \dots, t_n); 0 < t_1 < \cdots < t_m\}$. The probability density function of T_m is obtained by integrating out $t_1, \dots, t_{m-1}$ in $f_T(t_1, \dots, t_m)$, which gives $f_{T_m}(s) = (\lambda^m/(m-1)!)e^{-\lambda s}$. Therefore

$$\begin{aligned} f_T(t_1, \dots, t_m | T_{m+1} = s) &= \frac{f_{(T_1,\dots,T_{m+1})}(t_1, \dots, t_m, s)}{f_{T_m}(s)} \\ &= \frac{\lambda^{m+1} e^{-\lambda s}}{(\lambda^{m+1} s^m/m!) e^{-\lambda s}} 1_C(t_1, \dots, t_m) 1_{\{t_m < s\}} = \frac{m!}{s^m} 1_{t_1 < t_2 < \cdots < t_m < s} \end{aligned}$$

This is indeed the probability density function of the vector obtained by reordering $(sU_1, \dots, sU_m)$, where $(U_1, \dots, U_m)$ is a vector of m independent variables uniformly distributed on $[0, 1]$. In particular

$$\begin{aligned} E\left[e^{-\sum_{i=1}^m f(T_i)}\right] &= E\left[E^{T_{m+1}}\left[e^{-\sum_{i=1}^m f(T_i)}\right]\right] \\ &= \int_0^\infty E^{T_{m+1}=s}\left[e^{-\sum_{i=1}^m f(T_i)}\right] f_{T_{m+1}}(s)\, ds \\ &= \int_0^\infty E\left[e^{-\sum_{i=1}^m f(sU_i)}\right]) f_{T_{m+1}}(s)\, ds \\ &= \int_0^\infty \left(E\left[e^{-f(sU_1)}\right]\right)^m f_{T_{m+1}}(s)\, ds \\ &= E\left[\left(\int_0^{T_{m+1}} e^{-f(x)} \frac{dx}{T_{m+1}}\right)^m\right] \\ &= E\left[\left(1 - \frac{\int_0^{T_{m+1}} (1 - e^{-f(x)}) dx}{T_{m+1}}\right)^{T_{m+1} \frac{m}{T_{m+1}}}\right] \end{aligned}$$

By the law of large numbers, $\lim_{m \uparrow \infty} \frac{m}{T_{m+1}} = \lambda$. Therefore, passing to the limit as $m \uparrow \infty$ (dominated convergence),

$$E\left[e^{-\sum_{i=1}^\infty f(T_i)}\right] = e^{-\lambda \int_0^\infty \left(1 - e^{-f(x)}\right) dx}.$$

Solution (Exercise 5.7.17).
Letting Z be the (random) bias of the coin, and $X_n = 1$ if the sun rises on day n, we have to compute

$$\begin{aligned} &P(X_{n+1} = 1 \,|\, X_1 = 1, \dots, X_n = 1) \\ &\quad = P(X_1 = 1, \dots, X_n = 1, X_{n+1} = 1)/P(X_1 = 1, \dots, X_n = 1). \end{aligned}$$

But for all $k \geq 1$,

$$P(X_1 = 1, \ldots, X_k = 1) = \int_0^1 P(X_1 = 1, \ldots, X_k = 1 \,|\, Z = p)\, dp$$
$$= \int_0^1 p^k \, dp = \frac{1}{k+1},$$

and therefore

$$P(X_{n+1} = 1 \,|\, X_1 = 1, \ldots, X_n = 1) = \frac{n+1}{n+2}.$$

Solution (Exercise 5.7.18).
Both $E^Y[v(X)]$ and $E^Z[v(X)]$ are functions of Y (since a function of Z is a function of Y!). To prove equality, it suffices therefore to show that for all bounded φ,

$$E\left[E^Y[v(X)]\varphi(Y)\right] = E\left[E^Z[v(X)]\varphi(Y)\right].$$

Now

$$E\left[E^Y[v(X)]\varphi(Y)\right] = E\left[v(X)\varphi(Y)\right]$$

for all bounded φ. On the other hand

$$E\left[E^Z[v(X)]\psi(Z)\right] = E\left[v(X)\psi(Z)\right]$$

for all bounded ψ, and in particular

$$E\left[E^Z[v(X)]\psi(Z)\right] = E\left[v(X)\varphi(Y\right]$$

for all bounded φ (h is bijective).

Solution (Exercise 5.7.19).
Similarly to the unconditioned case

$$\text{Var}\,(X|Y) = E\left[X^2|Y\right] - E\left[X|Y\right]^2,$$

and therefore

$$E\left[\text{Var}\,(X|Y)\right] = E\left[X^2\right] - E\left[E\left[X|Y\right]^2\right].$$

On the other hand

$$\text{Var}\,(E[X|Y]) = E\left[E\left[X|Y\right]^2\right] - E\left[E\left[X|Y\right]\right]^2$$
$$= E\left[E\left[X|Y\right]^2\right] - E\left[X\right]^2.$$

Summing the last two equalities and using $\mathrm{Var}\,(X) = E\left[X^2\right] - E\left[X\right]^2$, we obtain the announced equality.

SOLUTION (Exercise 5.7.20).
Just imitate the proof of Theorem 2.1.25.

Chapter 6

Convergence Almost Sure

Order hidden in chaos: an erratic sequence of coin tosses exhibits a remarkable balance between heads and tails in the long run, at least "when the coin is fair and fairly tossed". This phenomenon is captured by the *strong law of large numbers*. The relevant mathematical notion, which is the object of this chapter, is that of *almost-sure convergence* of a sequence of random variables.

6.1 A Sufficient Condition and a Criterion

Consider a game of *heads or tails* with independent tosses of a single, possibly biased, coin. In other words, we have an IID sequence $\{X_n\}_{n\geq 1}$ of random variables taking two values, 1 (heads) and 0 (tails), with

$$P(X_n = 1) = p \in (0, 1) .$$

Let

$$S_n := X_1 + X_2 + \cdots + X_n .$$

The random variable S_n/n is the *empirical frequency* of heads after n tosses. We are interested in the limit of this quantity as $n \uparrow \infty$. As we know "from experience", the empirical frequency tends to p. In fact, this is a theorem: *Borel's strong law of large numbers*, which asserts that

$$P\left(\exists \lim_{n\uparrow\infty} \frac{S_n}{n} = p\right) = 1 . \tag{6.1}$$

More explicitly: the probability that there exists a limit of the sequence $\{\frac{S_n}{n}\}_{n\geq 1}$ and that this limit is p is equal to 1. We shall in general avoid in similar statements the use of the symbol $\exists$ and write (6.1) in the form $P\left(\lim_{n\uparrow\infty} \frac{S_n}{n} = p\right) = 1$, or $\lim_{n\uparrow\infty} \frac{S_n}{n} = p$, $P - a.s.$

P. Brémaud, *An Introduction to Applied Probability*, Texts in Applied Mathematics 77,
https://doi.org/10.1007/978-3-031-49306-5_6

Definition 6.1.1 *A sequence* $\{Z_n\}_{n\geq 1}$ *of random variables with values in* $\mathbb{C}$ *(resp. in* $\overline{\mathbb{R}}$*) is said to* converge P-almost surely *(P-a.s.) to the random variable* Z *with values in* $\mathbb{C}$ *(resp. in* $\overline{\mathbb{R}}$*) if*

$$P(\lim_{n\uparrow\infty} Z_n = Z) = 1\,. \tag{6.2}$$

This is also denoted by

$$Z_n \overset{\text{a.s.}}{\to} Z\,.$$

Paraphrasing: For all ω outside a negligible set, $\lim_{n\uparrow\infty} Z_n(\omega) = Z(\omega)$.

In the case where the sequence takes values in $\overline{\mathbb{R}}$, the limit may be infinite. Otherwise, when $P(Z < \infty) = 1$, one may add the precision: "converges to a *finite* limit".

The Borel–Cantelli Lemma

This is one of the fundamental tools in the study of almost sure convergence.

Consider a sequence of events $\{A_n\}_{n\geq 1}$. We are interested in the probability that A_n occurs infinitely often, that is, the probability of the event

$$\{A_n\, i.o.\} := \{\omega; \omega \in A_n \text{ for an infinity of indices } n\},$$

where *i.o.* abbreviates *infinitely often*. We have (Borel–Cantelli lemma):

Theorem 6.1.2 *For any sequence of events* $\{A_n\}_{n\geq 1}$,

$$\sum_{n=1}^{\infty} P(A_n) < \infty \implies P(A_n\, i.o.) = 0\,.$$

Proof. We first observe that

$$\{A_n \text{ i.o.}\} = \bigcap_{n=1}^{\infty} \bigcup_{k\geq n} A_k\,.$$

(Indeed, if ω belongs to the set on the right-hand side, then for *all* $n \geq 1$, ω belongs to at least one among $A_n, A_{n+1}, \ldots$, which implies that ω is in A_n for an infinite number of indices n. Conversely, if ω is in A_n for an infinite number of indices n, it is for *all* $n \geq 1$ in at least one of the sets $A_n, A_{n+1}, \ldots$.)

The set $\cup_{k\geq n}A_k$ decreases as n increases, so that by the sequential continuity property of probability,

$$P(A_n \text{ i.o.}) = \lim_{n\uparrow\infty} P\left(\bigcup_{k\geq n} A_k\right). \tag{6.3}$$

But by the sub-σ-additivity property of probability,

$$P\left(\bigcup_{k\geq n} A_k\right) \leq \sum_{k\geq n} P(A_k),$$

and by the summability assumption, the right-hand side of this inequality vanishes as $n \uparrow \infty$. □

The next result is usually called the *converse Borel–Cantelli lemma*. It is in fact a "pseudo-converse" since an additional assumption of independence is required.

Theorem 6.1.3 *Let $\{A_n\}_{n\geq 1}$ be a sequence of* independent *events. Then,*

$$\sum_{n=1}^{\infty} P(A_n) = \infty \implies P(A_n \ i.o.) = 1.$$

Proof. We may without loss of generality assume that $P(A_n) > 0$ for all $n \geq 1$. The divergence hypothesis implies, by the fundamental theorem of convergence of infinite products,[1] that for all $n \geq 1$,

$$\prod_{k=n}^{\infty}(1 - P(A_k)) = 0.$$

This infinite product equals, in view of the independence assumption,

$$\prod_{k=n}^{\infty} P\left(\overline{A_k}\right) = P\left(\bigcap_{k=n}^{\infty} \overline{A_k}\right) = 1 - P\left(\bigcup_{k=n}^{\infty} A_k\right).$$

Therefore,

$$P\left(\bigcup_{k=n}^{\infty} A_k\right) = 1$$

[1] Let $\{a_n\}_{n\geq 1}$ be a sequence of numbers in the interval $[0, 1)$. Then: (a) if $\sum_{n=1}^{\infty} a_n < \infty$, then $\lim_{n\uparrow\infty} \prod_{k=1}^{n}(1 - a_k) > 0$, and (b) if $\sum_{n=1}^{\infty} a_n = \infty$, then $\lim_{n\uparrow\infty} \prod_{k=1}^{n}(1 - a_k) = 0$.

and by (6.3),

$$P(A_n\, i.o.) = \lim_{n\uparrow\infty} P\left(\bigcup_{k=n}^{\infty} A_k\right) = 1\,.$$

□

EXAMPLE 6.1.4: BINARY SEQUENCE. Let $\{X_n\}_{n\geq 1}$ be a sequence of random variables with values in $\{0,1\}$, with $P(X_n = 1) = p_n$ $(n \geq 1)$.

If $\sum_n p_n < \infty$, then, by the direct Borel–Cantelli lemma, $P(X_n = 1 \text{ i.o.}) = 0$, and therefore $P(\lim_{n\uparrow\infty} X_n = 0) = 1$.

If $\sum_n p_n = \infty$, and if moreover the sequence is independent, then, by the converse Borel–Cantelli lemma, $P(X_n = 1 \text{ i.o.}) = 1$, and therefore the sequence cannot converge to 0. Therefore, in the independent case, a necessary and sufficient condition for convergence to 0 is $\sum_n p_n < \infty$.

A Sufficient Condition

The following *sufficient* condition guaranteeing almost-sure convergence is the most useful. It is a direct consequence of the Borel–Cantelli lemma.

Theorem 6.1.5 *Let $\{Z_n\}_{n\geq 1}$ and Z be complex random variables. If*

$$\sum_{n\geq 1} P(|Z_n - Z| \geq \varepsilon_n) < \infty \tag{6.4}$$

for some sequence of positive numbers $\{\varepsilon_n\}_{n\geq 1}$ converging to 0, then the sequence $\{Z_n\}_{n\geq 1}$ converges P-a.s. to Z.

Proof. Obviously, if $\{\varepsilon_n\}_{n\geq 1}$ is a sequence of positive real numbers converging to 0, then any sequence of non-negative real numbers $\{x_n\}_{n\geq 1}$ such that $x_n \geq \varepsilon_n$ for only a finite number of indices $n \geq 1$ also converges to 0. Therefore it suffices to prove that

$$P(|Z_n - Z| \geq \varepsilon_n \quad i.o.) = 0\,.$$

But this follows from hypothesis (6.4) and the Borel–Cantelli lemma. □

A Criterion

The result below is essentially of theoretical interest. It will be used later on for comparing convergence in probability and almost-sure convergence.

Theorem 6.1.6 *The sequence $\{Z_n\}_{n\geq 1}$ of complex random variables converges P-a.s. to the complex random variable Z if and only if for all $\epsilon > 0$,*

$$P(|Z_n - Z| \geq \epsilon \ i.o.) = 0 . \tag{6.5}$$

Proof. For the necessity, observe that

$$\{|Z_n - Z| \geq \epsilon \ i.o.\} \subseteq \overline{\{\omega; \lim_{n\uparrow\infty} Z_n(\omega) = Z(\omega)\}} ,$$

and therefore

$$P(|Z_n - Z| \geq \epsilon \ i.o.) \leq 1 - P(\lim_{n\uparrow\infty} Z_n = Z) = 0 .$$

For the sufficiency, let N_k be the last index n such that $|Z_n - Z| \geq \frac{1}{k}$ (letting $N_k := \infty$ if $|Z_n - Z| \geq \frac{1}{k}$ for an infinity of indices $n \geq 1$). By (6.5) with $\epsilon = \frac{1}{k}$, we have $P(N_k = \infty) = 0$. By sub-σ-additivity, $P(\cup_{k\geq 1}\{N_k = \infty\}) = 0$. Equivalently, $P(N_k < \infty, \quad \text{for all } k \geq 1) = 1$, which implies $P(\lim_{n\uparrow\infty} Z_n = Z) = 1$. □

6.2 The Strong Law of Large Numbers

In order to prove Borel's strong law of large numbers using Theorem 6.1.5, we must have some adequate upper bound for the general term of the series occurring in the left-hand side of (6.4). The basic tool for this is Markov's inequality.

The proof of (6.1) indeed relies on the Borel–Cantelli lemma and the Markov inequality. In fact, we shall apply Theorem 6.1.5, and for this we need to bound the probability that $\left|\frac{S_n}{n} - p\right|$ exceeds some $\varepsilon > 0$ where $p := E[X_1]$, which can be done by application of Markov's inequality as follows:

$$\begin{aligned} P\left(\left|\frac{S_n}{n} - p\right| \geq \varepsilon\right) &= P\left(\left(\frac{S_n}{n} - p\right)^4 \geq \varepsilon^4\right) \\ &\leq \frac{E\left[\left(\frac{S_n}{n} - p\right)^4\right]}{\varepsilon^4} \leq \frac{E\left[\left(\sum_{i=1}^n Y_i\right)^4\right]}{n^4\varepsilon^4} , \end{aligned}$$

where $Y_i := X_i - p$. In view of the independence hypothesis,

$$E[Y_1Y_2Y_3Y_4] = E[Y_1]E[Y_2]E[Y_3]E[Y_4] = 0 ,$$

$$E[Y_1Y_2^3] = E[Y_1]E[Y_2^3] = 0 ,$$

and the like. Finally, in the development

$$E\left[\left(\sum_{i=1}^{n} Y_i\right)^4\right] = \sum_{i,j,k,\ell=1}^{n} E[Y_iY_jY_kY_\ell],$$

only the terms of the form $E[Y_i^4]$ and $E[Y_i^2Y_j^2]$ $(i \neq j)$ remain. There are n terms of the first type and $3n(n-1)$ terms of the second type. Therefore, only $nE[Y_1^4] + 3n(n-1)E[Y_1^2Y_2^2]$ remains, which is less than Kn^2 for some finite K. Therefore

$$P\left(\left|\frac{S_n}{n} - p\right| \geq \varepsilon\right) \leq \frac{K}{n^2\varepsilon^4},$$

and in particular, with $\varepsilon = n^{-\frac{1}{8}}$,

$$P\left(\left|\frac{S_n}{n} - p\right| \geq n^{-\frac{1}{8}}\right) \leq \frac{K}{n^{\frac{3}{2}}},$$

from which it follows that

$$\sum_{n=1}^{\infty} P\left(\left|\frac{S_n}{n} - p\right| \geq n^{-\frac{1}{8}}\right) < \infty.$$

Therefore, by Theorem 6.1.5, $\left|\frac{S_n}{n} - p\right|$ converges almost surely to 0.

EXAMPLE 6.2.1: PATTERNS IN A BERNOULLI SEQUENCE. Let k be a positive integer. Let $\{n_i\}_{1\leq i\leq k}$ be a strictly increasing finite sequence of positive integers with $n_1 = 1$. Let $\{\varepsilon_i\}_{1\leq i\leq k}$ be a sequence of 0's and 1's. The sequence of pairs $\{(n_i, \varepsilon_i)\}_{1\leq i\leq k}$ is called a *k-pattern*. Patterns are represented by sequences of 0's, 1's and $\cdot$'s, where $\cdot$ is an "unspecified binary digit". For instance, the 4-pattern

$$(1,0), (3,1), (4,1), (6,0)$$

is represented by $0 \cdot 11 \cdot 0$, and this pattern is said *to occur at position n* in a sequence $x_1, x_2, \ldots$ of binary digits if and only if

$$x_n = 0, x_{n+2} = 1, x_{n+3} = 1, x_{n+5} = 0.$$

Let now $\{X_n\}_{n\geq 1}$ be an IID sequence of 0's and 1's such that $P(X_1 = 1) = p \in (0,1)$. Define for all $n \geq 1$ the random variable Y_n with values 0 or 1 by

$$Y_n := 1 \text{ iff } \quad X_{n+n_i} := \varepsilon_i \text{ for all } i \quad (1 \leq i \leq k),$$

that is, iff the pattern occurs at position n. Then (exercise):

$$\frac{Y_1 + \ldots + Y_n}{n} \stackrel{\text{a.s.}}{\to} p^h q^{k-h} \qquad \text{where } h := \sum_{i=1}^{k} \varepsilon_i \,. \qquad (\star).$$

In particular, the empirical frequency of any k-pattern in a fair game of heads or tails equals $\frac{1}{2^k}$. Since the Bernoulli sequence with $p = \frac{1}{2}$ (the random sequence "par excellence") satisfies $(\star)$ for all possible patterns, one is tempted to call a *deterministic* sequence $(x_n,\, n \geq 1)$ of 0's and 1's "random" if for all patterns

$$\lim_{n \uparrow \infty} \frac{y_1 + \cdots + y_n}{n} = \frac{1}{2^k}\,,$$

where the y_n's are defined in the same way as the Y_n's above. Although this definition seems reasonable, it is not satisfying. In fact, one can show that the (rather deterministic!) *Champernowne sequence*:

$$011011100101110111 1000 \ldots\,,$$

which consists of the succession of integers (starting with 0) written in base 2, is random in this sense.

Kolmogorov's Strong Law of Large Numbers

Borel's proof is easily adapted to the case where the X_n's are uniformly bounded. In 1933, Kolmogorov gave the following more general form of the strong law of large numbers that requires only that the X_n's be integrable.

Theorem 6.2.2 *Let $\{X_n\}_{n \geq 1}$ be an* IID *sequence of random variables such that*

$$E[|X_1|] < \infty\,. \tag{6.6}$$

Then,

$$P\left(\lim_{n \uparrow \infty} \frac{S_n}{n} = E[X_1]\right) = 1\,. \tag{6.7}$$

Proof. We may suppose, without loss of generality, that $E\left[X_1\right] = 0$. The proof is in two parts. In Part A the strong law is proved with the additional assumption that $\sigma^2 := E\left[X_1^2\right] < \infty$, and then Part B gets rid of this assumption.

A. Let

$$Z_n := \sup_{1 \leq k \leq 2m+1} \left(|X_{m^2+1} + \cdots + X_{m^2+k}|\right).$$

Defining for all $n \geq 1$ the integer $m(n)$ by

$$m(n)^2 < n \leq (m(n)+1)^2\,,$$

we have that

$$|\frac{S_n}{n}| \leq |\frac{S^2_{m(n)}}{m(n)^2}| + \frac{Z_{m(n)}}{m(n)^2}\,.$$

Since $\lim_{n\uparrow\infty} m(n) = +\infty$, it suffices to prove that

$$\lim_{n\uparrow\infty} |\frac{S^2_{m(n)}}{m(n)^2}| = 0 \tag{$\star$}$$

and

$$\lim_{n\uparrow\infty} \frac{Z_{m(n)}}{m(n)^2} = 0\,. \tag{$\star\star$}$$

For all $\varepsilon > 0$, by Chebyshev's inequality,

$$P\left(|\frac{S^2_m}{m^2}| \leq \varepsilon\right) \leq \frac{\text{Var}\,(S_{m^2})}{m^4\varepsilon^2} = \frac{m^2\sigma^2}{m^4\varepsilon^2} = \frac{\sigma^2}{m^2\varepsilon^2}\,.$$

Therefore $\sum_{m\geq 1} P\left(|\frac{S^2_m}{m^2}| \geq \varepsilon\right) < \infty$, which implies ($\star$) (by the Borel–Cantelli lemma and Theorem 6.1.6).

Let now

$$\xi_k := X_{m^2+1} + \cdots + X_{m^2+k}\,.$$

If $|Z_m| \geq m^2\varepsilon$, then for at least one k $(1 \leq k \leq 2m+1)$, $|\xi_k| \geq m^2\varepsilon$. In other words,

$$\left\{\frac{Z_m}{m^2} \geq \varepsilon\right\} \subseteq \bigcup_{k=1}^{2m+1} \left\{|\xi_k| \geq m^2\varepsilon\right\}\,,$$

so that

$$P\left(\frac{Z_m}{m^2} \geq \varepsilon\right) \leq P\left(\bigcup_{k=1}^{2m+1} \left\{|\xi_k| \geq m^2\varepsilon\right\}\right) \leq \sum_{k=1}^{2m+1} P\left(|\xi_k| \geq m^2\varepsilon\right)\,,$$

and therefore, by Chebyshev's inequality,

$$P\left(\frac{Z_m}{m^2} \geq \varepsilon\right) \leq \sum_{k=1}^{2m+1} \frac{\text{Var}\,(\xi_k)}{m^4\varepsilon^2}\,.$$

Now, when $k \le 2m+1$,

$$\text{Var}\,(\xi_k) = \sum_{i=1}^{k} \text{Var}\,(X_{m^2+i}) \le (2m+1)\sigma^2\,,$$

and therefore

$$P\left(\frac{Z_m}{m^2} \ge \varepsilon\right) \le \frac{(2m+1)^2\sigma^2}{m^4\varepsilon^2}\,,$$

so that

$$\sum_{m\ge 1} P\left(\frac{Z_m}{m^2} \ge \varepsilon\right) < \infty\,,$$

and then ($\star\star$) follows from the Borel–Cantelli lemma and the criterion of almost-sure convergence.

B. It remains to get rid of the assumption of finiteness of the second moment. The natural technique for this is truncation.

Let

$$\widetilde{X}_n := \begin{cases} X_n & \text{if } |X_n| \le n, \\ 0 & \text{otherwise.} \end{cases}$$

We proceed in three steps.

Step 1. We first show that

$$\lim_{n\uparrow\infty} \frac{1}{n} \sum_{k=1}^{n} (\widetilde{X}_k - E[\widetilde{X}_k]) = 0\,.$$

In view of Part A, it suffices to prove that

$$\sum_{n=1}^{\infty} \frac{E[(\widetilde{X}_n - E[\widetilde{X}_n])^2]}{n^2} < \infty\,.$$

But

$$E[(\widetilde{X}_n - E[\widetilde{X}_n])^2] \le E[\widetilde{X}_n^2] = E[X_1^2 1_{\{|X_1|\le n\}}]\,.$$

It is therefore enough to show that

$$\sum_{n=1}^{\infty} \frac{E[X_1^2 1_{\{|X_1|\le n\}}]}{n^2} < \infty\,.$$

The left-hand side of the above inequality is equal to

$$\sum_{n=1}^{\infty} \frac{1}{n^2} \sum_{k=1}^{n} E[X_1^2 1_{\{k-1<|X_1|\le k\}}] = \sum_{k=1}^{\infty} \sum_{n=k}^{\infty} \frac{1}{n^2} E[X_1^2 1_{\{k-1<|X_1|\le k\}}]\,.$$

Using the fact that

$$\sum_{n=k}^{\infty} \frac{1}{n^2} \leq \frac{1}{k^2} + \int_k^{\infty} \frac{1}{x^2}\mathrm{d}x = \frac{1}{k^2} + \frac{1}{k} \leq \frac{2}{k}$$

(draw the graph of $x \mapsto x^{-2}$), this quantity is less than or equal to

$$\begin{aligned}\sum_{k=1}^{\infty} \frac{2}{k} E[X_1^2 1_{\{k-1<|X_1|\leq k\}}] &= 2\sum_{k=1}^{\infty} E\left[\frac{X_1^2}{k} 1_{\{k-1<|X_1|\leq k\}}\right] \\ &\leq 2\sum_{k=1}^{\infty} E[|X_1| 1_{\{k-1<|X_1|\leq k\}}] = 2E[|X_1|] < \infty .\end{aligned}$$

Step 2. Since $E[|X_1|] < \infty$, we have by dominated convergence that

$$\lim_{n\uparrow\infty} E[X_1 1_{\{|X_1|\leq n\}}] = E[X_1] = 0 .$$

Since X_n has the same distribution as X_1,

$$\lim_{n\uparrow\infty} E[\widetilde{X}_n] = \lim_{n\uparrow\infty} E[X_n 1_{\{|X_n|\leq n\}}] = \lim_{n\uparrow\infty} E[X_1 1_{\{|X_1|\leq n\}}] = E[X_1] = 0 .$$

In particular, by Cesàro's lemma,[2]

$$\lim_{n\uparrow\infty} \frac{1}{n} \sum_{k=1}^{n} E[\widetilde{X}_k] = 0 .$$

Step 3. We have

$$\sum_{n=1}^{\infty} P(|X_n| > n) = \sum_{n=1}^{\infty} P(|X_1| > n) \leq E[|X_1|] < \infty ,$$

and therefore, by the Borel–Cantelli lemma,

$$P(\widetilde{X}_n \neq X_n \text{ i.o.}) = P(X_n > n \text{ i.o.}) = 0 ,$$

which implies that

$$\lim_{n\uparrow\infty} \frac{\widetilde{S}_n}{n} = \lim_{n\uparrow\infty} \frac{S_n}{n} .$$

[2] Let $\{b_n\}_{n\geq 0}$ be a sequence of real numbers such that $\lim_{n\uparrow\infty} b_n = 0$. Then $\lim_{n\uparrow\infty} \frac{b_1+\cdots+b_n}{n} = 0$.

□

The next result shows that the integrability condition is, in a sense, also necessary. See however Exercise 6.6.4.

Theorem 6.2.3 *Let $\{X_n\}_{n\geq 1}$ be a sequence of* IID *random variables such that*

$$\frac{S_n}{n} \to C < \infty \quad P-a.s.,$$

where $S_n := X_1 + \cdots X_n$. Then $E[|X_1|] < \infty$ and $C = E[X_1]$.

Proof. Under these circumstances,

$$\frac{X_n}{n} = \frac{S_n}{n} - \frac{n-1}{n}\frac{S_{n-1}}{n-1} \to 0$$

and therefore, $P(|X_n| > n\, i.o.) = 0$. By the converse Borel–Cantelli lemma,

$$\sum_{n=1}^{\infty} P(|X_n| > n) < \infty$$

or, since the distribution of any X_n does not depend on n,

$$\sum_{n=1}^{\infty} P(|X_1| > n) < \infty .$$

But, by the following inequalities concerning any non-negative random variable X (Exercise 5.7.2) and $|X_1|$ in particular,

$$\sum_{n=1}^{\infty} P(X \geq n) \leq E[X] \leq 1 + \sum_{n=1}^{\infty} P(X \geq n),$$

we have that $E[|X_1|] < \infty$. The identification of C and $E[|X_1|]$ is then just the strong law of large numbers. □

Large Deviations from the Strong Law of Large Numbers

The *large deviations theory* of random variables produces estimates for the deviation of such variables from their means. When applied to sums of IID variables $S_n = X_1 + \cdots X_n$, these estimates complement the strong law of large numbers.

The type of result produced by this theory is, in the case where the X_i's are IID and integrable, with common mean m,

$$\lim_{n\uparrow\infty} \frac{1}{n} \log P\left(\left|\frac{S_n}{n} - m\right| \geq a\right) = -h(a), \tag{$\star$}$$

where $a > 0$ and $h(a) > 0$. Such bounds have important theoretical implications, but they are somewhat imprecise in that the meaning of ($\star$) is

$$P\left(\left|\frac{S_n}{n} - m\right| \geq a\right) = g(n)\mathrm{e}^{-n(h(a))},$$

where $n^{-1}\log g(n)$ tends to 0 as $n \uparrow \infty$, but perhaps too slowly and in an uncontrolled manner.

To obtain practical (upper) bounds, it is often useful to look at specific cases, using the Chernoff bound below at the origin of the general abstract theory. These powerful bounds are easy consequences of the elementary Markov inequality.

Theorem 6.2.4 *Let X be a real-valued random variable and let $a \in \mathbb{R}$. Then* (Chernoff's bound)

$$P(X \geq a) \leq \min_{t>0} \frac{E\left[\mathrm{e}^{tX}\right]}{\mathrm{e}^{ta}}, \tag{6.8}$$

and

$$P(X \leq a) \leq \min_{t<0} \frac{E\left[\mathrm{e}^{tX}\right]}{\mathrm{e}^{ta}}. \tag{6.9}$$

Proof. By the $\uparrow$-monotony of $x \mapsto \mathrm{e}^x$ and Markov's inequality,

$$P(X \geq a) = P(\mathrm{e}^{tX} \geq \mathrm{e}^{ta}) \leq \frac{E\left[\mathrm{e}^{tX}\right]}{\mathrm{e}^{ta}} \quad (t > 0),$$

and

$$P(X \leq a) = P(\mathrm{e}^{tX} \geq \mathrm{e}^{ta}) \leq \frac{E\left[\mathrm{e}^{tX}\right]}{\mathrm{e}^{ta}} \quad (t < 0).$$

The result follows by minimizing the right-hand sides with respect to $t > 0$ and $t < 0$, respectively. □

EXAMPLE 6.2.5: LARGE DEVIATIONS FOR THE POISSON VARIABLE. Let X be a Poisson variable with mean θ and therefore $E\left[\mathrm{e}^{tX}\right] = \mathrm{e}^{\theta(\mathrm{e}^t - 1)}$. We prove that for $c \geq 0$

$$P(X \geq \theta + c) \leq \exp\left\{-\frac{1}{\mathrm{e}}\left(\frac{\theta + c}{\mathrm{e}\theta}\right)^{\theta+c}\right\}.$$

With $a = \theta + c$ in (6.8):

$$P(X \geq \theta + c) \leq \min_{t>0} \frac{\mathrm{e}^{\theta(\mathrm{e}^t-1)}}{\mathrm{e}^{t(\theta+c)}} = \mathrm{e}^{-\max_{t>0}\{t(\theta+c)-\theta(\mathrm{e}^t-1)\}}.$$

The derivative of the function $f : t \mapsto t(\theta+c)-\theta(\mathrm{e}^t-1)$ at $t \geq 0$ is $\theta+c-\theta\mathrm{e}^t$, and it is null for $\mathrm{e}^t = \frac{\theta+c}{\theta}$ or equivalently $t = \ln(\theta+c)-\ln(\theta)$, and this corresponds to a maximum since the second derivative $-\mathrm{e}^t$ is negative. Therefore

$$\max_{t>0}\left\{t(\theta+c)-\theta(\mathrm{e}^t-1)\right\} = \frac{1}{\mathrm{e}}\left(\frac{\theta+c}{\mathrm{e}\theta}\right)^{\theta+c}$$

and finally

$$P(X \geq \theta+c) \leq \exp\left\{-\frac{1}{\mathrm{e}}\left(\frac{\theta+c}{\mathrm{e}\theta}\right)^{\theta+c}\right\}.$$

Theorem 6.2.6 *Let $X_1, \ldots, X_n$ be* IID *real-valued random variables and let $a \in \mathbb{R}$. Then,*

$$P\left(\sum_{i=1}^{n} X_i \geq na\right) \leq \mathrm{e}^{-nh^+(a)},$$

where

$$h^+(a) = \sup_{t\geq 0}\{at - \ln E\left[\mathrm{e}^{tX_1}\right]\}. \tag{6.10}$$

Proof. For all $t \geq 0$, Markov's inequality gives

$$\begin{aligned}
P\left(\sum_{i=1}^{n} X_i \geq na\right) &= P\left(\exp\left\{t\sum_{i=1}^{n} X_i\right\} \geq \exp\{nta\}\right) \\
&\leq E\left[\exp\left\{t\sum_{i=1}^{n} X_i\right\}\right] \times \mathrm{e}^{-nta} \\
&\leq \exp\{-n\left(at - \ln E\left[\mathrm{e}^{tX_1}\right]\right)\},
\end{aligned}$$

from which the result follows by optimizing this bound over $t \geq 0$. $\square$

Suppose that $E\left[\mathrm{e}^{tX_1}\right] < \infty$ for all $t \geq 0$. Differentiating $t \mapsto at - \ln E\left[\mathrm{e}^{tX_1}\right]$ yields $a - \frac{E\left[X_1\mathrm{e}^{tX_1}\right]}{E\left[\mathrm{e}^{tX_1}\right]}$, and therefore the function $t \mapsto at - \ln E\left[\mathrm{e}^{tX_1}\right]$ is finite and differentiable on $\mathbb{R}$, with derivative at 0_+ equal to $a - E[X_1]$, which implies that when $a > E[X_1]$, $h^+(a)$ is positive.

Similarly to (6.10), we obtain that

$$P\left(\sum_{i=1}^{n} X_i \leq na\right) \leq \mathrm{e}^{-nh^-(a)},$$

where

$$h^-(a) := \sup_{t\le 0}\{at - \ln E\left[e^{tX_1}\right]\}.$$

Moreover, if $a < E[X_1]$, $h^-(a)$ is positive.

The Chernoff bound can be interpreted in terms of *large deviations* from the law of large numbers. Denote by μ the common mean of the X_n's, and define for $\varepsilon > 0$ the (positive) quantities

$$H^+(\varepsilon) = \sup_{t\ge 0}\left\{\varepsilon t - \ln E\left[e^{t(X_1-\mu)}\right]\right\},$$

$$H^-(\varepsilon) = \sup_{t\le 0}\left\{\varepsilon t - \ln E\left[e^{t(X_1-\mu)}\right]\right\}.$$

Then

$$P\left(\left|\frac{1}{n}\sum_{i=1}^{n} X_i\right| \ge +\varepsilon\right) \le e^{-nH^+(\varepsilon)} + e^{-nH^-(\varepsilon)}.$$

The computation of the supremum in (6.10) may be fastidious. There are short-cuts leading to practical bounds that are not as good but nevertheless satisfactory for certain applications.

EXAMPLE 6.2.7: LARGE DEVIATIONS FOR THE RANDOM WALK. Suppose for instance that $\{X_n\}_{n\ge 1}$ is IID, the X_n's taking the values -1 and $+1$ equiprobably so that $E\left[e^{tX}\right] = \frac{1}{2}e^{+t} + \frac{1}{2}e^{-t}$. Replacing $\frac{1}{2}e^{+t} + \frac{1}{2}e^{-t}$ by the upper bound $e^{\frac{t^2}{2}}$, we have that, for $a > 0$,

$$P\left(\sum_{i=1}^{n} X_i \ge na\right) \le e^{-n\left(at - \ln E\left[e^{tX_1}\right]\right)}$$
$$\le e^{-n\left(at - \frac{1}{2}t^2\right)},$$

and therefore, with $t = a$,

$$P\left(\sum_{i=1}^{n} X_i \ge na\right) \le e^{-n\frac{1}{2}a^2}.$$

By symmetry of the distribution of $\sum_{i=1}^{n} X_i$, we obtain for $a > 0$

$$P\left(\sum_{i=1}^{n} X_i \le -na\right) = P\left(\sum_{i=1}^{n} X_i \ge na\right) \le e^{-n\frac{1}{2}a^2},$$

and therefore, combining the two bounds,

$$P\left(\left|\sum_{i=1}^{n} X_i\right| \geq na\right) \leq 2e^{-n\frac{1}{2}a^2}.$$

6.3 Kolmogorov's Zero-one Law

(The result of this section will be used only in the chapter on martingales.)

Definition 6.3.1 *Let $\{X_n\}_{n\geq 1}$ be a sequence of random variables and let $\mathcal{F}_n^X := \sigma(X_1, \dots, X_n)$. The σ-field $\mathcal{T}^X := \cap_{n\geq 1}\sigma(X_n, X_{n+1}, \dots)$ is called the* tail σ-field *of this sequence.*

EXAMPLE 6.3.2: For any $a \in \mathbb{R}$, the event $\{\lim_{n\uparrow\infty} \frac{X_1+\cdots+X_n}{n} \leq a\}$ belongs to the tail σ-field, since the existence and the value of the limit of $\frac{X_1+\cdots+X_n}{n}$ does not depend on any fixed finite number of terms of the sequence. More generally, any event concerning $\lim_{n\uparrow\infty} \frac{X_1+\cdots+X_n}{n}$ such as, for instance, the event that such limit exists, is in the tail σ-field.

Recall the notation $\mathcal{F}_\infty^X := \vee_{n\geq 1}\mathcal{F}_n^X$.

Theorem 6.3.3 *The tail σ-field of a sequence $\{X_n\}_{n\geq 1}$ of independent random variables is trivial, that is, if $A \in \mathcal{T}^X$, then $P(A) = 0$ or 1.*

Proof. The σ-fields $\mathcal{F}_n^X$ and $\sigma(X_{n+k}, X_{n+k+1}, \dots)$ are independent for all $k \geq 1$ and therefore, since $\mathcal{T}^X = \cap_{k\geq 1}\sigma(X_{n+k}, X_{n+k+1})$, the σ-fields $\mathcal{F}_n^X$ and $\mathcal{T}^X$ are independent. Therefore the *algebra* $\cup_{n\geq 1}\mathcal{F}_n^X$ and $\mathcal{T}^X$ are independent, and consequently (Theorem 5.4.2) $\mathcal{F}_\infty^X$ and $\mathcal{T}^X$ are independent. But $\mathcal{F}_\infty^X \supseteq \mathcal{T}^X$, so that $\mathcal{T}^X$ is independent of itself. In particular, for all $A \in \mathcal{T}^X$, $P(A \cap A) = PA)P(A)$, that is $P(A) = P(A)^2$, which implies that $P(A) = 0$ or 1. □

6.4 Related Types of Convergence

Convergence in Probability

This type of convergence is closely related to almost-sure convergence, yet weaker, as we shall see.

Definition 6.4.1 *A sequence* $\{Z_n\}_{n\geq 1}$ *of complex random variables is said to* converge in probability *to the complex random variable* Z *if, for all* $\varepsilon > 0$,

$$\lim_{n\uparrow\infty} P(|Z_n - Z| \geq \varepsilon) = 0\,. \tag{6.11}$$

Theorem 6.4.2 A. *If the sequence* $\{Z_n\}_{n\geq 1}$ *of complex random variables converges almost surely to some complex random variable* Z, *it also converges in probability to the same random variable* Z.

B. *If the sequence of complex random variables* $\{X_n\}_{n\geq 1}$ *converges in probability to the complex random variable* X, *one can find a sequence of integers* $\{n_k\}_{k\geq 1}$, *strictly increasing, such that* $\{X_{n_k}\}_{k\geq 1}$ *converges almost surely to* X.

B says, in other words: From a sequence converging in probability to some random variable, one can extract a subsequence converging almost surely to the same random variable.

Proof. A. Suppose almost-sure convergence. By Theorem 6.1.6 , for all $\varepsilon > 0$,

$$P(|Z_n - Z| \geq \varepsilon \ i.o.) = 0\,,$$

that is

$$P(\cap_{n\geq 1} \cup_{k=n}^{\infty} (|Z_k - Z| \geq \varepsilon)) = 0\,,$$

or (sequential continuity of probability)

$$\lim_{n\uparrow\infty} P(\cup_{k=n}^{\infty} (|Z_k - Z| \geq \varepsilon)) = 0\,,$$

which in turn implies that

$$\lim_{n\uparrow\infty} P(|Z_n - Z| \geq \varepsilon) = 0\,.$$

B. By definition of convergence in probability, for all $\varepsilon > 0$,

$$\lim_{n\uparrow\infty} P\left(|X_n - X| \geq \varepsilon\right) = 0\,.$$

Therefore one can find n_1 such that

$$P\left(|X_{n_1} - X| \geq \frac{1}{1}\right) \leq \left(\frac{1}{2}\right)^1$$

Then, one can find $n_2 > n_1$ such that

$$P\left(|X_{n_2} - X| \geq \frac{1}{2}\right) \leq \left(\frac{1}{2}\right)^2$$

and so on, until we have a strictly increasing sequence of integers n_k ($k \geq 1$) such that

$$P\left(|X_{n_k} - X| \geq \frac{1}{k}\right) \leq \left(\frac{1}{2}\right)^k .$$

It then follows from Theorem 6.1.5 that

$$\lim_{k\uparrow\infty} X_{n_k} = X \quad \text{a.s.}$$

□

Exercise 6.6.6 gives an example of a sequence converging in probability, but not almost surely. Thus, convergence in probability is in general a notion strictly weaker than almost-sure convergence. However, Exercise 6.6.7 gives an important example where both convergences occur simultaneously.

There exists a distance between random variables that metrizes convergence in probability, namely

$$d(X,Y) := \mathrm{E}\left[|X - Y| \wedge 1\right] .$$

(The verification that d is indeed a metric is left as an exercise.) This means the following:

Theorem 6.4.3 *The sequence $\{X_n\}_{n\geq 1}$ converges in probability to the variable X if and only if*

$$\lim_{n\uparrow\infty} d(X_n, X) = 0 .$$

Proof. If: By Markov's inequality, for $\varepsilon \in (0, 1]$,

$$P(|X_n - X| \geq \varepsilon) = P(|X_n - X| \wedge 1 \geq \varepsilon) \leq \frac{d(X_n, X)}{\varepsilon} .$$

Only if: For all $\varepsilon > 0$,

$$\begin{aligned} d(X_n, X) &= \int_{\{|X_n - X| \geq \varepsilon\}} (|X_n - X| \wedge 1)\, \mathrm{d}P + \int_{\{|X_n - X| < \varepsilon\}} (|X_n - X| \wedge 1)\, \mathrm{d}P \\ &\leq P(|X_n - X| \geq \varepsilon) + \varepsilon . \end{aligned}$$

If the sequence converges in probability, there exists an n_0 such that for $n \geq n_0$, $P(|X_n - X| \geq \varepsilon) \leq \varepsilon$ and therefore $d(X_n, X) \leq 2\varepsilon$. Since $\varepsilon > 0$ is arbitrary, we have shown that $\lim_{n\uparrow\infty} d(X_n, X) = 0$. □

Convergence in the Quadratic Mean

This type of convergence concerns sequences of square-integrable random variables.

Definition 6.4.4 *A sequence $\{Z_n\}_{n\geq 1}$ of square-integrable complex random variables is said to converge in the quadratic mean to the square-integrable complex random variable Z if, for all $\varepsilon > 0$,*

$$\lim_{n\uparrow\infty} E[|Z_n - Z|^2] = 0 . \tag{6.12}$$

The next result follows from the fact that $L^2_{\mathbb{C}}(P)$, the collection of square-integrable complex-valued random variables, is a Hilbert space when endowed with the inner product

$$\langle X, Y\rangle := E\left[XY^*\right] .$$

(This is a particular case of Theorem 4.4.19.) In particular,

Theorem 6.4.5 *For the sequence $\{Z_n\}_{n\geq 1}$ of square-integrable complex random variables to converge in the quadratic mean to some square-integrable complex random variable Z, it is necessary and sufficient that*

$$\lim_{n,m\uparrow\infty} E[|Z_n - Z_m|^2] = 0 . \tag{6.13}$$

We now give the property of *continuity of the inner product.*

Theorem 6.4.6 *Let $\{X_n\}_{n\geq 1}$ $\{Y_n\}_{n\geq 1}$ be two sequences of square-integrable complex random variables that converge in the quadratic mean to the square-integrable complex random variables X and Y respectively. Then,*

$$\lim_{n,m\uparrow\infty} E[X_n Y_m^*] = E[XY^*] . \tag{6.14}$$

Proof. We have

$$\begin{aligned}
&|E[X_n Y_m^*] - E[XY^*]| \\
&\quad = |E[(X_n - X)(Y_m - Y)^*] + E[(X_n - X)Y^*] + E[X(Y_m - Y)^*]| \\
&\quad \leq |E[(X_n - X)(Y_m - Y)^*]| + |E[(X_n - X)Y^*]| + |E[X(Y_m - Y)^*]|
\end{aligned}$$

and the right-hand side of this inequality is, by Schwarz's inequality, less than or equal to

$$\begin{aligned}
&\left(E[|X_n - X|^2]\right)^{\frac{1}{2}} \left(E[|Y_m - Y|^2]\right)^{\frac{1}{2}} \\
&\quad + \left(E[|X_n - X|^2]\right)^{\frac{1}{2}} \left(E[|Y|^2]\right)^{\frac{1}{2}} \\
&\qquad + \left(E[|X|^2]\right)^{\frac{1}{2}} \left(E[|Y_m - Y|^2]\right)^{\frac{1}{2}} ,
\end{aligned}$$

which tends to 0 as $n, m \uparrow \infty$. □

Theorem 6.4.7 *If the sequence $\{Z_n\}_{n\geq 1}$ of square-integrable complex random variables converges in the quadratic mean to the complex random variable Z, it also converges in probability to the same random variable.*

Proof. It suffices to observe that, by Markov's inequality, for all $\varepsilon > 0$,

$$P(|Z_n - Z| \geq \varepsilon) \leq \frac{1}{\varepsilon^2} E[|Z_n - Z|^2].$$

□

EXAMPLE 6.4.8: CONVERGENCE IN QUADRATIC MEAN OF SERIES. Let $\{A_n\}_{n\in\mathbb{Z}}$ and $\{B_n\}_{n\in\mathbb{Z}}$ be two sequences of centered square-integrable complex random variables such that

$$\sum_{j\in\mathbb{Z}} E[|A_j|^2] < \infty, \sum_{j\in\mathbb{Z}} E[|B_j|^2] < \infty.$$

Suppose, moreover, that

$$E\left[A_i A_j^*\right] = E\left[B_i B_j^*\right] = E\left[A_i B_j^*\right] = 0 \quad (i \neq j).$$

Let

$$U_n := \sum_{j=-n}^{n} A_j, \; V_n := \sum_{j=-n}^{n} B_j.$$

Then $\{U_n\}_{n\geq 1}$ (resp., $\{V_n\}_{n\geq 1}$) converges in the quadratic mean to some square-integrable random variable U (resp., V) and

$$E[U] = E[V] = 0 \text{ and } E[UV^*] = \sum_{j\in\mathbb{Z}} E[A_j B_j^*].$$

Proof. We have

$$E[|U_n - U_m|^2] = E\left[\left|\sum_{j=n+1}^{m} A_j\right|^2\right] = \sum_{j=n+1}^{m}\sum_{i=n+1}^{m} E[A_j A_i^*] = \sum_{j=n+1}^{m} E[|A_j|^2]$$

since $E[A_j A_i^*] = 0$ when $i \neq j$. The conclusion then follows from the Cauchy criterion for convergence in the quadratic mean, since

$$\lim_{m,n\uparrow\infty} E[|U_n - U_m|^2] = \lim_{m,n\uparrow\infty} \sum_{j=n+1}^{m} E[|A_j|^2] = 0,$$

in view of hypothesis $\sum_{j\in\mathbb{Z}} E[|A_j|^2] < \infty$. By continuity of the inner product in $L^2_{\mathbb{C}}(P)$,

$$E[UV^*] = \lim_{n\uparrow\infty} E[U_n V_n^*] = \lim_{n\uparrow\infty} \sum_{j=1}^{n}\sum_{\ell=1}^{n} E[A_j B_\ell^*]$$

$$= \lim_{n\uparrow\infty} \sum_{j=1}^{n} E[A_j B_j^*] = \sum_{j\in\mathbb{Z}} E[A_j B_j^*]\,.$$

□

6.5 Uniform Integrability

The monotone and dominated convergence theorems are not all the tools that we have at our disposition giving conditions under which it is possible to exchange limits and expectations. Uniform integrability, which will be introduced now, is another such sufficient condition.

Definition 6.5.1 *A collection $\{X_i\}_{i\in I}$ (where I is an arbitrary index) of integrable random variables is called uniformly integrable if*

$$\lim_{c\uparrow\infty} \int_{\{|X_i|>c\}} |X_i| \,\mathrm{d}P = 0 \text{ uniformly in } i \in I\,.$$

EXAMPLE 6.5.2: COLLECTION DOMINATED BY AN INTEGRABLE VARIABLE. If, for some integrable random variable, $P(|X_i| \le X) = 1$ for all $i \in I$, then $\{X_i\}_{i\in I}$ is uniformly integrable. Indeed, in this case,

$$\int_{\{|X_i|>c\}} |X_i| \,\mathrm{d}P \le \int_{\{X>c\}} X \,\mathrm{d}P$$

and by monotone convergence the right-hand side of the above inequality tends to 0 as $c \uparrow \infty$.

Clearly, if one adds a finite number of integrable variables to a uniformly integrable collection, the augmented collection will also be uniformly integrable.

Theorem 6.5.3 *The collection $\{X_i\}_{i\in I}$ of integrable random variables is uniformly integrable if and only if*

(a) $\sup_i E[|X_i|] < \infty$, *and*

(b) *for every $\varepsilon > 0$, there exists a $\delta(\varepsilon) > 0$ such that*

$$\sup_n \int_A |X_i| \, \mathrm{d}P \le \varepsilon \text{ whenever } P(A) \le \delta(\varepsilon) .$$

(In other words, $\int_A |X_i| \, \mathrm{d}P \to 0$ uniformly in i as $P(A) \to 0$.)

Proof. Assume uniform integrability. For any $\varepsilon > 0$, there exists a c such that $\int_{\{X_i > c\}} |X_i| \, \mathrm{d}P \le \varepsilon$ for all $i \in I$. For all $A \in \mathcal{F}$, all $i \in I$,

$$\int_A |X_i| \, \mathrm{d}P \le cP(A) + \int_{\{|X_i|>c\}} |X_i| \, \mathrm{d}P \le cP(A) + \frac{1}{2}\varepsilon .$$

Therefore we have (b) by taking $\delta(\varepsilon) = \frac{\varepsilon}{2c}$ and (a) with $A = \Omega$.

Conversely, let $M := \sup_i E[|X_i|] < \infty$. Let ε and $\delta(\varepsilon)$ be as in (b). Let $c_0 := \frac{M}{\delta(\varepsilon)}$. For all $c \ge c_0$ and all $i \in I$, $P(|X_i| > c) \le \delta_\varepsilon$ (Markov's inequality). Apply (b) with $A = \{|X_c| > c\}$ to obtain that $\sup_n \int_{\{|X_c|>c\}} |X_i| \, \mathrm{d}P \le \varepsilon$. □

Since the "collection" consisting of a single integrable variable X is uniformly integrable, condition (b) of the theorem above reads

$$\sup_{A;\, P(A)<\delta} E[|X| 1_A] \to 0 \text{ as } \delta \to 0 . \tag{6.15}$$

This simple observation will be used in the proof of the next result.

Theorem 6.5.4 *Let Y be an integrable random variable and let $\{\mathcal{F}_i\}_{i\in I}$ be a collection of sub-σ fields of $\mathcal{F}$. The collection $X_i := E[Y \mid \mathcal{F}_i]$ $(i \in I)$ is uniformly integrable.*

Proof. By Jensen's inequality,

$$|X_i| = |E[Y \mid \mathcal{F}_i]| \le E[|Y| \mid \mathcal{F}_i]$$

and therefore, for all $a > 0$,

$$E\left[|X_i| \, 1_{\{|X_i|\ge a\}}\right] \le E[Z_i] \, 1_{\{Z_i \ge a\}} ,$$

where $Z_i := E[|Y| \mid \mathcal{F}_i]$. By definition of conditional expectation, since $\{Z_i \ge a\} \in \mathcal{F}_i$,

$$E\left[(|Y| - Z_i) \, 1_{\{Z_i\ge a\}}\right] = 0$$

and therefore

$$E\left[|X_i|\,1_{\{|X_i|\geq a\}}\right] \leq E\left[|Y|\,1_{\{Z_i\geq a\}}\right] . \qquad (\star)$$

By Markov's inequality,

$$P(Z_i \geq a) \leq \frac{E\left[Z_i\right]}{a} = \frac{E\left[|Y|\right]}{a} ,$$

and therefore $P(Z_i \geq a) \to 0$ as $a \to \infty$ uniformly in i. Use (6.15) to obtain that $E\left[|Y|\,1_{\{Z_i\geq a\}}\right] \to 0$ as $a \to \infty$ uniformly in i. Conclude with $(\star)$. □

Theorem 6.5.5 *A sufficient condition for the collection $\{X_i\}_{i\in I}$ of integrable random variables to be uniformly integrable is the existence of a non-negative non-decreasing function $G : \mathbb{R} \to \mathbb{R}$ such that*

$$\lim_{t\uparrow\infty} \frac{G(t)}{t} = +\infty$$

and

$$\sup_i E\left[G(|X_i|)\right] < \infty .$$

Proof. Fix $\varepsilon > 0$ and let $a = \frac{M}{\varepsilon}$ where $M := \sup_n(E\left[G(|X_i|)\right])$. Take c large enough so that $G(t)/t \geq a$ for $t \geq c$. In particular, $|X_i| \leq \frac{G(|X_i|)}{a}$ on $\{|X_i| > c\}$ and therefore

$$\int_{\{|X_i|>c\}} |X_i| \,\mathrm{d}P \leq \frac{1}{a} E\left[G(|X_i|)1_{\{|X_i|>c\}}\right] \leq \frac{M}{a} = \varepsilon$$

uniformly in i. □

EXAMPLE 6.5.6: TWO SUFFICIENT CONDITIONS FOR UNIFORM INTEGRABILITY. Two frequently used sufficient conditions guaranteeing uniform integrability are

$$\sup_i E\left[|X_i|^{1+\alpha}\right] < \infty \quad (\alpha > 1)$$

and

$$\sup_i E\left[|X_i| \log^+ |X_i|\right] < \infty .$$

Almost-sure convergence of a sequence of integrable random variables to an integrable random variable does not necessarily imply convergence in L^1. However:

Theorem 6.5.7 *Let $\{X_n\}_{n\geq 1}$ be a sequence of integrable random variables and let X be some random variable. The following are equivalent:*

(a) *$\{X_n\}_{n\geq 1}$ is uniformly integrable and $X_n \overset{Pr.}{\to} X$ as $n \to \infty$.*

(b) *X is integrable and $X_n \overset{L^1}{\to} X$ as $n \to \infty$.*

Proof. (a) implies (b): Since $X_n \overset{Pr}{\to} X$, there exists a subsequence $\{X_{n_k}\}_{k\geq 1}$ such that $X_{n_k} \overset{a.s.}{\to} X$. By Fatou's lemma,

$$E[|X|] \leq \liminf_k E[|X_{n_k}|] \leq \sup_{n_k} E[|X_{n_k}|] \leq \sup_n E[|X_n|] < \infty .$$

Therefore $X \in L^1_{\mathbb{R}}(P)$. Also for fixed $\varepsilon > 0$,

$$\begin{aligned} E[|X_n - X|] &\leq \int_{\{|X_n - X| < \varepsilon\}} |X_n - X| \, \mathrm{d}P + \cdots \\ &\quad \cdots + \int_{\{|X_n - X| \geq \varepsilon\}} |X_n| \, \mathrm{d}P + \int_{\{|X_n - X| \geq \varepsilon\}} |X| \, \mathrm{d}P \\ &\leq \varepsilon + \int_{\{|X_n - X| \geq \varepsilon\}} |X_n| \, \mathrm{d}P + \int_{\{|X_n - X| \geq \varepsilon\}} |X| \, \mathrm{d}P . \end{aligned}$$

Recall that adding an integrable random variable to a uniformly integrable collection retains uniformly integrability. Apply (b) of Theorem 6.5.3 to the uniformly integrable family $\{X_n\}_{n\geq 0}$ where $X_0 := X$, denoting by δ' the corresponding δ. By hypothesis, $P(|X_n - X| \geq \varepsilon) \leq \delta'$ for large enough n. By (b) of Theorem 6.5.3 with $A := \{|X_n - X| \geq \varepsilon\}$, for large enough n, $\int_{\{|X_n - X| \geq \varepsilon\}} |X_n| \, \mathrm{d}P \leq \varepsilon$ and $\int_{\{|X_n - X| \geq \varepsilon\}} |X| \, \mathrm{d}P \leq \varepsilon$. Therefore, $E[|X_n - X|] \leq 3\varepsilon$ for large enough n, thus proving convergence in L^1.

(b) implies (a): Let $\varepsilon > 0$ be given and let n_0 be such that $E[|X_n - X|] \leq \varepsilon$ for all $n \geq n_0$. The random variables $X, X_1, \ldots, X_{n_0}$ being integrable, there exists a $\delta > 0$ such that if $P(A) \leq \delta$, $\int_A |X| \, \mathrm{d}P \leq \frac{\varepsilon}{2}$ and $\int_A |X_n| \, \mathrm{d}P \leq \frac{\varepsilon}{2}$ for $n \leq n_0$. If $n \geq n_0$, by the triangle inequality,

$$\int_A |X_n| \, \mathrm{d}P \leq \int_A |X| \, \mathrm{d}P + \int_A |X_n - X| \, \mathrm{d}P \leq 2\varepsilon ,$$

and therefore (b) of Theorem 6.5.3 is satisfied. Whereas (a) of Theorem 6.5.3 is satisfied since $E[|X_n|] \leq E[|X_n - X|] + E[|X|]$. □

6.6 Exercises

Exercise 6.6.1. IN PROBABILITY BUT NOT ALMOST SURELY

Let $\{X_n\}_{n\geq 2}$ be an independent sequence of random variables such that

$$P(X_n = n) = P(X_n = -n) = \frac{1}{2n\ln n} \text{ and } P(X_n = 0) = 1 - \frac{1}{n\ln n} \quad (n \geq 2)\,.$$

Let $S_n := \sum_{i=2}^{n} X_i$. Prove that $\frac{S_n}{n} \to 0$ in probability but not almost surely.

Exercise 6.6.2. A RECURRENCE EQUATION, TAKE 2

Recall the notation $a^+ = \max(a, 0)$. Consider the recurrence equation,

$$X_{n+1} = (X_n - 1)^+ + Z_{n+1} \qquad (n \geq 0)\,,$$

where X_0 and Z_n $(n \geq 1)$ are integer-valued random variables, and $\{Z_n\}_{n\geq 1}$ is IID and independent of X_0.

(a) Show that $\lim_{n\uparrow\infty} X_n = +\infty$ if $E[Z_1] > 1$.

(b) Let T_0 be the first time $n \geq 1$ for which $X_n = 0$. Show that if $E[Z_1] < 1$, then $P(T_0 < \infty) = 1$.

Exercise 6.6.3. ASYMPTOTICS OF THE RENEWAL PROCESS

Let $\{S_n\}_{n\geq 1}$ be an IID sequence of real random variables such that

$$P(0 < S_1 < +\infty) = 1 \text{ and } E[S_1] < \infty\,,$$

and let for each $t \geq 0$, $N(t) = \sum_{n\geq 1} 1_{(0,t]}(T_n)$, where $T_n = S_1 + \cdots + S_n$. (The sequence $\{T_n\}_{n\geq 1}$ is called a *renewal process*.)

(a) Prove that P-almost surely $\lim_{n\uparrow\infty} T_n = \infty$ and $\lim_{t\uparrow\infty} N(t) = \infty$.

(b) Prove that P-almost surely $\lim_{t\to\infty} \frac{N(t)}{t} = \frac{1}{E[S_1]}$.

Exercise 6.6.4. SLLN FOR NON-NEGATIVE SEQUENCES

Let $\{X_n\}_{n\geq 1}$, be an IID sequence of non-negative random variables such that $E[X_1] = \infty$. Show that

$$\lim_{n\uparrow\infty} \frac{X_1 + \cdots + X_n}{n} = \infty \quad (= E[X_1]).$$

Exercise 6.6.5. A RESULT FROM ANALYSIS

Let $f : [0, 1] \to \mathbb{R}$ be a continuous function. Prove that

$$\lim_{n\uparrow\infty} \int_0^1 \cdots \int_0^1 f\left(\frac{x_1 + \cdots + x_n}{n}\right) \mathrm{d}x_1 \cdots \mathrm{d}x_n$$

exists, and find it. (A probabilistic proof is required.)

Exercise 6.6.6. CONVERGENCE IN PROBABILITY BUT NOT ALMOST-SURE, II
Let $\{X_n\}_{n\geq 1}$ be a sequence of independent random variables taking values in $\{0, 1\}$.

(A) Show that a necessary and sufficient condition for this sequence to converge almost surely to 0 is $\sum_{n\geq 1} P(X_n = 1) < \infty$.

(B) Show that a necessary and sufficient condition for this sequence to converge in probability to 0 is $\lim_{n\uparrow\infty} P(X_n = 1) = 0$.

(C) Deduce from the above that convergence in probability does not imply in general almost-sure convergence.

Exercise 6.6.7. WHEN CONVERGENCE IN PROBABILITY IMPLIES ALMOST-SURE CONVERGENCE
Let $\{X_n\}_{n\geq 1}$ be a sequence of *non-negative* random variables. Let $S_n := X_1 + \cdots + X_n$. Show that the convergence in probability of the sequence $\{S_n\}_{n\geq 1}$ implies its almost-sure convergence.

Exercise 6.6.8. IN PROBABILITY AND IN THE QUADRATIC MEAN
Let $\alpha > 0$, and let $\{Z_n\}_{n\geq 1}$ be a sequence of random variables such that

$$P(Z_n = 1) = 1 - \frac{1}{n^\alpha}, \ P(Z_n = n) = \frac{1}{n^\alpha}.$$

Show that $\{Z_n\}_{n\geq 1}$ converges in *probability* to some variable Z to be identified. For what values of α does $\{Z_n\}_{n\geq 1}$ converge to Z *in the quadratic mean*?

Exercise 6.6.9. CONTINUITY OF THE MEAN AND VARIANCE.
Prove the following: If the sequence $\{Z_n\}_{n\geq 1}$ of square-integrable complex random variables converges in the quadratic mean to the complex random variable Z, then

$$\lim_{n\uparrow\infty} E[Z_n] = E[Z] \text{ and } \lim_{n\uparrow\infty} E\left[|Z_n|^2\right] = E\left[|Z|^2\right].$$

Exercise 6.6.10. $g(Z_n)$
Suppose the sequence of random variables $\{Z_n\}_{n\geq 1}$ converges to a in probability. Let $g : \mathbb{R} \to \mathbb{R}$ be a *continuous* function. Show that $\{g(Z_n)\}_{n\geq 1}$ converges to $g(a)$ in probability.

Chapter 7

Convergence in Distribution

The next fundamental notion of convergence after almost-sure convergence is *convergence in distribution*, and the main result there is the *central limit theorem*, the heart of statistics, which is the art of assessing probability models (is this coin fair?). Although these notions are linked in various ways, they are fundamentally different.

7.1 Paul Lévy's Criterion

Let $\{X_n\}_{n\geq 1}$ and X be real random variables with respective cumulative distribution functions $\{F_n\}_{n\geq 1}$ and F. The "natural" definition of convergence in distribution of $\{X_n\}_{n\geq 1}$ to X could be the following:

$$\lim_{n\uparrow\infty} F_n(x) = F(x) \quad (x \in \mathbb{R})\,. \tag{$\star$}$$

In this provisional definition, there is no restriction on the x's in $\mathbb{R}$ for which $(\star)$ is required. However, if it was adopted, one could not say that the "random" (actually deterministic) sequence of random variables $X_n \equiv a + \frac{1}{n}$ where $a \in \mathbb{R}$ converges in distribution to $X \equiv a$. The following definition takes care of this anomaly.

Definition 7.1.1 *Let $\{X_n\}_{n\geq 1}$ and X be real random variables with respective cumulative distribution functions $\{F_n\}_{n\geq 1}$ and F. The sequence $\{X_n\}_{n\geq 1}$ is said to converge in distribution to X if*

$$\lim_{n\uparrow\infty} F_n(x) = F(x) \quad \textit{for all continuity points of } F\,, \tag{7.1}$$

where the point $x \in \mathbb{R}$ is called a continuity point *of the cumulative distribution function F on $\mathbb{R}$ if $F(x) = F(x-)$.*

P. Brémaud, *An Introduction to Applied Probability*, Texts in Applied Mathematics 77,
https://doi.org/10.1007/978-3-031-49306-5_7

This is denoted by:

$$X_n \xrightarrow{\mathcal{D}} X\,.$$

EXAMPLE 7.1.2: MAGNIFIED MINIMUM. Let $\{Y_n\}_{n\geq 1}$ be a sequence of IID random variables uniformly distributed on $[0,1]$. Then

$$X_n = n\min(Y_1,\dots,Y_n) \xrightarrow{\mathcal{D}} \mathcal{E}(1)\,,$$

(the exponential distribution with mean 1). In fact, for all $x \in [0,n]$,

$$P(X_n > x) = P\left(\min(Y_1,\dots,Y_n) > \frac{x}{n}\right) = \prod_{i=1}^{n} P\left(Y_i > \frac{x}{n}\right) = \left(1-\frac{x}{n}\right)^n,$$

and therefore $\lim_{n\uparrow\infty} P(X_n > x) = \mathrm{e}^{-x}1_{\mathbb{R}_+}(x)$.

For random vectors, another definition (which in the univariate case turns out to be equivalent; see Theorem 7.1.5) is needed.

Definition 7.1.3 *Let $\{X_n\}_{n\geq 1}$ and X be random vectors of $\mathbb{R}^d$. The sequence $\{X_n\}_{n\geq 1}$ is said to converge in distribution to X if for all continuous and bounded functions $f : \mathbb{R}^d \to \mathbb{R}$,*

$$\lim_{n\uparrow\infty} E[f(X_n)] = E[f(X)]\,.$$

The vectors X and X_n $(n \geq 1)$ need not be defined on the same probability space. Convergence in distribution concerns only probability distributions. As a matter of fact, very often, the X_n's are defined on the same probability space but there is no "visible" (that is, defined on the same probability space) limit random vector X. Therefore one sometimes denotes convergence in distribution as follows: $X_n \xrightarrow{\mathcal{D}} Q$, where Q is a probability distribution on $\mathbb{R}^d$. If Q is a "famous" probability distribution, for instance a standard Gaussian variable, we then say, that "$\{X_n\}_{n\geq 1}$ converges in distribution to a standard Gaussian distribution", and denote this by: $X_n \xrightarrow{\mathcal{D}} \mathcal{N}(0,1)$.

Theorem 7.1.4 *Let $\{X_n\}_{n\geq 1}$ be a sequence of random vectors of $\mathbb{R}^d$ with respective characteristic functions $\{\varphi_n\}_{n\geq 1}$.*

A. *Suppose that there exists a function φ such that*

$$\lim_{n\uparrow\infty} \varphi_n = \varphi. \tag{7.2}$$

If $\varphi(0) = 1$, this function is the characteristic function of a random vector X and $\{X_n\}_{n\geq 1}$ converges in distribution to X.

B. *In fact, a necessary and sufficient condition for $\{X_n\}_{n\geq 1}$ to converge in distribution to some random vector X with characteristic function φ is that (7.2) holds true.*

This result is the *Paul Lévy criterion for convergence in distribution.* Its (very technical) proof will be omitted as well as the proof of the next result.[1]

Theorem 7.1.5 *In the univariate case ($d = 1$), the conditions (7.2) and (7.1) are equivalent.*

The following result, *Slutsky's lemma*, is often used.

Theorem 7.1.6 *Let $\{X_n\}_{n\geq 1}$ and $\{Y_n\}_{n\geq 1}$ be sequences of real random variables such that $Y_n \overset{Pr}{\to} 0$ and $X_n \overset{\mathcal{D}}{\to} X$ for some real random variable X. Then $X_n + Y_n \overset{\mathcal{D}}{\to} X$.*

Proof. By Lévy's criterion, we must show that $\lim_{n\uparrow\infty} \psi_{X_n+Y_n}(u) \to \psi_X(u)$ for all $u \in \mathbb{R}$. Since

$$|\psi_{X_n+Y_n}(u) - \psi_X(u)| \leq |\psi_{X_n+Y_n}(u) - \psi_{X_n}(u)| + |\psi_{X_n}(u) - \psi_X(u)|,$$

and since by hypothesis ($X_n \overset{\mathcal{D}}{\to} X$) the second member of the right-hand side of the above inequality tends to 0, it remains to show that the first member tends to 0. But the latter equals

$$|E\left[e^{iuX_n}(e^{iuY_n} - 1)\right]| \leq |E\left[(e^{iuY_n} - 1)\right]|.$$

Now, for any $\varepsilon > 0$ there exists a $\delta > 0$ such that $|y| \leq \delta \Rightarrow |e^{iuy} - 1| \leq \varepsilon$. Therefore

$$\begin{aligned} |E\left[(e^{iuY_n} - 1)\right]| &= |E\left[(e^{iuY_n} - 1)1_{\{|Y_n|>\delta\}}\right]| + |E\left[(e^{iuY_n} - 1)1_{\{|Y_n|\leq\delta\}}\right]| \\ &\leq 2P(|Y_n| > \delta) + \varepsilon. \end{aligned}$$

[1] The classic reference is [2].

Since $Y_n \overset{Pr}{\to} 0$, $\lim_{n\uparrow\infty} P(|Y_n| > \delta) = 0$. Therefore

$$\limsup_{n\uparrow\infty} |E\left[(e^{iuY_n} - 1)\right]| \le \varepsilon\,,$$

and since ε is arbitrary, $\lim_{n\uparrow\infty} |E\left[(e^{iuY_n} - 1)\right]| = 0$. □

Bochner's Theorem

This result is of paramount importance in the theory of wide-sense stationary processes (Chapter 12).

The characteristic function φ of a real random variable X has the following properties:

A. it is hermitian symmetric (that is, $\varphi(-u) = \varphi(u)^*$) and uniformly bounded (in fact, $|\varphi(u)| \le \varphi(0)$);

B. it is uniformly continuous on $\mathbb{R}$; and

C. it is definite non-negative, in the sense that for all integers n, all $u_1, \ldots, u_n \in \mathbb{R}$, and all $z_1, \ldots, z_n \in \mathbb{C}$,

$$\sum_{j=1}^{n}\sum_{k=1}^{n} \varphi(u_j - u_k) z_j z_k^* \ge 0$$

(just observe that the left-hand side equals $E\left[\left|\sum_{j=1}^{n} z_j e^{iu_jX}\right|^2\right]$).

It turns out that Properties A, B and C characterize characteristic functions (up to a multiplicative constant). This is *Bochner's theorem*:

Theorem 7.1.7 *Let $\varphi : \mathbb{R} \to \mathbb{C}$ be a function satisfying properties A, B and C. Then there exists a constant $0 \le \beta < \infty$ and a real random variable X such that for all $u \in \mathbb{R}$,*

$$\varphi(u) = \beta E\left[e^{iuX}\right].$$

Proof. We henceforth eliminate the trivial case where $\varphi(0) = 0$ (implying, in view of condition A, that φ is the null function). For any continuous function $z : \mathbb{R} \to \mathbb{C}$ and any $A \ge 0$,

$$\int_0^A \int_0^A \varphi(u - v) z(u) z^*(v)\, du\, dv \ge 0\,. \qquad (\star)$$

Indeed, since the integrand is continuous, the integral is the limit as $n \uparrow \infty$ of

$$\frac{A^2}{4^n} \sum_{j=1}^{2^n} \sum_{k=1}^{2^n} \varphi\left(\frac{A(j-k)}{2^n}\right) z\left(\frac{Aj}{2^n}\right) z\left(\frac{Ak}{2^n}\right)^*,$$

a non-negative quantity by condition C. From $(\star)$ with $z(u) := \mathrm{e}^{-ixu}$, we have that

$$g(x, A) := \frac{1}{2\pi A} \int_0^A \int_0^A \varphi(u-v) \mathrm{e}^{-ix(u-v)} \, \mathrm{d}u \, \mathrm{d}v \geq 0 .$$

Changing variables, we obtain the alternative expression

$$\begin{aligned} g(x, A) :&= \frac{1}{2\pi} \int_{-A}^{A} \left(1 - \frac{|u|}{A}\right) \varphi(u) \mathrm{e}^{-iux} \, \mathrm{d}u \\ &= \frac{1}{2\pi} \int_{-\infty}^{+\infty} h\left(\frac{u}{A}\right) \varphi(u) \mathrm{e}^{-iux} \, \mathrm{d}u , \end{aligned}$$

where $h(u) = (1 - |u|) 1_{\{|u| \leq 1\}}$. Let $M > 0$. We have

$$\begin{aligned} &\int_{-\infty}^{+\infty} h\left(\frac{x}{2M}\right) g(x, A) \, \mathrm{d}x \\ &\qquad = \frac{1}{2\pi} \int_{-\infty}^{+\infty} h\left(\frac{u}{A}\right) \varphi(u) \left(\int_{-\infty}^{+\infty} h\left(\frac{x}{2M}\right) \mathrm{e}^{-iux} \, \mathrm{d}x\right) \mathrm{d}u \\ &\qquad = \frac{1}{\pi} M \int_{-\infty}^{+\infty} h\left(\frac{u}{A}\right) \varphi(u) \left(\frac{\sin Mu}{Mu}\right)^2 \mathrm{d}u . \end{aligned}$$

Therefore

$$\begin{aligned} \int_{-\infty}^{+\infty} h\left(\frac{x}{2M}\right) g(x, A) \, \mathrm{d}x &\leq \frac{1}{\pi} M \int_{-\infty}^{+\infty} h\left(\frac{u}{A}\right) |\varphi(u)| \left(\frac{\sin Mu}{Mu}\right)^2 \mathrm{d}u \\ &\leq \frac{1}{\pi} \varphi(0) \int_{-\infty}^{+\infty} \left(\frac{\sin u}{u}\right)^2 \mathrm{d}u = \varphi(0) . \end{aligned}$$

By monotone convergence,

$$\lim_{M \uparrow \infty} \int_{-\infty}^{+\infty} h\left(\frac{x}{2M}\right) g(x, A) \, \mathrm{d}x = \int_{-\infty}^{+\infty} g(x, A) \, \mathrm{d}x ,$$

and therefore

$$\int_{-\infty}^{+\infty} g(x, A) \, \mathrm{d}x \leq \varphi(0) .$$

The function $x \mapsto g(x,A)$ is therefore integrable and it is the Fourier transform of the integrable and continuous function $u \mapsto h\left(\frac{u}{A}\right)\varphi(u)$. Therefore, by the Fourier inversion formula:

$$h\left(\frac{u}{A}\right)\varphi(u) = \int_{-\infty}^{+\infty} g(x,A)e^{iux}\,dx\,.$$

In particular, with $u = 0$, $\int_{-\infty}^{+\infty} g(x,A)\,dx = \varphi(0)$. Therefore, $f(x,A) := \frac{g(x,A)}{\varphi(0)}$ is the probability density of some real random variable with characteristic function $h\left(\frac{u}{A}\right)\frac{\varphi(u)}{\varphi(0)}$. But

$$\lim_{A\uparrow\infty} h\left(\frac{u}{A}\right)\frac{\varphi(u)}{\varphi(0)} = \frac{\varphi(u)}{\varphi(0)}\,.$$

This limit of a sequence of characteristic functions is continuous at 0 and is therefore a characteristic function (Paul Lévy's criterion, Theorem 7.1.4). □

7.2 The Central Limit Theorem

This is the emblematic theorem of Statistics.

Theorem 7.2.1 *Let $\{X_n\}_{n\geq 1}$ be an* IID *sequence of real random variables such that*

$$E[X_1^2] < \infty\,. \tag{7.3}$$

(In particular, $E[|X_1|] < \infty$.) Then, for all $x \in \mathbb{R}$,

$$\frac{S_n - nE[X_1]}{\sigma\sqrt{n}} \xrightarrow{\mathcal{D}} \mathcal{N}(0,1)\,. \tag{7.4}$$

The random variable in the left of (7.4) is obtained by *centering* the sum S_n (subtracting its mean $nE[X_1]$) and then *normalizing* it (dividing by the square root of its variance so that the resulting variance equals 1).

Proof. Assume without loss of generality that $E[X_1] = 0$. Let σ^2 be the variance of X_1. By the characteristic function criterion for convergence in distribution, it suffices to show that

$$\lim_{n\uparrow\infty} \varphi_n(u) = e^{-\sigma^2 u^2/2},$$

where

$$\begin{aligned}\varphi_n(u) &= E\left[\exp\left\{iu\frac{\sum_{j=1}^n X_j}{\sqrt{n}}\right\}\right]\\ &= \prod_{j=1}^n E\left[\exp\left\{i\frac{u}{\sqrt{n}}X_j\right\}\right] = \psi\left(\frac{u}{\sqrt{n}}\right)^n,\end{aligned}$$

where ψ is the characteristic function of X_1. From the Taylor expansion of ψ about zero,

$$\psi(u) = 1 + \frac{\psi''(0)}{2!}u^2 + o(u^2)\,,$$

we have, for fixed $u \in \mathbb{R}$,

$$\psi\left(\frac{u}{\sqrt{n}}\right) = 1 - \frac{1}{n}\frac{\sigma^2 u^2}{2} + o\left(\frac{1}{n}\right),$$

and therefore

$$\lim_{n\uparrow\infty} \ln\{\varphi_n(u)\} = \lim_{n\uparrow\infty} n\left(\ln\left\{1 - \frac{\sigma^2u^2}{2n} + o\left(\frac{1}{n}\right)\right\}\right) = -\frac{1}{2}\sigma^2u^2\,.$$

The result then follows by Theorem 7.1.4. □

EXAMPLE 7.2.2: FAST SAMPLING OF THE POISSON DISTRIBUTION. In the case of a Poisson distribution with mean θ, the method of the inverse works as follows: letting $p_i := \mathrm{e}^{-\theta}\frac{\theta^i}{i!}$, sample a random variable U uniformly distributed on $[0,1]$ and set $T = k$ if U falls in the interval $I_i := [\sum_{i=0}^{k-1} p_i, \sum_{i=0}^{k} p_i]$. The crude version of this sampling algorithm consists in examining the intervals I_i sequentially until one is found that contains U. This would require on average $1 + E[T] = 1 + \theta$ trials. If θ is very large, a more economical procedure is available. It takes into account the fact that the probability mass of a Poisson variable is maximal at a value i_0 near the average value and decreases as one get farther away from this value. The exploration starts with the value i_0, and then proceeds to $i_0 - 1$, $i_0 + 1$, $i_0 - 2$, $i_0 + 2$, etc. The average number of trials is then roughly equal to

$$1 + E\left[|T - \theta|\right] = 1 + \sqrt{\theta}E\left[\frac{|T-\theta|}{\sqrt{\theta}}\right].$$

By the central limit theorem, $Z := \frac{T-\theta}{\sqrt{\theta}}$ is approximately distributed as a standard Gaussian variable. Therefore the average number of trials for large θ is approximately

$$1 + \sqrt{\theta}E\left[|\mathcal{N}(0,1)|\right] \simeq 1 + 0.82\sqrt{\theta}\,.$$

The central limit theorem admits a multidimensional version.

Theorem 7.2.3 *Let $\{X_n\}_{n\geq 1}$ be a sequence of independent random vectors of dimension d, and let $\{a_n\}_{n\geq 1}$ be a sequence of real numbers such that $\lim_{n\uparrow\infty} a_n = \infty$. Suppose that*

$$X_n \stackrel{a.s.}{\to} m$$

and

$$\sqrt{a_n}(X_n - m) \stackrel{\mathcal{D}}{\to} \mathcal{N}(0, \Gamma) .$$

Let $g : \mathbb{R}^d \to \mathbb{R}^q$ be a function twice continuously differentiable in a neighborhood U of m. Then

$$g(X_n) \stackrel{a.s.}{\to} g(m)$$

and

$$\sqrt{a_n}(g(X_n) - g(m)) \stackrel{\mathcal{D}}{\to} \mathcal{N}\left(0, J_g(m)^T \,\Gamma\, J_g(m)\right) ,$$

where $J_g(m)$ is the Jacobian matrix of g evaluated at m.

Proof. U can be chosen convex and compact. Let g_j denote the j-th coordinate of g, and let $D^2 g_j$ denote the second differential matrix of g_j. By Taylor's formula,

$$g_j(x) - g_j(m) = (x - m)^T (\text{grad } g_j(m)) + \frac{1}{2}(x - m)^T \, D^2 g_j(m^*) \, (x - m)$$

for some m^* in the closed segment linking m to x, denoted $[m, x]$. Therefore, if $X_n \in U$

$$\begin{aligned}\sqrt{a_n}(g_j(X_n) - g_j(m)) &= \sqrt{a_n}(X_n - m)^T (\text{grad } g_j(m)) \\ &\quad + \frac{1}{2} a_n (X_n - m)^T \frac{1}{\sqrt{a_n}} D^2 g_j(m_n^*) \, (X_n - m) ,\end{aligned}$$

where $m_n^* \in [m, X_n]$.

Suppose $X_n \in U$. Since U is convex and $m \in U$, also $m_n^* \in U$. Now since U is compact, the continuous function $D^2 g_j$ is bounded in U. Therefore, since $a_n \uparrow \infty$, $\frac{1}{\sqrt{a_n}} D^2 g_j(m_n^*) 1_U(X_n) \to 0$. Since $X_n \stackrel{a.s.}{\to} m$, we deduce from the above remarks that

$$\sqrt{a_n}(g_j(X_n) - g_j(m)) - \sqrt{a_n}(X_n - m)^T (\text{grad } g_j(m)) \stackrel{a.s.}{\to} 0 ,$$

and therefore

$$\sqrt{a_n}(g(X_n) - g(m)) - J_g(m)\sqrt{a_n}(X_n - m) \stackrel{a.s.}{\to} 0 .$$

But $\sqrt{a_n}(X_n - m) \stackrel{\mathcal{D}}{\to} \mathcal{N}(0, \Gamma)$ and therefore

$$\sqrt{a_n}(g(X_n) - g(m)) \stackrel{\mathcal{D}}{\to} J_g(m)\mathcal{N}(0, \Gamma) = \mathcal{N}\left(0, J_g(m)^T \,\Gamma\, J_g(m)\right) .$$

□

Confidence Intervals

We now briefly introduce a basic methodology of Statistics with the notion of *confidence interval.*

The central limit theorem (7.2.1) implies that for $x \geq 0$,

$$\lim_{n \uparrow \infty} P\left(E[X_1] - \frac{\sigma}{\sqrt{n}}x \leq \frac{S_n}{n} \leq E[X_1] + \frac{\sigma}{\sqrt{n}}x\right) = P(|\mathcal{N}(0;1)| \leq x).$$

Under the condition $E\left[|X_1|^3\right] < \infty$, this limit is uniform in $x \in \mathbb{R}$ (we shall admit this result, called the *Berry–Essen theorem*) and therefore, with $\frac{\sigma}{\sqrt{n}}x = a$,

$$\lim_{n \uparrow \infty} P\left(E[X_1] - a \leq \frac{S_n}{n} \leq E[X_1] + a\right) = P\left(|\mathcal{N}(0;1)| \leq \frac{a\sqrt{n}}{\sigma}\right).$$

That is, for large n,

$$P\left(E[X_1] - a \leq \frac{S_n}{n} \leq E[X_1] + a\right) \simeq P\left(|\mathcal{N}(0;1)| \leq \frac{a\sqrt{n}}{\sigma}\right).$$

In other words, for large n, the SLLN estimate of $E[X_1]$, that is $\frac{S_n}{n}$, lies within distance a of $E[X_1]$ with probability $P\left(|\mathcal{N}(0;1)| \leq \frac{a\sqrt{n}}{\sigma}\right)$.

In statistical practice, this result is used in two manners.

(1) One wishes to know the number n of experiments that guarantee that with probability, say 0.99, the estimation error is less than a. Choose n such that

$$P\left(|\mathcal{N}(0;1)| \leq \frac{a\sqrt{n}}{\sigma}\right) = 0.99.$$

Since

$$P(|\mathcal{N}(0;1)| \leq 2.58) = 0.99,$$

we have

$$2.58 = \frac{a\sqrt{n}}{\sigma}, \tag{7.5}$$

and therefore

$$n = \left(\frac{2.58a}{\sigma}\right)^2.$$

(2) The (usually large) number n of experiments is fixed. We want to determine the interval $\left[\frac{S_n}{n} - a, \frac{S_n}{n} + a\right]$ within which the mean $E[X_1]$ lies with probability at least 0.99. From (7.5):

$$a = \frac{2.58\sigma}{\sqrt{n}}.$$

If the standard deviation σ is unknown, it may be either replaced by an SLLN estimate of it (but then of course...), or the conservative method can be used, which consists of replacing σ by an upper bound.

EXAMPLE 7.2.4: TESTING A COIN. Consider the problem of estimating the bias p of a coin. Here, X_n takes two values, 1 and 0 with probability p and $1-p$ respectively, and in particular $E[X_1] = p$, $\text{Var}\,(X_1) = \sigma^2 = p(1-p)$. Clearly, since we are trying to estimate p, the standard deviation σ is unknown. Here the upper bound of σ is the maximum of $\sqrt{p(1-p)}$ for $p \in [0,1]$, which is attained for $p = \frac{1}{2}$. Thus $\sigma \leq \frac{1}{2}$.

Suppose the coin was tossed 10,000 times and that the experiment produced the estimate $\frac{S_n}{n} = 0.4925$. Can we "believe 99 percent" that the coin is unbiased? For this we would check that the corresponding confidence interval contains the value $\frac{1}{2}$. Using the conservative method (not a big problem since obviously the actual bias is not far from $\frac{1}{2}$), we have

$$a = \frac{\sigma 2.58}{\sqrt{n}} = 0.0129\,,$$

and indeed $\frac{1}{2} \in [0.4925 - 0.0129, 0.4925 - 0.0129]$, so that we are at least 99 percent confident that the coin is unbiased.

7.3 Convergence in Variation

This notion is introduced in the discrete time setting, since this will be sufficient for the study of convergence (in variation) of a Markov chain (see Chapter 9).

Definition 7.3.1 *Let E be a countable space. The* distance in variation *between two probability distributions α and β on E is the quantity*

$$d_V(\alpha,\beta) := \frac{1}{2}\sum_{i\in E} |\alpha(i) - \beta(i)|\,. \tag{7.6}$$

That d_V is indeed a distance is clear.

Lemma 7.3.2 *Let α and β be two probability distributions on the same countable space E. Then*

$$\begin{aligned} d_V(\alpha,\beta) &= \sup_{A\subseteq E}\{|\alpha(A) - \beta(A)|\} \\ &= \sup_{A\subseteq E}\{\alpha(A) - \beta(A)\}\,. \end{aligned}$$

Proof. For the second equality observe that for each subset A there is a subset B such that $|\alpha(A) - \beta(A)| = \alpha(B) - \beta(B)$ (take $B = A$ or $\bar{A}$). For the first equality, write

$$\alpha(A) - \beta(A) = \sum_{i \in E} 1_A(i)\{\alpha(i) - \beta(i)\}$$

and observe that the right-hand side is maximal for

$$A = \{i \in E;\ \alpha(i) > \beta(i)\}.$$

Therefore, with $g(i) = \alpha(i) - \beta(i)$,

$$\sup_{A \subseteq E}\{\alpha(A) - \beta(A)\} = \sum_{i \in E} g^+(i) = \frac{1}{2}\sum_{i \in E}|g(i)|$$

since $\sum_{i \in E} g(i) = 0$. □

The distance in variation *between two random variables* X and Y with values in E is the distance in variation between their probability distributions, and it is denoted (with a slight abuse of notation) by $d_V(X, Y)$. Therefore

$$d_V(X, Y) := \frac{1}{2}\sum_{i \in E}|P(X = i) - P(Y = i)|.$$

The distance in variation *between a random variable* X *with values in* E *and a probability distribution* α *on* E denoted (again with a slight abuse of notation) by $d_V(X, \alpha)$ is defined by

$$d_V(X, \alpha) := \frac{1}{2}\sum_{i \in E}|P(X = i) - \alpha(i)|.$$

We now introduce the notion of *coupling*.

Definition 7.3.3 *The coupling of two discrete probability distributions* π' *on* E' *and* π'' *on* E'' *consists, by definition, of the construction of a probability distribution* π *on* $E := E' \times E''$ *such that the marginal distributions of* π *on* E' *and* E'' *respectively are* π' *and* π''*, that is,*

$$\sum_{j \in E''} \pi(i, j) = \pi'(i) \text{ and } \sum_{i \in E'} \pi(i, j) = \pi''(j).$$

For two probability distributions α and β on the countable set E, let $\mathcal{D}(\alpha, \beta)$ be the collection of random vectors (X, Y) taking their values in $E \times E$ and with given marginal distributions α and β, that is,

$$P(X = i) = \alpha(i),\ P(Y = i) = \beta(i)\,. \tag{7.7}$$

Theorem 7.3.4 *For any pair $(X, Y) \in \mathcal{D}(\alpha, \beta)$, we have the fundamental* coupling inequality

$$d_V(\alpha, \beta) \leq P(X \neq Y)\,,$$

and equality is attained by some pair $(X, Y) \in \mathcal{D}(\alpha, \beta)$, which is then said to realize maximal coincidence.

Proof. For arbitrary $A \subset E$,

$$\begin{aligned} P(X \neq Y) &\geq P(X \in A, Y \in \bar{A}) \\ &= P(X \in A) - P(X \in A, Y \in A) \\ &\geq P(X \in A) - P(Y \in A)\,, \end{aligned}$$

and therefore

$$P(X \neq Y) \geq \sup_{A \subset E} \{P(X \in A) - P(Y \in A)\} = d_V(\alpha, \beta)\,.$$

We now construct $(X, Y) \in \mathcal{D}(\alpha, \beta)$ realizing equality. Let U, Z, V, and $\{W(t)\}_{t \in [0,1]}$ be independent random variables; U takes its values in $\{0, 1\}$, and Z, V, W take their values in E. The distributions of these random variables is given by

$$\begin{aligned} P(U = 1) &= 1 - d_V(\alpha, \beta)\,, \\ P(Z = i) &= \alpha(i) \wedge \beta(i) / \left(1 - d_V(\alpha, \beta)\right)\,, \\ P(V = i) &= (\alpha(i) - \beta(i))^+ / d_V(\alpha, \beta)\,, \\ P(W = i) &= (\beta(i) - \alpha(i))^+ / d_V(\alpha, \beta)\,. \end{aligned}$$

Observe that $P(V = W) = 0$. Defining

$$\begin{aligned} (X, Y) &= (Z, Z) \text{ if } U = 1 \\ &= (V, W) \text{ if } U = 0\,, \end{aligned}$$

we have

$$\begin{aligned} P(X = i) &= P(U = 1, Z = i) + P(U = 0, V = i) \\ &= P(U = 1) P(Z = i) + P(U = 0) P(V = i) \\ &= \alpha(i) \wedge \beta(i) + (\alpha(i) - \beta(i))^+ = \alpha(i)\,, \end{aligned}$$

and similarly, $P(Y = i) = \beta(i)$. Therefore, $(X, Y) \in \mathcal{D}(\alpha, \beta)$. Also, $P(X = Y) = P(U = 1) = 1 - d_V(\alpha, \beta)$. $\square$

EXAMPLE 7.3.5: POISSON'S LAW OF RARE EVENTS, TAKE 2. Let $Y_1, \ldots, Y_n$ be independent random variables taking their values in $\{0, 1\}$, with $P(Y_i = 1) = \pi_i$, $1 \leq i \leq n$. Let $X := \sum_{i=1}^n Y_i$ and $\lambda := \sum_{i=1}^n \pi_i$. Let p_λ be the Poisson distribution with mean λ. We wish to bound the variation distance between the distribution q of X and p_λ. For this we construct a coupling of the two distributions as follows. First we generate independent couples $(Y_1, Y_1'), \ldots, (Y_n, Y_n')$ such that

$$P(Y_i = j, Y_i' = k) = \begin{cases} 1 - \pi_i & \text{if } j = 0, k = 0, \\ \mathrm{e}^{-\pi_i} \frac{\pi_i^k}{k!} & \text{if } j = 1, k \geq 1, \\ \mathrm{e}^{-\pi_i} - (1 - \pi_i) & \text{if } j = 1, k = 0\,. \end{cases}$$

One verifies that for all $1 \leq i \leq n$, $P(Y_i = 1) = \pi_i$ and $Y_i' \sim \text{Poi}(\pi_i)$. In particular $X' := \sum_{i=1}^n Y_i'$ is a Poisson variable with mean λ. Now

$$\begin{aligned} P(X \neq X') &= P\left(\sum_{i=1}^n Y_i \neq \sum_{i=1}^n Y_i'\right) \\ &\leq P\left(Y_i \neq Y_i' \text{ for some } i\right) \leq \sum_{i=1}^n P\left(Y_i \neq Y_i'\right)\,. \end{aligned}$$

But

$$\begin{aligned} P\left(Y_i \neq Y_i'\right) &= \mathrm{e}^{-\pi_i} - (1 - \pi_i) + P(Y_1' \geq 1) \\ &= \pi_i\left(1 - \mathrm{e}^{-\pi_i}\right) \leq \pi_i^2\,. \end{aligned}$$

Therefore $P(X \neq X') \leq \sum_{i=1}^n \pi_i^2$ and by the coupling inequality

$$d_V(q, p_\lambda) \leq \sum_{i=1}^n \pi_i^2\,.$$

For instance, with $\pi_i = p := \frac{\lambda}{n}$, we have

$$d_V(q, p_\lambda) \leq \frac{\lambda^2}{n}\,.$$

In other terms the binomial distribution of size n and mean λ differs in variation by less than $\frac{\lambda^2}{n}$ from a Poisson variable with the same mean. This is obviously a refinement of the Poisson approximation theorem since it gives exploitable estimates for finite n.

Definition 7.3.6 *A sequence $\{X_n\}_{n\geq 1}$ of discrete random variables with values in E is said to converge in distribution to the probability distribution π on E if for all $i \in E$, $\lim_{n\uparrow\infty} P(X_n = i) = \pi(i)$. It is said to* converge in variation *to this distribution if*

$$\lim_{n\uparrow\infty} \sum_{i\in E} |P(X_n = i) - \pi(i)| = 0. \tag{7.8}$$

Observe that Definition 7.3.6 concerns only the marginal distributions of the stochastic process, not the stochastic process itself. Therefore, if there exists another stochastic process $\{X'_n\}_{n\geq 0}$ such that $X_n \stackrel{\mathcal{D}}{\sim} X'_n$ for all $n \geq 0$, and if there exists a third one $\{X''_n\}_{n\geq 0}$ such that $X''_n \stackrel{\mathcal{D}}{\sim} \pi$ for all $n \geq 0$, then (7.8) follows from

$$\lim_{n\uparrow\infty} d_V(X'_n, X''_n) = 0. \tag{7.9}$$

This trivial observation is useful because of the resulting freedom in the choice of $\{X'_n\}$ and $\{X''_n\}$. An interesting situation occurs when there exists a finite random time τ such that $X'_n = X''_n$ for all $n \geq \tau$.

Definition 7.3.7 *Two stochastic processes $\{X'_n\}_{n\geq 0}$ and $\{X''_n\}_{n\geq 0}$ taking their values in the same state space E are said to* couple *if there exists an almost surely* finite *random time τ such that*

$$n \geq \tau \Rightarrow X'_n = X''_n. \tag{7.10}$$

The random variable τ is called a coupling time *of the two processes.*

Theorem 7.3.8 *For any coupling time τ of $\{X'_n\}_{n\geq 0}$ and $\{X''_n\}_{n\geq 0}$, we have the* coupling inequality

$$d_V(X'_n, X''_n) \leq P(\tau > n). \tag{7.11}$$

Proof. For all $A \subseteq E$,

$$\begin{aligned} P(X'_n \in A) - P(X''_n \in A) &= P(X'_n \in A,\ \tau \leq n) + P(X'_n \in A,\ \tau > n) \\ &- P(X''_n \in A,\ \tau \leq n) - P(X''_n \in A, \tau > n) \\ &= P(X'_n \in A,\ \tau > n) - P(X''_n \in A,\ \tau > n) \\ &\leq P(X'_n \in A,\ \tau > n) \leq P(\tau > n). \end{aligned}$$

Inequality (7.11) then follows from Lemma 7.3.2. □

Therefore, if the coupling time is P-a.s. *finite*, that is $\lim_{n\uparrow\infty} P(\tau > n) = 0$,

$$\lim_{n\uparrow\infty} d_V(X_n, \pi) = \lim_{n\uparrow\infty} d_V(X'_n, X''_n) = 0.$$

Definition 7.3.9 (A) *A sequence $\{\alpha_n\}_{n\geq 0}$ of probability distributions on E is said to converge in variation to the probability distribution β on E if*

$$\lim_{n\uparrow\infty} d_V(\alpha_n, \beta) = 0\,.$$

(B) *An E-valued random sequence $\{X_n\}_{n\geq 0}$ such that for some probability distribution π on E,*

$$\lim_{n\uparrow\infty} d_V(X_n, \pi) = 0, \tag{7.12}$$

is said to converge in variation to π.

7.4 The Rank of Convergence in Distribution

Convergence in distribution is weaker than almost-sure convergence. This means the following.

Theorem 7.4.1 *If the sequence $\{X_n\}_{n\geq 1}$ of random vectors of $\mathbb{R}^d$ converges almost surely to some random vector X, it also converges in distribution to the same vector X.*

Proof. By dominated convergence, for all $u \in \mathbb{R}$,

$$\lim_{n\uparrow\infty} E\left[e^{i\langle u, X_n\rangle}\right] = E\left[e^{i\langle u, X\rangle}\right],$$

which implies, by Paul Lévy's theorem (Theorem 7.1.4), that $\{X_n\}_{n\geq 1}$ converges in distribution to X. □

In fact, convergence in distribution is even weaker than convergence in probability.

Theorem 7.4.2 *If the sequence $\{X_n\}_{n\geq 1}$ of random variables converges in probability to some random variable X, it also converges in distribution to X.*

Proof. If this were not the case, one could find a bounded continuous function f such that $E[f(X_n)]$ does not converge to $E[f(X)]$. In particular, there would exist a subsequence n_k and some $\varepsilon > 0$ such that $|E[f(X_{n_k})] - E[f(X)]| \geq \varepsilon$ for all k. As $\{X_{n_k}\}_{k\geq 1}$ converges in probability to X, one can extract from it a subsequence $\{X_{n_{k_\ell}}\}_{\ell\geq 1}$ converging almost surely to X. In particular, since f is bounded and continuous, $\lim_\ell E[f(X_{n_{k_\ell}}] = E[f(X)]$ by dominated convergence, a contradiction. □

Combining Theorems 6.4.7 and 7.4.2, we have that convergence in distribution is weaker than convergence in the quadratic mean:

Theorem 7.4.3 *If the sequence of real random variables $\{Z_n\}_{n\geq 1}$ converges in the quadratic mean to some random variable Z, it also converges in distribution to the same random variable Z.*

Convergence in distribution is weaker that convergence in variation:

Theorem 7.4.4 *If the sequence of real random variables $\{X_n\}_{n\geq 1}$ converges in variation to X, it converges in distribution to the same random variable.*

Proof. Indeed, for all x (not just the continuity points of the distribution of X),

$$|P(X_n \leq x) - P(X \leq x)| \leq d_V(X_n, X) \to 0\,.$$

□

A Stability Property of the Gaussian Distribution

Theorem 7.4.5 *Let $\{Z_n\}_{n\geq 1}$, where $Z_n = (Z_n^{(1)}, \ldots, Z_n^{(m)})$, be a sequence of Gaussian random vectors of fixed dimension m that converges componentwise in the quadratic mean to some vector $Z = (Z^{(1)}, \ldots, Z^{(m)})$. Then the latter vector is Gaussian.*

Proof. In fact, by continuity of the inner product in $L^2_{\mathbb{R}}(P)$, for all $1 \leq i, j \leq m$, $\lim_{n\uparrow\infty} E[Z_n^{(i)} Z_n^{(j)}] = E[Z^{(i)} Z^{(j)}]$ and $\lim_{n\uparrow\infty} E[Z_n^{(i)}] = E[Z^{(i)}]$, that is

$$\lim_{n\uparrow\infty} m_{Z_n} = m_Z, \qquad \lim_{n\uparrow\infty} \Gamma_{Z_n} = \Gamma_Z$$

and in particular, for all $u \in \mathbb{R}^m$,

$$\begin{aligned}\lim_{n\uparrow\infty} E\left[\mathrm{e}^{iu^T Z_n}\right] &= \lim_{n\uparrow\infty} \mathrm{e}^{iu^T \mu_{Z_n} - \frac{i}{2} u^T \Gamma_{Z_n} u} \\ &= \mathrm{e}^{iu^T \mu_Z - \frac{i}{2} u^T \Gamma_Z u}\,.\end{aligned}$$

The sequence $\{u^T Z_n\}_{n\geq 1}$ converges in the quadratic mean to $u^T Z$, and therefore it also converges in distribution to $u^T Z$. Therefore, $\lim_{n\uparrow\infty} E\left[\mathrm{e}^{iu^T Z_n}\right] = E[\mathrm{e}^{iu^T Z}]$, and finally

$$E[\mathrm{e}^{iu^T Z}] = \mathrm{e}^{iu^T \mu_Z - \frac{i}{2} u^T \Gamma_Z u}$$

for all $u \in \mathbb{R}^m$. This shows that Z is a Gaussian vector. □

Therefore, limits in the quadratic mean preserve the Gaussian nature of random vectors. This is the stability property referred to in the title of this subsection. Note that the Gaussian nature of random vectors is also preserved by linear transformations, as we already know.

Skorokhod's Theorem

There is an at first sight surprising connection between convergence in distribution and almost-sure convergence exemplified in the following example and generalized by the theorem following it.

EXAMPLE 7.4.6: CONVERGENCE IN DISTRIBUTION OF EXPONENTIAL RANDOM VARIABLES. Consider the sequence of exponential CDFs

$$F_n(x) = (1 - \mathrm{e}^{-\lambda_n x})1_{x\geq 0} \quad (n \geq 1),$$

where $\{\lambda_n\}_{n\geq 1}$ is a sequence of positive numbers converging to the finite positive number λ. Obviously $F_n \stackrel{\mathcal{D}}{\to} F$ where $F(x) = (1 - \mathrm{e}^{-\lambda x})1_{x\geq 0}$. Let now X be an exponential random variable with parameter λ. The random variables

$$X_n = \frac{\lambda}{\lambda_n} X \quad (n \geq 1)$$

have the respective CDF F_n $(n \geq 1)$ and obviously $X_n \stackrel{a.s.}{\to} X$.

The next result, *Skorokhod's theorem*, generalizes the previous example.

Theorem 7.4.7 *Let $\{F_n\}_{n\geq 1}$ be the CDFs of a sequence $\{X_n\}_{n\geq 1}$ of random variables converging in distribution to a random variable X with the CDF F. There exists a sequence $\{Y_n\}_{n\geq 1}$ of random variables with the CDF $\{F_n\}_{n\geq 1}$ that converges almost surely to a random variable X with the CDF F.*

Proof. Let $F_n^{\leftarrow}$ and $F^{\leftarrow}$ denote the generalized inverses of F_n and F respectively (see Theorem 3.2.30). The sequence we are looking for will be defined on the probability space $(\Omega, \mathcal{F}, P) := ((0,1), \mathcal{B}((0,1)), \ell)$, where ℓ is the Lebesgue measure. It is defined by $Y_n(\omega) := F_n^{\leftarrow}(\omega)$, that is[2]

$$Y_n(u) := F_n^{\leftarrow}(u)$$

and the putative limit is defined by

$$Y(u) := F^{\leftarrow}(u).$$

In order to prove that $Y_n \to Y$, it suffices to show that $F_n^{\leftarrow}(u) \to F^{\leftarrow}(u)$ for all $u \in (0,1)$ where $F^{\leftarrow}$ is continuous, since the complement of such points is of null Lebesgue measure.

---

[2] With a change of notation that will maybe avoid confusion.

Let u be such a point. Let $\mathcal{C}$ be the set of points of continuity of F. If $a, b \in \mathcal{C}$, $a < b$, are such that

$$a < F^{\leftarrow}(u) < b\,, \tag{$\star$}$$

then there exists a v, $u < v < 1$, such that $a < F^{\leftarrow}(u) \leq F^{\leftarrow}(v) \leq b$, that is, $F(a) < u < v \leq F(b)$.

Since $a, b \in \mathcal{C}$, $a < b$, for large enough n, $F_n(a) < u \leq F_n(b)$, that is,

$$a < F_n^{\leftarrow}(u) \leq b\,. \tag{$\star\star$}$$

The conclusion then follows from ($\star$) and ($\star\star$). □

7.5 Exercises

Exercise 7.5.1. AUTOREGRESSIVE GAUSSIAN MODEL, TAKE 2
This is a continuation of Exercise 3.6.32.

3. Show that X_n converges in distribution to a centered Gaussian variable of mean 0 and variance γ^2 to be computed.

4. Suppose now that X_0 is Gaussian with mean 0 and variance γ^2 as computed in the previous question. Show that $\{X_n\}_{n\geq 0}$ is a strictly stationary sequence, in the sense that for all n, $(X_k, X_{k+1}, \dots, X_{k+n})$ has a distribution independent of k.

Exercise 7.5.2. POISSON'S LAW OF RARE EVENTS IN THE PLANE
With A a positive real number, let $Z_1, \dots, Z_M$ be IID random vectors uniformly distributed in the square $\Gamma_A := [0, A] \times [0, A]$. Define for any set $C \subseteq \Gamma_A$, $N(C)$ to be the number of random vectors Z_i that fall in C. Let $C_1, \dots, C_K$ be *disjoint* bounded subsets of $\mathbb{R}^2$.

Let $M = M(A)$ be a function of A such that

$$\frac{M(A)}{A^2} = \lambda > 0.$$

Show that, as $A \uparrow \infty$, $(N(C_1), \dots, N(C_K))$ converges in distribution. Identify the limit distribution.

Exercise 7.5.3. A CHARACTERISTIC PROPERTY OF THE GAUSSIAN DISTRIBUTION
Let G be a cumulative distribution function on $\mathbb{R}$ such that

$$\int_{\mathbb{R}} x \, \mathrm{d}G(x) = 0 \text{ and } \int_{\mathbb{R}} x^2 \, \mathrm{d}G(x) = 1\,.$$

In addition, suppose that G has the following property: If X_1 and X_2 are independent random variables with the CDF G, then $\frac{X_1+X_2}{\sqrt{2}}$ also admits G as CDF. Prove that G is the CDF of a Gaussian variable with mean 0 and variance 1.

Exercise 7.5.4. MIXED MOMENTS OF A GAUSSIAN VECTOR
Let $X = (X_1, \ldots, X_n)^T$ be a centered (0-mean) n-dimensional Gaussian vector $X = (X_1, \ldots, X_n)^T$ with the covariance matrix $\Gamma = \{\sigma_{ij}\}$. Prove the following formula:

$$E[X_{i_1} X_{i_2}, \ldots, X_{i_{2k}}] = \sum_{\substack{(j_1,\ldots,j_{2k}) \\ j_1<j_2,\ldots,j_{2k-1}<j_{2k}}} \sigma_{j_1 j_2} \sigma_{j_3 j_4} \ldots \sigma_{j_{2k} j_{2k}}, \tag{7.13}$$

where the summation extends over all permutations $(j_1, \ldots, j_{2k})$ of $\{i_1, \ldots, i_{2k}\}$ such that $j_1 < j_2, \ldots, j_{2k-1} < j_{2k}$. There are $1 \cdot 3 \cdot 5 \ldots (2k-1)$ terms in the right-hand side of Eq. (7.13). The indices $i_1, \ldots, i_{2k}$ are in $\{1, \ldots, n\}$ and they may occur with repetitions. Also prove that the odd moments of a centered gaussian vector are null, that is:

$$E[X_{i_1} \ldots X_{i_{2k+1}}] = 0,$$

for all $(i_1, \ldots, i_{2k+1}) \in \{1, 2, \ldots, n\}^{2k+1}$. Apply the above to compute the quantities $E[X_1 X_2 X_3 X_4]$, $E[X_1^2 X_2^2]$, $E[X_1^4]$ and $E[X_1^{2k}]$.

Exercise 7.5.5. SERIES SUMMATION VIA THE CENTRAL LIMIT THEOREM
Prove, using the central limit theorem, that

$$\lim_{n\to\infty} \sum_{k=1}^{n} \mathrm{e}^{-n} \frac{n^k}{k!} = \frac{1}{2}.$$

Exercise 7.5.6. $g(X_n) \xrightarrow{\mathcal{D}} g(X)$
Let $\{X_n\}_{n\geq 1}$ and X be random variables such that $X_n \xrightarrow{\mathcal{D}} X$, and let $g : \mathbb{R} \to \mathbb{R}$ be a continuous function. Prove that $g(X_n) \xrightarrow{\mathcal{D}} g(X)$.

Exercise 7.5.7. CAUCHY TRICKS
Let $\{X_n\}_{n\geq 1}$ be a sequence of IID Cauchy random variables.

(a) What is the limit in distribution of $\frac{X_1+\cdots+X_n}{n}$?

(b) Does $\frac{X_1+\cdots+X_n}{n^2}$ converge in distribution?

(c) Does $\frac{X_1+\cdots+X_n}{n}$ converge almost surely to a (nonrandom constant)?

Exercise 7.5.8. CONVERGENCE IN DISTRIBUTION, BUT NOT IN PROBABILITY
Let Z be a random variable with a symmetric distribution (that is, Z and $-Z$ have the same distribution). Define the sequence $\{Z_n\}_{n\geq 1}$ as follows: $Z_n = Z$ if n is odd, $Z_n = -Z$ if n is even. In particular, $\{Z_n\}_{n\geq 1}$ converges in *distribution* to Z. Show that if Z is not the constant 0, then $\{Z_n\}_{n\geq 1}$ does NOT converge to Z *in probability*.

Exercise 7.5.9. CONVERGENCE IN PROBABILITY AND CONVERGENCE IN VARIATION
Let $\{Z_n\}_{n\geq 0}$ be a sequence of $\{0,1\}$-valued random variables. Show that it converges in variation to 0 if and only if it converges in probability to 0.

Exercise 7.5.10. CONVERGENCE IN PROBABILITY BUT NOT IN DISTRIBUTION
Give an example of a sequence of random variables that converges in probability but not in distribution.

Chapter 8

Martingales

A martingale is for the general public a clever way of gambling. In mathematics, it formalizes the notion of fair game and we shall see that martingale theory indeed has something to say about such games. However the interest and scope of martingale theory extends far beyond gambling and has become a fundamental tool of the theory of stochastic processes. The present chapter is an introduction to this topic, featuring the two main pillars on which it rests: the *optional sampling* theorem and the convergence theory of martingales.

8.1 The Martingale Property

Let $(\Omega, \mathcal{F}, P)$ be a probability space and let $\{\mathcal{F}_n\}_{n\geq 1}$ be a *history* (or *filtration*) defined on it, that is, a sequence of sub-σ-fields of $\mathcal{F}$ that is non-decreasing: $\mathcal{F}_n \subseteq \mathcal{F}_{n+1}$ $(n \geq 0)$. The *internal history* of a random sequence $\{X_n\}_{n\geq 0}$ is the filtration $\{\mathcal{F}_n^X\}_{n\geq 0}$ defined by $\mathcal{F}_n^X := \sigma(X_0, \ldots, X_n)$.

Definition 8.1.1 *A complex random sequence $\{Y_n\}_{n\geq 0}$ such that for all $n \geq 0$*

(i) *Y_n is $\mathcal{F}_n$-measurable and*

(ii) $E[|Y_n|] < \infty$

*is called a $(P, \mathcal{F}_n)$-*martingale *(resp.,* sub*martingale,* super*martingale) if, in addition, for all $n \geq 0$, P-almost surely,*

$$E[Y_{n+1} \mid \mathcal{F}_n] = Y_n \qquad (resp., \;\geq Y_n, \leq Y_n)\,. \tag{8.1}$$

When the context is clear as to the choice of the underlying probability measure P, we shall abbreviate, saying for instance, "$\mathcal{F}_n$-submartingale" instead of "$(P, \mathcal{F}_n)$-submartingale".

P. Brémaud, *An Introduction to Applied Probability*, Texts in Applied Mathematics 77,
https://doi.org/10.1007/978-3-031-49306-5_8

If the history is not mentioned, it is assumed to be the internal history. For instance, the phrase $\{Y_n\}_{n\geq 0}$ is a martingale means that it is an $\mathcal{F}_n^Y$-martingale.

Of course an $\mathcal{F}_n$-martingale is an $\mathcal{F}_n$-submartingale *and* an $\mathcal{F}_n$–supermartingale. Condition (8.1) implies that for all $k \geq 1$, all $n \geq 0$,

$$E[Y_{n+k} \mid \mathcal{F}_n] = Y_n \qquad (\text{resp.}, \ \geq Y_n, \leq Y_n).$$

Proof. In the martingale case, for instance, by the rule of successive conditioning

$$\begin{aligned} E[Y_{n+k} \mid \mathcal{F}_n] &= E[E[Y_{n+k} \mid \mathcal{F}_{n+k-1}] \mid \mathcal{F}_n] \\ &= E[Y_{n+k-1} \mid \mathcal{F}_n] = E[Y_{n+k-2} \mid \mathcal{F}_n] \\ &= \cdots = E[Y_n \mid \mathcal{F}_n] = Y_n \,. \end{aligned}$$

□

In particular, taking expectations and letting $n = 0$,

$$E[Y_k] = E[Y_0] \qquad (\text{resp.}, \ \geq E[Y_0], \leq E[Y_0])\,.$$

EXAMPLE 8.1.2: SUMS OF IID RANDOM VARIABLES. Let $\{X_n\}_{n\geq 0}$ be an IID sequence of *centered and integrable* random variables. The random sequence

$$Y_n := X_0 + X_1 + \cdots + X_n \quad (n \geq 0)$$

is an $\mathcal{F}_n^X$-martingale. Indeed, for all $n \geq 0$, Y_n is $\mathcal{F}_n^X$-measurable and

$$E[Y_{n+1} \mid \mathcal{F}_n^X] = E[Y_n \mid \mathcal{F}_n] + E[X_{n+1} \mid \mathcal{F}_n^X] = Y_n + E[X_{n+1}] = Y_n \,,$$

where the second equality is due to the fact that $\mathcal{F}_n^X$ and X_{n+1} are independent (Theorem 5.6.5).

EXAMPLE 8.1.3: PRODUCTS OF IIDS. Let $X = \{X_n\}_{n\geq 0}$ be an IID sequence of integrable random variables with mean 1. The random sequence

$$Y_n = \prod_{k=0}^{n} X_k \quad (n \geq 0)$$

is an $\mathcal{F}_n^X$-martingale. Indeed, for all $n \geq 0$, Y_n is $\mathcal{F}_n^X$-measurable and

$$\begin{aligned} E[Y_{n+1} \mid \mathcal{F}_n^X] &= E\left[X_{n+1} \prod_{k=0}^{n} X_k \mid \mathcal{F}_n^X\right] = E[X_{n+1} \mid \mathcal{F}_n^X] \prod_{k=0}^{n} X_k \\ &= E[X_{n+1}] \prod_{k=1}^{n} X_k = 1 \times Y_n = Y_n \,, \end{aligned}$$

where the second equality is due to the fact that $\mathcal{F}_n^X$ and X_{n+1} are independent (Theorem 5.6.5).

EXAMPLE 8.1.4: GAMBLING. Consider the random sequence $\{Y_n\}_{n\geq 0}$ with values in $\mathbb{R}_+$ defined by $Y_0 = a \in \mathbb{R}_+$ and

$$Y_{n+1} = Y_n + X_{n+1}\, b_{n+1}(X_0^n) \quad (n \geq 0)\,,$$

where $X_0^n := (X_0, \dots, X_n)$, $X_0 = Y_0$, $\{X_n\}_{n\geq 1}$ is an IID sequence of random variables taking the values $+1$ or -1 with equal probability, and the family of functions $b_n : \{0,1\}^n \to \mathbb{N}$ $(n \geq 1)$ is the *betting strategy*, that is, $b_{n+1}(X_0^n)$ is the stake at time $n+1$ of a gambler given the observed history $\mathcal{F}_n^X$ of the chance outcomes up to time n. Admissible bets must guarantee that the fortune Y_n remains non-negative at all times n, that is, $b_{n+1}(X_0^n) \leq Y_n$. The process so defined is an $\mathcal{F}_n^X$-martingale. Indeed, for all $n \geq 0$, Y_n is $\mathcal{F}_n^X$-measurable and

$$\begin{aligned} E\left[Y_{n+1} \mid \mathcal{F}_n^X\right] &= E\left[Y_n \mid \mathcal{F}_n^X\right] + E\left[X_{n+1} b_{n+1}(X_0^n) \mid \mathcal{F}_n^X\right] \\ &= Y_n + E\left[X_{n+1} \mid \mathcal{F}_n^X\right] b_{n+1}(X_0^n) = Y_n\,, \end{aligned}$$

where the second equality uses Theorem 5.6.9. The integrability condition should be checked on each application. It is satisfied if the stakes $b_n(X_0^n)$ are uniformly bounded.

EXAMPLE 8.1.5: HARMONIC FUNCTIONS OF AN HMC. Let $\{X_n\}_{n\geq 0}$ be an HMC with countable space E and transition matrix $\mathbf{P}$. A function $h : E \to \mathbb{R}$ is called *harmonic* (resp., *subharmonic, superharmonic*) if $\mathbf{P}h$ is well defined and

$$\mathbf{P}h = h \qquad (\text{resp., } \geq h, \leq h)\,, \tag{8.2}$$

that is,

$$\sum_{j\in E} p_{ij} h(j) = h(i) \qquad (\text{resp., } \geq h(i), \leq h(i)) \quad (i \in E)\,.$$

Superharmonic functions are also called *excessive* functions.

Equation (8.2) is equivalent, in the harmonic case for instance, to

$$E[h(X_{n+1}) \mid X_n = i] = h(i) \quad (i \in E)\,. \tag{$\star$}$$

In view of the Markov property, the left-hand side of the above equality is also equal to

$$E[h(X_{n+1}) \mid X_n = i, X_{n-1} = i_{n-1}, \dots, X_0 = i_0]\,,$$

and therefore ($\star$) is equivalent to

$$E[h(X_{n+1} \mid \mathcal{F}_n^X] = h(X_n)\,.$$

Therefore, if $E[|h(X_n)|] < \infty$ for all $n \geq 0$, the process $\{h(X_n)\}_{n\geq 0}$ is an $\mathcal{F}_n^X$-martingale. Similarly, for a subharmonic (*resp.* superharmonic) function h such that $E[|h(X_n)|] < \infty$ for all $n \geq 0$, the process $\{h(X_n)\}_{n\geq 0}$ is an $\mathcal{F}_n^X$-submartingale (*resp.* $\mathcal{F}_n^X$-supermartingale).

Definition 8.1.6 *Let $\{\mathcal{F}_n\}_{n\geq 0}$ be some filtration. A $(P, \mathcal{F}_n)$-*martingale difference *(resp.,* sub*martingale difference,* super*martingale difference) is, by definition, a complex random sequence $\{X_n\}_{n\geq 0}$ such that for all $n \geq 0$,*

(a) *X_n is $\mathcal{F}_n$-measurable,*

(b) *$E[|X_n|] < \infty$ and $E[X_n] = 0$, and*

(c) *$E[X_{n+1} \mid \mathcal{F}_n] = 0$ (resp. ≥ 0, ≤ 0).*

The notion of martingale difference generalizes that of centered IID sequences. Indeed for such IID sequences, X_n is independent of $\mathcal{F}_n^X$, and therefore (Theorem 5.6.5) $E[X_{n+1} \mid \mathcal{F}_n^X] = 0$.

Convex Functions of Martingales

Theorem 8.1.7 *Let $I \subseteq \mathbb{R}$ be an interval (closed, open, semi-closed, infinite, etc.) and let $\varphi : I \to \mathbb{R}$ be a convex function.*

A. *Let $\{Y_n\}_{n\geq 0}$ be an $\mathcal{F}_n$-martingale such that $P(Y_n \in I) = 1$ for all $n \geq 0$. Assume that $E[|\varphi(Y_n)|] < \infty$ for all $n \geq 0$. Then, the process $\{\varphi(Y_n)\}_{n\geq 0}$ is an $\mathcal{F}_n$-submartingale.*

B. *Assume moreover that φ is non-decreasing and suppose this time that $\{Y_n\}_{n\geq 0}$ is an $\mathcal{F}_n$-submartingale. Then, the process $\{\varphi(Y_n)\}_{n\geq 0}$ is an $\mathcal{F}_n$-submartingale.*

Proof. By Jensen's inequality for conditional expectations (Exercise 8.6.1),

$$E\left[\varphi(Y_{n+1})|\mathcal{F}_n\right] \geq \varphi(E\left[Y_{n+1}|\mathcal{F}_n\right])\,.$$

Therefore (case A)

$$E\left[\varphi(Y_{n+1})|\mathcal{F}_n\right] \geq \varphi(E\left[Y_{n+1}|\mathcal{F}_n\right]) = \varphi(Y_n)\,,$$

and (case B)

$$E[\varphi(Y_{n+1})|\mathcal{F}_n] \geq \varphi(E[Y_{n+1}|\mathcal{F}_n]) \geq \varphi(Y_n).$$

(For the last inequality, use the submartingale property $E[Y_{n+1}|\mathcal{F}_n] \geq Y_n$ and the hypothesis that φ is non-decreasing.) □

EXAMPLE 8.1.8: Let $\{Y_n\}_{n\geq 0}$ be an $\mathcal{F}_n$-martingale and let $p \geq 1$. As a special case of Theorem 8.1.7 with the convex function $x \to |x|^p$, we have that if $E[|Y_n|^p] < \infty$, $\{|Y_n|^p\}_{n\geq 0}$ is an $\mathcal{F}_n$-submartingale. Applying Theorem 8.1.7 with the convex function $x \mapsto x^+$, we have that $\{Y_n^+\}_{n\geq 0}$ is an $\mathcal{F}_n$-submartingale.

Martingale Transforms and Stopped Martingales

Let $\{\mathcal{F}_n\}_{n\geq 0}$ be some filtration. The complex stochastic process $\{H_n\}_{n\geq 1}$ is called *$\mathcal{F}_n$-predictable* if

$$H_n \text{ is } \mathcal{F}_{n-1}\text{-measurable} \quad \text{for all } n \geq 1.$$

Let $\{Y_n\}_{n\geq 0}$ be another complex stochastic process. The stochastic process

$$(H \circ Y)_n := \sum_{k=1}^{n} H_k(Y_k - Y_{k-1}) \quad (n \geq 1)$$

is called the *transform* of Y by H.

Theorem 8.1.9

(a) *Let $\{Y_n\}_{n\geq 0}$ be an $\mathcal{F}_n$-submartingale and let $\{H_n\}_{n\geq 0}$ be a bounded non-negative $\mathcal{F}_n$-predictable process. Then $\{(H\circ Y)_n\}_{n\geq 0}$ is an $\mathcal{F}_n$-submartingale.*

(b) *If $\{Y_n\}_{n\geq 0}$ is an $\mathcal{F}_n$-martingale and if $\{H_n\}_{n\geq 0}$ is bounded and $\mathcal{F}_n$-predictable, then $\{(H \circ Y)_n\}_{n\geq 0}$ is an $\mathcal{F}_n$-martingale.*

Proof. Conditions (i) and (ii) of (8.1.1) are obviously satisfied. Moreover,

$$\text{(a)} \quad E[(H\circ Y)_{n+1} - (H\circ Y)_n \,|\, \mathcal{F}_n] = E[H_{n+1}(Y_{n+1} - Y_n) \,|\, \mathcal{F}_n] = H_{n+1}E[Y_{n+1} - Y_n \,|\, \mathcal{F}_n] \geq 0,$$

using Theorem 5.6.9 for the second equality.

$$\text{(b)} \quad E[(H\circ Y)_{n+1} - (H\circ Y)_n \,|\, \mathcal{F}_n] = H_{n+1}E[Y_{n+1} - Y_n \,|\, \mathcal{F}_n] = 0,$$

by the same token. □

Definition 8.1.10 *Let $\{\mathcal{F}_n\}_{n\in\mathbb{N}}$ be a non-decreasing sequence of sub-σ-fields of $\mathcal{F}$. A random variable τ taking its values in $\overline{\mathbb{N}}$ and such that, for all $m \in \mathbb{N}$, the event $\{\tau = m\}$ is in $\mathcal{F}_m$ is called an $\mathcal{F}_n$-*stopping time.

In particular, τ is an $\mathcal{F}_n^X$-stopping time if, for all $m \in \mathbb{N}$, the event $\{\tau = m\}$ can be expressed as

$$1_{\{\tau=m\}} = \psi_m(X_0, \dots, X_m),$$

for some measurable function ψ_m with values in $\{0, 1\}$ (Theorem 5.6.3). This explains why stopping times are said to be *non anticipative.*

Theorem 8.1.11 *Let $\{\mathcal{F}_n\}_{n\geq 0}$ be a history and let $\mathcal{F}_\infty := \sigma(\cup_{n\geq 0}\mathcal{F}_n)$. Let τ be an $\mathcal{F}_n$-stopping time. The collection of events*

$$\mathcal{F}_\tau := \{A \in \mathcal{F}_\infty \mid A \cap \{\tau = n\} \in \mathcal{F}_n, \text{ for all } n \geq 1\}$$

is a σ-field, and τ is $\mathcal{F}_\tau$-measurable. Let $\{X_n\}_{n\geq 0}$ be an E-valued $\mathcal{F}_n$-adapted random sequence, and let τ be a finite $\mathcal{F}_n$-stopping time. Then $X(\tau)$ is $\mathcal{F}_\tau$-measurable.

The proof is left as an exercise.

If $\{\mathcal{F}_n\}_{n\geq 0}$ is the internal history of some random sequence $\{X_n\}_{n\geq 0}$, that is, if $\mathcal{F}_n = \mathcal{F}_n^X$ $(n \geq 0)$, one may interpret $\mathcal{F}_\tau^X$ as the collection of events that are determined by the observation of the random sequence up to time τ (included).

Theorem 8.1.9 immediately leads to the *stopped martingale* theorem:

Theorem 8.1.12 *Let $\{Y_n\}_{n\geq 0}$ be an $\mathcal{F}_n$-submartingale (resp., martingale) and let τ be an $\mathcal{F}_n$-stopping time. Then $\{Y_{n\wedge\tau}\}_{n\geq 0}$ is an $\mathcal{F}_n$-submartingale (resp., martingale). In particular,*

$$E[Y_{n\wedge\tau}] \geq E[Y_0] \quad (\textit{resp.}, \ = E[Y_0]) \quad (n \geq 0). \tag{8.3}$$

Proof. Let $H_n := 1_{\{n\leq\tau\}}$. The stochastic process H is $\mathcal{F}_n$–predictable since $\{H_n = 0\} = \{\tau \leq n-1\} \in \mathcal{F}_{n-1}$. We have

$$\begin{aligned} Y_{n\wedge\tau} &= Y_0 + \sum_{k=1}^{n\wedge\tau} (Y_k - Y_{k-1}) \\ &= Y_0 + \sum_{k=1}^{n} 1_{\{k\leq\tau\}} (Y_k - Y_{k-1}) . \end{aligned}$$

The result then follows by Theorem 8.1.9. □

8.2 Martingale Inequalities

Kolmogorov's Inequality

Theorem 8.2.1 *Let $\{S_n\}_{n\geq 0}$ be an $\mathcal{F}_n$-submartingale. Then, for all $\lambda \in \mathbb{R}_+$,*

$$\lambda P\left(\max_{0\leq i\leq n} S_i > \lambda\right) \leq E\left[S_n 1_{\{\max_{0\leq i\leq n} S_i > \lambda\}}\right]. \tag{8.4}$$

Proof. Define the random time

$$\tau = \inf\{n \geq 0\,;\, S_n > \lambda\}.$$

It is an $\mathcal{F}_n$-stopping time since

$$A_i := \{\tau = i\} = \left\{S_i > \lambda,\, \max_{0\leq j\leq i-1} S_j \leq \lambda\right\} \in \mathcal{F}_i.$$

The A_i's so defined are mutually disjoint and

$$A := \left\{\max_{0\leq i\leq n} S_i > \lambda\right\} = \bigcup_{i=1}^{n} A_i.$$

Since $\lambda 1_{A_i} \leq S_i 1_{A_i}$,

$$\lambda P(A) = \lambda \sum_{i=0}^{n} P(A_i) \leq \sum_{i=0}^{n} E[S_i 1_{A_i}].$$

For all $0 \leq i \leq n$, A_i being $\mathcal{F}_i$-measurable, we have by the submartingale property that $E[S_n \mid \mathcal{F}_i] \geq S_i$ and therefore $\int_{A_i} S_i \, dP \leq \int_{A_i} E[S_n \mid \mathcal{F}_i] \, dP$. Taking these observations into account,

$$\begin{aligned}
\lambda P(A) &\leq \sum_{i=0}^{n} E[S_i 1_{A_i}] \leq \sum_{i=0}^{n} E\left[E^{\mathcal{F}_i}[S_n] 1_{A_i}\right] \\
&= \sum_{i=0}^{n} E\left[E^{\mathcal{F}_i}[S_n 1_{A_i}]\right] = \sum_{i=0}^{n} E[S_n 1_{A_i}] \\
&= E\left[S_n \sum_{i=0}^{n} 1_{A_i}\right] = E[S_n 1_A].
\end{aligned}$$

□

Corollary 8.2.2 *Let $\{M_n\}_{n\geq 0}$ be an $\mathcal{F}_n$-martingale. Then, for all $p \geq 1$, all $\lambda \in \mathbb{R}$,*

$$\lambda^p P\left(\max_{0\leq i\leq n} |M_i| > \lambda\right) \leq E[|M_n|^p]\,. \tag{8.5}$$

Proof. Let $S_n = |M_n|^p$. This defines an $\mathcal{F}_n$-submartingale (Example 8.1.8) to which one may apply Kolmogorov's inequality with λ replaced by λ^p:

$$\lambda^p P\left(\max_{0\leq i\leq n} |M_i|^p > \lambda^p\right) \leq E\left[|M_n|^p 1_{\{\max_{0\leq i\leq n} |M_i|^p > \lambda^p\}}\right] \leq E[|M_n|^p]\,.$$

□

Doob's Inequality

Recall the notation $\| X \|_p := (E\left[|X|^p\right])^{1/p}$.

Theorem 8.2.3 *Let $\{M_n\}_{n\geq 0}$ be an $\mathcal{F}_n$-martingale. For all $p > 1$,*

$$\| M_n \|_p \leq \| \max_{0\leq i\leq n} |M_i| \|_p \leq q \| M_n \|_p\,, \tag{8.6}$$

where q (the "conjugate" of p) is defined by $\frac{1}{p} + \frac{1}{q} = 1$.

Proof. The first inequality is trivial. For the second inequality, observe that for all non-negative random variables X, by Fubini's theorem,

$$\begin{aligned} E[X^p] &= E\left[\int_0^X p x^{p-1}\,\mathrm{d}x\right] \\ &= E\left[\int_0^\infty p x^{p-1} 1_{\{x<X\}}\,\mathrm{d}x\right] = p\int_0^\infty x^{p-1} P(X > x)\,\mathrm{d}x\,. \end{aligned}$$

Therefore, applying this and Kolmogorov's inequality (8.4) to the submartingale

$S_n = |M_n|$,

$$\begin{aligned}
E\left[\max_{0\le i\le n} |M_i|^p\right] &\le E\left[\left(\max_{0\le i\le n} |M_i|\right)^p\right] \\
&= p\int_0^\infty x^{p-1} P\left(\max_{0\le i\le n} |M_i| > x\right) \mathrm{d}x \\
&\le p\int_0^\infty x^{p-2} E\left[|M_n|\, 1_{\{\max_{0\le i\le n} |M_i|>x\}}\right] \mathrm{d}x \\
&= pE\left[\int_0^\infty x^{p-2} |M_n|\, 1_{\{\max_{0\le i\le n} |M_i|>x\}}\, \mathrm{d}x\right] \\
&= pE\left[|M_n| \int_0^{\max_{0\le i\le n} |M_i|} x^{p-2}\, \mathrm{d}x\right] \\
&= \frac{p}{p-1} E\left[|M_n| \left(\max_{0\le i\le n} |M_i|\right)^{p-1}\right] \\
&= qE\left[|M_n| \left(\max_{0\le i\le n} |M_i|\right)^{p-1}\right].
\end{aligned}$$

By Hölder's inequality, and observing that $(p-1)q = p$,

$$\begin{aligned}
E\left[|M_n| \left(\max_{0\le i\le n} |M_i|\right)^{p-1}\right] &\le E[|M_n|^p]^{1/p}\; E\left[\left(\max_{0\le i\le n} |M_i|\right)^{(p-1)q}\right]^{1/q} \\
&= \| M_n \|_p \; E\left[\left(\max_{0\le i\le n} |M_i|\right)^p\right]^{1/q}.
\end{aligned}$$

Therefore

$$E\left[\max_{0\le i\le n} |M_i|^p\right] \le q \,\| M_n \|_p \; E\left[\left(\max_{0\le i\le n} |M_i|\right)^p\right]^{1/q},$$

or (eliminating the trivial case where $E\left[\max_{0\le i\le n} |M_i|^p\right] = \infty$)

$$E\left[\max_{0\le i\le n} |M_i|^p\right]^{1-\frac{1}{q}} \le q \,\| M_n \|_p \,,$$

that is, since $1 - \frac{1}{q} = \frac{1}{p}$,

$$\| \max_{0\le i\le n} |M_i| \|_p \le q \,\| M_n \|_p \,.$$

□

Hoeffding's Inequality

Theorem 8.2.4 *Let* $\{M_n\}_{n\geq 0}$ *be a real* $\mathcal{F}_n$*-martingale such that, for some sequence* $c_1, c_2, \dots$ *of real numbers,*

$$P(|M_n - M_{n-1}| \leq c_n) = 1 \quad (n \geq 1). \tag{8.7}$$

Then, for all $x \geq 0$ *and all* $n \geq 1$,

$$P(|M_n - M_0| \geq x) \leq 2 \exp\left(-\frac{1}{2}x^2 \Big/ \sum_{i=1}^{n} c_i^2\right). \tag{8.8}$$

Proof. By convexity of $z \mapsto \mathrm{e}^{az}$, for $|z| \leq 1$ and all $a \in \mathbb{R}$,

$$a^{az} \leq \frac{1}{2}(1-z)\mathrm{e}^{-a} + \frac{1}{2}(1+z)\mathrm{e}^{+a}.$$

In particular, if Z is a centered random variable such that $P(|Z| \leq 1) = 1$,

$$\begin{aligned} E[\mathrm{e}^{aZ}] &\leq \frac{1}{2}(1 - E[Z])\mathrm{e}^{-a} + \frac{1}{2}(1 + E[Z])\mathrm{e}^{+a} \\ &= \frac{1}{2}\mathrm{e}^{-a} + \frac{1}{2}\mathrm{e}^{+a} \leq \mathrm{e}^{a^2/2}. \end{aligned}$$

By similar arguments, for all $a \in \mathbb{R}$,

$$\begin{aligned} &E\left[\mathrm{e}^{a\left(\frac{M_n - M_{n-1}}{c_n}\right)} \middle| \mathcal{F}_{n-1}\right] \\ &\quad \leq \frac{1}{2}\left(1 - E\left[\frac{M_n - M_{n-1}}{c_n}\middle|\mathcal{F}_{n-1}\right]\right)\mathrm{e}^{-a} + \cdots \\ &\quad\quad \cdots + \frac{1}{2}\left(1 + E\left[\frac{M_n - M_{n-1}}{c_n}\middle|\mathcal{F}_{n-1}\right]\right)\mathrm{e}^{+a} \leq \mathrm{e}^{a^2/2}, \end{aligned}$$

and, with a replaced by $c_n a$,

$$E\left[\mathrm{e}^{a(M_n - M_{n-1})} \middle| \mathcal{F}_{n-1}\right] \leq \mathrm{e}^{a^2 c_n^2/2}.$$

Therefore,

$$\begin{aligned} E\left[\mathrm{e}^{a(M_n - M_0)}\right] &= E\left[\mathrm{e}^{a(M_{n-1} - M_0)}\mathrm{e}^{a(M_n - M_{n-1})}\right] \\ &= E\left[\mathrm{e}^{a(M_{n-1} - M_0)} E\left[\mathrm{e}^{a(M_n - M_{n-1})} \middle| \mathcal{F}_{n-1}\right]\right] \\ &\leq E\left[\mathrm{e}^{a(M_{n-1} - M_0)}\right] \times \mathrm{e}^{a^2 c_n^2/2}, \end{aligned}$$

and then by recurrence

$$E\left[e^{a(M_n-M_0)}\right] \le e^{\frac{1}{2}a^2\sum_{i=1}^n c_i^2}.$$

In particular, with $a > 0$, by Markov's inequality,

$$P(M_n - M_0 \ge x) \le e^{-ax}E\left[e^{a(M_n-M_0)}\right] \le e^{-ax+\frac{1}{2}a^2\sum_{i=1}^n c_i^2}.$$

Minimization of the right-hand side with respect to a gives

$$P(M_n - M_0 \ge x) \le e^{-\frac{1}{2}x^2/\sum_{i=1}^n c_i^2}.$$

The same argument with $M_0 - M_n$ instead of $M_n - M_0$ yields the bound

$$P(-(M_n - M_0) \ge x) \le e^{-\frac{1}{2}x^2/\sum_{i=1}^n c_i^2}.$$

The announced bound then follows from these two bounds since for any random variable X, and all $x \in \mathbb{R}_+$, $P(|X| \ge x) = P(X \ge x) + P(X \le -x)$. □

EXAMPLE 8.2.5: THE KNAPSACK. There are n objects, the i-th has a volume V_i and is worth W_i. All these non-negative random variables form an independent family, the V_i's have finite means and the means of the W_i's are bounded by $M < \infty$. You have to choose integers $z_1, \dots, z_n$ in such a way that the total volume $\sum_{i=1}^n z_iV_i$ does not exceed a given storage capacity c and that the total worth $\sum_{i=1}^n z_iV_i$ is maximized. Call this maximal worth Z. We shall see that

$$P\left(|Z - E[Z]| \ge x\right) \le 2\exp\left\{\frac{-x^2}{2nM^2}\right\} \quad (x \ge 0).$$

For this consider the variables Z_j which are the equivalent of Z when the j-th object has been removed. Let now $M_j := E[Z \mid \mathcal{F}_j]$, where $\mathcal{F}_j := \sigma\left((V_k, W_k); 1 \le k \le j\right)$. Note that in view of the independence assumptions $E[Z_j \mid \mathcal{F}_j] = E[Z_{j-1} \mid \mathcal{F}_j]$. Clearly $Z_j \le Z \le Z_j + M$. Taking conditional expectations given $\mathcal{F}_j$ and then $\mathcal{F}_{j-1}$ in this last chain of inequalities reveals that $|M_j - M_{j-1}| \le M$. The rest is then just Hoeffding's inequality.

We now give a general framework of application.

Let $\mathcal{X}$ be a finite set, and let $f : \mathcal{X}^N \to \mathbb{R}$ be a given function. We introduce the notation $x = (x_1, \dots, x_N)$ and $x_1^k = (x_1, \dots, x_k)$. In particular, $x = x_1^N$. For $x \in \mathcal{X}^N$, $z \in \mathcal{X}$ and $1 \le k \le N$, let

$$f_k(x, z) := f(x_1, \dots, x_{k-1}, z, x_{k+1}, \dots, x_N).$$

The function f is said to satisfy the *Lipschitz condition* with bound c if for all $x \in \mathcal{X}^N$, all $z \in \mathcal{X}$ and all $1 \le k \le N$,

$$|f_k(x, z) - f(x)| \le c\,.$$

Let $X_1, X_2, \dots, X_N$ be independent random variables with values in $\mathcal{X}$. Define the martingale

$$M_n = E\left[f(X) \mid X_1^n\right].$$

By the independence assumption, with obvious notations,

$$E\left[f(X) \mid X_1^n\right] = \sum_{x_{n+1}^N} f(X_1^{n-1}, X_n, x_{n+1}^N) P(X_{n+1}^N = x_{n+1}^N)$$

and

$$E\left[f(X) \mid X_1^{n-1}\right] = \sum_{x_{n+1}^N} \sum_{x_n} f(X_1^{n-1}, x_n, x_{n+1}^N) P(X_n = x_n) P(X_{n+1}^N = x_{n+1}^N)\,.$$

Therefore

$$|M_n - M_{n-1}| \le \sum_{x_{n+1}^N} \sum_{x_n} |f(X_1^{n-1}, x_n, x_{n+1}^N) - f(X_1^{n-1}, X_n, x_{n+1}^N)| P(X_n = x_n) P(X_{n+1}^N = x_{n+1}^N) \le c\,.$$

EXAMPLE 8.2.6: PATTERN MATCHING. Take $f(x)$ to be the number of occurrences of the fixed pattern $b = (b_1, \dots, b_k)$ ($k \le N$) in the sequence $x = (x_1, \dots, x_N)$, that is

$$f(x) = \sum_{i=1}^{N-k+1} 1_{\{x_i = b_1, \dots, x_{i+k-1} = b_k\}}\,.$$

The mean number of matches in an IID sequence $X = (X_1, \dots, X_N)$ with uniform distribution on $\mathcal{X}$ is therefore

$$E\left[f(X)\right] = \sum_{i=1}^{N-k+1} E\left[1_{\{X_i = b_1, \dots, X_{i+k-1} = b_k\}}\right] = \sum_{i=1}^{N-k+1} \left(\frac{1}{|\mathcal{X}|}\right)^k,$$

that is,

$$E\left[f(X)\right] = (N - k + 1) \left(\frac{1}{|\mathcal{X}|}\right)^k.$$

The martingale $M_n := E\left[f(X) \mid X_1^n\right]$ is such that $M_0 = E\left[f(X)\right]$. Changing the value of one coordinate of $x \in \mathcal{X}^N$ changes $f(x)$ by at most k, we can apply the bound of Theorem 8.8 with $c_i \equiv k$ to obtain the inequality

$$P(|f(X) - E\left[f(X)\right]| \ge \lambda) \le 2e^{-\frac{1}{2}\frac{\lambda^2}{Nk^2}}\,.$$

8.3 The Optional Sampling Theorem

Martingale theory rests on two pillars. The first pillar is the *Doob's optional sampling theorem.* The second pillar is the *martingale convergence theorem* (and its avatars).

The version of the optional sampling theorem given next is the most elementary one, sufficient for the elementary examples to be considered now. More general results are given later in this subsection.

Theorem 8.3.1 *Let $\{M_n\}_{n\geq 0}$ be an $\mathcal{F}_n$-martingale, and let τ be an $\mathcal{F}_n$-stopping time (see Definition 8.1.10). Suppose that at least one of the following conditions holds:*

(α) $P(\tau \leq n_0) = 1$ *for some* $n_0 \geq 0$, *or*

(β) $P(\tau < \infty) = 1$ *and* $|M_n| \leq K < \infty$ *when* $n \leq \tau$.

Then

$$E[M_\tau] = E[M_0]\,. \tag{8.9}$$

Proof. (α) Just apply Theorem 8.1.12 (Formula (8.3) with $n = n_0$).

(β) Apply the result of (α) to the $\mathcal{F}_n$-stopping time $\tau \wedge n_0$ to obtain

$$E[M_{\tau\wedge n_0}] = E[M_0]\,.$$

But, by dominated convergence,

$$\lim_{n_0\uparrow\infty} E[M_{\tau\wedge n_0}] = E[\lim_{n_0\uparrow\infty} M_{\tau\wedge n_0}] = E[M_\tau]\,.$$

□

EXAMPLE 8.3.2: THE RUIN PROBLEM VIA MARTINGALES. The symmetric random walk $\{X_n\}_{n\geq 0}$ on $\mathbb{Z}$ with initial state 0 is an $\mathcal{F}_n^X$-martingale (Example 8.1.2). Let τ be the first time n for which $X_n = -a$ or $+\,b$, where $a, b > 0$. This is an $\mathcal{F}_n^X$-stopping time and moreover $\tau < \infty$. Part (β) of the above result can be applied with $K = \sup(a, b)$ to obtain $0 = E[X_0] = E[X_\tau]$. Writing $v = P(-a$ is hit before $b)$, we have

$$E[X_\tau] = -av + b(1-v)\,,$$

and therefore

$$v = \frac{b}{a+b}.$$

EXAMPLE 8.3.3: A COUNTEREXAMPLE. Consider the symmetric random walk of the previous example, but now define τ to be the hitting time of $b > 0$, an almost surely finite time since the symmetric walk on $\mathbb{Z}$ is recurrent. If the optional sampling theorem applied, one would have

$$0 = E[X_0] = E[X_\tau] = b,$$

an obvious contradiction. Of course, neither condition (α) nor (β) is satisfied.

We are now ready for the statement and proof of Doob's optional sampling theorem generalizing the elementary results given at the beginning of the present section.

Theorem 8.3.4 *Let $\{Y_n\}_{n\geq 0}$ be an $\mathcal{F}_n$-submartingale (resp., martingale), and let τ_1, τ_2 be finite $\mathcal{F}_n$-stopping times such that $P(\tau_1 \leq \tau_2) = 1$. If for $i = 1, 2$,*

$$E\left[|Y_{\tau_i}|\right] < \infty, \tag{8.10}$$

and

$$\liminf_{n\uparrow\infty} E\left[|Y_n| 1_{\{\tau_i > n\}}\right] = 0, \tag{8.11}$$

then, P-a.s.

$$E[Y_{\tau_2} \mid \mathcal{F}_{\tau_1}] \geq Y_{\tau_1} \text{ (resp., } = Y_{\tau_1}). \tag{8.12}$$

In particular,

$$E[Y_{\tau_2}] \geq E[Y_{\tau_1}] \qquad (\text{resp., } = E[Y_{\tau_1}]). \tag{8.13}$$

More generally, if $\{\tau_n\}_{n\geq 1}$ is a non-decreasing sequence of finite $\mathcal{F}_n$-stopping times satisfying conditions (8.10) and (8.11), the sequence $\{Y_{\tau_n}\}_{n\geq 1}$ is an $\mathcal{F}_{\tau_n}$-submartingale (resp., martingale).

Proof. It suffices to give the proof for the submartingale case. The meaning of (8.12) is that, for all $A \in \mathcal{F}_{\tau_1}$,

$$E\left[1_A Y_{\tau_2}\right] \geq E\left[1_A Y_{\tau_1}\right].$$

It is sufficient to show that for all $n \geq 0$,

$$E\left[1_{A\cap\{\tau_1 = n\}} Y_{\tau_2}\right] \geq E\left[1_{A\cap\{\tau_1 = n\}} Y_{\tau_1}\right],$$

or, equivalently since $\tau_1 = n$ implies $\tau_2 \geq n$,

$$E\left[1_{A\cap\{\tau_1=n\}\cap\{\tau_2\geq n\}}Y_{\tau_2}\right] \geq E\left[1_{A\cap\{\tau_1=n\}\cap\{\tau_2\geq n\}}Y_{\tau_1}\right] = E\left[1_{A\cap\{\tau_1=n\}\cap\{\tau_2\geq n\}}Y_n\right].$$

Write this as

$$E\left[1_{B\cap\{\tau_2\geq n\}}Y_{\tau_2}\right] \geq E\left[1_{B\cap\{\tau_2\geq n\}}Y_n\right], \tag{$\star$}$$

where $B := A \cap \{\tau_1 = n\}$. By definition of $\mathcal{F}_{\tau_1}$, $B \in \mathcal{F}_n$. It is therefore sufficient to show that for all $n \geq 0$, all $B \in \mathcal{F}_n$, $(\star)$ holds. We have

$$\begin{aligned}
E\left[1_{B\cap\{\tau_2\geq n\}}Y_n\right] &= E\left[1_{B\cap\{\tau_2=n\}}Y_n\right] + E\left[1_{B\cap\{\tau_2\geq n+1\}}Y_n\right] \\
&\leq E\left[1_{B\cap\{\tau_2=n\}}Y_n\right] + E\left[1_{B\cap\{\tau_2\geq n+1\}}E[Y_{n+1}|\mathcal{F}_n]\right] \\
&= E\left[1_{B\cap\{\tau_2=n\}}Y_{\tau_2}\right] + E\left[1_{B\cap\{\tau_2\geq n+1\}}Y_{n+1}\right] \\
&\leq E\left[1_{B\cap\{n\leq\tau_2\leq n+1\}}Y_{\tau_2}\right] + E\left[1_{B\cap\{\tau_2\geq n+2\}}Y_{n+2}\right] \\
&\cdots \\
&\leq E\left[1_{B\cap\{n\leq\tau_2\leq m\}}Y_{\tau_2}\right] + E\left[1_{B\cap\{\tau_2>m\}}Y_m\right],
\end{aligned}$$

that is,

$$E\left[1_{B\cap\{n\leq\tau_2\leq m\}}Y_{\tau_2}\right] \geq E\left[1_{B\cap\{\tau_2\geq n\}}Y_n\right] - E\left[1_{B\cap\{\tau_2>m\}}Y_m\right]$$

for all $m \geq n$. Therefore, by dominated convergence and hypothesis (8.11)

$$\begin{aligned}
E\left[1_{B\cap\{\tau_2\geq n\}}Y_{\tau_2}\right] &= E\left[\lim_{m\uparrow\infty} 1_{B\cap\{n\leq\tau_2\leq m\}}Y_{\tau_2}\right] \\
&\geq E\left[1_{B\cap\{\tau_2\geq n\}}Y_n\right] - \liminf_{m\uparrow\infty} E\left[1_{B\cap\{\tau_2>m\}}Y_m\right] \\
&= E\left[1_{B\cap\{\tau_2\geq n\}}Y_n\right].
\end{aligned}$$

□

Corollary 8.3.5 *Let $\{Y_n\}_{n\geq 0}$ be an $\mathcal{F}_n$-submartingale (resp., martingale). Let τ_1, τ_2 be $\mathcal{F}_n$-stopping times such that $\tau_1 \leq \tau_2 \leq N$ a.s., for some constant $N < \infty$. Then (8.13) holds.*

Proof. This is an immediate consequence of Theorem 8.3.4. □

Corollary 8.3.6 *Let $\{Y_n\}_{n\geq 0}$ be a* uniformly integrable *$\mathcal{F}_n$-submartingale (resp., martingale). Let τ_1, τ_2 be* finite *$\mathcal{F}_n$-stopping times. Then (8.12) holds.*

Proof. In order to apply Theorem 8.3.4, we have to show that conditions (8.10) and (8.11) are satisfied when $\{Y_n\}_{n\geq 1}$ is uniformly integrable. Condition (8.11) follows from part (b) of Theorem 6.5.3 since the τ_i's are finite and therefore $P(\tau_i > n) \to 0$ as $n \uparrow \infty$. It remains to show that condition (8.10) is satisfied. Let $N < \infty$ be an integer. By Corollary 8.3.5, if τ is a stopping time (here τ_1 or τ_2),

$$E[Y_0] \le E[Y_{\tau\wedge N}]$$

and therefore

$$E[|Y_{\tau\wedge N}|] = 2E[Y^+_{\tau\wedge N}] - E[Y_{\tau\wedge N}] \le 2E[Y^+_{\tau\wedge N}] - E[Y_0]\,.$$

The submartingale $\{Y_n^+\}_{n\ge 0}$ satisfies

$$\begin{aligned} E[Y^+_{\tau\wedge N}] &= \sum_{j=0}^{N} E[1_{\{\tau\wedge N = j\}} Y_j^+] + E[1_{\{\tau > N\}} Y_N^+] \\ &\le \sum_{j=0}^{N} E[1_{\{\tau\wedge N = j\}} Y_N^+] + E[1_{\{\tau > N\}} Y_N^+] \\ &= E[Y_N^+] \le E[|Y_N|]\,. \end{aligned}$$

Therefore

$$E[|Y_{\tau\wedge N}|] \le 2E[|Y_N|] + E[|Y_0|] \le 3\sup_N E|Y_N|\,.$$

Since by Fatou's lemma $E[|Y_\tau|] \le \liminf_{N\uparrow\infty} E[|Y_{\tau\wedge N}|]$, we have

$$E[|Y_\tau|] \le 3\sup_N E[|Y_N|]\,,$$

a finite quantity since $\{Y_n\}_{n\ge 1}$ is uniformly integrable. □

Corollary 8.3.7 *Let $\{Y_n\}_{n\ge 0}$ be an $\mathcal{F}_n$-submartingale (resp., martingale) and let τ be an $\mathcal{F}_n$-stopping time such that*

$$E[\tau] < \infty.$$

Suppose moreover that there exists a constant $c < \infty$ such that, for all $n \ge 0$,

$$E[|Y_{n+1} - Y_n| \mid \mathcal{F}_n] \le c, \quad P\text{-a.s. on } \{\tau \ge n\}.$$

Then $E[|Y_\tau|] < \infty$ and

$$E[Y_\tau] \ge (\text{ resp., } =) \, E[Y_0]\,.$$

Proof. In order to apply Theorem 8.3.4 with $\tau_1 = 0$, $\tau_2 = \tau$, one just has to check conditions (8.10) and (8.11) for τ. Let $Z_0 := |Y_0|$. With $Z_n := |Y_n - Y_{n-1}|$ $(n \geq 1)$,

$$E\left[\sum_{j=0}^{\tau} Z_j\right] = \sum_{n=0}^{\infty} E\left[1_{\{\tau=n\}} \sum_{j=0}^{n} Z_j\right] = \sum_{n=0}^{\infty}\sum_{j=0}^{n} E\left[1_{\{\tau=n\}} Z_j\right]$$
$$= \sum_{j=0}^{\infty}\sum_{n=j}^{\infty} E\left[1_{\{\tau=n\}} Z_j\right] = \sum_{j=0}^{\infty} E\left[1_{\{\tau\geq j\}} Z_j\right].$$

For $j \geq 1$, $\{\tau \geq j\} = \overline{\{\tau < j-1\}} \in \mathcal{F}_{j-1}$ and therefore,

$$E\left[1_{\{\tau\geq j\}} Z_j\right] = E\left[1_{\{\tau\geq j\}} E\left[Z_j \mid \mathcal{F}_{j-1}\right]\right] \leq cP(\tau \geq j), \tag{$\star$}$$

and

$$E\left[\sum_{j=0}^{\tau} Z_j\right] \leq E[|Y_0|] + c\sum_{j=1}^{\infty} P(\tau \geq j) = E[|Y_0|] + cE[\tau] < \infty.$$

Therefore condition (8.10) is satisfied since $E\left[|Y_\tau|\right] \leq E\left[\sum_{j=0}^{\tau} Z_j\right]$. Moreover, if $\tau > n$,

$$\sum_{j=0}^{n} Z_j \leq \sum_{j=0}^{\tau} Z_j$$

and therefore

$$E\left[1_{\{\tau>n\}}|Y_n|\right] \leq E\left[1_{\{\tau>n\}} \sum_{j=0}^{\tau} Z_j\right].$$

But, by ($\star$), $E\left[\sum_{j=0}^{\tau} Z_j\right] < \infty$. Also, $\{\tau > n\} \downarrow \varnothing$ as $n \uparrow \infty$. Therefore, by dominated convergence

$$\liminf_{n\uparrow\infty} E[1_{\{\tau>n\}}|Y_n|] \leq \liminf_{n\uparrow\infty} E\left[1_{\{\tau>n\}} \sum_{j=0}^{\tau} Z_j\right] = 0.$$

This is condition (8.11). □

Wald's Formulas

Theorem 8.3.8 *Let $\{Z_n\}_{n\geq 1}$ be an* IID *sequence of real random variables such that $E[|Z_1|] < \infty$, and let τ be an $\mathcal{F}_n^Z$-stopping time with $E[\tau] < \infty$. Then*

$$E\left[\sum_{n=1}^{\tau} Z_n\right] = E[Z_1]E[\tau]. \tag{8.14}$$

If, moreover, $E[Z_1^2] < \infty$,

$$\text{Var}\left(\sum_{n=1}^{\tau} Z_n\right) = \text{Var}(Z_1)E[\tau]. \tag{8.15}$$

Proof. Let $X_0 := 0$, $X_n := (Z_1 + \cdots + Z_n) - nE[Z_1]$ $(n \geq 1)$. Then $\{X_n\}_{n\geq 1}$ is an $\mathcal{F}_n^Z$-martingale such that

$$\begin{aligned} E[|X_{n+1} - X_n| \mid \mathcal{F}_n^Z] &= E[|Z_{n+1} - E[Z_1]| \mid \mathcal{F}_n^Z] \\ &= E|Z_n - E[Z_1]| \leq 2E[|Z_1|] < \infty. \end{aligned}$$

Therefore Corollary 8.3.7 can be applied with $Y_n = \sum_{k=1}^{n}(Z_k - E[Z_1])$ to obtain (8.14). For the proof of (8.15), the same kind of argument works, this time with the martingale $Y_n = X_n^2 - n\,\text{Var}(Z_1)$. □

Theorem 8.3.9 *Let $\{Z_n\}_{n\geq 1}$ be* IID *real random variables and let $S_n = Z_1 + \cdots + Z_n$. Let $\varphi_Z(t) := E[e^{tZ_1}]$ and suppose that $\varphi_Z(t_0)$ exists and is greater than or equal to 1 for some $t_0 \neq 0$. Let τ be an $\mathcal{F}_n^Z$-stopping time such that $E[\tau] < \infty$ and $|S_n| \leq c$ on $\{\tau \geq n\}$ for some constant $c < \infty$. Then*

$$E\left[\frac{e^{t_0 S_\tau}}{\varphi_Z(t_0)^\tau}\right] = 1. \tag{8.16}$$

Proof. Let $Y_0 := 1$ and for $n \geq 1$,

$$Y_n := \frac{e^{t_0 S_n}}{\varphi_Z(t_0)^n}.$$

By application of the result of Example 8.1.3 with $X_i := \frac{e^{t_0 Z_i}}{\varphi_Z(t_0)}$, we have that the sequence $\{Y_n\}_{n\geq 0}$ is an $\mathcal{F}_n^Z$-martingale. Moreover, on $\{\tau \geq n\}$,

$$\begin{aligned} E[|Y_{n+1} - Y_n| \mid \mathcal{F}_n^Z] &= Y_n E\left[\left|\frac{e^{t_0 Z_{n+1}}}{\varphi_Z(t_0)} - 1\right| \mid \mathcal{F}_n^Z\right] \\ &= \frac{Y_n}{\varphi_Z(t_0)} E\left[|e^{t_0 Z_1} - \varphi_Z(t_0)|\right] \leq K < \infty \end{aligned}$$

since $\varphi_Z(t_0) \geq 1$ and

$$Y_n = \frac{e^{t_0 S_n}}{\varphi_Z(t_0)^n} \leq \frac{e^{|t_0|c}}{\varphi_Z(t_0)^n} \leq e^{|t_0|c}.$$

Therefore, Corollary 8.3.7 applies to give (8.16). □

8.4 The Martingale Convergence Theorem

The second pillar of martingale theory is the martingale convergence theorem. This result is the probabilistic counterpart of the convergence of a non-negative non-increasing, or bounded non-decreasing, sequence of real numbers to a finite limit. It says in particular (but we shall give a more complete result soon) that a non-negative supermartingale converges almost surely to a finite limit.

The Upcrossing Inequality

The proof of the martingale convergence theorem is based on the *upcrossing inequality*.

Theorem 8.4.1 *Let $\{S_n\}_{n\geq 0}$ be an $\mathcal{F}_n$-submartingale. Let $a, b \in \mathbb{R}$ with $a < b$, and let ν_n be the number of upcrossings of $[a, b]$ before ($\leq$) time n. Then*

$$(b - a)E[\nu_n] \leq E[(S_n - a)^+]. \tag{8.17}$$

(By definition, an upcrossing occurs at time ℓ if $S_k \leq a$ and if there exists $\ell > k$ such that $S_j < b$ for $j = 1, \ldots, \ell - 1$ and $S_\ell \geq b$.)

Proof. Since ν_n is the number of upcrossings of $[0, b - a]$ by the submartingale $\{(S_n - a)^+\}_{n\geq 1}$, we may suppose without loss of generality that $S_n \geq 0$ and take $a = 0$, and then prove that

$$bE[\nu_n] \leq E[S_n - S_0], \tag{8.18}$$

where $S_0 = 0$ and $\mathcal{F}_0$ is the gross σ-field. Define a sequence of $\mathcal{F}_n$-stopping times as follows

$$\begin{aligned}
\tau_0 &= 0 \\
\tau_1 &= \inf\{n > \tau_0 \, ; \, S_n = 0\} \\
\tau_2 &= \inf\{n > \tau_1 \, ; \, S_n \geq b\} \\
&\cdots \\
\tau_{2k+1} &= \inf\{n > \tau_{2k} \, ; \, S_n = 0\} \\
\tau_{2k} &= \inf\{n > \tau_{2k+1} \, ; \, S_n \geq b\} \\
&\cdots
\end{aligned}$$

For $i \geq 1$, let

$$\begin{aligned}
\varphi_i &= 1 \text{ if } \tau_m < i \leq \tau_{m+1} \text{ for some odd } m \\
&= 0 \text{ if } \tau_m < i \leq \tau_{m+1} \text{ for some even } m \, .
\end{aligned}$$

Observe that

$$\{\varphi_i = 1\} = \bigcup_{odd\, m} \left(\{\tau_m < i\} \cap \overline{\{\tau_{m+1} < i\}} \right) \in \mathcal{F}_{i-1}$$

and that

$$b\nu_n \leq \sum_{i=1}^{n} \varphi_i (S_i - S_{i-1}) \, .$$

Therefore

$$\begin{aligned}
bE[\nu_n] &\leq E[\sum_{i=1}^{n} \varphi_i (S_i - S_{i-1})] = \sum_{i=1}^{n} E[\varphi_i (S_i - S_{i-1})] \\
&= \sum_{i=1}^{n} E[\varphi_i E[(S_i - S_{i-1}) | \mathcal{F}_{i-1}]] = \sum_{i=1}^{n} E[\varphi_i (E[S_i | \mathcal{F}_{i-1}] - S_{i-1})] \\
&\leq \sum_{i=1}^{n} E[(E[S_i | \mathcal{F}_{i-1}] - S_{i-1})] \leq \sum_{i=1}^{n} (E[S_i] - E[S_{i-1}]) = E[S_n - S_0] \, .
\end{aligned}$$

□

We are now in a position to state and prove the fundamental *martingale convergence theorem.*

Theorem 8.4.2 *Let $\{S_n\}_{n \geq 0}$ be an $\mathcal{F}_n$-submartingale. Suppose moreover that it is L^1-bounded, that is,*

$$\sup_{n \geq 0} E[|S_n|] < \infty \, . \tag{8.19}$$

Then $\{S_n\}_{n \geq 0}$ converges P-a.s. to an integrable random variable S_∞.

Condition (8.19) can be replaced by the equivalent condition

$$\sup_{n\geq 0} E[S_n^+] < \infty .$$

Indeed, if $\{S_n\}_{n\geq 0}$ is an $\mathcal{F}_n$-submartingale,

$$E\left[S_n^+\right] \leq E\left[|S_n|\right] \leq 2E\left[S_n^+\right] - E\left[S_n\right] \leq 2E\left[S_n^+\right] - E\left[S_0\right] .$$

By changing signs, the same hypothesis leads to the same conclusion for a *supermartingale* $\{S_n\}_{n\geq 0}$. Similarly to the previous remark, condition (8.19) can be replaced by the equivalent condition

$$\sup_{n\geq 0} E[S_n^-] < \infty .$$

Proof. The proof is based on the following observation concerning any deterministic sequence $\{x_n\}_{n\geq 1}$. If this sequence does not converge, then it is possible to find two rational numbers a and b such that

$$\liminf_n x_n < a < b < \limsup_n x_n ,$$

which implies that the number of upcrossings of $[a, b]$ by this sequence is infinite. Therefore to prove convergence, it suffices to prove that any interval $[a, b]$ with rational extremities is crossed at most a finite number of times.

Let $\nu_n([a, b])$ be the number of upcrossings of an interval $[a, b]$ prior ($\leq$) to time n and let $\nu_\infty([a, b]) := \lim_{n\uparrow\infty} \nu_n([a, b])$. By (8.17),

$$\begin{aligned}(b-a)E[\nu_n([a, b])] &\leq E[(S_n - a)^+] \leq E[S_n^+] + |a| \\ &\leq \sup_{k\geq 0} E[S_k^+] + |a| \leq \sup_{k\geq 0} E[|S_k|] + |a| < \infty .\end{aligned}$$

Therefore, letting $n \uparrow \infty$,

$$(b-a)E[\nu_\infty([a, b])] < \infty .$$

In particular, $\nu_\infty([a, b]) < \infty$, P-a.s. Therefore, P-a.s. there is only a finite number of upcrossings of any *rational* interval $[a, b]$. Equivalently, in view of the observation made in the first lines of the proof, $\{S_n\}_{n\geq 0}$ converges P-a.s. to some random variable S_∞. Therefore (by Fatou's lemma for the second inequality):

$$E[|S_\infty|] = E[\lim_{n\uparrow\infty} |S_n|] \leq \liminf_{n\uparrow\infty} E|S_n| \leq \sup_{n\geq 0} E|S_n| < \infty .$$

□

Corollary 8.4.3

(a) *Any non-positive submartingale* $\{S_n\}_{n\geq 0}$ *almost surely converges to an integrable random variable.*

(b) *Any non-negative supermartingale almost surely converges to an integrable random variable.*

Proof. (b) follows from (a) by changing signs. For (a), we have

$$E[|S_n|] = -E[S_n] \leq -E[S_0] = E[|S_0|] < \infty .$$

Therefore (8.19) is satisfied and the conclusion then follows from Theorem 8.4.2. □

An immediate application of the martingale convergence theorem is to gambling. The next example teaches us that a gambler in a "fair game" is eventually ruined.

EXAMPLE 8.4.4: FAIR GAME NOT SO FAIR. Consider the situation in Example 8.1.4, assuming that the initial fortune a is a positive integer and that the bets are also positive integers (that is, the functions $b_{n+1}(X_0^n) \in \mathbb{N}_+$ except if $Y_n = 0$, in which case the gambler is not allowed to bet anymore, or equivalently $b_n(X_0^{n-1} * 0) := b_n(X_0, X_1, \dots, X_n, 0) = 0$). In particular, $Y_n \geq 0$ for all $n \geq 0$. Therefore the process $\{Y_n\}_{n\geq 0}$ is a non-negative $\mathcal{F}_n^X$-martingale and by the martingale convergence theorem it almost surely has a finite limit. Since the bets are assumed positive integers when the fortune of the player is positive, this limit cannot be other than 0. Since Y_n is a non-negative integer for all $n \geq 0$, this can happen only if the fortune of the gambler becomes null in finite time.

EXAMPLE 8.4.5: BRANCHING PROCESSES VIA MARTINGALES. The power of the concept of martingale will now be illustrated by revisiting the branching process. It is assumed that $P(Z = 0) < 1$ and $P(Z \geq 2) > 0$ (to get rid of trivialities). The stochastic process

$$Y_n = \frac{X_n}{m^n},$$

where m is the average number of sons of a given individual, is an $\mathcal{F}_n^X$-martingale. Indeed, since each one among the X_n members of the nth generation gives birth on average to m sons and does this independently of the rest of the population, $E[X_{n+1}|X_n] = mX_n$ and

$$E\left[\frac{X_{n+1}}{m^{n+1}}|\mathcal{F}_n^X\right] = E\left[\frac{X_{n+1}}{m^{n+1}}|X_n\right] = \frac{X_n}{m^n}.$$

By the martingale convergence theorem, almost surely

$$\lim_{n\uparrow\infty} \frac{X_n}{m^n} = Y < \infty .$$

In particular, if $m < 1$, then $\lim_{n\uparrow\infty} X_n = 0$ almost surely. Since X_n takes integer values, this implies that the branching process eventually becomes extinct.

If $m = 1$, then $\lim_{n\uparrow\infty} X_n = X_\infty < \infty$ and it is easily argued that this limit must be 0. Therefore, in this case as well the process eventually becomes extinct.

For the case $m > 1$, we consider the unique solution in $(0, 1)$ of $x = g(x)$ (g is the generating function of the typical progeny of a member of the population considered). Suppose we can show that $Z_n = x^{X_n}$ is a martingale. Then, by the martingale convergence theorem, Z_n converges to a finite limit and therefore X_n has a limit X_∞, which however can be infinite. One can easily argue that this limit cannot be other than 0 (extinction) or ∞ (non-extinction). Since $\{Z_n\}_{n\geq 0}$ is a martingale, $x = E[Z_0] = E[Z_n]$ and therefore, by dominated convergence, $x = E[Z_\infty] = E[x^{X_\infty}] = P(X_\infty = 0)$. Therefore x is the probability of extinction.

It remains to show that $\{Z_n\}_{n\geq 0}$ is an $\mathcal{F}_n^X$-martingale. For all $i \in \mathbb{N}$ and all $x \in [0, 1]$, $E[x^{X_{n+1}}|X_n = i] = x^i$. This is obvious if $i = 0$. If $i > 0$, X_{n+1} is the sum of i independent random variables with the same generating function g, and therefore, $E[x^{X_{n+1}}|X_n = i] = g(x)^i = x^i$. From this last result and the Markov property,

$$E[x^{X_{n+1}}|\mathcal{F}_n^X] = E[x^{X_{n+1}}|X_n] = x^{X_n}.$$

The following results are important refinements of the fundamental martingale convergence theorem.

Theorem 8.4.6 *Let $\{M_n\}_{n\geq 0}$ be an $\mathcal{F}_n$-martingale such that for some $p \in (1, \infty)$,*

$$\sup_{n\geq 0} E|M_n|^p < \infty . \tag{8.20}$$

Then $\{M_n\}_{n\geq 0}$ converges a.s. and in L^p to some finite variable M_∞.

Proof. By hypothesis, the martingale $\{M_n\}_{n\geq 0}$ is L^p-bounded and *a fortiori* L^1-bounded since $p > 1$. Therefore it converges almost surely. By Doob's inequality, $E[\max_{0\leq i\leq n} |M_i|^p] \leq q^p E|M_n|^p$ and in particular,

$$E[\max_{0\leq i\leq n} |M_i|^p] \leq q^p \sup_k E|M_k|^p < \infty .$$

Letting $n \uparrow \infty$, we have in view of condition (8.20) that

$$E[\sup_{n\geq 0} |M_n|^p] < \infty . \tag{8.21}$$

Therefore $\{|M_n|^p\}_{n\geq 0}$ is uniformly integrable (Theorem 6.5.5). In particular, since it converges almost surely, it also converges in L^1 (Theorem 6.5.7). In other words, $\{M_n\}_{n\geq 0}$ converges in L^p. □

The above result was proved for $p > 1$ (the proof depended on Doob's inequality, which is true for $p > 1$). For $p = 1$, a similar result holds with an additional assumption of uniform integrability. Note however that the next result also applies to submartingales.

Theorem 8.4.7 *A uniformly integrable $\mathcal{F}_n$-submartingale $\{S_n\}_{n\geq 0}$ converges a.s. and in L^1 to an integrable random variable S_∞ and $E[S_\infty \mid \mathcal{F}_n] \geq S_n$.*

Proof. By the uniform integrability hypothesis, $\sup_n E[|S_n|] < \infty$ and therefore, by Theorem 8.4.2, S_n converges almost surely to some integrable random variable S_∞. It also converges to this variable in L^1 since a uniformly integrable sequence that converges almost surely also converges in L^1 (Theorem 6.5.7).

By the submartingale property, for all $A \in \mathcal{F}_n$, all $m \geq n$,

$$E[1_A S_n] \leq E[1_A S_m] .$$

Since convergence is in L^1,

$$\lim_{m\uparrow\infty} E[1_A S_m] = E[1_A S_\infty] ,$$

so that finally $E[1_A S_n] \leq E[1_A S_\infty]$. This being true for all $A \in \mathcal{F}_n$, we have that $E[S_\infty \mid \mathcal{F}_n] \geq S_n$. □

The following result is *Lévy's continuity theorem for conditional expectations.*

Corollary 8.4.8 *Let $\{\mathcal{F}_n\}_{n\geq 1}$ be a filtration and let ξ be an integrable random variable. Let $\mathcal{F}_\infty := \sigma\big(\cup_{n\geq} \mathcal{F}_n\big)$. Then*

$$\lim_{n\uparrow\infty} E[\xi \mid \mathcal{F}_n] = E[\xi \mid \mathcal{F}_\infty] . \tag{8.22}$$

Proof. It suffices to treat the case where ξ is non-negative. The sequence $\{M_n = E[\xi \mid \mathcal{F}_n]\}_{n\geq 1}$ is a uniformly integrable $\mathcal{F}_n$-martingale (Theorem 6.5.4) and by Theorem 8.4.7, it converges almost surely and in L^1 to some integrable random variable M_∞. We have to show that $M_\infty = E[\xi \mid \mathcal{F}_\infty]$. For $m \geq n$ and $A \in \mathcal{F}_n$,

$$E[1_A M_m] = E[1_A M_n] = E[1_A E[\xi \mid \mathcal{F}_n]] = E[1_A \xi] .$$

Since convergence is also in L^1, $\lim_{m\uparrow\infty} E[1_A M_m] = E[1_A M_\infty]$. Therefore

$$E[1_A M_\infty] = E[1_A \xi] \tag{8.23}$$

for all $A \in \mathcal{F}_n$ and therefore for all $A \in \cup_n \mathcal{F}_n$. The σ-finite measures $A \mapsto E[1_A M_\infty]$ and $A \mapsto E[1_A \xi]$ agreeing on the algebra $\cup_n \mathcal{F}_n$ also agree on the smallest σ-algebra containing it, that is $\mathcal{F}_\infty$. Therefore (8.23) holds for all $A \in \mathcal{F}_\infty$ (Theorem 4.1.32) and this implies

$$E[1_A M_\infty] = E[1_A E[\xi \mid \mathcal{F}_\infty]] ,$$

and finally, since M_∞ is $\mathcal{F}_\infty$-measurable, $M_\infty = E[\xi \mid \mathcal{F}_\infty]$. □

Backwards (or Reverse) Martingales

In the following, pay attention to the indexation: the index set is the set of non-positive relative integers. Let $\{\mathcal{F}_n\}_{n\leq 0}$ be a non-decreasing family of σ-fields, that is, $\mathcal{F}_n \subseteq \mathcal{F}_{n+1}$ for all $n \leq -1$.

There is nothing new in the definition of "backwards" or "reverse" martingales or submartingales, except that the index set is now $\{\dots, -2, -1, 0\}$. For instance, $\{Y_n\}_{n\leq 0}$ is an $\mathcal{F}_n$-submartingale if $E[Y_n \mid \mathcal{F}_{n-1}] \geq Y_{n-1}$ for all $n \leq 0$. The term "backwards" in fact refers to one of the uses that is made of this notion, that of discussing the limit of Y_n as $n \downarrow -\infty$.

Reverse martingales or submartingales often appear in the following setting. Let $\{Z_k\}_{k\geq 0}$ be a sequence of integrable random variables. Suppose that

$$E[Z_{k-1} \mid Z_k, Z_{k+1}, Z_{k+2}, \dots] = Z_k \quad (k \geq 0) .$$

Clearly, the change of indexation $k \to -n$ gives a "backwards" martingale. The next example concerns that situation.

EXAMPLE 8.4.9: EMPIRICAL MEAN OF AN IID SEQUENCE. Let $\{X_n\}_{n\geq 1}$ be an IID sequence of integrable random variables and let

$$Z_k := \frac{1}{k} S_k ,$$

where $S_k := X_1 + \cdots + X_k$. We shall prove that

$$E[Z_{k-1} \mid \mathcal{G}_k] = Z_k ,$$

where $\mathcal{G}_k = \sigma(Z_k, Z_{k+1}, Z_{k+2}, \ldots)$. It suffices to prove that for all $k \geq 1$,

$$E[Z_1 \mid \mathcal{G}_k] = Z_k , \qquad (\star)$$

since it then follows that for $m \leq k$,

$$E[Z_m \mid \mathcal{G}_k] = E[E[Z_1 \mid \mathcal{G}_m] \mid \mathcal{G}_k] = E[Z_1 \mid \mathcal{G}_k] = Z_k .$$

By linearity,

$$S_k = E[S_k \mid \mathcal{G}_k] = \sum_{j=1}^{k} E[X_j \mid \mathcal{G}_k] .$$

¿From the fact that $\mathcal{G}_k = \sigma(Z_k, Z_{k+1}, Z_{k+2}, \ldots) = \sigma(S_k, X_{k+1}, X_{k+2}, \ldots)$ and by the IID assumption for $\{X_n\}_{n\geq 1}$,

$$S_k = \sum_{j=1}^{k} E[X_j \mid S_k, X_{k+1}, X_{k+2}, \ldots] = \sum_{j=1}^{k} E[X_j \mid S_k] .$$

But the pairs (X_j, S_k) $(1 \leq j \leq k)$ have the same distribution, and therefore

$$\sum_{j=1}^{k} E[X_j \mid S_k] = kE[X_1 \mid S_k] = kE[X_1 \mid \mathcal{G}_k] = kE[Z_1 \mid \mathcal{G}_k] ,$$

from which $(\star)$ follows.

Theorem 8.4.10 *Let $\{\mathcal{F}_n\}_{n\leq 0}$ be a non-decreasing family of σ-fields. Let $\{S_n\}_{n\leq 0}$ be an $\mathcal{F}_n$-submartingale. Then:*

A. *S_n converges P-a.s. and in L^1 as $n \downarrow -\infty$ to an integrable random variable $S_{-\infty}$, and*

B. *with $\mathcal{F}_{-\infty} := \cap_{n\leq 0}\mathcal{F}_n$,*

$$S_{-\infty} \leq E[S_0 \mid \mathcal{F}_{-\infty}] ,$$

with equality if $\{S_n\}_{n\leq 0}$ is an $\mathcal{F}_n$-martingale.

Proof. First note that by the submartingale property, $S_n \leq E[S_0 \mid \mathcal{F}_n]$ $(n \leq 0)$. In particular, $\{S_n\}_{n\leq 0}$ is not only L^1-bounded, but also uniformly integrable (Theorem 6.5.4).

A. Denoting by $\nu_m = \nu_m([a,b])$ the number of upcrossings of $[a,b]$ by $\{S_n\}_{n\leq 0}$ in the integer interval $[-m,0]$ and by $\nu = \nu([a,b])$ the total number of upcrossings of $[a,b]$, the upcrossing inequality yields

$$(b-a)E[\nu_m] \leq E\left[(S_0-a)^+\right] < \infty,$$

and letting $m \uparrow \infty$, $E[\nu] < \infty$. Almost-sure convergence to an integrable random variable $S_{-\infty}$ is then proved as in Theorem 8.4.2. Since $\{S_n\}_{n\leq 0}$ is uniformly integrable, convergence to $S_{-\infty}$ is also in L^1.

B. Clearly, $S_{-\infty}$ is $\mathcal{F}_{-\infty}$-measurable. Also, by the submartingale property, $S_n \leq E[S_0 \mid \mathcal{F}_n]$ $(n \leq -1)$, that is, for all $n \leq -1$ and all $A \in \mathcal{F}_n$,

$$\int_A S_n \, \mathrm{d}P \leq \int_A S_0 \, \mathrm{d}P.$$

This is true for any $A \in \mathcal{F}_{-\infty}$ because $\mathcal{F}_{-\infty} \subseteq \mathcal{F}_n$ for all $n \leq -1$. Since S_n converges to $S_{-\infty}$ in L^1 as $n \downarrow -\infty$, $\int_A S_n \, \mathrm{d}P \to \int_A S_{-\infty} \, \mathrm{d}P$ and therefore

$$\int_A S_{-\infty} \, \mathrm{d}P \leq \int_A S_0 \, \mathrm{d}P \quad (A \in \mathcal{F}_{-\infty}),$$

which implies that $S_{-\infty} \leq E[S_0 \mid \mathcal{F}_{-\infty}]$.

The martingale case is obtained using the same proof with each $\leq$ symbol replaced by $=$. □

Statement B says that $\{S_n\}_{n\in -\mathbb{N}\cup\{-\infty\}}$ is a submartingale relatively to the history $\{\mathcal{F}_n\}_{n\in -\mathbb{N}\cup\{-\infty\}}$.

EXAMPLE 8.4.11: THE STRONG LAW OF LARGE NUMBERS. The situation is that of Example 8.4.9. By Theorem 8.4.10, $S_k/k \to$ converges almost surely. By Kolmogorov's zero-one law (Theorem 6.3.3), $S_k/k \to a$, a deterministic number. It remains to identify a with $E[X_1]$. We know from the first lines of the proof of Theorem 8.4.10 that $\{S_k/k\}_{k\geq 1}$ is uniformly integrable. Therefore, by Theorem 6.5.7,

$$\lim_{k\uparrow\infty} E\left[\frac{S_k}{k}\right] = a.$$

But for all $k \geq 1$, $E[S_k/k] = E[X_1]$.

The uniform integrability of the backwards submartingale in Theorem 8.4.10 followed directly from the submartingale property. This is not the case for a *supermartingale* unless one adds a condition.

Theorem 8.4.12 *Let $\{\mathcal{F}_n\}_{n\leq 0}$ be a filtration and let $\{S_n\}_{n\leq 0}$ be an $\mathcal{F}_n$-supermartingale such that*

$$\sup_{n\leq 0} E\left[S_n\right] < \infty\,. \tag{8.24}$$

Then

A. *S_n converges P-a.s. and in L^1 as $n \downarrow -\infty$ to an integrable random variable $S_{-\infty}$, and*

B. *with $\mathcal{F}_{-\infty} := \cap_{n\leq 0}\mathcal{F}_n$,*

$$S_{-\infty} \geq E\left[S_0 \mid \mathcal{F}_{-\infty}\right] \quad P\text{-a.s.}$$

Proof. It suffices to prove uniform integrability, since the rest of the proof then follows the same lines as in Theorem 8.4.7.

Fix $\varepsilon > 0$ and select $k \leq 0$ such that

$$\lim_{i\downarrow -\infty} E\left[S_i\right] - E\left[S_k\right] \leq \varepsilon\,. \tag{$\star$}$$

Then $0 \leq E\left[S_n\right] - E\left[S_k\right] \leq \varepsilon$ for all $n \leq k$. We first show that for sufficiently large $\lambda > 0$,

$$\int_{\{|S_n|>\lambda\}} |S_n|\,\mathrm{d}P \leq \varepsilon\,.$$

It is enough to prove this for sufficiently large $-n$, here for $-n \geq -k$. The previous integral is equal to

$$-\int_{\{S_n<-\lambda\}} S_n\,\mathrm{d}P + E\left[S_n\right] - \int_{\{S_n\leq\lambda\}} S_n\,\mathrm{d}P\,.$$

By the supermartingale hypothesis, this quantity is

$$\leq -\int_{\{S_k<-\lambda\}} S_n\,\mathrm{d}P + E\left[S_n\right] - \int_{\{S_n\leq\lambda\}} S_k\,\mathrm{d}P\,.$$

In view of ($\star$), this is less than or equal to

$$-\int_{\{S_n<-\lambda\}} S_k\,\mathrm{d}P + E\left[S_k\right] - \int_{\{S_n\leq\lambda\}} S_k\,\mathrm{d}P + \varepsilon\,,$$

which is equal to

$$\int_{\{|S_n|>\lambda\}} |S_k|\,\mathrm{d}P + \varepsilon\,.$$

Since ε is an arbitrary positive quantity, it remains to show that $\int_{\{|S_n|>\lambda\}} |S_k| \, \mathrm{d}P$ tends to 0 uniformly in $n \leq 0$ as $\lambda \uparrow \infty$. But since $\{S_n^-\}_{n\geq 1}$ is a supermartingale

$$E[|S_n|] = E[S_n] + 2E[S_n^-] \leq \sup_{n\leq 0} E[S_n] + 2E[S_0^-] .$$

Therefore, in view of hypothesis (8.24),

$$P(|S_n| > \lambda) \leq \frac{E[|S_n|]}{\lambda} \to 0$$

uniformly in $n \leq 0$, and therefore

$$\int_{\{|S_n|>\lambda\}} |S_k| \, \mathrm{d}P \to 0$$

uniformly in n. □

The following result is the *backwards Lévy's continuity theorem for conditional expectations.*

Corollary 8.4.13 *Let $\{\mathcal{F}_n\}_{n\leq 0}$ be a history and let ξ be an integrable random variable. Then, with $\mathcal{F}_{-\infty} := \cap_{n\leq 0}\mathcal{F}_n$,*

$$\lim_{n\downarrow -\infty} E[\xi \,|\, \mathcal{F}_n] = E[\xi \,|\, \mathcal{F}_{-\infty}] . \tag{8.25}$$

Proof. $M_n := E[\xi \,|\, \mathcal{F}_n]$ $(n \leq 0)$ is an $\mathcal{F}_n$-martingale and therefore by the backwards martingale convergence theorem, it converges as $n \downarrow -\infty$ almost surely and in L^1 to some integrable variable $M_{-\infty}$ and

$$M_{-\infty} = E[M_0 \mid \mathcal{F}_{-\infty}] = E[E[\xi \mid \mathcal{F}_0] \mid \mathcal{F}_{-\infty}] = E[\xi \mid \mathcal{F}_{-\infty}]$$

since $\mathcal{F}_{-\infty} \subseteq \mathcal{F}_0$. □

The Robbins–Sigmund Theorem

In applications, one often encounters random sequences that are not quite martingales, submartingales or supermartingales, but "nearly" so, up to "perturbations". The statement of the result below will make this precise.

Theorem 8.4.14 *Let $\{V_n\}_{n\geq 1}$, $\{\beta_n\}_{n\geq 1}$, $\{\gamma_n\}_{n\geq 1}$ and $\{\delta_n\}_{n\geq 1}$ be real non-negative sequences of random variables adapted to some filtration $\{\mathcal{F}_n\}_{n\geq 1}$ and such that*

$$\mathrm{E}[V_{n+1} \,|\, \mathcal{F}_n] \leq V_n(1 + \beta_n) + \gamma_n - \delta_n \quad (n \geq 1) . \tag{8.26}$$

Then, on the set

$$\Gamma = \left\{ \sum_{n\geq 1} \beta_n < \infty \right\} \cap \left\{ \sum_{n\geq 1} \gamma_n < \infty \right\} \tag{8.27}$$

the sequence $\{V_n\}_{n\geq 1}$ *converges almost surely to a finite random variable and moreover* $\sum_{n\geq 1} \delta_n < \infty$ *P-almost surely.*

Proof. 1. Let $\alpha_0 := 0$ and

$$\alpha_n := \left(\prod_{k=1}^{n} (1+\beta_k) \right)^{-1} \quad (n \geq 1),$$

and let

$$V'_n := \alpha_{n-1} V_n, \quad \gamma'_n := \alpha_n \gamma_n, \quad \delta'_n := \alpha_n \delta_n \quad (n \geq 1).$$

Then

$$\mathrm{E}[V'_{n+1} \mid \mathcal{F}_n] = \alpha_n \mathrm{E}[V_{n+1} \mid \mathcal{F}_n] \leq \alpha_n V_n (1+\beta_n) + \alpha_n \gamma_n - \alpha_n \delta_n ,$$

that is, since $\alpha_n V_n (1+\beta_n) = \alpha_{n-1} V_n$,

$$\mathrm{E}[V'_{n+1} \mid \mathcal{F}_n] \leq V'_n + \gamma'_n - \delta'_n .$$

Therefore, the random sequence $\{Y_n\}_{n\geq 1}$ defined by

$$Y_n := V'_n - \sum_{k=1}^{n-1} (\gamma'_k - \delta'_k)$$

is an $\mathcal{F}_n$-supermartingale.

2. For $a > 0$, let

$$T_a := \inf \left\{ n \geq 1 \,;\, \sum_{k=1}^{n-1} (\gamma'_k - \delta'_k) \geq a \right\} .$$

The sequence $\{Y_{n\wedge T_a}\}_{n\geq 1}$ is an $\mathcal{F}_n$-supermartingale bounded from below by $-a$. It therefore converges to a finite limit. Therefore, on $\{T_a = \infty\}$, $\{Y_n\}_{n\geq 1}$ converges to a finite limit.

3. On Γ, $\prod_{k=1}^{\infty}(1+\beta_k)$ converges almost surely to a positive limit and therefore $\lim_{n\uparrow\infty} \alpha_n > 0$. Therefore, condition $\sum_{n\geq 1} \gamma_n < \infty$ implies $\sum_{n\geq 1} \gamma'_n < \infty$.

4. By definition of Y_n,

$$Y_n + \sum_{k=1}^{n-1} \gamma'_k = V'_n + \sum_{k=1}^{n-1} \delta'_k \geq \sum_{k=1}^{n-1} \delta'_k ,$$

But on $\Gamma \cap \{T_a = \infty\}$, $\{Y_n\}_{n\geq 1}$ converges to a finite random variable, and therefore $\sum_{n\geq 1} \delta'_n < \infty$.

5. Since on $\Gamma \cap \{T_a = \infty\}$, $\sum_{n\geq 1} \gamma'_n < \infty$, $\sum_{n\geq 1} \delta'_n < \infty$ and $\{Y_n\}_{n\geq 1}$ converges to a finite random variable, it follows that $\{V'_n\}_{n\geq 1}$ converges to a finite limit. Since $\lim_{n\uparrow\infty} \alpha_n > 0$, it follows in turn that $\{V_n\}_{n\geq 1}$ converges to a finite limit and $\sum_{n\geq 1} \delta_n < \infty$ on $\Gamma \cap \{T_a = \infty\}$, and therefore on $\Gamma \cap (\cup_a \{T_a = \infty\}) = \Gamma$. □

Corollary 8.4.15 *Let $\{V_n\}_{n\geq 1}$, $\{\gamma_n\}_{n\geq 1}$ and $\{\delta_n\}_{n\geq 1}$ be real non-negative sequences of random variables adapted to some filtration $\{\mathcal{F}_n\}_{n\geq 1}$. Suppose that for all $n \geq 1$*

$$\mathrm{E}[V_{n+1} \mid \mathcal{F}_n] \leq V_n + \gamma_n - \delta_n . \tag{8.28}$$

Let $\{a_n\}_{n\geq 1}$ be a random sequence that is strictly positive and strictly increasing and let

$$\widetilde{\Gamma} := \left\{ \sum_{n\geq 1} \frac{\gamma_n}{a_n} < \infty \right\} . \tag{8.29}$$

Then, almost-surely:

1. *on $\widetilde{\Gamma}$, the series $\sum_{n\geq 1} \frac{V_{n+1}-V_n}{a_n}$ is convergent and $\sum_{n\geq 1} \frac{\delta_n}{a_n} < \infty$,*
2. *on $\widetilde{\Gamma} \cap \{\lim_{n\uparrow\infty} a_n < \infty\}$, $\{V_n\}_{n\geq 1}$ converges almost surely, and*
3. *on $\widetilde{\Gamma} \cap \{\lim_{n\uparrow\infty} a_n = \infty\}$, $\lim_{n\uparrow\infty} \frac{V_n}{a_n} = 0$ and $\lim_{n\uparrow\infty} \frac{V_{n+1}}{a_n} = 0$.*

Proof. 1. Let for $n \geq 1$

$$Z_n := \sum_{k=1}^{n-1} \frac{V_{k+1} - V_k}{a_k} + \frac{V_1}{a_0} = \sum_{k=1}^{n} V_k \left(\frac{1}{a_{k-1}} - \frac{1}{a_k} \right) + \frac{V_n}{a_n} .$$

Since $\frac{1}{a_{k-1}} - \frac{1}{a_k} > 0$, we have that $Z_n \geq 0$ $(n \geq 1)$. Also

$$\mathrm{E}[Z_{n+1} \mid \mathcal{F}_n] \leq Z_n + \frac{\gamma_n}{a_n} - \frac{\delta_n}{a_n}.$$

Therefore, by Theorem 8.4.14, on $\widetilde{\Gamma}$, $\{Z_n\}_{n\geq 1}$ converges and $\sum_{n\geq 1} \frac{\delta_n}{a_n} < \infty$. Note that in particular

$$\lim_{n\uparrow\infty} \frac{V_{n+1} - V_n}{a_n} = 0 \text{ on } \widetilde{\Gamma} . \tag{8.30}$$

2. If moreover $\lim_{n\uparrow\infty} a_n = a_\infty < \infty$, the convergence of $\sum_{n\geq 1} \frac{V_{n+1}-V_n}{a_n}$ implies that of $\frac{1}{a_\infty} \sum_{n\geq 1} (V_{n+1} - V_n)$, and therefore $\{V_n\}_{n\geq 1}$ converges.

3. If on the contrary $\lim_{n\uparrow\infty} a_n = \infty$, the convergence of $\sum_{n\geq 1} \frac{V_{n+1}-V_n}{a_n}$ implies that of $\frac{V_{n+1}}{a_n}$ (and therefore that of $\frac{V_n}{a_n}$, by (8.30)) to 0 (recall Kronecker's lemma: if $a_n > 0$ and $a_n \uparrow \infty$, the convergence of $\sum_{n\geq 1} \frac{x_n}{a_n}$ implies that $\lim_{n\uparrow\infty} \frac{1}{a_n}\sum_{k=1}^{n} x_k = 0$). □

8.5 Square-integrable Martingales

Let $\{\mathcal{F}_n\}_{n\geq 0}$ be a filtration. Recall that a process $\{H_n\}_{n\geq 0}$ is called $\mathcal{F}_n$-*predictable* if for all $n \geq 1$, H_n is $\mathcal{F}_{n-1}$-measurable.

Doob's decomposition

Theorem 8.5.1 *Let $\{S_n\}_{n\geq 0}$ be an $\mathcal{F}_n$-submartingale. Then there exists a P-a.s. unique non-decreasing $\mathcal{F}_n$-predictable process $\{A_n\}_{n\geq 0}$ with $A_0 \equiv 0$ and a unique $\mathcal{F}_n$-martingale $\{M_n\}_{n\geq 0}$ such that for all $n \geq 0$,*

$$S_n = M_n + A_n\,.$$

Proof. Existence is proved by explicit construction. Let $M_0 := S_0$, $A_0 = 0$ and, for $n \geq 1$,

$$M_n := S_0 + \sum_{j=0}^{n-1} \{S_{j+1} - E[S_{j+1}|\mathcal{F}_j]\}\,,$$

$$A_n := \sum_{j=0}^{n-1} (E[S_{j+1}|\mathcal{F}_j] - S_j)\,.$$

Clearly, $\{M_n\}_{n\geq 0}$ and $\{A_n\}_{n\geq 0}$ have the announced properties. In order to prove uniqueness, let $\{M'_n\}_{n\geq 0}$ and $\{A'_n\}_{n\geq 0}$ be another such decomposition. In particular, for $n \geq 1$,

$$A'_{n+1} - A'_n = (A_{n+1} - A_n) + (M_{n+1} - M_n) - (M'_{n+1} - M'_n)\,.$$

Therefore

$$E[A'_{n+1} - A'_n \mid \mathcal{F}_n] = E[A_{n+1} - A_n \mid \mathcal{F}_n]\,,$$

and, since $A'_{n+1} - A'_n$ and $A_{n+1} - A_n$ are $\mathcal{F}_n$-measurable,

$$A'_{n+1} - A'_n = A_{n+1} - A_n, \qquad \text{P-a.s.} \quad (n \geq 1)\,,$$

from which it follows that $A'_n = A_n$ a.s. for all $n \geq 0$ (recall that $A'_0 = A_0$) and then $M'_n = M_n$ a.s. for all $n \geq 0$. □

Definition 8.5.2 *The sequence* $\{A_n\}_{n\geq 0}$ *in Theorem 8.5.1 is called the* compensator *of* $\{S_n\}_{n\geq 0}$.

Definition 8.5.3 *Let* $\{M_n\}_{n\geq 0}$ *be a square-integrable* $\mathcal{F}_n$*-martingale (that is,* $E[M_n^2] < \infty$ *for all* $n \geq 0$*). The compensator of the* $\mathcal{F}_n$*-submartingale* $\{M_n^2\}_{n\geq 0}$ *is denoted by* $\{\langle M\rangle_n\}_{n\geq 0}$ *and is called the* bracket process *of* $\{M_n\}_{n\geq 0}$.

By the explicit construction in the proof of Theorem 8.5.1, $\langle M\rangle_0 := 0$ and for $n \geq 1$,

$$\langle M\rangle_n := \sum_{j=0}^{n-1} \{E[M_{j+1}^2 \mid \mathcal{F}_j] - M_j^2\} = \sum_{j=0}^{n-1} \{E[(M_{j+1}^2 - M_j^2) \mid \mathcal{F}_j]\} . \tag{8.31}$$

Also, for all $0 \leq k \leq n$,

$$E[(M_n - M_k)^2 \mid \mathcal{F}_k] = E[M_n^2 - M_k^2 \mid \mathcal{F}_k] = E[\langle M\rangle_n - \langle M\rangle_k \mid \mathcal{F}_k] .$$

Therefore, $\{M_n^2 - \langle M\rangle_n\}_{n\geq 0}$ is an $\mathcal{F}_n$-martingale. In particular, if $M_0 = 0$, $E[M_n^2] = E[\langle M\rangle_n]$.

EXAMPLE 8.5.4: Let $\{Z_n\}_{n\geq 0}$ be a sequence of IID centered random variables of finite variance. Let $M_0 := 0$ and $M_n := \sum_{j=1}^{n} Z_j$ for $n \geq 1$. Then, for $n \geq 1$,

$$\langle M\rangle_n = \sum_{j=1}^{n} \operatorname{Var}(Z_j) .$$

Theorem 8.5.5 *If* $E[\langle M\rangle_\infty] < \infty$*, the square-integrable martingale* $\{M_n\}_{n\geq 0}$ *converges almost surely to a finite limit, and convergence takes place also in* L^2.

Proof. This is Theorem 8.4.6 for the particular case $p = 2$. In fact, condition (8.20) thereof is satisfied since

$$\sup_{n\geq 1} E\left[M_n^2\right] = \sup_{n\geq 1} E[\langle M\rangle_n] = E\left[\langle M\rangle_\infty\right] < \infty .$$

□

The Martingale Law of Large Numbers

Theorem 8.5.6 *Let $\{M_n\}_{n\geq 0}$ be a square-integrable $\mathcal{F}_n$-martingale. Then:*

A. *On $\{\langle M\rangle_\infty < \infty\}$, M_n converges to a finite limit.*

B. *On $\{\langle M\rangle_\infty = \infty\}$, $M_n/\langle M\rangle_n \to 0$.*

Proof. A. Let $K > 0$ be fixed, the random time

$$\tau_K := \inf\{n \geq 0 : \ \langle M\rangle_{n+1} > K\}$$

is an $\mathcal{F}_n$-stopping time since the bracket process is $\mathcal{F}_n$-predictable. Also $\langle M\rangle_{n\wedge\tau_K} \leq K$ and therefore by Theorem 8.5.5, $\{M_{n\wedge\tau_K}\}_{n\geq 0}$ converges to a finite limit. Therefore $\{M_n\}_{n\geq 0}$ converges to a finite limit on the set $\{\langle M\rangle_\infty < K\}$ contained in $\{\tau_K = \infty\}$. Hence the result since

$$\{\langle M\rangle_\infty < \infty\} = \bigcup_{K\geq 1}\{\tau_K = \infty\}\,.$$

B. Note that

$$E[M_{n+1}^2 \,|\, \mathcal{F}_n] = M_n^2 + \langle M\rangle_{n+1} - \langle M\rangle_n.$$

Define

$$V_n = M_n^2, \quad \gamma_n = \langle M\rangle_{n+1} - \langle M\rangle_n, \quad a_n = \langle M\rangle_{n+1}^2\,.$$

The result then follows from Part 3 of Corollary 8.4.15 (observe that there exists a k_0 such that $a_k \geq 1$ for $k \geq k_0$ and

$$\sum_{k=k_0}^{\infty}\gamma_k/a_k = \sum_{k=k_0}^{\infty}(\langle M\rangle_{k+1} - \langle M\rangle_k)/\langle M\rangle_{k+1}^2 \leq \int_1^\infty x^{-2}\,\mathrm{d}x < \infty)\,,$$

which says, in particular, that $\sqrt{V_{n+1}/a_n} = M_{n+1}/\langle M\rangle_{n+1}$ converges to 0. $\square$

We do not have in general $\{\langle M\rangle_\infty < \infty\} = \{\{M_n\}_{n\geq 0} \text{ converges}\}$.

The following is a *conditioned version of the Borel–Cantelli lemma.* Note that, in this form, we have a necessary and sufficient condition.

Corollary 8.5.7 *Let $\{\mathcal{F}_n\}_{n\geq 1}$ be a filtration and let $\{A_n\}_{n\geq 1}$ be a sequence of events such that $A_n \in \mathcal{F}_n$ $(n \geq 1)$. Then*

$$\left\{\sum_{n\geq 1} P(A_n \,|\, \mathcal{F}_{n-1}) = \infty\right\} \equiv \left\{\sum_{n\geq 1} 1_{A_n} = \infty\right\}.$$

Proof. Define $\{M_n\}_{n\geq 0}$ by $M_0 := 0$ and for $n \geq 1$,

$$M_n := \sum_{k=1}^{n} (1_{A_k} - P(A_k \,|\, \mathcal{F}_{k-1}))\,.$$

This is a square-integrable $\mathcal{F}_n$-martingale, with bracket process

$$\langle M \rangle_n = \sum_{k=1}^{n} P(A_k \,|\, \mathcal{F}_{k-1})(1 - P(A_k \,|\, \mathcal{F}_{k-1}))\,.$$

In particular,

$$\langle M \rangle_n \leq \sum_{k=1}^{n} P(A_k \,|\, \mathcal{F}_{k-1})\,.$$

A. Suppose that $\sum_{k=1}^{\infty} P(A_k \,|\, \mathcal{F}_{k-1}) < \infty$. Then, by the above inequality, $\langle M \rangle_\infty < \infty$, and therefore, by Part A of Theorem 8.5.6, M_n converges. Since by hypothesis, $\sum_{k=1}^{\infty} P(A_k \,|\, \mathcal{F}_{k-1}) < \infty$, this implies that $\sum_{k=1}^{\infty} 1_{A_k} < \infty$.

B. Suppose that $\sum_{k=1}^{\infty} P(A_k \,|\, \mathcal{F}_{k-1}) = \infty$ and $\langle M \rangle_\infty < \infty$. Then M_n converges to a finite random variable and therefore

$$\frac{M_n}{\sum_{k=1}^{n} P(A_k \,|\, \mathcal{F}_{k-1})} = \frac{\sum_{k=1}^{n} 1_{A_k}}{\sum_{k=1}^{n} P(A_k \,|\, \mathcal{F}_{k-1})} - 1 \to 0\,.$$

C. Suppose that $\sum_{k=1}^{\infty} P(A_k \,|\, \mathcal{F}_{k-1}) = \infty$ and $\langle M \rangle_\infty = \infty$. Then $\frac{M_n}{\langle M \rangle_n} \to 0$ and *a fortiori*,

$$\frac{M_n}{\sum_{k=1}^{n} P(A_k \,|\, \mathcal{F}_{k-1})} \to 0\,,$$

that is,

$$\frac{\sum_{k=1}^{n} 1_{A_k}}{\sum_{k=1}^{n} P(A_k \,|\, \mathcal{F}_{k-1})} \to 1\,.$$

□

8.6 Exercises

Exercise 8.6.1. CONDITIONAL JENSEN'S INEQUALITY
Let I be a general interval of $\mathbb{R}$ (closed, open, semi-closed, infinite, etc.) and let (a, b) be its interior, assumed non-empty. Let $\varphi : I \to \mathbb{R}$ be a convex function. Let X be an integrable real-valued random variable such that $P(X \in I) = 1$. Assume

moreover that either φ is non-negative, or that $\varphi(X)$ is integrable. Prove that for any sub-σ-field $\mathcal{G} \subseteq \mathcal{F}$

$$E\left[\varphi(X) \mid \mathcal{G}\right] \geq \varphi(E\left[X \mid \mathcal{G}\right]).$$

Exercise 8.6.2. DISCOUNTED PRODUCT
Let $\{X_n\}_{n\geq 1}$ be a sequence of independent integrable random variables with a common mean $m \neq 0$. Show that

$$Y_n := m^{-n} X_1 X_2 \cdots X_n \quad (n \geq 1)$$

is an $\mathcal{F}_n^X$-martingale.

Exercise 8.6.3. MEAN HITTING TIME VIA MARTINGALES
Let $\{X_n\}_{n\geq 0}$ be a symmetric random walk on $\mathbb{Z}$. Show that $\{X_n\}_{n\geq 0}$ and $\{X_n^2 - n\}_{n\geq 0}$ are $\mathcal{F}_n^X$-martingales. Deduce from this the mean of T of the hitting time of $\{-a, b\}$, where a and b are positive integers.

Exercise 8.6.4. PROBABILITY OF HIT
Let $\{X_n\}_{n\geq 0}$ be a HMC with state space E, and let B be a closed subset of states, that is,

$$i \in B \Rightarrow \sum_{j\in B} p_{ij} = 1.$$

Let T be the hitting time of B, and let for $i \in E$,

$$h(i) := P_i(T < \infty).$$

Show that $\{h(X_n)\}_{n\geq 0}$ is a $\mathcal{F}_n^X$-martingale.

Exercise 8.6.5. RUINED AGAIN!
Show that the function $h(i) = \left(\frac{q}{p}\right)^i$ is harmonic for the nonsymmetric random walk on $\mathbb{Z}$ (with $p_{i,i+1} = p, p_{i,i-1} = q = 1-p,\ p \neq \frac{1}{2}$), where $p \in (0,1), p \neq \frac{1}{2}$. Apply the optional sampling theorem to obtain the ruin probability in the ruin problem of Example 9.1.10.

Exercise 8.6.6. THE LÉVY MARTINGALE
Let $\{X_n\}_{n\geq 0}$ be an HMC with state space E and transition matrix $\mathbf{P}$, and let $f : E \to \mathbb{R}$ be a bounded function. Show that the process

$$M_n^f = f(X_n) - f(X_0) - \sum_{k=0}^{n-1} (\mathbf{P} - I) f(X_k)$$

is an $\mathcal{F}_n^X$-martingale.

Exercise 8.6.7. MARTINGALE CHARACTERIZATION OF AN HMC
Let $\{X_n\}_{n\geq 0}$ be a stochastic process with values in the countable space E. It is not assumed to be an HMC. Let $\mathbf{P}$ be some transition matrix on E. Prove that if for all bounded $f : E \to \mathbb{R}$, $\{M_n^f\}_{n\geq 0}$ defined in Exercise 8.6.6 is a martingale with respect to $\{X_n\}_{n\geq 0}$, then $\{X_n\}_{n\geq 0}$ is an HMC with transition matrix $\mathbf{P}$.

Exercise 8.6.8. A MARTINGALE REPRESENTATION THEOREM
Let $\{X_n\}_{n\geq 0}$ be a sequence of $\{0,1\}$-valued random variables and let $\lambda_{n-1} := E\left[X_n \mid \mathcal{F}_{n-1}^X\right]$ $(n \geq 0)$, where $\mathcal{F}_{-1} := \{\varnothing, \Omega\}$. Show that any $\mathcal{F}_n^X$-martingale $\{M_n\}_{n\geq 0}$ is of the form

$$M_n = M_0 + \sum_{j=0}^{n} H_j(X_j - \lambda_{j-1}),$$

where $\{H_n\}_{n\geq 0}$ is an $\mathcal{F}_n^X$-predictable sequence.

Exercise 8.6.9. A MARTINGALE BUILT ON A PERMUTATION
Let $a_1, \dots, a_k \in \mathbb{R}$ be such that $\sum_{j=1}^k a_j = 0$. Let π be a completely random permutation of $\{1, \dots, k\}$, that is, $P(\pi = \pi_0) = \frac{1}{k!}$ for all permutations π_0 of $\{1, \dots, k\}$. Let $\mathcal{F}_n := \sigma(\pi(1), \dots, \pi(n))$ $(1 \leq n \leq k)$ and

$$X_n := \frac{k}{k-n} \sum_{j=1}^{n} a_{\pi(j)}.$$

Show that $\{X_n\}_{1\leq n\leq k}$ is an $\mathcal{F}_n$-martingale.

Exercise 8.6.10. $\mathcal{F}_\tau$
Prove Theorem 8.1.11.

Exercise 8.6.11. RUINED AGAIN
Show that the function $h(i) = \left(\frac{1-p}{p}\right)^i$ is harmonic for the nonsymmetric random walk on $\mathbb{Z}$ (with $p_{i,i+1} = p, p_{i,i-1} = 1-p$, where $p \in (0,1)$ and $p \neq \frac{1}{2}$). Apply the optional sampling theorem to obtain the ruin probability in Example 9.1.10.

Exercise 8.6.12. ABSORPTION PROBABILITY
Consider the HMC $\{X_n\}_{n\geq 0}$ with state space $E = \{0, 1, \dots, m\}$ and transition probabilities

$$p_{ij} = \binom{m}{j} \left(\frac{i}{m}\right)^j \left(1 - \frac{i}{m}\right)^{m-j}.$$

In particular, 0 and m are absorbing states.

(a) Show that $\{X_n\}_{n\geq 0}$ is a martingale.

(b) Compute the probability of absorption by state 0.

Exercise 8.6.13. UPCROSSINGS
Let $\{M_n\}_{n\geq 0}$ be an $\mathcal{F}_n$-martingale. Let $a, b \in \mathbb{R}$ with $a < b$, and let ν_n be the number of upcrossings of $[a, b]$ before ($\leq$) time n. For $k \geq 1$, let A_k be the event that there are exactly $k - 1$ upcrossings of $[a, b]$ before ($\leq$) time n. Show that

$$(b - a)P(\nu_n \geq k) \leq E\left[(a - M_n)1_{A_k}\right] .$$

Exercise 8.6.14. $E\left[X \mid \mathcal{F}_\tau\right] = X(\tau)$
Let τ be a stopping time for the filtration $\{\mathcal{F}_n\}_{n\geq 1}$. Let $X_n := E\left[X \mid \mathcal{F}_n\right]$ $(n \geq 1)$ where X is an integrable random variable. Prove that $E\left[X \mid \mathcal{F}_\tau\right] = X(\tau)$.

Exercise 8.6.15. MARTINGALE BOUNDED BY AN INTEGRABLE RANDOM VARIABLE
Let $\{X_n\}_{n\geq 1}$ be an $\mathcal{F}_n$-martingale and let Z be an integrable random variable such that $X_n \leq Z$ $(n \geq 1)$. Prove that $\{X_n\}_{n\geq 1}$ converges almost surely.

Exercise 8.6.16. THE GAMBLER WITH UNLIMITED CREDIT
Consider the gambling situation of Example 8.1.4 when the stakes are bounded, say by M, and when the initial fortune of the gambler is a. But we suppose that the gambler can borrow whatever amount he needs, so that his "fortune" Y_n at any time n can take arbitrary values. Prove that

$$P(|Y_n - a| \geq \lambda \leq 2\exp\left(-\frac{\lambda^2}{2nM^2}\right) .$$

Exercise 8.6.17. FAIR COIN TOSSES
Consider a Bernoulli sequence of parameter $\frac{1}{2}$ representing a fair game of HEADS and TAILS. Let X_n be the number of HEADS after n tosses. Use Hoeffding's inequality to prove that

$$P(|X_n - E[X_n]| \geq \lambda) \leq 2\exp\left(-\frac{\lambda^2}{n}\right) .$$

Exercise 8.6.18. KRICKEBERG'S DECOMPOSITION
Prove that an $\mathcal{F}_n$-martingale $\{M_n\}_{n\geq 0}$ such that $\sup_{n\geq 0} E[|M_n|] < \infty$ is the difference of two non-negative $\mathcal{F}_n$-martingales. (Hint: Doob's decomposition applied to $|M_n|$.)

Exercise 8.6.19. POLYA'S URN
At time 0 an urn contains exactly one black ball and one white ball. At time $n \geq 0$, a ball is drawn at random and then at time $n+1$ this ball is put back into the urn together with another ball of the same color. In particular, there are at time n exactly $n+2$ balls in the urn. Let B_n be the number of black balls in the urn. Let $X_n := \frac{B_n}{n+2}$ be the proportion of black balls at time n. Show that $\{X_n\}_{n\geq 0}$ is a martingale and that the ratio of the number of black balls to the number of white balls converges.

Exercise 8.6.20. RECORDS
Let $\{X_n\}_{n\geq 1}$ be an IID sequence of random variables with a common cumulative distribution F that is continuous. For $1 \leq i \leq n$, let $Y_i := 1$ if and only if $X_i = \max(X_1, \dots, X_i)$. We shall admit that X_i is uniformly distributed on $\{1, \dots, i\}$ and that $\{Y_i\}_{1\leq i\leq n}$ is IID. Let $Z_n := \sum_{i=1}^n 1_{\{Y_i=1\}}$ (the number of times a record is broken, that is, the number of i's such that $X_i > \max(X_1, \dots, X_{i-1})$). Prove that $\frac{Z_n}{\ln n} \to 1$ almost surely.

Exercise 8.6.21. A MAXIMAL INEQUALITY
Let $\{X_n\}_{n\geq 0}$ be a centered square-integrable martingale. Let $\lambda > 0$. Prove the following inequality:

$$P\left(\max_{0\leq k\leq n} X_k > \lambda\right) \leq \frac{E\left[X_n^2\right]}{E\left[X_n^2\right] + \lambda^2}.$$

Hint: With $c > 0$, work with the sequence $\{(X_n + c)^2\}_{n\geq 0}$ and then select an appropriate c.

Exercise 8.6.22. AN EXTENSION OF HOEFFDING'S INEQUALITY
Let M be a real $\mathcal{F}_n^X$-martingale such that, for some sequence $d_1, d_2, \dots$ of real numbers,

$$P(B_n \leq M_n - M_{n-1} \leq B_n + d_n) = 1 \quad (n \geq 1),$$

where for each $n \geq 1$, B_n is a function of X_0^{n-1}. Prove that, for all $x \geq 0$,

$$P(|M_n - M_0| \geq x) \leq 2\exp\left(-2x^2 \Big/ \sum_{i=1}^n d_i^2\right).$$

Exercise 8.6.23. THE DERIVATIVE OF A LIPSCHITZ CONTINUOUS FUNCTION
Let $f : [0,1) \to \mathbb{R}$ satisfy a Lipschitz condition, that is,

$$|f(x) - f(y)| \leq M|x - y| \quad (x, y \in [0,1)),$$

where $M < \infty$. Let $\Omega = [0,1)$, $\mathcal{F} = \mathcal{B}([0,1))$ and let P be the Lebesgue measure on $[0,1)$. Let for all $n \geq 1$

$$\xi_n(\omega) := \sum_{k=1}^{2^n} 1_{\{[(k-1)2^{-n}, k2^{-n})\}}(\omega)$$

and

$$\mathcal{F}_n = \sigma(\xi_k \,;\, 1 \leq k \leq n).$$

(i) Show that $\mathcal{F}_n = \sigma(\xi_n)$ and $\vee_n \mathcal{F}_n = \mathcal{B}([0,1))$.

(ii) Let

$$X_n := \frac{f(\xi_n + 2^{-n}) - f(\xi_n)}{2^{-n}}.$$

Show that $\{X_n\}_{n\geq 1}$ is a uniformly integrable $\mathcal{F}_n$-martingale.

(iii) Show that there exists a measurable function $g : [0,1) \to \mathbb{R}$ such that $X_n \to g$ P-almost surely and that $X_n = E[g \mid \mathcal{F}_n]$.

(iv) Show that for all $n \geq 1$ and all k $(1 \leq k \leq 2^n)$

$$f(k2^{-n}) - f(0) = \int_0^{k2^{-n}} g(x)\, dx$$

and deduce from this that

$$f(x) - f(0) = \int_0^x g(y)\, dy \quad (x \in [0,1)).$$

Exercise 8.6.24. A NON-UNIFORMLY INTEGRABLE MARTINGALE
Let $\{X_n\}_{n\geq 0}$ be a sequence of IID random variables such that $P(X_n = 0) = P(X_n = 2) = \frac{1}{2}$ $(n \geq 0)$. Define

$$Z_n := \prod_{j=1}^{n} X_j \quad (n \geq 0).$$

Show that $\{Z_n\}_{n\geq 0}$ is an $\mathcal{F}_n^X$-martingale and prove that it is not uniformly integrable.

Exercise 8.6.25. THE BALLOT PROBLEM VIA MARTINGALES
This exercise proposes an alternative proof for the ballot problem. Let $k := a + b$ and let D_n be the difference between the number of votes for A and the number of votes for B at time $n \geq 1$. Prove that

$$X_n = \frac{D_{k-n}}{k-n} \quad (1 \leq n \leq k)$$

is a martingale. Deduce from this that the probability that A leads throughout the voting process is $(a-b)/(a+b)$. Hint: $\tau := \inf\{n\,;\, X_n = 0\} \wedge (k-1)$.

Exercise 8.6.26. A VOTING MODEL
Let $G = (V, \mathcal{E})$ be a finite graph. Each vertex v shelters a random variable $X_n(v)$ representing the opinion (0 or 1) at time n of the voter located at this vertex. At each time n, an edge $\langle v, w \rangle$ is chosen at random, and one of the two vertices, again chosen at random (say v), reconsiders his opinion passing from $X_n(v)$ to $X_{n+1}(v) = X_n(w)$. The initial opinions at time 0 are given. Let Z_n be the total number of votes for 1 at time n. Show that $\{Z_n\}_{n\geq 1}$ is a martingale that converges in finite random time to a random variable Z_∞ taking the values 0 or $|V|$, the probability that all opinions are eventually 1 being equal to the initial proportion of 1's.

Chapter 9

Markov Chains

Discrete-time homogeneous Markov chains are sequences $\{X_n\}_{n\geq 0}$ of random variables with values in some denumerable set E, that can always be represented (in a sense to be made precise) by a recurrence equation $X_{n+1} = f(X_n, Z_{n+1})$, where $\{Z_n\}_{n\geq 1}$ is an IID sequence independent of the initial state X_0. The probabilistic dependence on the past is only through the previous state, but this limited amount of memory suffices to produce enough varied and complex behavior to make Markov chains the most important source of stochastic models in the applied sciences.

9.1 The Transition Matrix

A particle moves on a denumerable set E. If at time n, the particle is in position $X_n = i$, it will be at time $n+1$ in a position $X_{n+1} = j$ chosen independently of the past trajectory X_{n-1}, X_{n-2} with probability p_{ij}. This can be represented by a labeled directed graph, called the *transition graph*, whose set of vertices is E, and for which there is a directed edge from $i \in E$ to $j \in E$ with label p_{ij} if and only the latter quantity is positive. Note that there may be "self-loops", corresponding to positions i such that $p_{ii} > 0$.

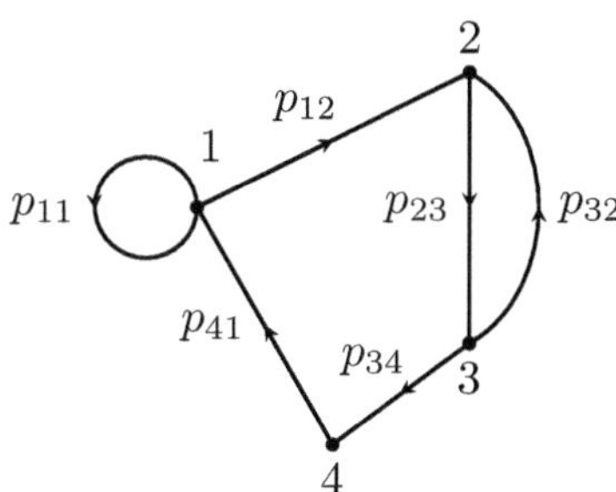

P. Brémaud, *An Introduction to Applied Probability*, Texts in Applied Mathematics 77,
https://doi.org/10.1007/978-3-031-49306-5_9

This graphical interpretation of a Markov chain in terms of a "random walk" on a set E is adapted to the study of random walks on graphs. Since the interpretation of a Markov chain in such terms is not always the natural one, we proceed to give a more formal definition.

Definition 9.1.1 *If for all integers $n \geq 0$ and all states $i_0, i_1, \dots, i_{n-1}, i, j$,*

$$P(X_{n+1} = j \,|\, X_n = i, X_{n-1} = i_{n-1}, \dots, X_0 = i_0) = P(X_{n+1} = j \,|\, X_n = i)\,,$$

this stochastic process is called a Markov chain, *and a* homogeneous *Markov chain* (HMC) *if, in addition, the right-hand side is independent of n.*

The matrix $\mathbf{P} = \{p_{ij}\}_{i,j \in E}$, where

$$p_{ij} = P(X_{n+1} = j \,|\, X_n = i)\,,$$

is called the *transition matrix* of the HMC. Since the entries are probabilities, and since a transition from any state i must be to some state, it follows that

$$p_{ij} \geq 0, \text{ and } \sum_{k \in E} p_{ik} = 1$$

for all states i, j. A matrix $\mathbf{P}$ indexed by E and satisfying the above properties is called a *stochastic matrix*. The state space may be infinite, and therefore such a matrix is in general not of the kind studied in linear algebra. However, the basic operations of addition and multiplication will be defined by the same formal rules. The notation $x = \{x(i)\}_{i \in E}$ formally represents a column vector, and x^T is the corresponding row vector.

The Markov property easily extends (Exercise 9.7.2) to

$$P(A \,|\, X_n = i, B) = P(A \,|\, X_n = i)\,,$$

where

$$A = \{X_{n+1} = j_1, \dots, X_{n+k} = j_k\}, B = \{X_0 = i_0, \dots, X_{n-1} = i_{n-1}\}\,.$$

This is in turn equivalent to

$$P(A \cap B \,|\, X_n = i) = P(A \,|\, X_n = i) P(B \,|\, X_n = i)\,.$$

That is, A and B are conditionally independent given $X_n = i$.

In other words, the future at time n and the past at time n are conditionally independent given the present state $X_n = i$. In particular, the Markov property is independent of the direction of time.

Notation. We shall from now on abbreviate $P(A \mid X_0 = i)$ as $P_i(A)$. Also, if μ is a probability distribution on E, then $P_\mu(A)$ is the probability of A given that the initial state X_0 is distributed according to μ.

The distribution at time n of the chain is the vector $\nu_n := \{\nu_n(i)\}_{i\in E}$, where

$$\nu_n(i) := P(X_n = i)\,.$$

From the Bayes rule of total causes, $\nu_{n+1}(j) = \sum_{i\in E} \nu_n(i)p_{ij}$, that is, in matrix form, $\nu_{n+1}^T = \nu_n^T \mathbf{P}$. Iteration of this equality yields

$$\nu_n^T = \nu_0^T \mathbf{P}^n\,. \tag{9.1}$$

The matrix $\mathbf{P}^m$ is called the *m-step transition matrix* because its general term is

$$p_{ij}(m) = P(X_{n+m} = j \mid X_n = i)\,.$$

In fact, by the Bayes sequential rule and the Markov property, the right-hand side equals $\sum_{i_1,\dots,i_{m-1}\in E} p_{ii_1}p_{i_1i_2}\cdots p_{i_{m-1}j}$, which is the general term of the m-th power of $\mathbf{P}$.

The probability distribution ν_0 of the *initial state* X_0 is called the *initial distribution.* ¿From the Bayes sequential rule and in view of the homogeneous Markov property and the definition of the transition matrix,

$$P(X_0 = i_0, X_1 = i_1, \dots, X_k = i_k) = \nu_0(i_0)p_{i_0i_1}\cdots p_{i_{k-1}i_k}\,.$$

Therefore,

Theorem 9.1.2 *The distribution of a discrete-time* HMC *is uniquely determined by its initial distribution and its transition matrix.*

Many HMCs receive a natural description in terms of a recurrence equation.

Theorem 9.1.3 *Let $\{Z_n\}_{n\geq 1}$ be an* IID *sequence of random variables with values in an arbitrary space F. Let E be a countable space, and $f : E \times F \to E$ be some function. Let X_0 be a random variable with values in E, independent of $\{Z_n\}_{n\geq 1}$. The recurrence equation*

$$X_{n+1} = f(X_n, Z_{n+1}) \tag{9.2}$$

then defines an HMC.

Proof. Iteration of recurrence (9.2) shows that for all $n \geq 1$, there is a function g_n such that $X_n = g_n(X_0, Z_1, \dots, Z_n)$, and therefore $P(X_{n+1} = j \mid X_n = i, X_{n-1} = i_{n-1}, \dots, X_0 = i_0) = P(f(i, Z_{n+1}) = j \mid X_n = i, X_{n-1} = i_{n-1}, \dots, X_0 = i_0) = P(f(i, Z_{n+1}) = j)$, since the event $\{X_0 = i_0, \dots, X_{n-1} = i_{n-1}, X_n = i\}$ is expressible in terms of $X_0, Z_1, \dots, Z_n$ and is therefore independent of Z_{n+1}. Similarly,

$P(X_{n+1} = j \,|\, X_n = i) = P(f(i, Z_{n+1}) = j)$. We therefore have a Markov chain, and it is homogeneous since the right-hand side of the last equality does not depend on n. Explicitly:

$$p_{ij} = P(f(i, Z_1) = j)\,. \tag{9.3}$$

□

EXAMPLE 9.1.4: 1-D RANDOM WALK, TAKE 1. Let X_0 be a random variable with values in $\mathbb{Z}$. Let $\{Z_n\}_{n\geq 1}$ be a sequence of IID random variables, independent of X_0, taking the values $+1$ or -1, and with the probability distribution $P(Z_n = +1) = p$, where $p \in (0, 1)$. The process $\{X_n\}_{n\geq 1}$ defined by

$$X_{n+1} = X_n + Z_{n+1}$$

is, in view of Theorem 9.1.3, an HMC, called a *random walk* on $\mathbb{Z}$. It is called a "symmetric" random walk if $p = \frac{1}{2}$.

EXAMPLE 9.1.5: THE REPAIR SHOP, TAKE 1. During day n, Z_{n+1} machines break down, and they enter the repair shop on day $n + 1$. Every day one machine among those waiting for service is repaired. Therefore, denoting by X_n the number of machines in the shop on day n,

$$X_{n+1} = (X_n - 1)^+ + Z_{n+1}\,, \tag{9.4}$$

where $a^+ = \max(a, 0)$. In particular, if $\{Z_n\}_{n\geq 1}$ is an IID sequence independent of the initial state X_0, then $\{X_n\}_{n\geq 0}$ is a homogeneous Markov chain. In terms of the probability distribution $P(Z_1 = k) = a_k \quad (k \geq 0)$, its transition matrix is

$$\mathbf{P} = \begin{pmatrix} a_0 & a_1 & a_2 & a_3 & \cdots \\ a_0 & a_1 & a_2 & a_3 & \cdots \\ 0 & a_0 & a_1 & a_2 & \cdots \\ 0 & 0 & a_0 & a_1 & \cdots \\ \vdots & \vdots & \vdots & \vdots & \end{pmatrix}.$$

Indeed, from (9.3),

$$p_{ij} = P((i-1)^+ + Z_1 = j) = P(Z_1 = j - (i-1)^+) = a_{j-(i-1)^+}\,.$$

EXAMPLE 9.1.6: STOCHASTIC AUTOMATA. A finite automaton $(E, \mathcal{A}, f)$ can read sequences of letters from a finite alphabet $\mathcal{A}$ written on some infinite tape. It can be in any state of a finite set E, and its evolution is governed by a function $f : E \times \mathcal{A} \to E$, as follows. When the automaton is in state $i \in E$ and reads letter $a \in \mathcal{A}$, it switches from state i to state $j = f(i, a)$ and then reads on the tape the next letter to the right.

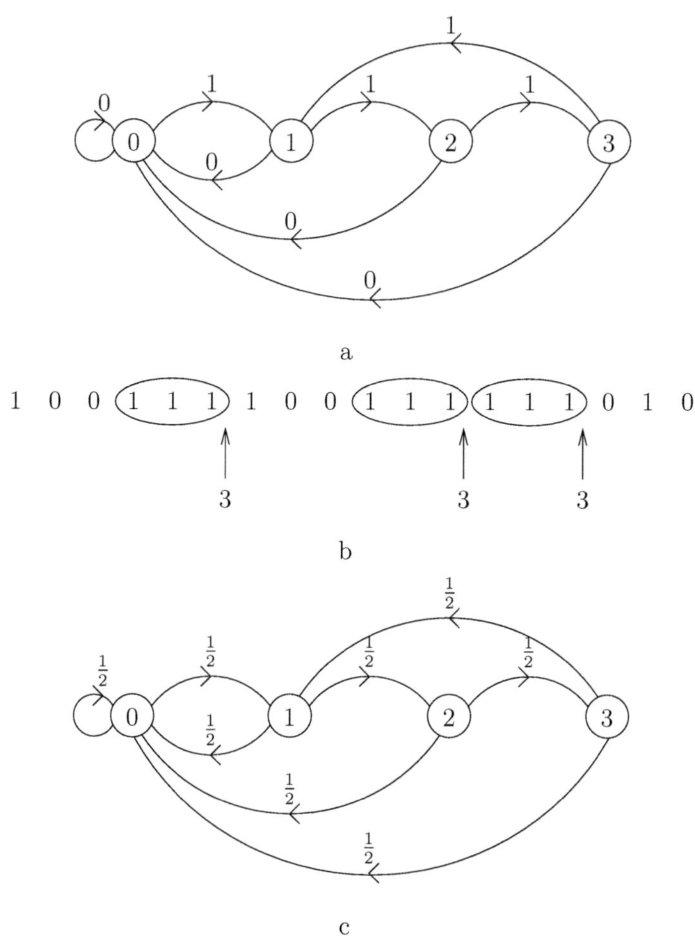

Figure 9.1: The automaton: the recognition process and the Markov chain.

An automaton can be represented by its transition graph G having for nodes the states of E. There is an oriented edge from the node (state) i to the node j if and only if there exists an $a \in \mathcal{A}$ such that $j = f(i, a)$, and this edge then receives label a. If $j = f(i, a_1) = f(i, a_2)$ for $a_1 \neq a_2$, then there are two edges from i to j with labels a_1 and a_2, or, more economically, one such edge with label (a_1, a_2). More generally, a given oriented edge can have multiple labels of any order.

Consider, for instance, the automaton with alphabet $\mathcal{A} = \{0, 1\}$ corresponding to the transition graph of Figure 9.1a. As the automaton, initialized in state 0, reads the sequence of Figure 9.1b from left to right, it passes successively through the states (including the initial state 0)

$$0\,1\,0\,0\,1\,2\,3\,1\,0\,0\,1\,2\,3\,1\,2\,3\,0\,1\,0\,.$$

Rewriting the sequence of states below the sequence of letters, it appears that the automaton is in state 3 after it has seen three consecutive 1's. This automaton is therefore able to recognize and count such blocks of 1's. However, it does not take into account overlapping blocks (see Figure 9.1b).

If the sequence of letters read by the automaton is $\{Z_n\}_{n\geq 1}$, the sequence of states $\{X_n\}_{n\geq 0}$ is then given by the recurrence equation $X_{n+1} = f(X_n, Z_{n+1})$ and therefore, if $\{Z_n\}_{n\geq 1}$ is IID and independent of the initial state X_0, then $\{X_n\}_{n\geq 1}$ is, according to Theorem 9.2, an HMC.

Not all homogeneous Markov chains receive a "natural" description of the type featured in Theorem 9.1.3. However, it is always possible to find a "theoretical" description of this kind.

Theorem 9.1.7 *For any transition matrix* $\mathbf{P}$ *on* E*, there exists a homogeneous Markov chain with this transition matrix and with a representation such as in Theorem 9.1.3.*

Proof. Define

$$X_{n+1} := j \text{ if } \sum_{k=0}^{j-1} p_{X_n k} \leq Z_{n+1} < \sum_{k=0}^{j} p_{X_n k}\,,$$

where $\{Z_n\}_{n\geq 1}$ is IID, uniform on $[0, 1]$. By application of Theorem 9.1.3 and of formula (9.3), we check that this HMC has the announced transition matrix. $\square$

As we already mentioned, not all homogeneous Markov chains are *naturally* described by the model of Theorem 9.1.3. A slight modification of this result considerably enlarges its scope.

Theorem 9.1.8 *Let things be as in Theorem 9.1.3 except for the joint distribution of $X_0, Z_1, Z_2, \ldots$. Suppose instead that for all $n \geq 0$, Z_{n+1} is conditionally independent of $Z_n, \ldots, Z_1, X_{n-1}, \ldots, X_0$ given X_n, and that for all $i, j \in E$, $P(Z_{n+1} = k \mid X_n = i)$ is independent of n. Then $\{X_n\}_{n \geq 0}$ is an HMC, with transition probabilities*

$$p_{ij} = P(f(i, Z_1) = j \mid X_0 = i) .$$

Proof. The proof is quite similar to that of Theorem 9.1.3 and is left as an exercise. □

EXAMPLE 9.1.9: THE EHRENFEST URN, TAKE 1. This idealized model of diffusion through a porous membrane, proposed in 1907 by the Austrian physicists Tatiana and Paul Ehrenfest to describe in terms of statistical mechanics the exchange of heat between two systems at different temperatures, considerably helped our understanding of the phenomenon of thermodynamic irreversibility. It features N particles that can be either in compartment A or in compartment B.

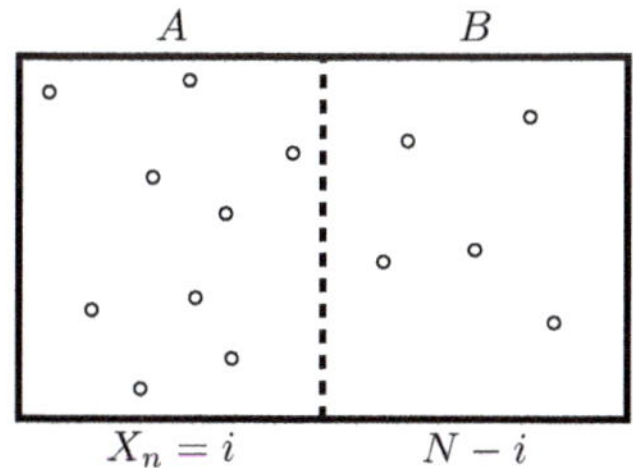

Suppose that at time $n \geq 0$, $X_n = i$ particles are in A. One then chooses a particle at random, and this particle is moved at time $n + 1$ from where it is to the other compartment. Thus, the next state X_{n+1} is either $i - 1$ (the displaced particle was found in compartment A) with probability $\frac{i}{N}$, or $i + 1$ (it was found in B) with probability $\frac{N-i}{N}$. This model pertains to Theorem 9.1.8. For all $n \geq 0$,

$$X_{n+1} = X_n + Z_{n+1} ,$$

where $Z_n \in \{-1, +1\}$ and $P(Z_{n+1} = -1 \mid X_n = i) = \frac{i}{N}$. The nonzero entries of the transition matrix are therefore

$$p_{i,i+1} = \frac{N-i}{N} , \quad p_{i,i-1} = \frac{i}{N} .$$

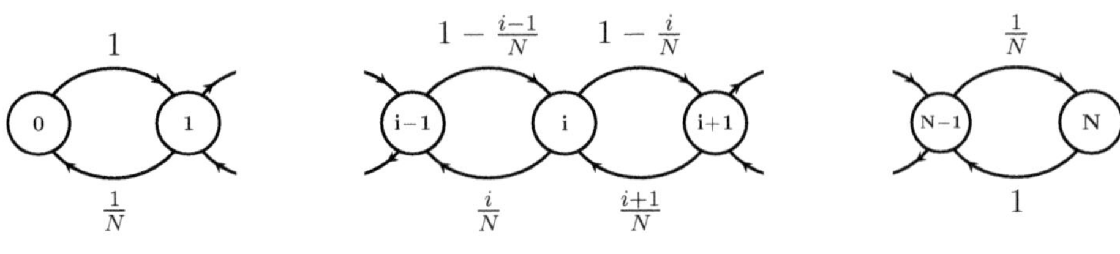

First-step Analysis

Some functionals of homogeneous Markov chains such as probabilities of absorption by a closed set and average times before absorption can be evaluated by a technique called *first-step analysis*.

EXAMPLE 9.1.10: THE GAMBLER'S RUIN, TAKE 1. Two players A and B play "heads or tails", where heads occur with probability $p \in (0,1)$, and the successive outcomes form an IID sequence. Calling X_n the fortune in dollars of player A at time n, then $X_{n+1} = X_n + Z_{n+1}$, where $Z_{n+1} = +1$ (resp., -1) with probability p (resp., $q := 1-p$), and $\{Z_n\}_{n\geq 1}$ is IID. In other words, A bets \$1 on heads at each toss, and B bets \$1 on tails. The respective initial fortunes of A and B are a and b (positive integers). The game ends when a player is ruined, and therefore the process $\{X_n\}_{n\geq 1}$ is a random walk as described in Example 9.1.4, except that it is restricted to $E = \{0, \dots, a, a+1, \dots, a+b = c\}$. The duration of the game is T, the first time n at which $X_n = 0$ or c, and the probability of winning for A is $u(a) = P(X_T = c \,|\, X_0 = a)$.

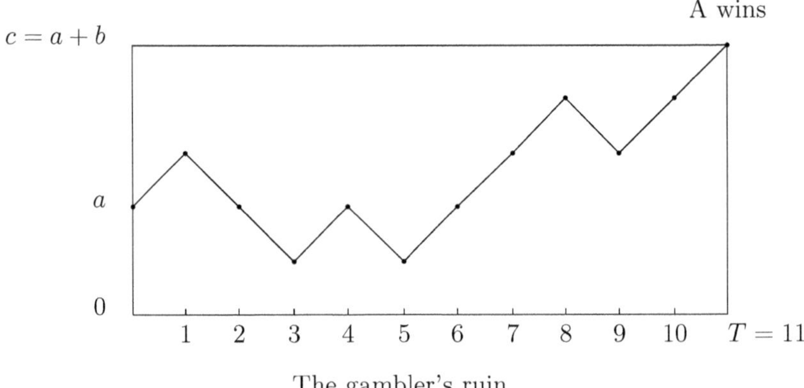

The gambler's ruin

Instead of computing $u(a)$ alone, first-step analysis computes

$$u(i) = P(X_T = c \mid X_0 = i)$$

for all states i, $0 \le i \le c$, and for this, it first generates a recurrence equation for $u(i)$ by breaking down event "A wins" according to what can happen after the first step (the first toss) and using the rule of total causes. If $X_0 = i$, $1 \le i \le c-1$, then $X_1 = i+1$ (resp., $X_1 = i-1$) with probability p (resp., q), and the probability of winning for A with updated initial fortune $i+1$ (resp., $i-1$) is $u(i+1)$ (resp., $u(i-1)$). Therefore, for i, $1 \le i \le c-1$,

$$u(i) = pu(i+1) + qu(i-1)\,,$$

with the boundary conditions $u(0) = 0$, $u(c) = 1$.

The characteristic equation associated with this linear recurrence equation is $pr^2 - r + q = 0$. It has two distinct roots, $r_1 = 1$ and $r_2 = \frac{q}{p}$, if $p \neq \frac{1}{2}$, and a double root, $r_1 = 1$, if $p = \frac{1}{2}$. Therefore, the general solution is $u(i) = \lambda r_1^i + \mu r_2^i = \lambda + \mu \left(\frac{q}{p}\right)^i$ when $p \neq q$, and $u(i) = \lambda r_1^i + \mu i r_1^i = \lambda + \mu i$ when $p = q = \frac{1}{2}$. Taking into account the boundary conditions, one can determine the values of λ and μ. The result is, for $p \neq q$,

$$u(i) = \frac{1 - (\frac{q}{p})^i}{1 - (\frac{q}{p})^c}\,,$$

and for $p = q = \frac{1}{2}$,

$$u(i) = \frac{i}{c}\,.$$

In the case $p = q = \frac{1}{2}$, the probability $v(i)$ that B wins when the initial fortune of B is $c-i$ is obtained by replacing i by $c-i$ in the expression for $u(i)$: $v(i) = \frac{c-i}{c} = 1 - \frac{i}{c}$. One checks that $u(i) + v(i) = 1$, which means in particular that the probability that the game lasts forever is null. The reader is invited to check that the same is true in the case $p \neq q$.

First-step analysis can also be used to compute average times before absorption (Exercise 9.7.5).

Communication and Period

These two concepts are *topological* in the sense that they concern only the *naked* transition graph (with only the arrows, without the labels).

Definition 9.1.11 *State j is said to be* accessible *from state i if there exists an $M \geq 0$ such that $p_{ij}(M) > 0$. States i and j are said to* communicate *if i is accessible from j* and *j is accessible from i, and this is denoted by $i \leftrightarrow j$.*

In particular, a state i is always accessible from itself, since $p_{ii}(0) = 1$ ($\mathbf{P}^0 = I$, the identity).

For $M \geq 1$, $p_{ij}(M) = \sum_{i_1,\ldots,i_{M-1}} p_{ii_1} \cdots p_{i_{M-1}j}$, and therefore $p_{ij}(M) > 0$ if and only if there exists at least one path $i, i_1, \ldots, i_{M-1}, j$ from i to j such that

$$p_{ii_1} p_{i_1 i_2} \cdots p_{i_{M-1}j} > 0\,,$$

or, equivalently, if there is a directed path from i to j in the transition graph G. Clearly,

$$\begin{aligned} i &\leftrightarrow i && \text{(reflexivity)},\\ i \leftrightarrow j &\Rightarrow j \leftrightarrow i && \text{(symmetry)},\\ i \leftrightarrow j, j \leftrightarrow k &\Rightarrow i \leftrightarrow k && \text{(transitivity)}. \end{aligned}$$

Therefore, the communication relation ($\leftrightarrow$) is an equivalence relation, and it generates a partition of the state space E into disjoint equivalence classes called *communication classes*.

Definition 9.1.12 *A* closed state *i is one such that $p_{ii} = 1$. More generally, a* closed set *C of states is one such that for all $i \in C$, $\sum_{j \in C} p_{ij} = 1$.*

Definition 9.1.13 *If there exists only one communication class, then the chain, its transition matrix, and its transition graph are said to be* irreducible.

EXAMPLE 9.1.14: THE REPAIR SHOP, TAKE 2. Recall that this Markov chain satisfies the recurrence equation

$$X_{n+1} = (X_n - 1)^+ + Z_{n+1}\,, \tag{9.5}$$

where $a^+ = \max(a, 0)$. The sequence $\{Z_n\}_{n \geq 1}$ is assumed to be IID, independent of the initial state X_0, and with common probability distribution

$$P(Z_1 = k) = a_k,\ k \geq 0$$

of generating function g_Z.

This chain is irreducible if and only if $P(Z_1 = 0) > 0$ and $P(Z_1 \geq 2) > 0$ as we now prove formally. Looking at (9.14), we make the following observations. If

$P(Z_{n+1} = 0) = 0$, then $X_{n+1} \geq X_n$ a.s. and there is no way of going from i to $i-1$. If $P(Z_{n+1} \leq 1) = 1$, then $X_{n+1} \leq X_n$, and there is no way of going from i to $i+1$. Therefore, the two conditions $P(Z_1 = 0) > 0$ and $P(Z_2 \geq 2) > 0$ are *necessary* for irreducibility. They are also sufficient. Indeed if there exists an integer $k \geq 2$ such that $P(Z_{n+1} = k) > 0$, then one can jump with positive probability from any $i > 0$ to $i + k - 1 > i$ or from $i = 0$ to $k > 0$. Also if $P(Z_{n+1} = 0) > 0$, one can step down from $i > 0$ to $i - 1$ with positive probability. In particular, one can go from i to $j < i$ with positive probability. Therefore, one way to travel from i to $j \geq i$ is by taking several successive steps of height at least $k - 1$ in order to reach a state $l \geq i$, and then (in the case of $l > i$) stepping down one stair at a time from l to i. All this with positive probability.

Consider the random walk on $\mathbb{Z}$ (Example 9.1.4). Since $0 < p < 1$, it is irreducible. Observe that $E = C_0 + C_1$, where C_0 and C_1, the set of even and odd relative integers respectively, have the following property. If you start from $i \in C_0$ (resp., C_1), then in one step you can go only to a state $j \in C_1$ (resp., C_0). The chain $\{X_n\}$ passes alternately from one cyclic class to the other. In this sense, the chain has a periodic behavior, corresponding to the period 2. More generally, for any *irreducible* Markov chain, one can find a *unique partition* of E into d classes C_0, C_1, ..., C_{d-1} such that for all $k, i \in C_k$,

$$\sum_{j \in C_{k+1}} p_{ij} = 1\,,$$

where by convention $C_d = C_0$, and where d is maximal (that is, there is no other such partition C'_0, $C'_1, \ldots, C'_{d'-1}$ with $d' > d$). The proof follows directly from Theorem 9.1.17 below.

The number $d \geq 1$ is called the *period* of the chain (resp., of the transition matrix, of the transition graph). The classes $C_0, C_1, \ldots, C_{d-1}$ are called the *cyclic classes*. The chain therefore moves from one class to the other at each transition, and this cyclically.

We now give the formal definition of period. It is based on the notion of *greatest common divisor* of a set of positive integers.

Definition 9.1.15 *The period d_i of state $i \in E$ is, by definition,*

$$d_i = \text{GCD}\{n \geq 1\,;\, p_{ii}(n) > 0\}\,,$$

with the convention $d_i = +\infty$ if there is no $n \geq 1$ with $p_{ii}(n) > 0$. If $d_i = 1$, the state i is called aperiodic.

Theorem 9.1.16 *If states i and j communicate, then they have the same period.*

Proof. As i and j communicate, there exist integers N and M such that $p_{ij}(M) > 0$ and $p_{ji}(N) > 0$. For any $k \geq 1$,

$$p_{ii}(M + nk + N) \geq p_{ij}(M)(p_{jj}(k))^n p_{ji}(N)$$

(indeed, the path $X_0 = i, X_M = j, X_{M+k} = j, \ldots, X_{M+nk} = j, X_{M+nk+N} = i$ is just one way of going from i to i in $M + nk + N$ steps). Therefore, for any $k \geq 1$ such that $p_{jj}(k) > 0$, we have $p_{ii}(M + nk + N) > 0$ for all $n \geq 1$. Therefore, d_i divides $M+nk+N$ for all $n \geq 1$, and in particular, d_i divides k. We have therefore shown that d_i divides all k such that $p_{jj}(k) > 0$, and in particular, d_i divides d_j. By symmetry, d_j divides d_i, and therefore, finally, $d_i = d_j$. □

We may therefore speak of the period of a communication class or of an irreducible chain.

The important result concerning periodicity is the following.

Theorem 9.1.17 *Let $\mathbf{P}$ be an irreducible stochastic matrix with period d. Then for all states i, j there exist $m \geq 0$ and $n_0 \geq 0$ (m and n_0 possibly depending on i, j) such that*

$$p_{ij}(m + nd) > 0, \quad \textit{for all } n \geq n_0.$$

Proof. It suffices to prove the theorem for $i = j$. Indeed, there exists an m such that $p_{ij}(m) > 0$, because j is accessible from i, the chain being irreducible, and therefore, if for some $n_0 \geq 0$ we have $p_{jj}(nd) > 0$ for all $n \geq n_0$, then $p_{ij}(m + nd) \geq p_{ij}(m)p_{jj}(nd) > 0$ for all $n \geq n_0$. The rest of the proof is an immediate consequence of a classical result of number theory.[1] Indeed, the GCD of the set $A = \{k \geq 1; p_{jj}(k) > 0\}$ is d, and A is closed under addition. The set A therefore contains all but a finite number of the positive multiples of d. In other words, there exists an n_0 such that $n > n_0$ implies $p_{jj}(nd) > 0$. □

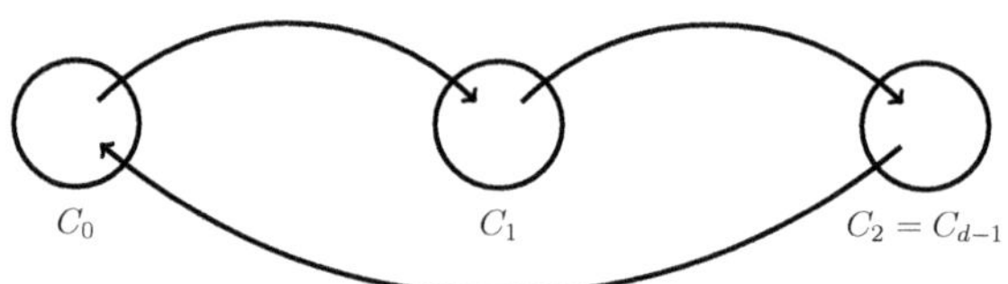

Behavior of a Markov chain with period 3

[1] Let d be the g.c.d of $A = \{a_n ; n \geq 1\}$, a set of positive integers that is closed under addition. Then A contains all but a finite number of the positive multiples of d.

Stationary Distributions

The central notion of the stability theory of discrete-time HMCs is that of a stationary distribution.

Definition 9.1.18 *A probability distribution π satisfying*

$$\pi^T = \pi^T \mathbf{P} \tag{9.6}$$

is called a stationary distribution of the transition matrix $\mathbf{P}$, or of the corresponding HMC.

The *global balance equation* (9.6) says that for all states i,

$$\pi(i) = \sum_{j \in E} \pi(j) p_{ji} \,.$$

Iteration of (9.6) gives $\pi^T = \pi^T \mathbf{P}^n$ for all $n \geq 0$, and therefore, in view of (9.1), if the initial distribution $\nu = \pi$, then $\nu_n = \pi$ for all $n \geq 0$. Thus, if a chain is started with a stationary distribution, it keeps the same distribution forever. But there is more, because then,

$$\begin{aligned} P(X_n = i_0, X_{n+1} = i_1, \ldots, X_{n+k} = i_k) &= P(X_n = i_0) p_{i_0 i_1} \ldots p_{i_{k-1} i_k} \\ &= \pi(i_0) p_{i_0 i_1} \ldots p_{i_{k-1} i_k} \end{aligned}$$

does not depend on n. In this sense the chain is *stationary*. One also says that the chain is in a *stationary regime*, or in *steady state*. In summary:

Theorem 9.1.19 *An* HMC *whose initial distribution is a stationary distribution is stationary.*

The balance equation $\pi^T \mathbf{P} = \pi^T$, together with the requirement that π be a probability vector, that is, $\pi^T \mathbf{1} = 1$ (where $\mathbf{1}$ is a column vector with all its entries equal to 1), constitute when E is finite, $|E|+1$ equations for $|E|$ unknown variables. One of the $|E|$ equations in $\pi^T \mathbf{P} = \pi^T$ is superfluous given the constraint $\pi^T \mathbf{1} = 1$. Indeed, summing up all equalities of $\pi^T \mathbf{P} = \pi^T$ yields the equality $\pi^T \mathbf{P} \mathbf{1} = \pi^T \mathbf{1}$, that is, $\pi^T \mathbf{1} = 1$.

EXAMPLE 9.1.20: TWO-STATE MARKOV CHAIN. Take $E = \{1, 2\}$ and define the transition matrix

$$\mathbf{P} = \begin{pmatrix} 1-\alpha & \alpha \\ \beta & 1-\beta \end{pmatrix},$$

where $\alpha, \beta \in (0,1)$. The global balance equations are

$$\pi(1) = \pi(1)(1-\alpha) + \pi(2)\beta\,, \qquad \pi(2) = \pi(1)\alpha + \pi(2)(1-\beta)\,.$$

These two equations are dependent and reduce to the single equation $\pi(1)\alpha = \pi(2)\beta$, to which must be added the constraint $\pi(1) + \pi(2) = 1$ expressing that π is a probability vector. We obtain

$$\pi(1) = \frac{\beta}{\alpha+\beta}, \quad \pi(2) = \frac{\alpha}{\alpha+\beta}\,.$$

EXAMPLE 9.1.21: THE EHRENFEST URN, TAKE 2. The global balance equations are, for $i \in [1, N-1]$,

$$\pi(i) = \pi(i-1)\left(1 - \frac{i-1}{N}\right) + \pi(i+1)\frac{i+1}{N}$$

and, for the boundary states,

$$\pi(0) = \pi(1)\frac{1}{N}, \quad \pi(N) = \pi(N-1)\frac{1}{N}\,.$$

Leaving $\pi(0)$ undetermined, one can solve the balance equations for $i = 0, 1, \dots, N$ successively, to obtain $\pi(i) = \pi(0)\binom{N}{i}$. The value of $\pi(0)$ is then determined by writing that π is a probability vector: $1 = \sum_{i=0}^{N} \pi(i) = \pi(0)\sum_{i=0}^{N}\binom{N}{i} = \pi(0)2^N$. This gives for π the binomial distribution of size N and parameter $\frac{1}{2}$:

$$\pi(i) = \frac{1}{2^N}\binom{N}{i}\,.$$

This is the distribution one would obtain by placing independently each particle in the compartments, with probability $\frac{1}{2}$ for each compartment.

Stationary distributions may be many. Take the identity as transition matrix. Then any probability distribution on the state space is a stationary distribution. Also there may well not exist any stationary distribution. See Exercise 9.7.10.

Reversible Chains

Let $\{X_n\}_{n\geq 0}$ be an HMC with transition matrix $\mathbf{P}$ and admitting a stationary distribution $\pi > 0$ (meaning $\pi(i) > 0$ for all states i). Define the matrix $\mathbf{Q}$, indexed by E, by

$$\pi(i)q_{ij} = \pi(j)p_{ji}\,. \tag{9.7}$$

This is a stochastic matrix since

$$\sum_{j\in E} q_{ij} = \sum_{j\in E} \frac{\pi(j)}{\pi(i)} p_{ji} = \frac{1}{\pi(i)} \sum_{j\in E} \pi(j) p_{ji} = \frac{\pi(i)}{\pi(i)} = 1,$$

where the third equality uses the global balance equations. Its interpretation is the following: Suppose that the initial distribution of the chain is π, in which case for all $n \geq 0$, all $i \in E$, $P(X_n = i) = \pi(i)$. Then, from Bayes' retrodiction formula,

$$P(X_n = j \mid X_{n+1} = i) = \frac{P(X_{n+1} = i \mid X_n = j) P(X_n = j)}{P(X_{n+1} = i)},$$

that is, in view of (9.7),

$$P(X_n = j \mid X_{n+1} = i) = q_{ji}.$$

We see that $\mathbf{Q}$ is the transition matrix of the initial chain when time is reversed.

The following is a very simple observation that will be promoted to the rank of a theorem in view of its usefulness.

Theorem 9.1.22 *Let $\mathbf{P}$ be a stochastic matrix indexed by a countable set E, and let π be a probability distribution on E. Define the matrix $\mathbf{Q}$ indexed by E by (9.7). If $\mathbf{Q}$ is a stochastic matrix, then π is a stationary distribution of $\mathbf{P}$.*

Proof. For fixed $i \in E$, sum equalities (9.7) with respect to $j \in E$ to obtain

$$\sum_{j\in E} \pi(i) q_{ij} = \sum_{j\in E} \pi(j) p_{ji}.$$

This is the global balance equation since the left-hand side is equal to $\pi(i) \sum_{j\in E} q_{ij} = \pi(i)$. □

Definition 9.1.23 *One calls* reversible *a stationary Markov chain with initial distribution π (a stationary distribution) if for all $i, j \in E$, we have the so-called* detailed balance equations

$$\pi(i) p_{ij} = \pi(j) p_{ji}. \tag{9.8}$$

We then say: the pair $(\mathbf{P}, \pi)$ is reversible.

In this case, $q_{ij} = p_{ij}$, and therefore the chain and the time-reversed chain are statistically the same, since the distribution of a homogeneous Markov chain is entirely determined by its initial distribution and its transition matrix.

The next result is an immediate corollary of Theorem 9.1.22.

Theorem 9.1.24 *Let* $\mathbf{P}$ *be a transition matrix on the countable state space* E*, and let* π *be some probability distribution on* E*. If for all* $i, j \in E$*, the detailed balance equations (9.8) are satisfied, then* π *is a stationary distribution of* $\mathbf{P}$.

EXAMPLE 9.1.25: THE EHRENFEST URN, TAKE 3. The verification of the detailed balance equations $\pi(i)p_{i,i+1} = \pi(i+1)p_{i+1,i}$ is immediate.

The Strong Markov Property

The Markov property, that is, the independence of past and future given the present state, extends to the situation where the present time is a *stopping time*, a notion which we now introduce.

Let $\{X_n\}_{n\geq 0}$ be a stochastic process with values in the denumerable set E. For an event A, the notation $A \in \mathcal{X}_0^n$ means that there exists a function $\varphi : E^{n+1} \mapsto \{0,1\}$ such that

$$1_A(\omega) = \varphi(X_0(\omega), \dots, X_n(\omega)) .$$

In other terms, this event is expressible in terms of $X_0(\omega), \dots, X_n(\omega)$. Let now τ be a random variable with values in $\overline{\mathbb{N}}$. It is called a X_0^n-stopping time if for all $m \in \mathbb{N}$, $\{\tau = m\} \in X_0^m$. In other words, it is a non-anticipative random time with respect to $\{X_n\}_{n\geq 0}$, since in order to check if $\tau = m$, one need only observe the process up to time m and not beyond. It is immediate to check that if τ is a X_0^n-stopping time, then so is $\tau + n$ for all $n \geq 1$.

EXAMPLE 9.1.26: RETURN TIME. Let $\{X_n\}_{n\geq 0}$ be an HMC with state space E. Define for $i \in E$ the *return time* to i by

$$T_i := \inf\{n \geq 1 \,;\, X_n = i\}$$

using the convention $\inf \varnothing = \infty$ for the empty set of $\mathbb{N}$. This is a X_0^n-stopping time since for all $m \in \mathbb{N}$,

$$\{T_i = m\} = \{X_1 \neq i, X_2 \neq i, \dots, X_{m-1} \neq i, X_m = i\} .$$

Note that $T_i \geq 1$. It is a "return" time, not to be confused with the closely related "hitting" time of i, defined as $S_i := \inf\{n \geq 0 \,;\, X_n = i\}$, which is also a X_0^n-stopping time, equal to T_i if and only if $X_0 \neq i$.

EXAMPLE 9.1.27: SUCCESSIVE RETURN TIMES. This continues the previous example. Let us fix a state, conventionally labeled 0, and let T_0 be the return time to 0. We define the successive return times to 0, τ_k, $k \geq 1$ by $\tau_1 = T_0$ and for $k \geq 1$,

$$\tau_{k+1} := \inf\{n \geq \tau_k + 1\,;\, X_n = 0\}$$

with the above convention that $\inf \varnothing = \infty$. In particular, if $\tau_k = \infty$ for some k, then $\tau_{k+\ell} = \infty$ for all $\ell \geq 1$. The identity

$$\{\tau_k = m\} \equiv \left\{ \sum_{n=1}^{m-1} 1_{\{X_n=0\}} = k-1\,,\, X_m = 0 \right\}$$

for $m \geq 1$ shows that τ_k is a X_0^n-stopping time.

Let $\{X_n\}_{n\geq 0}$ be a stochastic process with values in the countable set E and let τ be a random time taking its values in $\overline{\mathbb{N}} := \mathbb{N} \cup \{+\infty\}$. In order to define X_τ when $\tau = \infty$, one must decide how to define X_∞. This is done by taking some arbitrary element Δ not in E, and setting

$$X_\infty = \Delta\,.$$

By definition, the "process after τ" is the stochastic process

$$\{S_\tau X_n\}_{n\geq 0} := \{X_{n+\tau}\}_{n\geq 0}\,.$$

The "process before τ," or the "process stopped at τ," is the process

$$\{X_n^\tau\}_{n\geq 0} := \{X_{n\wedge\tau}\}_{n\geq 0}\,,$$

which freezes at time τ at the value X_τ.

Theorem 9.1.28 *Let $\{X_n\}_{n\geq 0}$ be an* HMC *with state space E and transition matrix $\mathbf{P}$. Let τ be a X_0^n-stopping time. Then for any state $i \in E$,*

(α) *Given that $X_\tau = i$, the process after τ and the process before τ are independent.*

(β) *Given that $X_\tau = i$, the process after τ is an* HMC *with transition matrix $\mathbf{P}$.*

Proof. (α) We have to show that for all times $k \geq 1$, $n \geq 0$, and all states $i_0, \dots, i_n, i, j_1, \dots, j_k$,

$$P(X_{\tau+1} = j_1, \dots, X_{\tau+k} = j_k \mid X_\tau = i, X_{\tau\wedge 0} = i_0, \dots, X_{\tau\wedge n} = i_n)$$
$$= P(X_{\tau+1} = j_1, \dots, X_{\tau+k} = j_k \mid X_\tau = i)\,.$$

We shall prove a simplified version of the above equality, namely

$$P(X_{\tau+k} = j \mid X_\tau = i, X_{\tau\wedge n} = i_n) = P(X_{\tau+k} = j \mid X_\tau = i) . \qquad (\star)$$

The general case is obtained by the same arguments. The left-hand side of $(\star)$ equals

$$\frac{P(X_{\tau+k} = j, X_\tau = i, X_{\tau\wedge n} = i_n)}{P(X_\tau = i, X_{\tau\wedge n} = i_n)} .$$

The numerator of the above expression can be developed as

$$\sum_{r\in\mathbb{N}} P(\tau = r, X_{r+k} = j, X_r = i, X_{r\wedge n} = i_n) . \qquad (\star\star)$$

(The sum is over $\mathbb{N}$ because $X_\tau = i \neq \Delta$ implies that $\tau < \infty$.) But

$$\begin{aligned}&P(\tau = r, X_{r+k} = j, X_r = i, X_{r\wedge n} = i_n)\\ &\quad = P(X_{r+k} = j \mid X_r = i, X_{r\wedge n} = i_n, \tau = r)P(\tau = r, X_{r\wedge n} = i_n, X_r = i) ,\end{aligned}$$

and since $r \wedge n \le r$ and $\{\tau = r\} \in X_0^r$, the event $B := \{X_{r\wedge n} = i_n, \tau = r\}$ is in X_0^r. Therefore, by the Markov property, $P(X_{r+k} = j \mid X_r = i, X_{r\wedge n} = i_n, \tau = r\} = P(X_{r+k} = j \mid X_r = i) = p_{ij}(k)$. Finally, expression $(\star\star)$ reduces to

$$\sum_{r\in\mathbb{N}} p_{ij}(k)P(\tau = r, X_{r\wedge n} = i_n, X_r = i) = p_{ij}(k)P(X_{\tau=i}, X_{\tau\wedge n} = i_n) .$$

Therefore, the left-hand side of $(\star)$ is just $p_{ij}(k)$. Similar computations show that the right-hand side of $(\star)$ is also $p_{ij}(k)$, so that (α) is proven.

(β) We must show that for all states $i, j, k, i_{n-1}, \dots, i_1$,

$$\begin{aligned}&P(X_{\tau+n+1} = k \mid X_{\tau+n} = j, X_{\tau+n-1} = i_{n-1}, \dots, X_\tau = i)\\ &\qquad\qquad = P(X_{\tau+n+1} = k \mid X_{\tau+n} = j) = p_{jk} .\end{aligned}$$

But the first equality follows from the fact proven in (α) that for the stopping time $\tau' = \tau + n$, the processes before and after τ' are independent given $X_{\tau'} = j$. The second equality is obtained by the same calculations as in the proof of (α). □

The Cycle Independence Property

Consider a Markov chain with a state conventionally denoted by 0 such that $P_0(T_0 < \infty) = 1$. In view of the strong Markov property, the chain starting from state 0 will return infinitely often to this state. Let $\tau_1 = T_0, \tau_2, \dots$ be the successive return times to 0, and set $\tau_0 \equiv 0$.

By the strong Markov property, for any $k \geq 1$, the process after τ_k is independent of the process before τ_k (observe that condition $X_{\tau_k} = 0$ is always satisfied), and the process after τ_k is a Markov chain with the same transition matrix as the original chain, and with initial state 0, by construction. Therefore, the successive times of visit to 0, the pieces of trajectory

$$\{X_{\tau_k}, X_{\tau_k+1}, \dots, X_{\tau_{k+1}-1}\}, \quad k \geq 0,$$

are independent and identically distributed. Such pieces are called the *regenerative cycles* of the chain between visits to state 0. Each random time τ_k is a *regeneration time*, in the sense that $\{X_{\tau_k+n}\}_{n\geq 0}$ is independent of the past $X_0, \dots, X_{\tau_k-1}$ and has the same distribution as $\{X_n\}_{n\geq 0}$. In particular, the sequence $\{\tau_k - \tau_{k-1}\}_{k\geq 1}$ is IID.

EXAMPLE 9.1.29: RETURNS TO ZERO OF THE 1-D SYMMETRIC WALK. Let $\tau_1 = T_0, \tau_2, \dots$ be the successive return times to state 0 of the random walk on $\mathbb{Z}$ of Example 9.1.4 with $p = \frac{1}{2}$. We shall admit that $P_0(T_0 < \infty) = 1$, a fact that will be proved in the next section, and obtain the probability distribution of T_0 given $X_0 = 0$.

Observe that for $n \geq 1$,

$$P_0(X_{2n} = 0) = \sum_{k\geq 1} P_0(\tau_k = 2n)\,,$$

and therefore, for all $z \in \mathbb{C}$ such that $|z| < 1$,

$$\sum_{n\geq 1} P_0(X_{2n} = 0) z^{2n} = \sum_{k\geq 1} \sum_{n\geq 1} P_0(\tau_k = 2n) z^{2n} = \sum_{k\geq 1} E_0[z^{\tau_k}]\,.$$

But $\tau_k = \tau_1 + (\tau_2 - \tau_1) + \cdots + (\tau_k - \tau_{k-1})$ and therefore, since $\tau_1 = T_0$,

$$E_0[z^{\tau_k}] = (E_0[z^{T_0}])^k\,.$$

In particular,

$$\sum_{n\geq 0} P_0(X_{2n} = 0) z^{2n} = \frac{1}{1 - E_0[z^{T_0}]}$$

(note that the latter sum includes the term for $n = 0$, that is, 1). Direct evaluation of the left-hand side yields

$$\sum_{n\geq 0} \frac{1}{2^{2n}} \frac{(2n)!}{n!n!} z^{2n} = \frac{1}{\sqrt{1 - z^2}}\,.$$

Therefore, the generating function of the return time to 0 given $X_0 = 0$ is

$$E_0[z^{T_0}] = 1 - \sqrt{1 - z^2}\,.$$

Its first derivative

$$\frac{z}{\sqrt{1 - z^2}}$$

tends to ∞ as $z \to 1$ from below via real values. Therefore, by Abel's theorem,

$$E_0[T_0] = \infty\,.$$

We see that although given $X_0 = 0$ the return time is almost surely finite, it has an infinite expectation.

9.2 Recurrence

In the theory of Markov chains, recurrence refers to the possibility of an infinite number of visits to a given state. The basic definition is in terms of return times.

Recall that T_i denotes the *return* time to state i.

Definition 9.2.1 *State $i \in E$ is called* recurrent *if*

$$P_i(T_i < \infty) = 1\,,$$

and otherwise it is called transient*. A recurrent state $i \in E$ such that*

$$E_i[T_i] < \infty$$

is called positive *recurrent, and otherwise it is called* null *recurrent.*

The definition in terms of return times will now be connected to that in terms of the number of visits.

Theorem 9.2.2 *The distribution given $X_0 = j$ of $N_i = \sum_{n\geq 1} 1_{\{X_n = i\}}$, the number of visits to state i strictly after time 0, is*

$$P_j(N_i = r) = f_{ji} f_{ii}^{r-1}(1 - f_{ii})\ (r \geq 1)$$
$$P_j(N_i = 0) = 1 - f_{ji}\,,$$

where $f_{ji} = P_j(T_i < \infty)$ and T_i is the return time to i.

Proof. We first go from j to i (probability f_{ji}) and then, $r - 1$ times in succession, from i to i (each time with probability f_{ii}), and the last time, that is the $r + 1$-st time, we leave i never to return to it (probability $1-f_{ii}$). By the cycle independence property, all these "cycles" are independent, so that the successive probabilities multiply. □

The distribution of N_i given $X_0 = j$ and given $N_i \geq 1$ is geometric. This has two main consequences. Firstly, $P_i(T_i < \infty) = 1 \Longleftrightarrow P_i(N_i = \infty) = 1$. In words: starting from i, the chain almost surely returns to i, and will then visit i infinitely often. Secondly,

$$E_i[N_i] = \sum_{r=1}^{\infty} r P_i(N_i = r) = \sum_{r-1}^{\infty} r f_{ii}^r (1 - f_{ii}) = \frac{f_{ii}}{1 - f_{ii}} .$$

In particular, $P_i(T_i < \infty) < 1 \Longleftrightarrow E_i[N_i] < \infty$.

We collect these results for future reference. For any state $i \in E$,

$$P_i(T_i < \infty) = 1 \Longleftrightarrow P_i(N_i = \infty) = 1$$

and

$$P_i(T_i < \infty) < 1 \Longleftrightarrow P_i(N_i = \infty) = 0 \Longleftrightarrow E_i[N_i] < \infty . \tag{9.9}$$

In particular, the event $\{N_i = \infty\}$ has P_i-probability 0 or 1.

The Potential Matrix Criterion

The *potential matrix* $\mathbf{G}$ associated with the transition matrix $\mathbf{P}$ is defined by

$$\mathbf{G} = \sum_{n \geq 0} \mathbf{P}^n .$$

Its general term

$$g_{ij} = \sum_{n=0}^{\infty} p_{ij}(n) = \sum_{n=0}^{\infty} P_i(X_n = j) = \sum_{n=0}^{\infty} E_i[1_{\{X_n=j\}}] = E_i \left[\sum_{n=0}^{\infty} 1_{\{X_n=j\}} \right]$$

is the average number of visits to state j, given that the chain starts from state i.

Although the next criterion of recurrence is of theoretical rather than practical interest, it can be helpful in a few situations, for instance in the study of recurrence of random walks (see the examples below).

Theorem 9.2.3 *State $i \in E$ is recurrent if and only if*

$$\sum_{n=0}^{\infty} p_{ii}(n) = \infty .$$

Proof. This merely rephrases Eqn. (9.9). □

EXAMPLE 9.2.4: 1-D RANDOM WALK. The state space of this Markov chain is $E := \mathbb{Z}$ and the non-null terms of its transition matrix are $p_{i,i+1} = p$, $p_{i,i-1} = 1-p$, where $p \in (0,1)$. Since this chain is irreducible, it suffices to elucidate the nature (recurrent or transient) of any one of its states, say, 0. We have $p_{00}(2n+1) = 0$ and

$$p_{00}(2n) = \frac{(2n)!}{n!n!} p^n (1-p)^n .$$

By Stirling's equivalence formula $n! \sim (n/\mathrm{e})^n \sqrt{2\pi n}$, the above quantity is equivalent to

$$\frac{[4p(1-p)]^n}{\sqrt{\pi n}} \tag{$\star$}$$

and the nature of the series $\sum_{n=0}^{\infty} p_{00}(n)$ (convergent or divergent) is that of the series with general term ($\star$). If $p \neq \frac{1}{2}$, in which case $4p(1-p) < 1$, the latter series converges, and if $p = \frac{1}{2}$, in which case $4p(1-p) = 1$, it diverges. In summary, the states of the 1-D random walk are transient if $p \neq \frac{1}{2}$, recurrent if $p = \frac{1}{2}$.

EXAMPLE 9.2.5: 3-D RANDOM WALK. The state space of this HMC is $E = \mathbb{Z}^3$. Denoting by e_1, e_2, and e_3 the canonical basis vectors of $\mathbb{R}^3$ (respectively $(1,0,0)$, $(0,1,0)$, and $(0,0,1)$), the nonnull terms of the transition matrix of the 3-D symmetric random walk are given by

$$p_{x,x\pm e_i} = \frac{1}{6} .$$

We elucidate the nature of state, say, $0 = (0,0,0)$. Clearly, $p_{00}(2n+1) = 0$ for all $n \geq 0$, and (exercise)

$$p_{00}(2n) = \sum_{0\leq i+j\leq n} \frac{(2n)!}{(i!j!(n-i-j)!)^2} \left(\frac{1}{6}\right)^{2n} .$$

This can be rewritten as

$$p_{00}(2n) = \sum_{0\leq i+j\leq n} \frac{1}{2^{2n}} \binom{2n}{n} \left(\frac{n!}{i!j!(n-i-j)!}\right)^2 \left(\frac{1}{3}\right)^{2n} .$$

Using the *trinomial formula*

$$\sum_{0\leq i+j\leq n} \frac{n!}{i!j!(n-i-j)!} \left(\frac{1}{3}\right)^n = 1 ,$$

we obtain the bound

$$p_{00}(2n) \leq K_n \frac{1}{2^{2n}} \binom{2n}{n} \left(\frac{1}{3}\right)^n ,$$

where

$$K_n = \max_{0 \leq i+j \leq n} \frac{n!}{i!j!(n-i-j)!} .$$

For large values of n, K_n is bounded as follows. Let i_0 and j_0 be the values of i, j that maximize $n!/(i!j!(n+-i-j)!)$ in the domain of interest $0 \leq i+j \leq n$. ¿From the definition of i_0 and j_0, the quantities

$$\frac{n!}{(i_0-1)!j_0!(n-i_0-j_0+1)!} ,$$

$$\frac{n!}{(i_0+1)!j_0!(n-i_0-j_0-1)!} ,$$

$$\frac{n!}{i_0!(j_0-1)!(n-i_0-j_0+1)!} ,$$

$$\frac{n!}{i_0!(j_0+1)!(n-i_0-j_0-1)!} ,$$

are bounded by

$$\frac{n!}{i_0!j_0!(n-i_0-j_0)!} .$$

The corresponding inequalities reduce to

$$n-i_0-1 \leq 2j_0 \leq n-i_0+1 \text{ and } n-j_0-1 \leq 2i_0 \leq n-j_0+1 ,$$

and this shows that for large n, $i_0 \sim n/3$ and $j_0 \sim n/3$. Therefore, for large n,

$$p_{00}(2n) \sim \frac{n!}{(n/3)!(n/3)!2^{2n}e^n} \binom{2n}{n} .$$

By Stirling's equivalence formula, the right-hand side of the latter equivalence is in turn equivalent to

$$\frac{3\sqrt{3}}{2(\pi n)^{3/2}} ,$$

the general term of a convergent series. State 0 is therefore transient.

One might wonder at this point about the symmetric random walk on $\mathbb{Z}^2$, which moves at each step northward, southward, eastward and westward equiprobably. Exercise 9.7.25 asks you to show that it is null recurrent. Exercise 9.7.26 asks you to prove that the symmetric random walks on $\mathbb{Z}^p$, $p \geq 4$, are transient.

A theoretical application of the potential matrix criterion is to the proof that recurrence is a (communication) class property.

Theorem 9.2.6 *If i and j communicate, then they are either both recurrent or both transient.*

Proof. By definition, i and j communicate if and only if there exist integers M and N such that $p_{ij}(M) > 0$ and $p_{ji}(N) > 0$. Going from i to j in M steps, then from j to j in n steps, then from j to i in N steps, is just one way of going from i back to i in $M + n + N$ steps. Therefore, $p_{ii}(M + n + N) \geq p_{ij}(M) \times p_{jj}(n) \times p_{ji}(N)$. Similarly, $p_{jj}(N + n + M) \geq p_{ji}(N) \times p_{ii}(n) \times p_{ij}(M)$. Therefore, with $\alpha := p_{ij}(M)\, p_{ji}(N)$ (a strictly positive quantity), we have $p_{ii}(M + N + n) \geq \alpha\, p_{jj}(n)$ and $p_{jj}(M + N + n) \geq \alpha\, p_{ii}(n)$. This implies that the series $\sum_{n=0}^{\infty} p_{ii}(n)$ and $\sum_{n=0}^{\infty} p_{jj}(n)$ either both converge or both diverge. The potential matrix criterion concludes the proof. □

Invariant Measure

This notion extends that of a stationary distribution and plays a central role in the recurrence theory of Markov chains.

Definition 9.2.7 *A non-trivial (that is, non-null) vector x (indexed by E) of non-negative real numbers (notation: $0 \leq x < \infty$) is called an* invariant measure *of the stochastic matrix $\mathbf{P}$ (indexed by E) if*

$$x^T = x^T \mathbf{P}\,. \tag{9.10}$$

Theorem 9.2.8 *Let $\mathbf{P}$ be the transition matrix of an irreducible recurrent* HMC *$\{X_n\}_{n\geq 0}$. Let 0 be an arbitrary state and let T_0 be the return time to 0. Define for all $i \in E$*

$$x_i = E_0 \left[\sum_{n=1}^{T_0} 1_{\{X_n = i\}} \right]. \tag{9.11}$$

(For $i \neq 0$, x_i is the expected number of visits to state i before returning to 0.) Then, $0 < x < \infty$ and x is an invariant measure of $\mathbf{P}$.

Proof. We make three preliminary observations. First, it will be convenient to rewrite (9.11) as

$$x_i = E_0 \left[\sum_{n \geq 1} 1_{\{X_n = i\}} 1_{\{n \leq T_0\}} \right].$$

Next, when $1 \leq n \leq T_0$, $X_n = 0$ if and only if $n = T_0$. Therefore,

$$x_0 = 1\,.$$

Also,

$$\sum_{i\in E}\sum_{n\geq 1} 1_{\{X_n=i\}}1_{\{n\leq T_0\}} = \sum_{n\geq 1}\left(\sum_{i\in E} 1_{\{X_n=i\}}\right)1_{\{n\leq T_0\}} = \sum_{n\geq 1} 1_{\{n\leq T_0\}} = T_0\,,$$

and therefore

$$\sum_{i\in E} x_i = E_0[T_0]\,. \tag{9.12}$$

We introduce the quantity

$$_0p_{0i}(n) := E_0[1_{\{X_n=i\}}1_{\{n\leq T_0\}}] = P_0(X_1 \neq 0, \cdots, X_{n-1} \neq 0, X_n = i)\,.$$

This is the probability, starting from state 0, of visiting i at time n before returning to 0. From the definition of x,

$$x_i = \sum_{n\geq 1} {}_0p_{0i}(n)\,. \tag{†}$$

We first prove (9.10). Observe that $_0p_{0i}(1) = p_{0i}$, and, by first-step analysis, for all $n \geq 2$, $_0p_{0i}(n) = \sum_{j\neq 0} {}_0p_{0j}(n-1)p_{ji}$. Summing up all the above equalities, and taking (†) into account, we obtain

$$x_i = p_{0i} + \sum_{j\neq 0} x_j p_{ji}\,,$$

that is, (9.10), since $x_0 = 1$.

Next we show that $x_i > 0$ for all $i \in E$. Indeed, iterating (9.10), we find $x^T = x^T\mathbf{P}^n$, that is, since $x_0 = 1$,

$$x_i = \sum_{j\in E} x_j p_{ji}(n) = p_{0i}(n) + \sum_{j\neq 0} x_j p_{ji}(n)\,.$$

If x_i were null for some $i \in E$, $i \neq 0$, the latter equality would imply that $p_{0i}(n) = 0$ for all $n \geq 0$, which means that 0 and i do not communicate, in contradiction to the irreducibility assumption.

It remains to show that $x_i < \infty$ for all $i \in E$. As before, we find that

$$1 = x_0 = \sum_{j\in E} x_j p_{j0}(n)$$

for all $n \geq 1$, and therefore if $x_i = \infty$ for some i, necessarily $p_{i0}(n) = 0$ for all $n \geq 1$, and this also contradicts irreducibility. □

Theorem 9.2.9 *The invariant measure of an irreducible recurrent* HMC *is unique up to a multiplicative factor.*

Proof. In the proof of Theorem 9.2.8, we showed that for an invariant measure y of an irreducible chain, $y_i > 0$ for all $i \in E$, and therefore, one can define, for all $i, j \in E$, the matrix $\mathbf{Q}$ by

$$q_{ji} = \frac{y_i}{y_j} p_{ij} \,. \tag{$\star$}$$

It is a transition matrix, since $\sum_{i\in E} q_{ji} = \frac{1}{y_j}\sum_{i\in E} y_i p_{ij} = \frac{y_j}{y_j} = 1$. The general term of $\mathbf{Q}^n$ is

$$q_{ji}(n) = \frac{y_i}{y_j} p_{ij}(n) \,. \tag{$\star\star$}$$

Indeed, supposing ($\star\star$) true for n,

$$\begin{aligned} q_{ji}(n+1) &= \sum_{k\in E} q_{jk} q_{ki}(n) = \sum_{k\in E} \frac{y_k}{y_j} p_{kj} \frac{y_i}{y_k} p_{ik}(n) \\ &= \frac{y_i}{y_j} \sum_{k\in E} p_{ik}(n) p_{kj} = \frac{y_i}{y_j} p_{ij}(n+1) \,, \end{aligned}$$

and ($\star\star$) follows by induction.

Clearly, $\mathbf{Q}$ is irreducible, since $\mathbf{P}$ is irreducible (just observe that $q_{ji}(n) > 0$ if and only if $p_{ij}(n) > 0$ in view of ($\star\star$)). Also, $p_{ii}(n) = q_{ii}(n)$, and therefore $\sum_{n\geq 0} q_{ii}(n) = \sum_{n\geq 0} p_{ii}(n)$, and therefore $\mathbf{Q}$ is recurrent by the potential matrix criterion. Call $g_{ji}(n)$ the probability, relative to the chain governed by the transition matrix $\mathbf{Q}$, of returning to state i for the first time at step n when starting from j. First-step analysis gives

$$g_{i0}(n+1) = \sum_{j\neq 0} q_{ij} g_{j0}(n) \,,$$

that is, using ($\star$),

$$y_i g_{i0}(n+1) = \sum_{j\neq 0} (y_j g_{j0}(n)) p_{ji} \,.$$

Recall that ${}_0p_{0i}(n+1) = \sum_{j\neq 0} {}_0p_{0j}(n) p_{ji}$, or, equivalently,

$$y_0 \, {}_0p_{0i}(n+1) = \sum_{j\neq 0} (y_0 \, {}_0p_{0j}(n)) p_{ji} \,.$$

We therefore see that the sequences $\{y_0 \, {}_0p_{0i}(n)\}$ and $\{y_i g_{i0}(n)\}$ satisfy the same recurrence equation. Their first terms ($n = 1$), respectively $y_0 \, {}_0p_{0i}(1) = y_0 p_{0i}$ and $y_i g_{i0}(1) = y_i q_{i0}$, are equal in view of ($\star$). Therefore, for all $n \geq 1$,

$$ {}_0p_{0i}(n) = \frac{y_i}{y_0} g_{i0}(n)\,. $$

Summing up with respect to $n \geq 1$ and using $\sum_{n\geq 1} g_{i0}(n) = 1$ ($\mathbf{Q}$ is recurrent), we obtain that $x_i = \frac{y_i}{y_0}$. □

Equality (9.12) and the definition of positive recurrence give the following.

Theorem 9.2.10 *An irreducible recurrent* HMC *is positive recurrent if and only if its invariant measures x satisfy*

$$ \sum_{i\in E} x_i < \infty\,. $$

The Stationary Distribution Criterion of Positive Recurrence

An HMC may well be irreducible and possess an invariant measure, and yet not be recurrent. The simplest example is the 1-D non-symmetric random walk, which was shown to be transient and yet admits $x_i = 1$ ($i \in \mathbb{Z}$) for invariant measure. However, it turns out that the existence of a stationary probability distribution is necessary and sufficient for an irreducible chain (not a priori assumed recurrent) to be recurrent positive.

Theorem 9.2.11 *An irreducible* HMC *is positive recurrent if and only if there exists a stationary distribution. Moreover, the stationary distribution π is, when it exists, unique, and $\pi > 0$.*

Proof. The direct part follows from Theorems 9.2.8 and 9.2.10. For the converse part, assume the existence of a stationary distribution π. Iterating $\pi^T = \pi^T\mathbf{P}$, we obtain $\pi^T = \pi^T\mathbf{P}^n$, that is, for all $i \in E$, $\pi(i) = \sum_{j\in E} \pi(j)p_{ji}(n)$. If the chain were transient, then, for all states i, j,

$$ \lim_{n\uparrow\infty} p_{ji}(n) = 0\,. $$

The following is a formal proof:[2]

$$\begin{aligned}\sum_{n\geq 1} p_{ji}(n) &= \sum_{n\geq 1}\sum_{k\geq 1} P_j(T_i = k)p_{ii}(n-k) \\ &= \sum_{k\geq 1} P_j(T_i = k)\sum_{n\geq 1} p_{ii}(n-k) \\ &\leq \left(\sum_{k\geq 1} P_j(T_i = k)\right)\left(\sum_{n\geq 1} p_{ii}(n)\right) \\ &= P_j(T_i < \infty)\left(\sum_{n\geq 1} p_{ii}(n)\right) \leq \sum_{n\geq 1} p_{ii}(n) < \infty .\end{aligned}$$

In particular, $\lim_n p_{ji}(n) = 0$. Since $p_{ji}(n)$ is bounded uniformly in j and n by 1, by the dominated convergence theorem for series:[3]

$$\pi(i) = \lim_{n\uparrow\infty}\sum_{j\in E}\pi(j)p_{ji}(n) = \sum_{j\in E}\pi(j)\left(\lim_{n\uparrow\infty} p_{ji}(n)\right) = 0 .$$

This contradicts the assumption that π is a stationary distribution ($\sum_{i\in E}\pi(i) = 1$). The chain must therefore be recurrent, and by Theorem 9.2.10, it is positive recurrent.

The stationary distribution π of an irreducible positive recurrent chain is unique (use Theorem 9.2.9 and the fact that there is no choice for a multiplicative factor but 1). Also recall that $\pi(i) > 0$ for all $i \in E$ (see Theorem 9.2.8). □

Theorem 9.2.12 *Let π be the unique stationary distribution of an irreducible positive recurrent* HMC, *and let T_i be the return time to state i. Then*

$$\pi(i)E_i[T_i] = 1 . \tag{9.13}$$

Proof. This equality is a direct consequence of expression (9.11) for the invariant measure. Indeed, π is obtained by normalization of x: for all $i \in E$,

$$\pi(i) = \frac{x_i}{\sum_{j\in E} x_j},$$

and in particular, for $i = 0$, recalling that $x_0 = 1$ and using (9.12),

$$\pi(0) = \frac{1}{E_0[T_0]} .$$

[2] Rather awkward, but using only the elementary tools available.

[3] Let $\{a_{nk}\}_{n\geq 1,k\geq 1}$ be an array of real numbers such that, for some sequence $\{b_k\}_{k\geq 1}$ of non-negative numbers satisfying $\sum_{k=1}^{\infty} b_k < \infty$, it holds that for all $n \geq 1$, $k \geq 1$, $|a_{nk}| \leq b_k$. If moreover for all $k \geq 1$, $\lim_{n\uparrow\infty} a_{nk} = a_k$, then $\lim_{n\uparrow\infty}\sum_{k=1}^{\infty} a_{nk} = \sum_{k=1}^{\infty} a_k$. (Note that this result is a particular case of the dominated convergence theorem, Theorem 5.1.3.)

Since state 0 does not play a special role in the analysis, (9.13) is true for all $i \in E$. □

The situation is extremely simple when the state space is finite.

Theorem 9.2.13 *An irreducible* HMC *with* finite *state space is positive recurrent.*

Proof. We first show recurrence. We have

$$\sum_{j \in E} p_{ij}(n) = 1 \,,$$

and in particular, the limit of the left-hand side is 1. If the chain were transient, then, as we saw in the proof of Theorem 9.2.11, for all $i, j \in E$,

$$\lim_{n \uparrow \infty} p_{ij}(n) = 0 \,,$$

and therefore, since the state space is finite

$$\lim_{n \uparrow \infty} \sum_{j \in E} p_{ij}(n) = 0 \,,$$

a contradiction. Therefore, the chain is recurrent. By Theorem 9.2.8 it has an invariant measure x. Since E is finite, $\sum_{i \in E} x_i < \infty$, and therefore the chain is positive recurrent, by Theorem 9.2.10. □

EXAMPLE 9.2.14: THE REPAIR SHOP, TAKE 2. Recall that this Markov chain satisfies the recurrence equation

$$X_{n+1} = (X_n - 1)^+ + Z_{n+1} \,, \tag{9.14}$$

where $a^+ = \max(a, 0)$. The sequence $\{Z_n\}_{n \geq 1}$ is assumed to be IID, independent of the initial state X_0, and with common probability distribution

$$P(Z_1 = k) = a_k, \; k \geq 0$$

of generating function g_Z.

This chain is irreducible if and only if $P(Z_1 = 0) > 0$ and $P(Z_1 \geq 2) > 0$ as we now prove formally. Looking at (9.14), we make the following observations. If $P(Z_{n+1} = 0) = 0$, then $X_{n+1} \geq X_n$ a.s. and there is no way of going from i to $i-1$. If $P(Z_{n+1} \leq 1) = 1$, then $X_{n+1} \leq X_n$, and there is no way of going from i to $i+1$. Therefore, the two conditions $P(Z_1 = 0) > 0$ and $P(Z_2 \geq 2) > 0$ are *necessary* for irreducibility. They are also sufficient. Indeed if there exists an integer $k \geq 2$

such that $P(Z_{n+1}=k)>0$, then one can jump with positive probability from any $i>0$ to $i+k-1>i$ or from $i=0$ to $k>0$. Also if $P(Z_{n+1}=0)>0$, one can step down from $i>0$ to $i-1$ with positive probability. In particular, one can go from i to $j<i$ with positive probability. Therefore, one way to travel from i to $j\geq i$ is by taking several successive steps of height at least $k-1$ in order to reach a state $l\geq i$, and then (in the case of $l>i$) stepping down one stair at a time from l to i. All this with positive probability.

EXAMPLE 9.2.15: THE REPAIR SHOP, TAKE 3. Assuming irreducibility (see Example 9.2.14), we now seek a necessary and sufficient condition for positive recurrence. For any complex number z with modulus not larger than 1, it follows from the recurrence equation (9.14) that

$$z^{X_{n+1}+1}=\left(z^{(X_n-1)^++1}\right)z^{Z_{n+1}}=\left(z^{X_n}-1_{\{X_n=0\}}+z1_{\{X_n=0\}}\right)z^{Z_{n+1}},$$

and therefore $zz^{X_{n+1}}-z^{X_n}z^{Z_{n+1}}=(z-1)1_{\{X_n=0\}}z^{Z_{n+1}}$. ¿From the independence of X_n and Z_{n+1}, $E[z^{X_n}z^{Z_{n+1}}]=E[z^{X_n}]g_Z(z)$, and $E[1_{\{X_n=0\}}z^{Z_{n+1}}]=\pi(0)g_Z(z)$, where $\pi(0)=P(X_n=0)$. Therefore, $zE[z^{X_{n+1}}]-g_Z(z)E[z^{X_n}]=(z-1)\pi(0)g_Z(z)$. But in steady state, $E[z^{X_{n+1}}]=E[z^{X_n}]=g_X(z)$, and therefore

$$g_X(z)\left(z-g_Z(z)\right)=\pi(0)(z-1)g_Z(z)\,. \tag{9.15}$$

This gives the generating function $g_X(z)=\sum_{i=0}^{\infty}\pi(i)z^i$, as long as $\pi(0)$ is available. To obtain $\pi(0)$, differentiate (9.15):

$$g_X'(z)\left(z-g_Z(z)\right)+g_X(z)\left(1-g_Z'(z)\right)=\pi(0)\left(g_Z(z)+(z-1)g_Z'(z)\right),$$

and let $z=1$, to obtain, taking into account the equalities $g_X(1)=g_Z(1)=1$ and $g_Z'(1)=E[Z]$,

$$\pi(0)=1-E[Z]\,. \tag{9.16}$$

But the stationary distribution of an irreducible HMC is positive, hence the necessary condition of positive recurrence:

$$E[Z_1]<1\,.$$

It turns out that this condition is also sufficient for positive recurrence.

From (9.15) and (9.16), we have the generating function of the stationary distribution:

$$\sum_{i=0}^{\infty}\pi(i)z^i=(1-E[Z])\frac{(z-1)g_Z(z)}{z-g_Z(z)}\,. \tag{9.17}$$

If $E[Z_1] > 1$, the chain is transient, as a simple argument based on the strong law of large numbers shows. In fact, $X_n = X_0 + \sum_{k=1}^{n} Z_k - n + \sum_{k=1}^{n} 1_{\{X_k=0\}}$, and therefore

$$X_n \geq \sum_{k=1}^{n} Z_k - n = \sum_{k=1}^{n} (Z_k - 1),$$

which tends to ∞ because, by the strong law of large numbers,

$$\frac{\sum_{k=1}^{n}(Z_k - 1)}{n} \to E[Z] - 1 > 0.$$

This is of course incompatible with recurrence.

In the case $E[Z_1] = 1$, there are only two possibilities left: transient or null recurrent. It turns out that the chain is null recurrent in this case.

EXAMPLE 9.2.16: THE PURE RANDOM WALK ON A GRAPH. Consider a finite non-directed *connected* graph $G = (V, \mathcal{E})$ where V is the set of vertices, or nodes, and $\mathcal{E}$ is the set of edges. Let d_i be the *index* of vertex i (the number of edges "adjacent" to vertex i). Since there are no isolated nodes (a consequence of the connectedness assumption), $d_i > 0$ for all $i \in V$. Transform this graph into a directed graph by splitting each edge into two directed edges of opposite directions, and make it a transition graph by associating to the directed edge from i to j the transition probability $\frac{1}{d_i}$ (see the figure below). Note that $\sum_{i \in V} d_i = 2|\mathcal{E}|$.

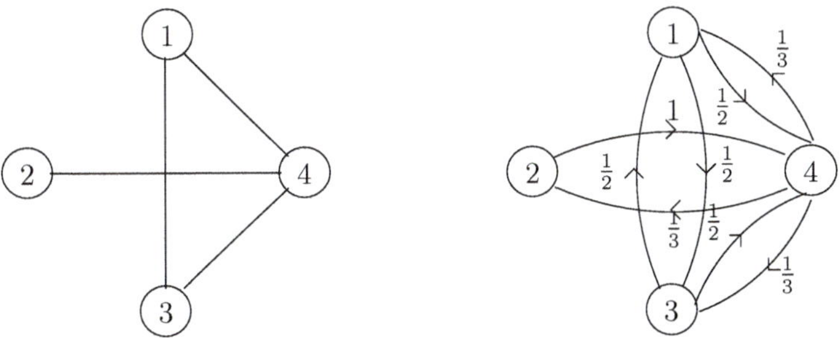

A random walk on a graph

The corresponding HMC with state space $E \equiv V$ is irreducible (G is connected). It therefore admits a unique stationary distribution π, which we attempt to find via Theorem 9.1.24. Let i and j be connected by an edge, and therefore $p_{ij} = \frac{1}{d_i}$

and $p_{ji} = \frac{1}{d_j}$, so that the detailed balance equation between these two states is

$$\pi(i)\frac{1}{d_i} = \pi(j)\frac{1}{d_j}\,.$$

This gives $\pi(i) = Kd_i$, where K is obtained by normalization: $K = \left(\sum_{j\in E} d_j\right)^{-1} = (2|\mathcal{E}|)^{-1}$. Therefore

$$\pi(i) = \frac{d_i}{2|\mathcal{E}|}\,.$$

The lazy random walk on the graph is, by definition, the Markov chain on V with the transition probabilities $p_{ii} = \frac{1}{2}$ and for $i, j \in V$ such that i and j are connected by an edge of the graph, $p_{i,i} = \frac{1}{2d_i}$. This modified chain admits the same stationary distribution as the original random walk. The difference is that the lazy version is always aperiodic, whereas the original version may be periodic.

The stationary distribution criterion can also be used to prove instability.

Birth-and-Death Markov Chain

We first define the *birth-and-death process with a bounded population*. The state space of such a chain is $E = \{0, 1, \dots, N\}$ and its transition matrix is

$$\mathbf{P} = \begin{pmatrix} r_0 & p_0 & & & & & & \\ q_1 & r_1 & p_1 & & & & & \\ & q_2 & r_2 & p_2 & & & & \\ & & & \ddots & & & & \\ & & & q_i & r_i & p_i & & \\ & & & & \ddots & \ddots & \ddots & \\ & & & & & q_{N-1} & r_{N-1} & p_{N-1} \\ & & & & & & p_N & r_N \end{pmatrix},$$

where $p_i > 0$ for all $i \in E\backslash\{N\}$, $q_i > 0$ for all $i \in E\backslash\{0\}$, $r_i \geq 0$ for all $i \in E$, and $p_i + q_i + r_i = 1$ for all $i \in E$. The positivity conditions placed on the p_i's and q_i's guarantee that the chain is irreducible. Since the state space is finite, it is positive recurrent (Theorem 9.2.13), and it has a unique stationary distribution. Motivated by the Ehrenfest HMC, which is reversible in the stationary state, we make the educated guess that the birth and death process considered has the same property. This will be the case if and only if there exists a probability distribution π

on E satisfying the detailed balance equations, that is, such that for all $1 \le i \le N$, $\pi(i-1)p_{i-1} = \pi(i)q_i$. Letting $w_0 = 1$ and for all $1 \le i \le N$,

$$w_i = \prod_{k=1}^{i} \frac{p_{k-1}}{q_k}$$

we find that

$$\pi(i) = \frac{w_i}{\sum_{j=0}^{N} w_j} \tag{9.18}$$

indeed satisfies the detailed balance equations and is therefore the (unique) stationary distribution of the chain.

We now consider the *unbounded birth-and-death process*. This chain has the state space $E = \mathbb{N}$ and its transition matrix is as in the previous example (only, it is unbounded on the right). In particular, we assume that the p_i's and q_i's are positive in order to guarantee irreducibility. The same reversibility argument as above applies with a little difference. In fact we can show that the w_i's defined above satisfy the detailed balance equations and therefore the global balance equations. Therefore the vector $\{w_i\}_{i \in E}$ is the unique, up to a multiplicative factor, invariant measure of the chain. It can be normalized to a probability distribution if and only if

$$\sum_{j=0}^{\infty} w_j < \infty \,.$$

Therefore, in this case and in this case only there exists a (unique) stationary distribution, also given by (9.18).

Note that the stationary distribution, when it exists, does not depend on the r_i's. The recurrence properties of the above unbounded birth-and-death process are therefore the same as those of the chain below, which is however not aperiodic. For aperiodicity, it suffices to suppose at least one of the r_i's is positive.

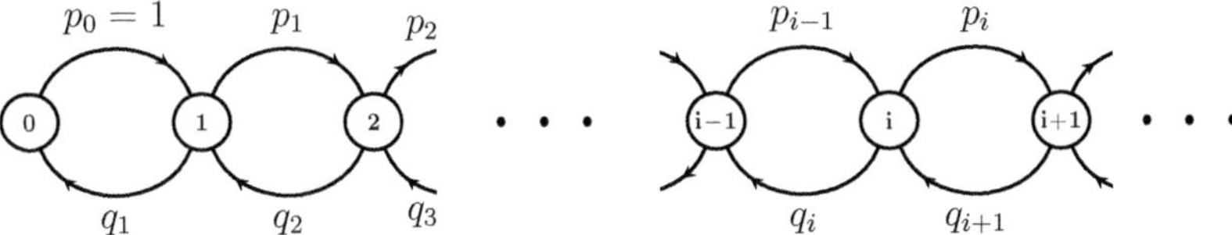

We now compute, for the (bounded or unbounded) irreducible birth-and death process, the average time it takes to reach a state b from a state $a < b$. In fact, we shall prove that

$$E_a\left[T_b\right] = \sum_{k=a+1}^{b} \frac{1}{q_k w_k} \sum_{j=0}^{k-1} w_j \,. \tag{9.19}$$

Since obviously $E_a\left[T_b\right] = \sum_{k=a+1}^{b} E_{k-1}\left[T_k\right]$, it suffices to prove that

$$E_{k-1}\left[T_k\right] = \frac{1}{q_k w_k} \sum_{j=0}^{k-1} w_j \,. \tag{$\star$}$$

For this, consider for any given $k \in \{0, 1, \ldots, N\}$ the truncated chain, which moves on the state space $\{0, 1, \ldots, k\}$ as the original chain, except in state k where it moves one step down with probability q_k and stays still with probability $p_k + r_k$. Write $\widetilde{E}$ for expectations of the modified chain. The unique stationary distribution of this chain is given by

$$\widetilde{\pi}_\ell = \frac{w_\ell}{\sum_{j=0}^{k} w_\ell}$$

for all $0 \le \ell \le k$. First-step analysis shows that $\widetilde{E}_k\left[T_k\right] = (r_k + p_k) \times 1 + q_k\left(1 + \widetilde{E}_{k-1}\left[T_k\right]\right)$, that is

$$\widetilde{E}_k\left[T_k\right] = 1 + q_k \widetilde{E}_{k-1}\left[T_k\right] \,.$$

Also

$$\widetilde{E}_k\left[T_k\right] = \frac{1}{\widetilde{\pi}_k} = \frac{1}{w_k} \sum_{j=0}^{k} w_j \,,$$

and therefore, since $\widetilde{E}_{k-1}\left[T_k\right] = E_{k-1}\left[T_k\right]$, we have ($\star$).

In the special case where $(p_j, q_j, r_j) = (p, q, r)$ for all $j \neq 0, N$, $(p_0, q_0, r_0) = (p, q+r, 0)$ and $(p_N, q_N, r_N) = (0, p+r, q)$, we have $w_i = \left(\frac{p}{q}\right)^i$, and for $1 \le k \le N$,

$$E_{k-1}\left[T_k\right] = \frac{1}{q\left(\frac{p}{q}\right)^k} \sum_{j=0}^{k-1} \left(\frac{p}{q}\right)^j = \frac{1}{p-q}\left(1 - \left(\frac{q}{p}\right)^k\right) .$$

In the further particularization where $p = q$, $w_i = 1$ for all i and

$$E_{k-1}\left[T_k\right] = \frac{k}{p} \,.$$

Foster's Theorem

The stationary distribution criterion of positive recurrence of an irreducible chain requires solving the balance equation, an often hopeless enterprise. The following *sufficient* condition is more tractable and indeed quite powerful.

Theorem 9.2.17 *Let* $\mathbf{P}$ *be an irreducible transition matrix on the countable state space* E. *Suppose that there exists a function* $h : E \to \mathbb{R}$ *such that* $\inf_i h(i) > -\infty$,

$$\sum_{k\in E} p_{ik}h(k) < \infty \quad (i \in F), \tag{9.20}$$

and

$$\sum_{k\in E} p_{ik}h(k) \le h(i) - \epsilon \quad (i \notin F), \tag{9.21}$$

for some finite set F *and some* $\epsilon > 0$. *Then the corresponding* HMC *is positive recurrent.*

Proof. Recall the notation X_0^n for $(X_0, \dots, X_n)$. Since $\inf_i h(i) > -\infty$, one may assume without loss of generality that $h \ge 0$, by adding a constant if necessary. Call τ the return time to F and let $Y_n := h(X_n)1_{\{n<\tau\}}$. Equality (9.21) implies that $E[h(X_{n+1}) \mid X_n = i] \le h(i) - \epsilon$ for all $i \notin F$. For $i \notin F$,

$$\begin{aligned} E_i[Y_{n+1} \mid X_0^n] &= E_i[Y_{n+1}1_{\{n<\tau\}} \mid X_0^n] + E_i(Y_{n+1}1_{\{n\ge\tau\}} \mid X_0^n] \\ &= E_i[Y_{n+1}1_{\{n<\tau\}} \mid X_0^n] \le E_i[h(X_{n+1})1_{\{n<\tau\}} \mid X_0^n] \\ &= 1_{\{n<\tau\}}E_i[h(X_{n+1}) \mid X_0^n] = 1_{\{n<\tau\}}E_i[h(X_{n+1}) \mid X_n] \\ &\le 1_{\{n<\tau\}}h(X_n) - \epsilon 1_{\{n<\tau\}}, \end{aligned}$$

where the third *equality* comes from the fact that $1_{\{n<\tau\}}$ is a function of X_0^n (Theorem 2.4.6), the fourth *equality* is the Markov property and the last *inequality* is true because P_i-a.s., $X_n \notin F$ on $n < \tau$. Therefore, P_i-a.s.,

$$E_i[Y_{n+1} \mid X_0^n] \le Y_n - \epsilon 1_{\{n<\tau\}}$$

and, taking expectations,

$$0 \le E_i[Y_{n+1}] \le E_i[Y_n] - \epsilon P_i(\tau > n).$$

Iterating the above equality and taking into account the fact that Y_n is non-negative, we obtain

$$0 \le E_i[Y_0] - \epsilon \sum_{k=0}^{n} P_i(\tau > k).$$

But $Y_0 = h(i)$, P_i-a.s., and $\sum_{k=0}^{\infty} P_i(\tau > k) = E_i[\tau]$. Therefore, for all $i \notin F$,

$$E_i[\tau] \le \epsilon^{-1} h(i) \,.$$

For $j \in F$, by first-step analysis

$$E_j[\tau] = 1 + \sum_{i \notin F} p_{ji} E_i[\tau] \,.$$

Therefore $E_j[\tau] \le 1 + \epsilon^{-1} \sum_{i \notin F} p_{ji} h(i)$, a finite quantity in view of assumption (9.20): the return time to F starting anywhere in F has finite expectation. Since F is a finite set, this implies positive recurrence in view of the following lemma. □

Lemma 9.2.18 *Let $\{X_n\}_{n \ge 0}$ be an irreducible* HMC, *let F be a finite subset of the state space E and let $\tau(F)$ be the return time to F. If $E_j[\tau(F)] < \infty$ for all $j \in F$, the chain is positive recurrent.*

Proof. Exercise 9.7.15. □

The function h in Foster's theorem is called a *Lyapunov function* because it plays a role similar to the Lyapunov functions in the stability theory of ordinary differential equations. It has a tendency to decrease along the trajectories of the process, at least outside a finite set of states, called the *refuge*. Since it is non-negative, it cannot decrease forever and therefore it eventually enters the refuge.

The following corollary of Foster's theorem is sometimes referred to as *Pakes' lemma*.

Corollary 9.2.19 *Let $\{X_n\}_{n \ge 0}$ be an irreducible* HMC *on $E = \mathbb{N}$ such that for all $n \ge 0$ and all $i \in E$,*

$$E[X_{n+1} \mid X_n = i] < \infty \tag{9.22}$$

and

$$\limsup_{i \uparrow \infty} E[X_{n+1} - X_n \mid X_n = i] < 0 \,. \tag{9.23}$$

Such an HMC *is positive recurrent.*

Proof. Let -2ϵ be the left-hand side of (9.23). In particular, $\epsilon > 0$. By (9.23), for i sufficiently large, say $i > i_0$, $E[X_{n+1} - X_n \mid X_n = i] < -\epsilon$, and therefore the conditions of Foster's theorem are satisfied with $h(i) = i$ and $F = \{i; i \le i_0\}$. □

EXAMPLE 9.2.20: A RANDOM WALK ON $\mathbb{N}$. Let $\{Z_n\}_{n\geq 1}$ be an IID sequence of integrable random variables with values in $\mathbb{Z}$ such that

$$E[Z_1] < 0\,,$$

and define $\{X_n\}_{n\geq 0}$, an HMC with state space $E = \mathbb{N}$, by

$$X_{n+1} = (X_n + Z_{n+1})^+\,,$$

where X_0 is independent of $\{Z_n\}_{n\geq 1}$. Assume irreducibility (the reader is invited to find a necessary and sufficient condition for this). Here

$$\begin{aligned} E[X_{n+1} - i \mid X_n = i] &= E[(i + Z_{n+1})^+ - i] \\ &= E[-i 1_{\{Z_{n+1} \leq -i\}} + Z_{n+1} 1_{\{Z_{n+1} > -i\}}] \leq E[Z_1 1_{\{Z_1 > -i\}}]\,. \end{aligned}$$

By dominated convergence, the limit of $E[Z_1 1_{\{Z_1 > -i\}}]$ as i tends to ∞ is $E[Z_1] < 0$ and therefore, by Pakes' lemma, the HMC is positive recurrent.

EXAMPLE 9.2.21: THE REPAIR SHOP, TAKE 4. Continuation of Example 9.2.15. Arguments very similar to those of the previous example show that in the repair shop HMC (assumed irreducible), condition $E[Z_1] < 1$ implies positive recurrence.

9.3 Long-run Behavior

The Markov Chain Ergodic Theorem

The ergodic theorem for Markov chains gives conditions guaranteeing that empirical averages of the type

$$\frac{1}{N} \sum_{k=1}^{N} f(X_k, \ldots, X_{k+L})$$

converge to the corresponding probabilistic averages. This result is an almost immediate application of the strong law of large numbers.

Proposition 9.3.1 *Let $\{X_n\}_{n\geq 0}$ be an irreducible recurrent* HMC *and let x denote the canonical invariant measure associated with state $0 \in E$, which is given by (9.11). Define for $n \geq 1$*

$$\nu(n) = \sum_{k=1}^{n} 1_{\{X_k = 0\}}\,. \tag{9.24}$$

Let $f : E \to \mathbb{R}$ be such that

$$\sum_{i \in E} |f(i)| x_i < \infty . \tag{9.25}$$

Then, for any initial distribution μ, P_μ-a.s.,

$$\lim_{N \uparrow \infty} \frac{1}{\nu(N)} \sum_{k=1}^{N} f(X_k) = \sum_{i \in E} f(i) x_i . \tag{9.26}$$

Before the proof, we shall harvest the most interesting consequences.

Theorem 9.3.2 *Let $\{X_n\}_{n \geq 0}$ be an irreducible positive recurrent Markov chain with the stationary distribution π, and let $f : E \to \mathbb{R}$ be such that*

$$\sum_{i \in E} |f(i)| \pi(i) < \infty . \tag{9.27}$$

Then for any initial distribution μ, P_μ-a.s.,

$$\lim_{n \uparrow \infty} \frac{1}{N} \sum_{k=1}^{N} f(X_k) = \sum_{i \in E} f(i) \pi(i) . \tag{9.28}$$

Proof. Apply Proposition 9.3.1 to $f \equiv 1$. Condition (9.25) is satisfied, since in the positive recurrent case, $\sum_{i \in E} x_i < \infty$. Therefore, P_μ-a.s.,

$$\lim_{N \uparrow \infty} \frac{N}{\nu(N)} = \sum_{j \in E} x_j .$$

Now, f satisfying (9.27) also satisfies (9.25), since x and π are proportional, and therefore, P_μ-a.s.,

$$\lim_{N \uparrow \infty} \frac{1}{\nu(N)} \sum_{k=1}^{N} f(X_k) = \sum_{i \in E} f(i) x_i .$$

Combining the above equalities gives, P_μ-a.s.,

$$\lim_{N \to \infty} \frac{1}{N} \sum_{k=1}^{N} f(X_k) = \lim_{N \to \infty} \frac{\nu(N)}{N} \frac{1}{\nu(N)} \sum_{k=1}^{N} f(X_k) = \frac{\sum_{i \in E} f(i) x_i}{\sum_{j \in E} x_j} ,$$

from which (9.28) follows, since π is obtained by normalization of x. □

Corollary 9.3.3 *Let $\{X_n\}_{n\geq 1}$ be an irreducible positive recurrent Markov chain with the stationary distribution π, and let $g : E^{L+1} \to \mathbb{R}$ be such that*

$$\sum_{i_0,\dots,i_L} |g(i_0,\dots,i_L)|\pi(i_0)p_{i_0 i_1}\cdots p_{i_{L-1}i_L} < \infty .$$

Then for all initial distributions μ, P_μ-a.s.

$$\lim \frac{1}{N}\sum_{k=1}^{N} g(X_k, X_{k+1},\dots,X_{k+L}) = \sum_{i_0,i_1,\dots,i_L} g(i_0,i_1,\dots,i_L)\pi(i_0)p_{i_0 i_1}\cdots p_{i_{L-1}i_L} .$$

Proof. Apply Theorem 9.3.2 to the "snake chain" $\{(X_n, X_{n+1},\dots,X_{n+L})\}_{n\geq 0}$, which is (see Exercise 9.7.13) irreducible recurrent and admits the stationary distribution

$$\pi(i_0)p_{i_0 i_1}\cdots p_{i_{L-1}i_L} .$$

□

Note that

$$\sum_{i_0,i_1,\dots,i_L} g(i_0,i_1,\dots,i_L)\pi(i_0)p_{i_0 i_1}\cdots p_{i_{L-1}i_L} = E_\pi[g(X_0,\dots,X_L)] .$$

Proof. (*of Proposition 9.3.1.*) Let $T_0 = \tau_1, \tau_2, \tau_3, \dots$ be the successive return times to state 0, and define

$$U_p = \sum_{n=\tau_p+1}^{\tau_{p+1}} f(X_n) .$$

In view of the regenerative cycle theorem, $\{U_p\}_{p\geq 1}$ is an IID sequence. Moreover, assuming $f \geq 0$ and using the strong Markov property,

$$\begin{aligned} E[U_1] &= E_0\left[\sum_{n=1}^{T_0} f(X_n)\right] = E_0\left[\sum_{n=1}^{T_0}\sum_{i\in E} f(i)1_{\{X_n=i\}}\right] \\ &= \sum_{i\in E} f(i)E_0\left[\sum_{n=1}^{T_0} 1_{\{X_n=i\}}\right] = \sum_{i\in E} f(i)x_i . \end{aligned}$$

This quantity is finite by hypothesis and therefore the strong law of large numbers applies to give

$$\lim_{n\uparrow\infty} \frac{1}{n}\sum_{p=1}^{n} U_p = \sum_{i\in E} f(i)x_i ,$$

that is,

$$\lim_{n\uparrow\infty} \frac{1}{n} \sum_{k=T_0+1}^{\tau_{n+1}} f(X_k) = \sum_{i\in E} f(i)x_i \,. \tag{9.29}$$

Observing that

$$\tau_{\nu(n)} \le n < \tau_{\nu(n)+1} \,,$$

we have

$$\frac{\sum_{k=1}^{\tau_{\nu(n)}} f(X_k)}{\nu(n)} \le \frac{\sum_{k=1}^{n} f(X_k)}{\nu(n)} \le \frac{\sum_{k=1}^{\tau_{\nu(n)+1}} f(X_i)}{\nu(n)} \,.$$

Since the chain is recurrent, $\lim_{n\uparrow\infty} \nu(n) = \infty$, and therefore, from (9.29), the extreme terms of the above chain of inequalities tend to $\sum_{i\in E} f(i)x_i$ as n goes to ∞, and this implies (9.26). The case of a function f of arbitrary sign is obtained by considering (9.26) written separately for $f^+ = \max(0, f)$ and $f^- = \max(0, -f)$, and then taking the difference of the two equalities obtained in this way. The difference is not an undetermined form $\infty - \infty$ due to hypothesis (9.25). □

The version of the ergodic theorem for Markov chains featured in Theorem 9.3.2 is a kind of strong law of large numbers, and it can be used in simulations to compute, when π is unknown, quantities of the type $E_\pi[f(X_0)]$.

The Markov Chain Convergence Theorem

This is one of the fundamental theoretical results of Markov chain theory. The proof will be given in terms of convergence in variation and is based on coupling.

Definition 9.3.4 (A) *A sequence $\{\alpha_n\}_{n\ge 0}$ of probability distributions on E is said to converge in variation to the probability distribution β on E if*

$$\lim_{n\uparrow\infty} d_V(\alpha_n, \beta) = 0 \,.$$

(B) *An E-valued random sequence $\{X_n\}_{n\ge 0}$ such that for some probability distribution π on E,*

$$\lim_{n\uparrow\infty} d_V(X_n, \pi) = 0 \,, \tag{9.30}$$

is said to converge in variation to π.

Observe that Definition 9.3.4 concerns only the marginal distributions of the stochastic process, not the stochastic process itself. Therefore, if there exists another stochastic process $\{X'_n\}_{n\ge 0}$ such that $X_n \overset{\mathcal{D}}{\sim} X'_n$ for all $n \ge 0$, and if there exists a third one $\{X''_n\}_{n\geq 0}$ such that $X''_n \stackrel{\mathcal{D}}{\sim} \pi$ for all $n \geq 0$, then (9.30) follows from

$$\lim_{n\uparrow\infty} d_V(X'_n, X''_n) = 0\,. \tag{9.31}$$

This trivial observation is useful because of the resulting freedom in the choice of $\{X'_n\}$ and $\{X''_n\}$. An interesting situation occurs when there exists a finite random time τ such that $X'_n = X''_n$ for all $n \geq \tau$.

Definition 9.3.5 *Two stochastic processes $\{X'_n\}_{n\geq 0}$ and $\{X''_n\}_{n\geq 0}$ taking their values in the same state space E are said to* couple *if there exists an almost surely* finite *random time τ such that*

$$n \geq \tau \Rightarrow X'_n = X''_n\,. \tag{9.32}$$

The random variable τ is called a coupling time *of the two processes.*

Theorem 9.3.6 *For any coupling time τ of $\{X'_n\}_{n\geq 0}$ and $\{X''_n\}_{n\geq 0}$, we have the* coupling inequality

$$d_V(X'_n, X''_n) \leq P(\tau > n)\,. \tag{9.33}$$

Proof. For all $A \subseteq E$,

$$\begin{aligned} P(X'_n \in A) - P(X''_n \in A) &= P(X'_n \in A,\ \tau \leq n) + P(X'_n \in A,\ \tau > n) \\ &\quad - P(X''_n \in A,\ \tau \leq n) - P(X''_n \in A, \tau > n) \\ &= P(X'_n \in A,\ \tau > n) - P(X''_n \in A,\ \tau > n) \\ &\leq P(X'_n \in A,\ \tau > n) \leq P(\tau > n)\,. \end{aligned}$$

Inequality (9.33) then follows from Lemma 7.3.2. □

Therefore, if the coupling time is P-a.s. *finite*, that is $\lim_{n\uparrow\infty} P(\tau > n) = 0$,

$$\lim_{n\uparrow\infty} d_V(X_n, \pi) = \lim_{n\uparrow\infty} d_V(X'_n, X''_n) = 0\,.$$

Consider an HMC that is irreducible and positive recurrent. If its initial distribution is the stationary distribution, it keeps the same distribution at all times. The chain is then said to be in the *stationary regime*, or in *equilibrium*, or in *steady state.*

A question arises naturally: What is the long-run behavior of the chain when the initial distribution μ is *arbitrary*? For instance, will it *converge to equilibrium*? In what sense?

The classical form of the result is that for arbitrary states i and j,

$$\lim_{n\uparrow\infty} p_{ij}(n) = \pi(j)\,, \tag{9.34}$$

if the chain is *ergodic*, according to the following definition:

Definition 9.3.7 *An irreducible positive recurrent and* aperiodic HMC *is called* ergodic.

In fact, (9.34) can be drastically improved:

Theorem 9.3.8 *Let $\{X_n\}_{n\geq 0}$ be an ergodic* HMC *on the countable state space E with transition matrix $\mathbf{P}$ and stationary distribution π, and let μ be an arbitrary initial distribution. Then*

$$\lim_{n\uparrow\infty} \sum_{i\in E} |P_\mu(X_n = i) - \pi(i)| = 0\,,$$

and in particular, for all $j \in E$,

$$\lim_{n\uparrow\infty} \sum_{i\in E} |p_{ji}(n) - \pi(i)| = 0\,.$$

In fact, for all *probability distributions μ and ν on E,*

$$\lim_{n\uparrow\infty} d_V(\mu^T\mathbf{P}^n, \nu^T\mathbf{P}^n) = 0\,.$$

Proof. (The first two statements correspond to the particular case where ν is the stationary distribution π, and particularizing further, $\mu = \delta_j$.) The proof will be given *via* the coupling method.[4] From the discussion preceding Definition 9.3.5, it suffices to construct two coupling chains with initial distributions μ and ν, respectively. This is done in the next lemma. □

Lemma 9.3.9 *Let $\{X_n^{(1)}\}_{n\geq 0}$ and $\{X_n^{(2)}\}_{n\geq 0}$ be two independent* ergodic HMCs *with the same transition matrix $\mathbf{P}$ and initial distributions μ and ν, respectively. Let $\tau = \inf\{n \geq 0;\ X_n^{(1)} = X_n^{(2)}\}$, with $\tau = \infty$ if the chains never intersect. Then τ is, in fact, almost surely finite. Moreover, the process $\{X'_n\}_{n\geq 0}$ defined by*

$$X'_n = \begin{cases} X_n^{(1)} & \text{if } n \leq \tau, \\ X_n^{(2)} & \text{if } n \geq \tau \end{cases} \tag{9.35}$$

is an HMC *with transition matrix $\mathbf{P}$.*

[4] For the general theory of coupling and its numerous applications, see [14].

Proof. Step 1. Consider the product HMC $\{Z_n\}_{n\geq 0}$ defined by $Z_n = (X_n^{(1)}, X_n^{(2)})$. It takes values in $E \times E$, and the probability of transition from (i,k) to (j,ℓ) in n steps is $p_{ij}(n)p_{k\ell}(n)$. We first show that this chain is irreducible. The probability of transition from (i,k) to (j,ℓ) in n steps is $p_{ij}(n)p_{k\ell}(n)$. Since $\mathbf{P}$ is irreducible and *aperiodic*, by Theorem 9.1.17, there exists an m such that for all pairs (i,j) and (k,ℓ), $n \geq m$ implies $p_{ij}(n)p_{k\ell}(n) > 0$. This implies irreducibility. (Note the essential role of aperiodicity. A simple counterexample is that of the symmetric random walk on $\mathbb{Z}$, which is irreducible but of period 2. The product of two independent such HMCs is the symmetric random walk on $\mathbb{Z}^2$, which has two communications classes.)

Step 2. Next we show that the two independent chains meet in finite time. Clearly, the distribution $\tilde{\sigma}$ defined by $\tilde{\sigma}(i,j) := \pi(i)\pi(j)$ is a stationary distribution for the product chain, where π is the stationary distribution of $\mathbf{P}$. Therefore, by the stationary distribution criterion, the product chain is positive recurrent. In particular, it reaches the diagonal of E^2 in finite time, and consequently, $P(\tau < \infty) = 1$.

It remains to show that $\{X'_n\}_{n\geq 0}$ given by (9.35) is an HMC with transition matrix $\mathbf{P}$. For this we use the following lemma.

Lemma 9.3.10 *Let $X_0^1, X_0^2, Z_n^1, Z_n^2$ $(n \geq 1)$ be independent random variables, and suppose moreover that Z_n^1, Z_n^2 $(n \geq 1)$ are identically distributed. Let τ be a non-negative integer-valued random variable such that for all $m \in \mathbb{N}$, the event $\{\tau = m\}$ is expressible in terms of $X_0^1, X_0^2, Z_n^1, Z_n^2$ $(n \leq m)$. Define the sequence $\{Z_n\}_{n\geq 1}$ by*

$$Z_n = \begin{cases} Z_n^1 & \text{if } n \leq \tau, \\ Z_n^2 & \text{if } n > \tau. \end{cases}$$

Then, $\{Z_n\}_{n\geq 1}$ has the same distribution as $\{Z_n^1\}_{n\geq 1}$ and is independent of X_0^1, X_0^2.

Proof. For any sets C_1, C_2, A_1, $\ldots$, A_k in the appropriate spaces,

$$\begin{aligned}
&P(X_0^1 \in C_1, X_0^2 \in C_2, Z_\ell \in A_\ell, 1 \leq \ell \leq k) \\
&\quad = \textstyle\sum_{m=0}^{k} P(X_0^1 \in C_1, X_0^2 \in C_2, Z_\ell \in A_\ell, 1 \leq \ell \leq k, \tau = m) \\
&\qquad + P(X_0^1 \in C_1, X_0^2 \in C_2, Z_1 \in A_1, \ldots, Z_k \in A_k, \tau > k) \\
&\quad = \textstyle\sum_{m=0}^{k} P(X_0^1 \in C_1, X_0^2 \in C_2, Z_\ell^1 \in A_\ell, 1 \leq \ell \leq m, \tau = m, Z_r^2 \in A_r, m+1 \leq r \leq k) \\
&\qquad + P(X_0^1 \in C_1, X_0^2 \in C_2, Z_\ell^1 \in A_\ell, 1 \leq \ell \leq k, \tau > k)\,.
\end{aligned}$$

Since the event $\{\tau = m\}$ is independent of $Z^2_{m+1} \in A_{m+1}, \dots, Z^2_k \in A_k$ $(k \geq m)$,

$$
\begin{aligned}
&= \sum_{m=0}^{k} P(X_0^1 \in C_1, X_0^2 \in C_2, Z_\ell^1 \in A_\ell, 1 \leq \ell \leq m, \tau = m) P(Z_r^2 \in A_r, m+1 \leq r \leq k) \\
&\quad + P(X_0^1 \in C_1, X_0^2 \in C_2, Z_\ell^1 \in A_\ell, 1 \leq \ell \leq k, \tau > k) \\
&= \sum_{m=0}^{k} P(X_0^1 \in C_1, X_0^2 \in C_2, Z_\ell^1 \in A_\ell, 1 \leq \ell \leq m, \tau = m, Z_r^1 \in A_r, m+1 \leq r \leq k) \\
&\quad + P(X_0^1 \in C_1, X_0^2 \in C_2, Z_\ell^1 \in A_\ell, 1 \leq \ell \leq k, \tau > k) \\
&= P(X_0^1 \in C_1, X_0^2 \in C_2, Z_1^1 \in A_1, \dots, Z_k^1 \in A_k) .
\end{aligned}
$$

□

Step 3. We now complete the proof. The statement of the theorem concerns only the distributions of $\{X_n^\ell\}_{n\geq 0}$ $(\ell = 1, 2)$, and therefore we may assume a representation

$$X_{n+1}^\ell = f(X_n^\ell, Z_{n+1}^\ell) \quad (n \geq 1, \ell = 1, 2) ,$$

where X_0^ℓ, Z_n^ℓ $(n \geq 1, \ell = 1, 2)$ satisfy the conditions in Lemma 9.3.10. The random time τ satisfies the condition of Lemma 9.3.10. Defining $\{Z_n\}_{n\geq 1}$ in the same manner as in this lemma, we therefore have

$$X_{n+1} = f(X_n, Z_{n+1}) ,$$

which proves the announced result. □

9.4 Absorption

The special nature of the branching process allowed for a simple and elegant computation of the probability of absorption into state 0. We now consider the absorption problem for HMCs with no special structure,[5] based only on the transition matrix $\mathbf{P}$, not necessarily assumed irreducible. The state space E is then decomposable as $E = T + \sum_j R_j$, where $R_1, R_2, \dots$ are the disjoint recurrent classes and T is the collection of transient states. (Note that the number of recurrent classes as well as the number of transient states may be infinite.) The transition matrix

[5] Such as those occurring in sociology, for instance in models describing migration (whether geographical or sociological) of populations, for which the transition matrix is obtained empirically.

can therefore be block-partitioned as

$$\mathbf{P} = \begin{pmatrix} \mathbf{P}_1 & 0 & \cdots & 0 \\ 0 & \mathbf{P}_2 & \cdots & 0 \\ \vdots & \vdots & \ddots & \vdots \\ B(1) & B(2) & \cdots & \mathbf{Q} \end{pmatrix}$$

or in condensed notation,

$$\mathbf{P} = \begin{pmatrix} D & 0 \\ B & \mathbf{Q} \end{pmatrix}. \tag{9.36}$$

This structure of the transition matrix accounts for the fact that one cannot go from a state in a given recurrent class to any state not belonging to this recurrent class. In other words, a recurrent class is closed.

What is the probability of being absorbed by a given recurrent class when starting from a given transient state? This kind of problem was already addressed when the first-step analysis method was introduced. This method leads to a system of linear equations with boundary conditions, for which the solution is unique, due to the finiteness of the state space. With an infinite state space, the uniqueness issue cannot be overlooked, and the absorption problem will be reconsidered with this in mind, and also with the intention of finding general matrix-algebraic expressions for the solutions. Another phenomenon not manifesting itself in the finite case is the possibility, when the set of transient states is infinite, of never being absorbed by the recurrent set. We shall consider this problem first, and then proceed to derive the distribution of the time to absorption by the recurrent set, and the probability of being absorbed by a given recurrent class.

Before Absorption

Let A be a subset of the state space E (typically the set of transient states, but not necessarily). We aim at computing for any initial state $i \in A$ the probability of remaining forever in A,

$$v(i) = P_i(X_r \in A;\ r \geq 0).$$

Defining $v_n(i) := P_i(X_1 \in A, \ldots, X_n \in A)$, we have, by monotone sequential continuity,

$$\lim_{n\uparrow\infty} \downarrow v_n(i) = v(i).$$

But for $j \in A$,

$$P_i(X_1 \in A, \ldots, X_{n-1} \in A, X_n = j) = \sum_{i_1\in A} \cdots \sum_{i_{n-1}\in A} p_{ii_1} \cdots p_{i_{n-1}j}$$

is the general term $q_{ij}(n)$ of the n-th iterate of the restriction $\mathbf{Q}$ of $\mathbf{P}$ to the set A. Therefore $v_n(i) = \sum_{j\in A} q_{ij}(n)$, that is, in vector notation,

$$v_n = \mathbf{Q}^n \mathbf{1}_A\,,$$

where $\mathbf{1}_A$ is the column vector indexed by A with all entries equal to 1. From this equality we obtain

$$v_{n+1} = \mathbf{Q} v_n\,,$$

and by dominated convergence $v = \mathbf{Q}v$. Moreover, $\mathbf{0}_A \le v \le \mathbf{1}_A$, where $\mathbf{0}_A$ is the column vector indexed by A with all entries equal to 0. The above result can be refined as follows:

Theorem 9.4.1 *The vector v is the* maximal *solution of*

$$v = \mathbf{Q}v, \quad \mathbf{0}_A \le v \le \mathbf{1}_A\,.$$

Moreover, either $v = \mathbf{0}_A$ or $\sup_{i\in A} v(i) = 1$. In the case of a finite transient set T, the probability of infinite sojourn in T is null.

Proof. Only maximality and the last statement remain to be proved. To prove maximality consider a vector u indexed by A such that $u = \mathbf{Q}u$ and $\mathbf{0}_A \le u \le \mathbf{1}_A$. Iteration of $u = \mathbf{Q}u$ yields $u = \mathbf{Q}^n u$, and $u \le \mathbf{1}_A$ implies that $\mathbf{Q}^n u \le \mathbf{Q}^n \mathbf{1}_A = v_n$. Therefore $u \le v_n$, which gives $u \le v$ by passage to the limit.

To prove the last statement of the theorem, let $c = \sup_{i\in A} v(i)$. ¿From $v \le c\mathbf{1}_A$, we obtain $v \le cv_n$ as above, and therefore, at the limit, $v \le cv$. This implies either $v = \mathbf{0}_A$ or $c = 1$.

When the set T is *finite*, the probability of infinite sojourn in T is null, because otherwise at least one transient state would be visited infinitely often. □

Equation $v = \mathbf{Q}v$ reads

$$v(i) = \sum_{j\in A} p_{ij} v(j)\ (i \in A)\,.$$

First-step analysis gives this equality as a necessary condition. However, it does not help to determine which solution to choose, in case there are several.

EXAMPLE 9.4.2: REPAIR SHOP, TAKE 5. We shall prove in a different way a result already obtained previously, that is: the repair shop chain is recurrent if and only if $\rho \le 1$. Observe that the restriction of $\mathbf{P}$ to $A_i := \{i+1, i+2, \ldots\}$, namely

$$\mathbf{Q} = \begin{pmatrix} a_1 & a_2 & a_3 & \cdots \\ a_0 & a_1 & a_2 & \cdots \\ & a_0 & a_1 & \cdots \\ & & & \cdots \end{pmatrix},$$

does not depend on $i \geq 0$. In particular, the maximal solution of $v = \mathbf{Q}v$, $\mathbf{0}_A \leq v \leq \mathbf{1}_A$ when $A \equiv A_i$ has, in view of Theorem 9.4.1, the following two interpretations. Firstly, for $i \geq 1$, $1 - v(i)$ is the probability of visiting 0 when starting from $i \geq 1$. Secondly, $(1 - v(1))$ is the probability of visiting $\{0, 1, \dots, i\}$ when starting from $i + 1$. But when starting from $i + 1$, the chain visits $\{0, 1, \dots, i\}$ if and only if it visits i, and therefore $(1 - v(1))$ is also the probability of visiting i when starting from $i + 1$. The probability of visiting 0 when starting from $i + 1$ is

$$1 - v(i+1) = (1 - v(1))(1 - v(i)),$$

because in order to go from $i + 1$ to 0 one must first reach i, and then go to 0. Therefore, for all $i \geq 1$,

$$v(i) = 1 - \beta^i,$$

where $\beta = 1 - v(1)$. To determine β, write the first equality of $v = \mathbf{Q}v$:

$$v(1) = a_1 v(1) + a_2 v(2) + \cdots,$$

that is,

$$(1 - \beta) = a_1(1 - \beta) + a_2(1 - \beta^2) + \cdots.$$

Since $\sum_{i \geq 0} a_i = 1$, this reduces to

$$\beta = g(\beta), \tag{$\star$}$$

where g is the generating function of the probability distribution $(a_k, k \geq 0)$. Also, all other equations of $v = \mathbf{Q}v$ reduce to $(\star)$.

Under the irreducibility assumptions $a_0 > 0$, $a_0 + a_1 < 1$, $(\star)$ has only one solution in $[0, 1]$, namely $\beta = 1$ if $\rho \leq 1$, whereas if $\rho > 1$, it has two solutions in $[0, 1]$, this probability is $\beta = 1$ and $\beta = \beta_0 \in (0, 1)$. We must take the smallest solution. Therefore, if $\rho > 1$, the probability of visiting state 0 when starting from state $i \geq 1$ is $1 - v(i) = \beta_0^i < 1$, and therefore the chain is transient. If $\rho \leq 1$, the latter probability is $1 - v(i) = 1$, and therefore the chain is recurrent.

EXAMPLE 9.4.3: 1-D RANDOM WALK, TAKE 3. The transition matrix of the random walk on $\mathbb{N}$ with a reflecting barrier at 0,

$$\mathbf{P} = \begin{pmatrix} 0 & 1 & & & \\ q & 0 & p & & \\ & q & 0 & p & \\ & & q & 0 & p \\ & & & \cdot & \cdot & \cdot \end{pmatrix},$$

where $p \in (0,1)$, is clearly irreducible. Intuitively, if $p > q$, there is a drift to the right, and one expects the chain to be transient. This will be proved formally by showing that the probability $v(i)$ of never visiting state 0 when starting from state $i \geq 1$ is strictly positive. In order to apply Theorem 9.4.1 with $A = \mathbb{N} - \{0\}$, we must find the general solution of $u = \mathbf{Q}u$. This equation reads

$$\begin{aligned} u(1) &= pu(2)\,, \\ u(2) &= qu(1) + pu(3)\,, \\ u(3) &= qu(2) + pu(4)\,, \\ \cdots \end{aligned}$$

and its general solution is $u(i) = u(1)\sum_{j=0}^{i-1}\left(\frac{q}{p}\right)^j$. The largest value of $u(1)$ respecting the constraint $u(i) \in [0,1]$ is $u(1) = 1 - \left(\frac{q}{p}\right)$. The solution $v(i)$ is therefore

$$v(i) = 1 - \left(\frac{q}{p}\right)^i.$$

Time to Absorption

We now turn to the determination of the distribution of τ, the time of exit from the transient set T. Theorem 9.4.1 says that $v = \{v(i)\}_{i \in T}$, where $v(i) = P_i(\tau = \infty)$, is the largest solution of $v = \mathbf{Q}v$ subject to the constraints $\mathbf{0}_T \leq v \leq \mathbf{1}_T$, where $\mathbf{Q}$ is the restriction of $\mathbf{P}$ to the transient set T. The probability distribution of τ when the initial state is $i \in T$ is readily computed starting from the identity

$$P_i(\tau = n) = P_i(\tau \geq n) - P_i(\tau \geq n+1)$$

and the observation that for $n \geq 1$, $\{\tau \geq n\} = \{X_{n-1} \in T\}$, from which we obtain, for $n \geq 1$,

$$P_i(\tau = n) = P_i(X_{n-1} \in T) - P(X_n \in T) = \sum_{j \in T}(p_{ij}(n-1) - p_{ij}(n))\,.$$

Now, $p_{ij}(n)$ $(i, j \in T)$ is the general term of the matrix $\mathbf{Q}^n$, and therefore:

Theorem 9.4.4

$$P_i(\tau = n) = \{(\mathbf{Q}^{n-1} - \mathbf{Q}^n)\mathbf{1}_T\}_i . \tag{9.37}$$

In particular, if $P_i(\tau = \infty) = 0$,

$$P_i(\tau > n) = \{\mathbf{Q}^n \mathbf{1}_T\}_i .$$

Proof. Only the last statement remains to be proved. From (9.37),

$$\begin{aligned} P_i(n < \tau \leq n+m) &= \sum_{j=0}^{m-1} \{(\mathbf{Q}^{n+j} - \mathbf{Q}^{n+j-1})\mathbf{1}_T\}_i \\ &= \left\{\left(\mathbf{Q}^n - \mathbf{Q}^{n+m}\right)\mathbf{1}_T\right\}_i , \end{aligned}$$

and therefore, if $P_i(\tau = \infty) = 0$, we obtain (9.37) by letting $m \uparrow \infty$. □

Final Destination

We seek to compute the probability of absorption by a given recurrent class when starting from a given transient state. As we shall see later, it suffices for the theory to treat the case where the recurrent classes are singletons. We therefore suppose that the transition matrix has the form

$$\mathbf{P} = \begin{pmatrix} I & 0 \\ B & \mathbf{Q} \end{pmatrix} . \tag{9.38}$$

Let f_{ij} be the probability of absorption by recurrent class $R_j = \{j\}$ when starting from the transient state i. We have

$$\mathbf{P}^n = \begin{pmatrix} I & 0 \\ L_n & \mathbf{Q}^n \end{pmatrix} ,$$

where $L_n = (I + \mathbf{Q} + \cdots + \mathbf{Q}^n)B$. Therefore, $\lim_{n\uparrow\infty} L_n = SB$. For $i \in T$, the (i, j) term of L_n is

$$L_n(i, j) = P(X_n = j | X_0 = i) .$$

Now, if T_{R_j} is the first time of visit to R_j after time 0, then

$$L_n(i, j) = P_i(T_{R_j} \leq n) ,$$

since R_j is a closed state. Letting n go to ∞ gives the following:

Theorem 9.4.5 *For an* HMC *with transition matrix* $\mathbf{P}$ *of the form (9.38), the probability of absorption by recurrent class* $R_j = \{j\}$ *starting from transient state* i *is*

$$P_i(T_{R_j} < \infty) = (SB)_{i,R_j} .$$

The general case, where the recurrence classes are not necessarily singletons, can be reduced to the singleton case as follows. Let $\mathbf{P}^*$ be the matrix obtained from the transition matrix $\mathbf{P}$, by grouping for each j the states of recurrent class R_j into a single state $\hat{\jmath}$:

$$\mathbf{P}^* = \begin{pmatrix} 1 & 0 & 0 & 0 \\ 0 & 1 & 0 & 0 \\ & & & \\ 0 & 0 & \ddots & 0 \\ b_{\hat{1}} & b_{\hat{2}} & \cdots & \mathbf{Q} \end{pmatrix} \tag{9.39}$$

where $b_{\hat{\jmath}} = B(j)\mathbf{1}_T$ is obtained by summation of the columns of $B(j)$, the matrix consisting of the columns $i \in R_j$ of B. The probability f_{iR_j} of absorption by class R_j when starting from $i \in T$ equals $\hat{f}_{i\hat{\jmath}}$, the probability of ever visiting $\hat{\jmath}$ when starting from i, computed for the chain with transition matrix $\mathbf{P}^*$.

EXAMPLE 9.4.6: SIBMATING. In the reproduction model called *sibmating* (sister-brother mating), two individuals are mated and two individuals from their offspring are chosen at random to be mated, and this incestuous process goes on through the subsequent generations.

Denote by X_n the genetic type of the mating pair at the nth generation. Clearly, $\{X_n\}_{n\geq 0}$ is an HMC with six states representing the different pairs of genotypes $AA \times AA$, $aa \times aa$, $AA \times Aa$, $Aa \times Aa$, $Aa \times aa$, $AA \times aa$, denoted respectively 1, 2, 3, 4, 5, 6. The following table gives the probabilities of occurrence of the three possible genotypes in the descent of a mating pair:

parents' genotype	AA	Aa	aa
AA AA	1	0	0
aa aa	0	0	1
AA Aa	1/2	1/2	0
Aa Aa	1/4	1/2	1/4
Aa aa	0	1/2	1/2
AA aa	0	1	0

descendant's genotype

The transition matrix of $\{X_n\}_{n\geq 0}$ is then easily deduced:

$$\mathbf{P} = \begin{pmatrix} 1 & & & & & \\ & 1 & & & & \\ 1/4 & & 1/2 & 1/4 & & \\ 1/16 & 1/16 & 1/4 & 1/4 & 1/4 & 1/8 \\ & 1/4 & & 1/4 & 1/2 & \\ & & & 1 & & \end{pmatrix}.$$

The set $R = \{1, 2\}$ is absorbing, and the restriction of the transition matrix to the transient set $T = \{3, 4, 5, 6\}$ is

$$\mathbf{Q} = \begin{pmatrix} 1/2 & 1/4 & 0 & 0 \\ 1/4 & 1/4 & 1/4 & 1/8 \\ 0 & 1/4 & 1/2 & 0 \\ 0 & 1 & 0 & 0 \end{pmatrix}.$$

We find

$$S = (1 - \mathbf{Q})^{-1} = \frac{1}{6}\begin{pmatrix} 16 & 8 & 4 & 1 \\ 8 & 16 & 8 & 2 \\ 4 & 8 & 16 & 1 \\ 8 & 16 & 8 & 8 \end{pmatrix},$$

and the absorption probability matrix is

$$SB = S\begin{pmatrix} 1/4 & 0 \\ 1/16 & 1/16 \\ 0 & 1/4 \\ 0 & 0 \end{pmatrix} = \begin{pmatrix} 3/4 & 1/4 \\ 1/2 & 1/2 \\ 1/4 & 3/4 \\ 1/2 & 1/2 \end{pmatrix}.$$

For instance, the $(3, 2)$ entry, $\frac{3}{4}$, is the probability that when starting from a couple of ancestors of type $Aa \times aa$, the race will end up in genotype $aa \times aa$.

9.5 The Markov Property on Graphs

This section introduces the Markov fields on a graph, a notion of special interest in Physics.

Let $G = (V, \mathcal{E})$ be a finite graph, and let $v_1 \sim v_2$ denote the fact that $\langle v_1, v_2 \rangle$ is an edge of the graph.[6] Such vertices are called *neighbors* (one of the other). One sometimes refers to vertices of V as *sites*. The *boundary* with respect to $\sim$ of a set $A \subset V$ is the set

$$\partial A := \{v \in V \backslash A \, ; \, v \sim w \text{ for some } w \in A\}.$$

Let Λ be a finite set, called the phase space. A *random field* on V with phases in Λ is a collection $X = \{X(v)\}_{v \in V}$ of random variables with values in Λ. A random field can be regarded as a random variable taking its values in the *configuration*

[6] Recall that, in the definition of an edge $\langle v_1, v_2 \rangle$, v_1 and v_2 are distinct vertices.

space $E := \Lambda^V$, where a configuration is a function $x : v \in V \mapsto x(v) \in \Lambda$. For a given configuration x and a given subset $A \subseteq V$, let

$$x(A) := (x(v), v \in A)$$

denote the restriction of x to A. If $V \backslash A$ denotes the complement of A in V, one writes $x = (x(A), x(V \backslash A))$. In particular, for fixed $v \in V$, $x = (x(v), x(V \backslash v))$, where $V \backslash v$ is a shorter way of writing $V \backslash \{v\}$, the complement of the singleton $\{v\}$ in V.

Of special interest are the random fields characterized by local interactions. This leads to the notion of a Markov random field. The "locality" is in terms of the neighborhood structure inherited from the graph structure. More precisely, for any $v \in V$, $N_v := \{w \in V; w \sim v\}$ is the neighborhood of v. In the following, $\widetilde{\mathcal{N}}_v$ denotes the set $\mathcal{N}_v \cup \{v\}$.

Definition 9.5.1 *The random field X is called a* Markov random field (MRF) *with respect to $\sim$ if for all sites $v \in V$, the random elements $X(v)$ and $X(V \backslash \widetilde{\mathcal{N}}_v)$ are independent given $X(\mathcal{N}_v)$.*

In symbols:

$$P(X(v) = x(v) \mid X(V \backslash v) = x(V \backslash v)) = P(X(v) = x(v) \mid X(\mathcal{N}_v) = x(\mathcal{N}_v)) \quad (9.40)$$

for all $x \in \Lambda^V$ and all $v \in V$. Property (9.40) is of the Markov type in the sense that the distribution of the phase at a given site is directly influenced only by the phases of the neighboring sites.

Note that any random field is Markovian with respect to the trivial topology, where the neighborhood of any site v is $V \backslash v$. However, the interesting Markov fields (from the point of view of modeling, simulation and optimization) are those with relatively small neighborhoods.

EXAMPLE 9.5.2: MARKOV CHAIN AS MARKOV FIELD. The Markov property of a stochastic sequence $\{X_n\}_{n \geq 0}$ implies (Exercise 9.7.18) that for all $n \geq 1$, X_n is independent of $(X_k, k \notin \{n-1, n, n+1\})$ given (X_{n-1}, X_{n+1}). Calling n a vertex, X_n the value of the process at vertex n and the set $\{n-1, n+1\}$ the neighborhood of vertex n, the above property can be rephrased as: For all $n \geq 1$, the value at vertex n is independent of the values at vertices $k \notin \{n-1, n, n+1\}$ given the values in the neighborhood of vertex n.

Definition 9.5.3 *The* local characteristic *of the* MRF *at site* v *is the function* $\pi^v : \Lambda^V \to [0,1]$ *defined by*

$$\pi^v(x) := P(X(v) = x(v) \mid X(\mathcal{N}_v) = x(\mathcal{N}_v)) .$$

The family $\{\pi^v\}_{v \in V}$ *is called the* local specification *of the* MRF.

One sometimes writes $\pi^v(x) := \pi(x(v) \mid x(\mathcal{N}_v))$.

Theorem 9.5.4 *Two positive distributions of a random field with a finite configuration space* Λ^V *that have the same local specification are identical.*

Proof. Enumerate V as $\{1, 2, \dots, K\}$. Therefore a configuration $x \in \Lambda^V$ is represented as $x = (x_1, \dots, x_{K-1}, x_K)$ where $x_i \in \Lambda$ $(1 \le i \le K)$. The following identity

$$\pi(z_1, z_2, \dots, z_k) = \prod_{i=1}^{K} \frac{\pi(z_i \mid z_1, \dots, z_{i-1}, y_{i+1}, \dots, y_K)}{\pi(y_i \mid z_1, \dots, z_{i-1}, y_{i+1}, \dots, y_K)} \pi(y_1, y_2, \dots, y_k) \qquad (\star)$$

holds for any $z, y \in \Lambda^K$. For the proof, write

$$\pi(z) = \prod_{i=1}^{K} \frac{\pi(z_1, \dots, z_{i-1}, z_i, y_{i+1}, \dots, y_K)}{\pi(z_1, \dots, z_{i-1}, y_i, y_{i+1}, \dots, y_K)} \pi(y)$$

and use Bayes' rule to obtain for each i $(1 \le i \le K)$:

$$\frac{\pi(z_1, \dots, z_{i-1}, z_i, y_{i+1}, \dots, y_K)}{\pi(z_1, \dots, z_{i-1}, y_i, y_{i+1}, \dots, y_K)} = \frac{\pi(z_i \mid z_1, \dots, z_{i-1}, y_{i+1}, \dots, y_K)}{\pi(y_i \mid z_1, \dots, z_{i-1}, y_{i+1}, \dots, y_K)} .$$

Let now π and π' be two positive probability distributions on V with the same local specification. Choose any $y \in \Lambda^V$. Identity $(\star)$ shows that for all $z \in \Lambda^V$,

$$\frac{\pi'(z)}{\pi(z)} = \frac{\pi'(y)}{\pi(y)} .$$

Therefore $\frac{\pi'(z)}{\pi(z)}$ is a constant, necessarily equal to 1 since π and π' are probability distributions. □

Gibbs Distributions

Consider the probability distribution

$$\pi_T(x) = \frac{1}{Z_T} \mathrm{e}^{-\frac{1}{T} U(x)} \qquad (9.41)$$

on the configuration space Λ^V, where $T > 0$ is a "*temperature*", $U(x)$ is the "*energy*" of configuration x and Z_T is the normalizing constant, called the *partition function*. Since $\pi_T(x)$ takes its values in $[0,1]$, necessarily $-\infty < U(x) \leq +\infty$. Note that $U(x) < +\infty$ if and only if $\pi_T(x) > 0$. One of the challenges associated with Gibbs models is obtaining explicit formulas for averages, considering that it is generally hard to compute the partition function. (This is however feasible in exceptional cases; see Exercise 9.7.19.)

Such distributions are of interest to physicists when the energy is expressed in terms of a potential function describing the local interactions. The notion of clique then plays a central role.

Definition 9.5.5 *Any singleton $\{v\} \subset V$ is a clique. A subset $C \subseteq V$ with more than one element is called a* clique *(with respect to $\sim$) if and only if any two distinct sites of C are mutual neighbors. A clique C is called* maximal *if for any site $v \notin C$, $C \cup \{v\}$ is not a clique.*

The collection of cliques will be denoted by $\mathcal{C}$.

Definition 9.5.6 *A* Gibbs potential *on Λ^V relative to $\sim$ is a collection $\{V_C\}_{C \subseteq V}$ of functions $V_C : \Lambda^V \to \mathbb{R} \cup \{+\infty\}$ such that*

(i) *$V_C \equiv 0$ if C is not a clique, and*

(ii) *for all $x, x' \in \Lambda^V$ and all $C \subseteq V$,*

$$x(C) = x'(C) \Rightarrow V_C(x) = V_C(x') .$$

The energy function U is said to derive from the potential $\{V_C\}_{C \subseteq V}$ if

$$U(x) = \sum_C V_C(x) .$$

The function V_C depends only on the phases at the sites inside subset C. One could write more explicitly $V_C(x(C))$ instead of $V_C(x)$, but this notation will not be used.

In this context, the distribution in (9.41) is called a Gibbs distribution (with respect to $\sim$).

Example 9.5.7: Ising Model, take 1. In statistical physics, the following model is regarded as a qualitatively correct idealization of a piece of ferromagnetic material. Here $V = \mathbb{Z}_m^2 = \{(i,j) \in \mathbb{Z}^2, \quad (1 \leq i,j \leq m)\}$ and $\Lambda = \{+1,-1\}$, where ± 1 is the orientation of the magnetic spin at a given site. The neighbor of a site consists of its four closest sites. The Gibbs potential is

$$\begin{aligned} V_{\{v\}}(x) &= -\frac{H}{k}x(v)\,, \\ V_{\langle v,w\rangle}(x) &= -\frac{J}{k}x(v)x(w)\,, \end{aligned}$$

where $\langle v, w\rangle$ is the 2-element clique ($v \sim w$). For physicists, k is the *Boltzmann constant*, H is the *external magnetic field*, and J is the *internal energy* of an elementary magnetic dipole. The energy function corresponding to this potential is therefore

$$U(x) = -\frac{J}{k}\sum_{\langle v,w\rangle} x(v)x(w) - \frac{H}{k}\sum_{v\in V} x(v)\,.$$

The Hammersley–Clifford Theorem

Gibbs distributions with an energy deriving from a Gibbs potential relative to a neighborhood system are distributions of Markov fields relative to the same neighborhood system.

Theorem 9.5.8 *If X is a random field with a distribution π of the form $\pi(x) = \frac{1}{Z}\mathrm{e}^{-U(x)}$, where the energy function U derives from a Gibbs potential $\{V_C\}_{C\subseteq V}$ relative to $\sim$, then X is a Markov random field with respect to $\sim$. Moreover, its local specification is given by the formula*

$$\pi^v(x) = \frac{\mathrm{e}^{-\sum_{C\ni v} V_C(x)}}{\sum_{\lambda\in\Lambda} \mathrm{e}^{-\sum_{C\ni v} V_C(\lambda, x(V\backslash v))}}\,, \tag{9.42}$$

where the notation $\sum_{C\ni v}$ means that the sum extends over the sets C that contain the site v.

Proof. First observe that the right-hand side of (9.42) depends on x only through $x(v)$ and $x(\mathcal{N}_v)$. Indeed, $V_C(x)$ depends only on $(x(w), w \in C)$, and for a clique C, if $w \in C$ and $v \in C$, then either $w = v$ or $w \sim v$. Therefore, if it can be shown that $P(X(v) = x(v)|X(V\backslash v) = x(V\backslash v))$ equals the right-hand side of (9.42), then (Theorem 2.1.14) the Markov property is proved. By definition of conditional probability,

$$P(X(v) = x(v) \mid X(V\backslash v) = x(V\backslash v)) = \frac{\pi(x)}{\sum_{\lambda\in\Lambda} \pi(\lambda, x(V\backslash v))}\,. \tag{†}$$

But

$$\pi(x) = \frac{1}{Z}\mathrm{e}^{-\sum_{C\ni v} V_C(x) - \sum_{C\not\ni v} V_C(x)}\,,$$

and similarly,

$$\pi(\lambda, x(V\backslash v)) = \frac{1}{Z} e^{-\sum_{C \ni v} V_C(\lambda, x(V\backslash v)) - \sum_{C \not\ni v} V_C(\lambda, x(V\backslash v))}.$$

If C is a clique and v is not in C, then $V_C(\lambda, x(V\backslash v)) = V_C(x)$ and is therefore independent of $\lambda \in \Lambda$. Therefore, after factoring out $\exp\left\{-\sum_{C \not\ni v} V_C(x)\right\}$, the right-hand side of (†) is found to be equal to the right-hand side of (9.42). □

The *local energy* at site v of configuration x is

$$U_v(x) = \sum_{C \ni v} V_C(x).$$

With this notation, (9.42) becomes

$$\pi^v(x) = \frac{e^{-U_v(x)}}{\sum_{\lambda \in \Lambda} e^{-U_v(\lambda, x(V\backslash v))}}.$$

EXAMPLE 9.5.9: ISING MODEL, TAKE 2. The local characteristics in the Ising model are

$$\pi_T^v(x) = \frac{e^{\frac{1}{kT}\left\{J \sum_{w; w \sim v} x(w) + H\right\} x(v)}}{e^{+\frac{1}{kT}\left\{J \sum_{w; w \sim v} x(w) + H\right\}} + e^{-\frac{1}{kT}\left\{J \sum_{w; w \sim v} x(w) + H\right\}}}.$$

Theorem 9.5.8 above is the direct part of the Gibbs–Markov equivalence theorem: A Gibbs distribution relative to a neighborhood system is the distribution of a Markov field with respect to the same neighborhood system. The converse part (Hammersley–Clifford theorem) is important from a theoretical point of view, since together with the direct part it concludes that Gibbs distributions and MRFs are essentially the same objects.

Theorem 9.5.10 *Let $\pi > 0$ be the distribution of a Markov random field with respect to $\sim$. Then*

$$\pi(x) = \frac{1}{Z} e^{-U(x)}$$

for some energy function U deriving from a Gibbs potential $\{V_C\}_{C \subseteq V}$ with respect to $\sim$.

The proof is omitted,[7] since in practice, the potential as well as the topology of V can be obtained directly from the expression of the energy, as the following example shows.

[7] See for instance Theorem 10.1.11 of [4].

EXAMPLE 9.5.11: MARKOV CHAINS AS MARKOV FIELDS. Let $V = \{0, 1, \dots N\}$ and $\Lambda = E$, a finite space. A random field X on V with phase space Λ is therefore a vector X with values in E^{N+1}. Suppose that $X_0, \dots, X_N$ is a homogeneous Markov chain with transition matrix $\mathbf{P} = \{p_{ij}\}_{i,j \in E}$ and initial distribution $\nu = \{\nu_i\}_{i \in E}$. In particular, with $x = (x_0, \dots, x_N)$,

$$\pi(x) = \nu_{x_0} p_{x_0 x_1} \cdots p_{x_{N-1} x_N} ,$$

that is,

$$\pi(x) = \mathrm{e}^{-U(x)} ,$$

where

$$U(x) = -\log \nu_{x_0} - \sum_{n=0}^{N-1} (\log p_{x_n x_{n+1}}) .$$

Clearly, this energy derives from a Gibbs potential associated with the nearest-neighbor topology for which the cliques are, besides the singletons, the pairs of adjacent sites. The potential functions are:

$$V_{\{0\}}(x) = -\log \nu_{x_0}, \quad V_{\{n,n+1\}}(x) = -\log p_{x_n x_{n+1}} .$$

The local characteristic at site n, $2 \le n \le N-1$, can be computed from formula (9.42), which gives

$$\pi^n(x) = \frac{\exp(\log p_{x_{n-1} x_n} + \log p_{x_n x_{n+1}})}{\sum_{y \in E} \exp(\log p_{x_{n-1} y} + \log p_{y x_{n+1}})} ,$$

that is,

$$\pi^n(x) = \frac{p_{x_{n-1} x_n} p_{x_n x_{n+1}}}{p^{(2)}_{x_{n-1} x_{n+1}}} ,$$

where $p^{(2)}_{ij}$ is the general term of the two-step transition matrix $\mathbf{P}^2$. Similar computations give $\pi^0(x)$ and $\pi^N(x)$. We note that, in view of the neighborhood structure, for $2 \le n \le N-1$, X_n is independent of $X_0, \dots, X_{n-2}, X_{n+2}, \dots, X_N$ given X_{n-1} and X_{n+1}.

9.6 Monte Carlo Markov Chains

Let us return to the problem of generating a random variable with a given distribution (see Section 3.2).

Both the inverse method and the acceptance–rejection method apply *in principle* when Z is a discrete random variable with values in a finite space $E = \{1, 2, ..., r\}$. Denote by π the distribution of Z. The inverse method is in this case always theoretically feasible. It consists in generating a random variable U uniformly distributed on $[0, 1]$ and letting $Z = i$ if and only if $\sum_{\ell=1}^{i-1} \pi(\ell) \leq U < \sum_{\ell=1}^{i} \pi(\ell)$. When the size r of the state space E is large, problems arise that are due to the small size of the intervals partitioning $[0, 1]$ and to the cost of precision in computing.

Another difficulty with the classical methods, besides the usual round-off errors, is that the probability π is in important applications known only up to a normalizing factor, that is, $\pi = K\tilde{\pi}$, and then, the integral that gives the normalizing factor K is difficult or impossible to compute. In physics, this is frequently the case, because the partition function of a Gibbs distribution is usually unavailable in closed form.

In random field simulation, another, maybe more important, reason is the necessity to enumerate the configurations, which implies coding and decoding of a mapping from the integers to the configuration space. The decoding part is usually very difficult and a small error may lead to a far-out sample (the configurations corresponding to close integers may be very different, which is a problem in image processing).

The *Monte Carlo Markov chain* (MCMC) method for sampling a probability distribution π on the finite space E partially avoids the problems just enumerated, but at the cost of obtaining only an approximate sample.

The basic methodology is as follows. One constructs an irreducible aperiodic HMC $\{X_n\}_{n\geq 0}$ with state space E admitting π as stationary distribution. Since E is finite, the chain is ergodic and therefore, for any initial distribution μ,

$$\lim_{n\to\infty} P_\mu(X_n = i) = \pi(i) \quad (i \in E)$$

and for any non-negative function $\varphi : E \to \mathbb{R}$,

$$\lim_{N\to\infty} \frac{1}{N} \sum_{n=1}^{N} \varphi(X_n) = E_\pi[\varphi(X)] \,.$$

When n is "large," we can consider that X_n has a distribution "close" to π. Of course, one would like to know how accurately X_n imitates an E-valued random variable Z with distribution π. For this we need estimates of the form

$$|\mu \mathbf{P}^n - \pi| \leq A\alpha^n \,,$$

where $\alpha < 1$. This issue will not be treated in this book, and only the basic problem, that of designing the MCMC algorithm, is considered. One looks for an ergodic transition matrix $\mathbf{P}$ on E whose stationary distribution is the *target distribution* π. There are infinitely many such transition matrices, and among them there are infinitely many that correspond to a reversible chain, that is, such that

$$\pi(i)p_{ij} = \pi(j)p_{ji}\,. \tag{9.43}$$

We seek solutions of the form

$$p_{ij} = q_{ij}\alpha_{ij} \tag{9.44}$$

for $j \neq i$, where $Q = \{q_{ij}\}_{i,j\in E}$ is an arbitrary irreducible transition matrix on E, called the *candidate-generating* matrix: When the present state is i, the next *tentative* state j is chosen with probability q_{ij}. When $j \neq i$, this new state is accepted with probability α_{ij}. Otherwise, the next state is the same state i. Hence, the resulting probability of moving from i to j when $i \neq j$ is given by (9.44). It remains to select the *acceptance probabilities* α_{ij}.

EXAMPLE 9.6.1: THE METROPOLIS ALGORITHM. In this example, the candidate-generation mechanism is purely random, that is, q_{ij} = constant, and

$$\alpha_{ij} = \min\left(1, \frac{\pi(j)}{\pi(i)}\right).$$

EXAMPLE 9.6.2: BARKER'S SAMPLER. In the special case of a purely random selection of the candidate,

$$\alpha_{ij} = \frac{\pi(j)}{\pi(i)+\pi(j)}\,.$$

In each case, the reversibility condition (9.43) is satisfied and therefore π is the stationary distribution (by Theorem 9.1.24).

Simulation of Random Fields

Consider a random field that changes randomly with time. In other words, we have a stochastic process $\{X_n\}_{n\geq 0}$ where

$$X_n = (X_n(v), v \in V)$$

and $X_n(v) \in \Lambda$. The state at time n of this process is a random field on V with phases in Λ, or equivalently, a random variable with values in the state space $E = \Lambda^V$, which for simplicity we assume finite. The stochastic process $\{X_n\}_{n\geq 0}$ will be called a *dynamical random field*.

Our purpose now is to show how a given random field with probability distribution

$$\pi(x) = \frac{1}{Z}e^{-\mathcal{E}(x)} \tag{9.45}$$

can arise as the stationary distribution of a field-valued Markov chain.

The *Gibbs sampler* uses a strictly positive probability distribution $(q_v, v \in V)$ on V, and the transition from $X_n = x$ to $X_{n+1} = y$ is made according to the following rule.

The new state y is obtained from the old state x by changing (or not) the value of the phase at *one site only*. The site v to be changed (or not) at time n is chosen independently of the past with probability q_v. When site v has been selected, the current configuration x is changed into y as follows: $y(V\backslash v) = x(V\backslash v)$, and the new phase $y(v)$ at site v is selected with probability $\pi(y(v) \mid x(V\backslash v))$. Thus, configuration x is changed into $y = (y(s), x(S\backslash s))$ with probability $\pi(y(v) \mid x(V\backslash v)$, according to the local specification at site v. This gives for the non-null entries of the transition matrix

$$P(X_{n+1} = y \mid X_n = x) = q_v \pi(y(v) \mid x(V\backslash v)) 1_{y(V\backslash v)=x(V\backslash v)} \,. \tag{9.46}$$

The corresponding chain is irreducible and aperiodic if $q_v > 0$ $(v \in V)$. To prove that π is the stationary distribution, we use the detailed balance test. For this, we have to check that for all $x, y \in \Lambda^V$,

$$\pi(x)P(X_{n+1} = y \mid X_n = x) = \pi(y)P(X_{n+1} = x \mid X_n = y)\,,$$

that is, in view of (9.46), for all $v \in V$,

$$\pi(x)q_v\pi(y(v) \mid y(V\backslash v)) = \pi(y)q_v\pi(x(v) \mid x(V\backslash v))\,.$$

But the last equality is just

$$\pi(x)q_v \frac{\pi(y(v), x(V\backslash v)}{P(X(V\backslash v) = x(V\backslash v))} = \pi(y(v), x(V\backslash v))q_v \frac{\pi(x)}{P(X(V\backslash v) = x(V\backslash v))}\,.$$

EXAMPLE 9.6.3: SIMULATION OF THE ISING MODEL. The local specification at site v depends only on the local configuration $x(\mathcal{N}_v)$. Note that small neighborhoods speed up computations. Note also that the Gibbs sampler is a natural sampler, in the sense that in a piece of ferromagnetic material, for instance, the spins are randomly changed according to the local specification. When nature decides to update the orientation of a dipole, it does so according to the law of statistical mechanics. It computes the local energy for each of the two possible spins, $E_+ = E(+1, x(\mathcal{N}_v))$ and $E_- = E(-1, x(\mathcal{N}_v))$, and takes the corresponding orientation with a probability proportional to e^{E_+} and e^{E_-}, respectively.

EXAMPLE 9.6.4: GIBBS SAMPLER FOR RANDOM VECTORS. Clearly, Gibbs sampling applies to any multivariate probability distribution

$$\pi(x(1), \ldots, x(N))$$

on a set $E = \Lambda^N$, where Λ is countable (but this restriction is not essential).

The basic step of the Gibbs sampler for the multivariate distribution π consists in selecting a coordinate number i ($1 \le i \le N$) at random, and then choosing the new value $y(i)$ of the corresponding coordinate, given the present values $x(1), \ldots, x(i-1), x(i+1), \ldots, x(N)$ of the other coordinates, with probability

$$\pi(y(i) \mid x(1), \ldots, x(i-1), x(i+1), \ldots, x(N)) .$$

One checks as above that π is the stationary distribution of the corresponding chain.

The Propp–Wilson Algorithm

We now present the basic idea of a theoretical method for obtaining an *exact sample* of a given distribution π on a finite state space E, that is, a random variable Z such that $P(Z = i) = \pi(i)$ for all $i \in E$. The following algorithm, the *Propp–Wilson algorithm*, is based on a coupling idea. One starts from an *ergodic* transition matrix $\mathbf{P}$ with stationary distribution π, just as in the classical MCMC method.

The algorithm is based on a representation of $\mathbf{P}$ in terms of a recurrence equation, that is, for given a function f and an IID sequence $\{Z_n\}_{n\ge1}$ independent of the initial state, the chain satisfies the recurrence

$$X_{n+1} = f(X_n, Z_{n+1}) . \tag{9.47}$$

The algorithm constructs a family of HMCs with this transition matrix with the help of a unique IID sequence of random vectors $\{Y_n\}_{n\in\mathbb{Z}}$, called the *updating sequence*, where $Y_n = (Z_{n+1}(1), \cdots, Z_{n+1}(r))$ is an r-dimensional random vector, and where the coordinates $Z_{n+1}(i)$ have a common distribution, that of Z_1. For each $N \in \mathbb{Z}$ and each $k \in E$, a process $\{X_n^N(k)\}_{n \geq N}$ is defined recursively by:

$$X_N^N(k) = k\,,$$

and, for $n \geq N$,

$$X_{n+1}^N(k) = f(X_n^N(k), Z_{n+1}(X_n^N(k))\,.$$

(Thus, if the chain is in state i at time n, it will be at time $n+1$ in state $j = f(i, Z_{n+1}(i)$.) Each of these processes is therefore an HMC with the transition matrix $\mathbf{P}$. Note that for all $k, \ell \in E$, and all $M, N \in \mathbb{Z}$, the HMCs $\{X_n^N(k)\}_{n \geq N}$ and $\{X_n^M(\ell)\}_{n \geq M}$ use at any time $n \geq \max(M, N)$ the same updating random vector Y_{n+1}.

If, in addition to the independence of $\{Y_n\}_{n \in \mathbb{Z}}$, the components $Z_{n+1}(1)$, $Z_{n+1}(2)$, ..., $Z_{n+1}(r)$ are, for each $n \in \mathbb{Z}$, independent, we say that the updating is *componentwise independent*.

Definition 9.6.5 *The random time*

$$\tau^+ = \inf\{n \geq 0; X_n^0(1) = X_n^0(2) = \cdots = X_n^0(r)\}$$

is called the forward coupling time (Figure 9.2). The random time

$$\tau^- = \inf\{n \geq 1; X_0^{-n}(1) = X_0^{-n}(2) = \cdots = X_0^{-n}(r)\}$$

is called the backward coupling time (Figure 9.2).

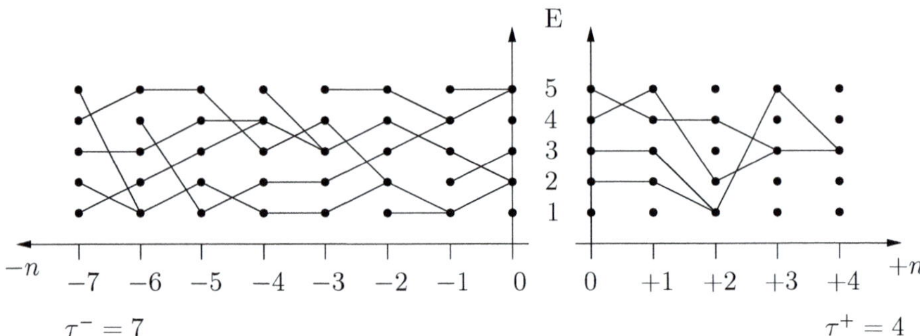

Figure 9.2: Backward and forward coupling

Thus, τ^+ is the first time at which the chains $\{X_n^0(i)\}_{n \geq 0}$, $1 \leq i \leq r$, coalesce.

Lemma 9.6.6 *When the updating is componentwise independent, the forward coupling time τ^+ is almost surely finite.*

Proof. Consider the (immediate) extension of Theorem 9.3.8 to the case of r independent HMCs with the same transition matrix. It cannot be applied directly to our situation, because the chains are not independent. However, the probability of coalescence in our situation is bounded below by the probability of coalescence in the completely independent case. To see this, first construct the independent chains model, using r independent IID componentwise independent updating sequences. The difference with our model is that we use too many updates. In order to construct from this a set of r chains as in our model, it suffices to use for two chains the same updates as soon as they meet. Clearly, the forward coupling time of the so modified model is smaller than or equal to that of the initial completely independent model. □

For a simpler notation, let $\tau^- := \tau$. Let

$$Z = X_0^{-\tau}(i) .$$

(This random variable is independent of i. In Figure 9.2, $Z = 2$.) Then,

Theorem 9.6.7 *With a componentwise independent updating sequence, the backward coupling time τ is almost surely finite. Also, the random variable Z has the distribution π.*

Proof. We shall show at the end of the current proof that for all $k \in \mathbb{N}$, $P(\tau \le k) = P(\tau^+ \le k)$, and therefore the finiteness of τ follows from that of τ^+ proven in the last lemma. Now, since for $n \ge \tau$, $X_0^{-n}(i) = Z$,

$$\begin{aligned} P(Z = j) &= P(Z = j, \tau > n) + P(Z = j, \tau \le n) \\ &= P(Z = j, \tau > n) + P(X_0^{-n}(i) = j, \tau \le n) \\ &= P(Z = j, \tau > n) - P(X_0^{-n}(i) = j, \tau > n) + P(X_0^{-n}(i) = j) \\ &= P(Z = j, \tau > n) - P(X_0^{-n}(i) = j, \tau > n) + p_{ij}(n) \\ &= A_n - B_n + p_{ij}(n) . \end{aligned}$$

But A_n and B_n are bounded above by $P(\tau > n)$, a quantity that tends to 0 as $n \uparrow \infty$ since τ is almost surely finite. Therefore

$$P(Z = j) = \lim_{n \uparrow \infty} p_{ij}(n) = \pi(j) .$$

It remains to prove the equality of the distributions of the forwards and backwards coupling time. For this, select an arbitrary integer $k \in \mathbb{N}$. Consider an updating

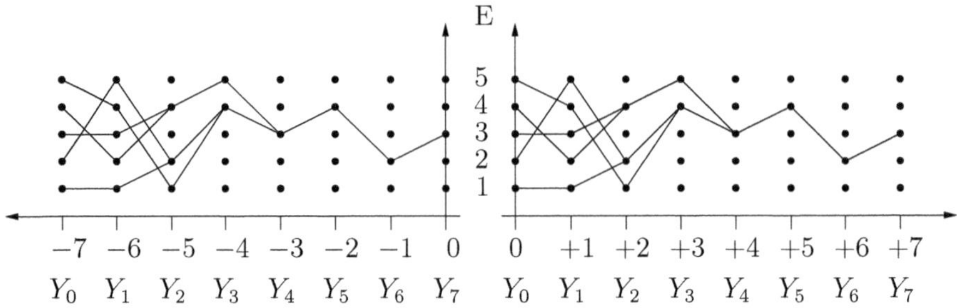

Figure 9.3: $\tau^+ \leq k$ implies $\tau' \leq k$

sequence constructed from a *bona fide* updating sequence $\{Y_n\}_{n\in\mathbb{Z}}$, by replacing $Y_{-k+1}, Y_{-k+2}, \ldots, Y_0$ by $Y_1, Y_2, \ldots, Y_k$. Call τ' the backwards coupling time in the modified model. Clearly τ and τ' have the same distribution.

Suppose that $\tau^+ \leq k$. Consider in the modified model the chains starting at time $-k$ from states $1, \ldots, r$. They coalesce at time $-k + \tau^+ \leq 0$ (see Figure 9.3), and consequently $\tau' \leq k$. Therefore $\tau^+ \leq k$ implies $\tau' \leq k$, so that

$$P(\tau^+ \leq k) \leq P(\tau' \leq k) = P(\tau \leq k)\,.$$

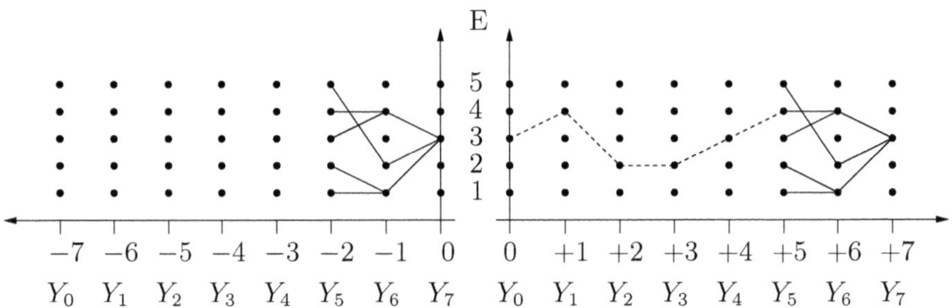

Figure 9.4: $\tau' \leq k$ implies $\tau^+ \leq k$

Now, suppose that $\tau' \leq k$. Then, in the modified model, the chains starting at time $k - \tau'$ from states $1, \ldots, r$ must at time $-k + \tau^+ \leq 0$ coalesce at time k. Therefore (Figure 9.4), $\tau^+ \leq k$. Therefore $\tau' \leq k$ implies $\tau^+ \leq k$, so that

$$P(\tau \leq k) = P(\tau' \leq k) \leq P(\tau^+ \leq k)\,.$$

□

Note that the coalesced value at the forward coupling time is not a sample of π (see Exercise 9.7.21).

The above exact sampling algorithm is often prohibitively time-consuming when the state space is large. However, if the algorithm required the coalescence of *two*, instead of r processes, then it would take less time. The Propp and Wilson algorithm does this in a special, yet not rare, case, which we now describe.

It is now assumed that there exists a partial order relation on E, denoted by $\preceq$, with a minimal and a maximal element (say, respectively, 1 and r), and that we can perform the updating in such a way that for all $i, j \in E$, all $N \in \mathbb{Z}$, and all $n \geq N$,

$$i \preceq j \Rightarrow X_n^N(i) \preceq X_n^N(j)\,.$$

However we do not require componentwise independent updating (but the updating vectors sequence remains IID). The corresponding sampling procedure is called the *monotone Propp–Wilson algorithm.*

Define the backwards *monotone* coupling time

$$\tau_m = \inf\{n \geq 1; X_0^{-n}(1) = X_0^{-n}(r)\}\,.$$

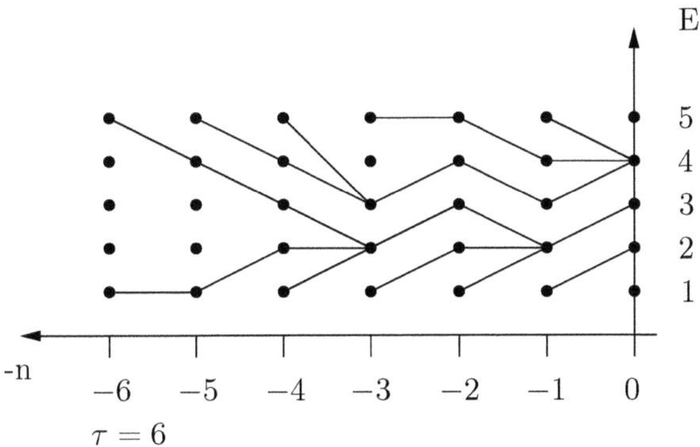

Figure 9.5: The Monotone Propp–Wilson algorithm

Theorem 9.6.8 *The monotone backwards coupling time τ_m is almost surely finite. Also, the random variable $X_0^{-\tau_m}(1) = X_0^{-\tau_m}(r)$ has the distribution π.*

Proof. We can use most of the proof of Theorem 9.6.7. We need only to prove independently that τ^+ is finite. It is so because τ^+ is dominated by the first time

$n \geq 0$ such that $X_n^0(r) = 1$, and the latter is finite in view of the recurrence assumption. □

Monotone coupling will occur with representations of the form (9.47) such that for all z,

$$i \preceq j \Rightarrow f(i,z) \preceq f(j,z),$$

and if for all $n \in \mathbb{Z}$, all $i \in \{1, \dots, r\}$,

$$Z_{n+1}(i) = Z_{n+1}.$$

EXAMPLE 9.6.9: A DAM MODEL. We consider the following model of a dam reservoir. The corresponding HMC, with values in $E = \{0, 2, \dots, r\}$, satisfies the recurrence equation

$$X_{n+1} = \min\{r, \max(0, X_n + Z_{n+1})\},$$

where, as usual, $\{Z_n\}_{n\geq 1}$ is IID. In this specific model, X_n is the content at time n of a dam reservoir with maximum capacity r, and $Z_{n+1} = A_{n+1} - c$, where A_{n+1} is the input into the reservoir during the time period from n to $n+1$, and c is the maximum release during the same period. The updating rule is then monotone.

In practical implementations, instead of trying the times -1, -2, etc., one may use successive starting times of the form $\alpha^r T_0$. Let k be the first k for which $\alpha^k T_0 \geq \tau_-$. The number of simulation steps used is $2\left(T_0 + \alpha T_0 + \cdots + \alpha^k T_0\right)$ (the factor 2 accounts for the fact that we are running two chains), that is,

$$2T_0\left(\frac{\alpha^{k+1}-1}{\alpha-1}\right) < 2T_0\left(\frac{\alpha^2}{\alpha-1}\right)\alpha^{k-1} \leq 2\tau_-\frac{\alpha^2}{\alpha-1}$$

steps, where we have assumed that $T_0 \leq \tau_-$. In the best case, supposing we are informed of the exact value of τ_- by some oracle, the number of steps is $2\tau_-$. The ratio of the worst to best cases is $\frac{\alpha^2}{\alpha-1}$, which is minimized for $\alpha = 2$. This is why it is usually suggested to start the successive attempts of backward coalescence at times of the form $-2^k T_0$ $(k \geq 0)$.

9.7 Exercises

Exercise 9.7.1. A COUNTEREXAMPLE
Find a simple example of an HMC $\{X_n\}_{n\geq 0}$ with state space $E = \{1, 2, 3, 4, 5, 6\}$ such that

$$P(X_2 = 6 \,|\, X_1 \in \{3,4\}, X_0 = 2) \neq P(X_2 = 6 \,|\, X_1 \in \{3,4\}).$$

Exercise 9.7.2. PAST, PRESENT, FUTURE
For an HMC $\{X_n\}_{n\geq 0}$ with state space E, prove that for all $n \in \mathbb{N}$, and all states $i_0, i_1, \ldots, i_{n-1}, i, j_1, j_2, \ldots, j_k \in E$,

$$P(X_{n+1} = j_1, \ldots, X_{n+k} = j_k \mid X_n = i, X_{n-1} = i_{n-1}, \ldots, X_0 = i_0)$$
$$= P(X_{n+1} = j_1, \ldots, X_{n+k} = j_k \mid X_n = i)\,.$$

Exercise 9.7.3. GIVEN ADJACENT STATES
Let $\{X_n\}_{n\geq 0}$ be an HMC with state space E and transition matrix $\mathbf{P}$. Show that for all $n \geq 1$ and all $k \geq 2$, X_n is conditionally independent of X_0, ..., X_{n-2}, X_{n+2}, ..., X_{n+k} given X_{n-1}, X_{n+1}. Compute the conditional distribution of X_n given X_{n-1}, X_{n+1}.

Exercise 9.7.4. STREET GANGS
Three characters, A, B, and C, armed with guns, suddenly meet at the corner of a Washington D.C. street, whereupon they naturally start shooting at one another. Each street gang kid shoots every tenth second, as long as he is still alive. The probabilities of a hit for A, B, and C are α, β, and γ respectively. A is the most hated, and therefore, as long as he is alive, B and C ignore each other and shoot at A. For historical reasons not developed here, A cannot stand B, and therefore he shoots only at B while the latter is still alive. Lucky C is shot at if and only if he is in the presence of A alone or B alone. What are the survival probabilities of A, B, and C, respectively?

Exercise 9.7.5. THE GAMBLER'S RUIN
(This exercise continues Example 9.1.10.) Compute the average duration of the game when $p = \frac{1}{2}$.

Exercise 9.7.6. RECORDS
Let $\{Z_n\}_{n\geq 1}$ be an IID sequence of geometric random variables: For $k \geq 0$, $P(Z_n = k) = (1-p)^k p$, where $p \in (0,1)$. Let $X_n = \max(Z_1, \ldots, Z_n)$ be the *record value* at time n, and suppose X_0 is an integer-valued random variable independent of the sequence $\{Z_n\}_{n\geq 1}$. Show that $\{X_n\}_{n\geq 0}$ is an HMC and give its transition matrix.

Exercise 9.7.7. AGGREGATION OF STATES
Let $\{X_n\}_{n\geq 0}$ be a HMC with state space E and transition matrix $\mathbf{P}$, and let $(A_k, k \geq 1)$ be a countable partition of E. Define the process $\{\hat{X}_n\}_{n\geq 0}$ with state space $\hat{E} = \{\hat{1}, \hat{2}, \ldots\}$ by $\hat{X}_n = \hat{k}$ if and only if $X_n \in A_k$. Show that if $\sum_{j\in A_\ell} p_{ij}$

is independent of $i \in A_k$ for all k, ℓ, then $\{\hat{X}_n\}_{n\geq 0}$ is an HMC with transition probabilities $\hat{p}_{\hat{k}\hat{\ell}} = \sum_{j\in A_\ell} p_{ij}$ (any $i \in A_k$).

Exercise 9.7.8. Truncated HMC

Let $\mathbf{P}$ be a transition matrix on the countable state space E, with the positive stationary distribution π. Let A be a subset of the state space, and define the truncation of $\mathbf{P}$ on A to be the transition matrix $\mathbf{Q}$ indexed by A and given by

$$q_{ij} = p_{ij} \text{ if } i, j \in A,\ i \neq j \text{ and } q_{ii} = p_{ii} + \sum_{k\in \bar{A}} p_{ik}\,.$$

Show that if $(\mathbf{P}, \pi)$ is reversible, then so is $(\mathbf{Q}, \frac{\pi}{\pi(A)})$.

Exercise 9.7.9. Moving stones

Stones $S_1, \dots, S_M$ are placed in line. At each time n a stone is selected at random, and this stone and the one ahead of it in the line exchange positions. If the selected stone is at the head of the line, nothing is changed. For instance, with $M = 5$: Let the current configuration be $S_2S_3S_1S_5S_4$ (S_2 is at the head of the line). If S_5 is selected, the new situation is $S_2S_3S_5S_1S_4$, whereas if S_2 is selected, the configuration is not altered. At each step, stone S_i is selected with probability $\alpha_i > 0$. Call X_n the situation at time n, for instance $X_n = S_{i_1}\cdots S_{i_M}$, meaning that stone S_{i_j} is in the jth position. Show that $\{X_n\}_{n\geq 0}$ is an irreducible HMC and that it has a stationary distribution given by the formula

$$\pi(S_{i_1}\cdots S_{i_M}) = C\alpha_{i_1}^M \alpha_{i_2}^{M-1}\cdots \alpha_{i_M}\,,$$

for some normalizing constant C.

Exercise 9.7.10. No stationary distribution

Show that the symmetric random walk on $\mathbb{Z}$ cannot have a stationary distribution.

Exercise 9.7.11. An interpretation of invariant measure

A countable number of particles move independently in the countable space E, each according to a Markov chain with the transition matrix $\mathbf{P}$. Let $A_n(i)$ be the number of particles in state $i \in E$ at time $n \geq 0$, and suppose that the random variables $A_0(i)$, $i \in E$, are independent Poisson random variables with respective means $\mu(i)$, $i \in E$, where $\mu = \{\mu(i)\}_{i\in E}$ is an invariant measure of $\mathbf{P}$. Show that for all $n \geq 1$, the random variables $A_n(i)$, $i \in E$, are independent Poisson random variables with respective means $\mu(i)$, $i \in E$.

Exercise 9.7.12. Return time to the initial state

Let τ be the first return time to initial state of an irreducible positive recurrent HMC $\{X_n\}_{n\geq 0}$, that is,

$$\tau = \inf\{n \geq 1; X_n = X_0\},$$

with $\tau = +\infty$ if $X_n \neq X_0$ for all $n \geq 1$. Compute the expectation of τ when the initial distribution is the stationary distribution π. Conclude that it is finite if and only if E is finite.

Exercise 9.7.13. THE SNAKE CHAIN
Let $\{X_n\}_{n\geq 0}$ be an HMC with state space E and transition matrix $\mathbf{P}$. Let for $L \geq 1, Y_n := (X_n, X_{n+1}, \dots, X_{n+L})$.

(a) The process $\{Y_n\}_{n\geq 0}$ takes its values in $F = E^{L+1}$. Prove that $\{Y_n\}_{n\geq 0}$ is an HMC and give the general entry of its transition matrix. (The chain $\{Y_n\}_{n\geq 0}$ is called the *snake chain* of length $L+1$ associated with $\{X_n\}_{n\geq 0}$.)

(b) Show that if $\{X_n\}_{n\geq 0}$ is irreducible, then so is $\{Y_n\}_{n\geq 0}$ if we restrict the state space of the latter to be $F = \{(i_0, \dots, i_L) \in E^{L+1}; p_{i_0 i_1} p_{i_1 i_2} \cdots p_{i_{L-1} i_L} > 0\}$. Show that if the original chain is irreducible aperiodic, so is the snake chain.

(c) Show that if $\{X_n\}_{n\geq 0}$ has a stationary distribution π, then $\{Y_n\}_{n\geq 0}$ also has a stationary distribution. Which one?

Exercise 9.7.14. PRODUCT MARKOV CHAIN
Let $\{X_n^{(1)}\}_{n\geq 0}$ and $\{X_n^{(2)}\}_{n\geq 0}$ be two *independent* irreducible and aperiodic HMCs with the same transition matrix $\mathbf{P}$. Define the *product* HMC $\{Z_n\}_{n\geq 0}$ taking its values in $E \times E$ by $Z_n = (X_n^{(1)}, X_n^{(2)})$. Prove that it is indeed a HMC. What is its n-step transition matrix? Prove that it is irreducible. Give a counterexample if the hypothesis of aperiodicity is omitted.

Exercise 9.7.15. PROOF OF LEMMA 9.2.18
Prove Lemma 9.2.18.

Exercise 9.7.16. IID RANDOM FIELDS
A. Let $(Z(v), v \in V)$ be a family of IID random variables indexed by a finite set V, with $P(Z(v) = -1) = p$, $P(Z(v) = +1) = q = 1 - p$. Show that

$$P(Z = z) = Ke^{\gamma \sum_{v\in V} z(s)},$$

for some constants γ and K to be identified.

B. Do the same with $P(Z(v) = 0) = p$, $P(Z(v) = +1) = q = 1 - p$.

Exercise 9.7.17. TWO-STATE HMC AS GIBBS FIELD

Consider an HMC $\{X_n\}_{n\geq 0}$ with state space $E = \{-1, 1\}$ and transition matrix

$$\mathbf{P} = \begin{pmatrix} 1-\alpha & \alpha \\ \beta & 1-\beta \end{pmatrix},$$

where $\alpha, \beta \in (0,1)$, and with the stationary initial distribution

$$(\nu_0, \nu_1) = \frac{1}{\alpha+\beta}(\beta, \alpha).$$

Give a representation of $Z := (X_0, \dots, X_N)$ as a Markov random field, that is, give its local characteristics.

Exercise 9.7.18. Markov Chain as Markov field
Let $\{X_n\}_{n\geq 0}$ be an HMC. Prove that for all $n \geq 1$, X_n is independent of $(X_k, k \notin \{n-1, n, n+1\})$ given (X_{n-1}, X_{n+1}).

Exercise 9.7.19. Ising on the Torus
Consider the classical Ising model of Example 9.5.7, except that the site space $V = \{1, 2, \dots, N\}$ consists of N points arranged in this order on a circle. The neighbors of site i are $i+1$ and $i-1$, with the convention that site $N+1$ is site 1. The phase space is $\Lambda = \{+1, -1\}$. Compute the partition function. Hint: express the normalizing constant Z_N in terms of the N-th power of the matrix

$$R = \begin{pmatrix} R(+1,+1) & R(+1,-1) \\ R(-1,+1) & R(-1,-1) \end{pmatrix} = \begin{pmatrix} \mathrm{e}^{K+h} & \mathrm{e}^{-K} \\ \mathrm{e}^{-K} & \mathrm{e}^{K-h} \end{pmatrix},$$

where $K := \frac{J}{kT}$ and $h := \frac{H}{kT}$.

Exercise 9.7.20. Monotonicity of the Gibbs Sampler
Let μ be an arbitrary probability measure on Λ^V and let ν be the probability measure obtained by applying the Gibbs sampler at an arbitrary site $v \in V$. Show that $d_V(\nu, \pi) \leq d_V(\mu, \pi)$.

Exercise 9.7.21. A Counterexample
Let Z^+ be the common value of the coalesced chains at the forwards coupling time τ^+ for the usual two-state ergodic HMC. Is the distribution of Z^+ the stationary distribution?

Exercise 9.7.22. Aperiodicity
a. Show that an irreducible transition matrix $\mathbf{P}$ with at least one state $i \in E$ such that $p_{ii} > 0$ is aperiodic.

b. Let $\mathbf{P}$ be an irreducible transition matrix on the *finite* state space E. Show that a necessary and sufficient condition for $\mathbf{P}$ to be aperiodic is the existence of an integer m such that $\mathbf{P}^m$ has all its entries positive.

c. Consider an HMC that is irreducible with period $d \geq 2$. Show that the restriction of the transition matrix to any cyclic class is irreducible. Show that the restriction of $\mathbf{P}^d$ to any cyclic class is aperiodic.

Exercise 9.7.23. DOUBLY STOCHASTIC TRANSITION MATRIX
A stochastic matrix $\mathbf{P}$ on the state space E is called *doubly stochastic* if for all states i, $\sum_{j \in E} p_{ji} = 1$. Suppose in addition that $\mathbf{P}$ is irreducible, and that E is *infinite*. Find the invariant measure of $\mathbf{P}$. Show that $\mathbf{P}$ cannot be positive recurrent.

Exercise 9.7.24. RETURNS TO A GIVEN SET
Let $\{X_n\}_{n \geq 0}$ be an HMC on the state space E with transition matrix $\mathbf{P}$. Let $\{\tau_k\}_{k \geq 1}$ be the successive return times to a given subset $F \subset E$. Assume these times are almost surely finite. Let $X_0 \equiv 0 \in F$, and define $Y_n = X(\tau_n)$. Show that $\{Y_n\}_{n \geq 0}$ is an HMC with state space F.

Exercise 9.7.25. NULL RECURRENCE OF THE 2-D SYMMETRIC RANDOM WALK
Show that the 2-D symmetric random walk on $\mathbb{Z}^2$ is null recurrent.

Exercise 9.7.26. TRANSIENCE OF THE 4-D SYMMETRIC RANDOM WALK
Show that the projection of the 4-D symmetric random walk on $\mathbb{Z}^3$ is a lazy symmetric random walk on $\mathbb{Z}^3$. Deduce from this that the 4-D symmetric random walk is transient. More generally, show that the symmetric random walk on $\mathbb{Z}^p$, $p \geq 5$, is transient.

Exercise 9.7.27. COUPLING TIME
Let $\mathbf{P}$ be an ergodic transition matrix on the *finite* state space E. Prove that for any initial distributions μ and ν, one can construct two HMCs $\{X_n\}_{n \geq 0}$ and $\{Y_n\}_{n \geq 0}$ on E with the same transition matrix $\mathbf{P}$, and the respective initial distributions μ and ν, in such a way that they couple at a finite time τ such that $E[e^{\alpha \tau}] < \infty$ for some $\alpha > 0$.

Exercise 9.7.28. THE LAZY RANDOM WALK ON THE CIRCLE
The state space $E := \{0, 1, \ldots, N-1\}$ consists of a succession of N equidistant points on a circle in such a way that two points i and j such that $j = i \pm 1$ *modulo* N are neighbors. Consider the Markov chain $\{(X_n, Y_n)\}_{n \geq 0}$ with state space $E \times E$ and representing two particles moving on E as follows. At each time n choose X_n

or Y_n with probability $\frac{1}{2}$ and move the corresponding particle to the left or to the right, equiprobably while the other particle remains still. The initial positions of the particles are a and b respectively. Compute the average time it takes until the two particles collide (the average coupling time of two lazy random walks).

Exercise 9.7.29. COUPLING TIME FOR THE 2-STATE HMC
Find the distribution of the first meeting time of two independent HMCs with state space $E = \{1, 2\}$ and transition matrix

$$\mathbf{P} = \begin{pmatrix} 1-\alpha & \alpha \\ \beta & 1-\beta \end{pmatrix},$$

where $\alpha, \beta \in (0, 1)$, when their initial states are different.

Exercise 9.7.30. EXTENSION TO NEGATIVE TIMES
Let $\{X_n\}_{n\geq 0}$ be an HMC with state space E, transition matrix $\mathbf{P}$, and suppose that there exists a stationary distribution $\pi > 0$. Suppose moreover that the initial distribution is π. Define the matrix $\mathbf{Q} = \{q_{ij}\}_{i,j\in E}$ by (9.7). Construct $\{X_{-n}\}_{n\geq 1}$, independent of $\{X_n\}_{n\geq 1}$ given X_0, as follows:

$$\begin{aligned} &P(X_{-1} = i_1, X_{-2} = i_2, \ldots, X_{-k} = i_k \mid X_0 = i, X_1 = j_1, \ldots, X_n = j_n) \\ &\quad = P(X_{-1} = i_1, X_{-2} = i_2, \ldots, X_{-k} = i_k \mid X_0 = i) = q_{ii_1} q_{i_1 i_2} \cdots q_{i_{k-1} i_k} \end{aligned}$$

for all $k \geq 1, n \geq 1, i, i_1, \ldots, i_k, j_1, \ldots, j_n \in E$. Prove that $\{X_n\}_{n\in\mathbb{Z}}$ is an HMC with transition matrix $\mathbf{P}$ and $P(X_n = i) = \pi(i)$, for all $i \in E$, all $n \in \mathbb{Z}$.

Chapter 10

Poisson Processes

Poisson processes are particular types of random point processes. A random point process on the line (*resp.* in space) is, roughly speaking, a countable random set of points of the real line (*resp.* in some space[1]).

In most applications to engineering and operations research, a *point* of a point process on the line is the time of occurrence of some event, and this is why points are also called *events*. For instance, the arrival times of customers at the desk of a post office or of jobs at the central processing unit of a computer are point process events. In biology the time of birth of an organism and in physiology the firing time of a neuron are events. In applications to ecology, a point of a spatial point process could be the location of a tree in a forest, or of a source of pollution. In a communications context, it may represent the position of a cellphone or of a relay antenna.

10.1 Poisson Processes on the Line

This section introduces the *homogeneous Poisson process*, the simplest example of a *random point process* on the line.

Definition 10.1.1 *A* random point process *on the line is a sequence* $\{T_n\}_{n\in\mathbb{Z}}$ *of real random variables such that, almost surely,*

(i) $\cdots \leq T_{-2} \leq T_{-1} \leq T_0 \leq 0 < T_1 < T_2 < \cdots$, *and*

(ii) $\lim_{|n|\uparrow\infty} T_{|n|} = +\infty$.

The usual definition of a random point process is less restrictive. In particular, condition (i) is relaxed in the more general definition, where multiple points

[1] In this book, $\mathbb{R}^m$ for $m \geq 2$.

P. Brémaud, *An Introduction to Applied Probability*, Texts in Applied Mathematics 77,
https://doi.org/10.1007/978-3-031-49306-5_10

(simultaneous arrivals to a ticket booth, for instance) are allowed. When condition (i) holds, one speaks of a *simple point process.* Also, condition (ii) is not required in the more general definition which allows with positive probability an *explosion*, that is, an accumulation of events in finite time. However, conditions (i) and (ii) fit the special case of homogeneous Poisson processes, the center of interest in this section.

The sequence $\{T_n - T_{n-1}\}_{n\in\mathbb{Z}}$ is called the *inter-event* sequence or, in the appropriate context, the *inter-arrival* sequence. For any interval $(a,b]$ in $\mathbb{R}$,

$$N((a,b]) := \sum_{n\in\mathbb{Z}} 1_{(a,b]}(T_n)$$

is an integer-valued random variable counting the events occurring in the time interval $(a,b]$. For typographical simplicity, it will be occasionally denoted by $N(a,b]$, omitting the external parentheses. If $t \geq 0$, we sometimes let $N(t) := N(0,t]$.

Since the interval $(a,t]$ $(t \geq 0)$ is closed on the right, the trajectories (or sample paths) $t \mapsto N((a,t],\omega)$ are right-continuous. They are non-decreasing, have limits on the left at every time t and jump one unit upwards at each event of the point process.

The family of random variables $N := \{N(a,b]\}_{(a,b]\subset\mathbb{R}}$ is called the *counting process* of the point process $\{T_n\}_{n\in\mathbb{Z}}$. Since the sequence of events can be recovered from N, the latter also receives the appellation "point process."

The Counting Process of an HPP

There exist several equivalent definitions of a Poisson process. The one adopted here is the most practical.

Definition 10.1.2 *A point process N on the positive half-line is called a* homogeneous Poisson process (HPP) *with* intensity $\lambda > 0$ *if*

- (α) *for all times $t_i \in \mathbb{R}$ $(1 \leq i \leq k)$ such that $t_1 \leq t_2 \leq \cdots \leq t_k$, the random variables $N(t_i, t_{i+1}]$ $(1 \leq i \leq k)$ are independent, and*
- (β) *for any interval $(a,b] \subset \mathbb{R}$, $N(a,b]$ is a Poisson random variable with mean $\lambda(b-a)$.*

In particular,

$$P(N(a,b] = k) = \mathrm{e}^{-\lambda(b-a)} \frac{[\lambda(b-a)]^k}{k!} \quad (k \geq 0)$$

and

$$E[N(a,b]] = \lambda(b-a)\,.$$

In this sense, λ is the average density of points.

Condition (α) is the property of *independence of increments* of Poisson processes. It implies in particular that for any interval $(a,b]$, the random variable $N(a,b]$ is independent of $(N(c,d]$ $(c \le d \le a)$. For this reason, Poisson processes are sometimes called *memoryless*. A more precise statement is "the *increments* of homogeneous Poisson processes have no memory of the past".

The definition adopted for random point processes does not allow for multiple points or explosions. But suppose it did. It turns out that requirements (α) and (β) in Definition 10.1.2 suffice to prevent such occurrences.

A proof is as follows. Since $E[N(a)] = \lambda a < \infty$, $N(a) < \infty$ almost surely. Since this is true for all $a \ge 0$, $\lim_{n\uparrow\infty} T_n = \infty$ almost surely. Simplicity will follow from $P(D(a)) = 0$ for all $a \ge 0$, where

$$D(a) := \{\text{there exists multiple points in } \ (0,a]\}\,.$$

We prove this for $D = D(1)$ (without loss of generality). The event

$$D_n := \left\{N\left(\frac{i}{2^n}, \frac{i+1}{2^n}\right] \ge 2 \text{ for some } i \ (1 \le i \le 2^n - 1)\right\}$$

decreases to D as n tends to infinity and therefore, by the monotone sequential continuity of probability,

$$P(D) = \lim_{n\uparrow\infty} P(D_n) = 1 - \lim_{n\uparrow\infty} P(\overline{D}_n)\,.$$

But

$$\begin{aligned} P(\overline{D}_n) &= P\left(\bigcap_{i=0}^{2^n-1}\left\{N\left(\frac{i}{2^n}, \frac{i+1}{2^n}\right] \le 1\right\}\right) = \prod_{i=0}^{2^n-1} P\left(N\left(\frac{i}{2^n}, \frac{i+1}{2^n}\right] \le 1\right) \\ &= \prod_{i=0}^{2^n-1} \mathrm{e}^{-\lambda 2^{-n}}(1+\lambda 2^{-n}) = \mathrm{e}^{-\lambda}(1+\lambda 2^{-n})^{2^n}. \end{aligned}$$

The limit of the latter quantity is 1 as $n \uparrow \infty$, and therefore, $P(D) = 0$.

Let

$$S_1 := T_1 \text{ and } S_n := T_n - T_{n-1} (n \ge 2)\,.$$

Theorem 10.1.3 *The sequence $\{S_n\}_{n\ge 1}$ of an* HPP *with intensity $\lambda > 0$ is* IID *with a common exponential distribution of parameter λ.*

The cumulative distribution function of an arbitrary inter-event time is therefore

$$P(S_n \leq t) = 1 - \mathrm{e}^{-\lambda t}.$$

Recall that

$$E[S_n] = \lambda^{-1},$$

that is, the average number of events per unit of time equals the inverse average inter-event time.

Proof. Suppose we can show that for any $n \geq 1$, the random vector $T := (T_1, \ldots, T_n)$ admits the probability density function

$$f_T(t_1, \ldots, t_n) = \lambda^n \mathrm{e}^{-\lambda t_n} 1_C(t_1, \ldots, t_n), \tag{10.1}$$

where $C := \{(t_1, \ldots, t_n); 0 < t_1 < \cdots < t_n\}$. Since

$$S_1 = T_1, \quad S_2 = T_2 - T_1, \ldots, \quad S_n = T_n - T_{n-1},$$

the formula of smooth change of variables gives for the probability density function of $S = (S_1, \ldots, S_n)$

$$f_S(s_1, \ldots, s_n) = f_T(s_1, s_1 + s_2, \ldots, s_1 + \cdots + s_n) = \prod_{i=1}^{n} \{\lambda \mathrm{e}^{-\lambda s_i} 1_{\{s_i > 0\}}\}.$$

It remains to prove (10.1).

The proof that we now give is somewhat heuristic (and we let the reader discover why) but most convincing.

The probability density function of T at $t = (t_1, \ldots, t_n)$ is obtained as the limit as $h_1, \ldots, h_n \in \mathbb{R}_+$ tend to 0 of the quantity

$$\frac{P(\cap_{i=1}^n \{T_i \in (t_i, t_i + h_i]\})}{\prod_{i=1}^n h_i}, \tag{10.2}$$

where it suffices to consider those $(t_1, \ldots, t_n)$ inside C since the points $T_1, \ldots, T_n$ are strictly ordered in increasing order. For sufficiently small $h_1, \ldots, h_n$, the event $\cap_{i=1}^n \{T_i \in (t_i, t_i + h_i]\}$ is the intersection of the events $\{N(0, t_1] = 0\}$, $\cap_{i=1}^{n-1} \{N(t_i, t_i + h_i] = 1, N(t_i + h_i, t_{i+1}] = 0\}$ and $\{N(t_n, t_n + h_n] \geq 1\}$, and therefore the numerator of (10.2) equals

$$P(N(0, t_1] = 0) \left(\prod_{i=1}^{n-1} P(N(t_i, t_i + h_i] = 1, N(t_i + h_i, t_{i+1}] = 0) \right) \times \cdots$$
$$\cdots \times P(N(t_n, t_n + h_n] \geq 1)$$
$$= \mathrm{e}^{-\lambda t_1} \prod_{i=1}^{n-1} \left(\mathrm{e}^{-\lambda h_i} \lambda h_i \mathrm{e}^{-\lambda(t_{i+1} - t_i - h_i)}\right) \left(1 - \mathrm{e}^{-\lambda h_n}\right) = \lambda^{n-1} \mathrm{e}^{-\lambda t_n} h_1 \cdots h_{n-1} (1 - \mathrm{e}^{-\lambda h_n}).$$

Dividing by $h_1 \cdots h_n$ and taking the limit as $h_1, \dots, h_n$ tend to 0, we obtain $\lambda^n e^{-\lambda t_n}$. □

Competing Poisson Processes

Let $\{T_n^1\}_{n\geq 1}$ and $\{T_n^2\}_{n\geq 1}$ be two *independent* HPPs on $\mathbb{R}_+$ with respective intensities $\lambda_1 > 0$ and $\lambda_2 > 0$. Their *superposition* is defined to be the sequence $\{T_n\}_{n\geq 1}$ formed by merging the two sequences $\{T_n^1\}_{n\geq 1}$ and $\{T_n^2\}_{n\geq 1}$ (see Figure 10.1). We shall prove that

(i) the point processes $\{T_n^1\}_{n\geq 1}$ and $\{T_n^2\}_{n\geq 1}$ have no points in common, and

(ii) the point process $\{T_n\}_{n\geq 1}$ is an HPP with intensity $\lambda = \lambda_1 + \lambda_2$.

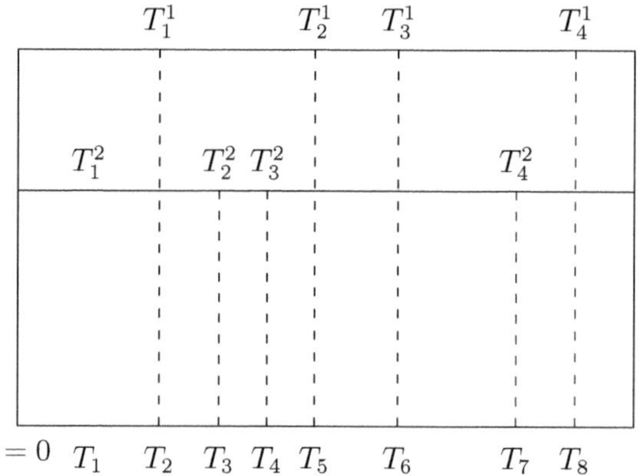

Figure 10.1: Superposition, or sum, of two point processes

Indeed, defining N by

$$N(a,b] := N_1(a,b] + N_2(a,b]\,,$$

we see that condition (α) of Definition 10.1.2 is satisfied, in view of the independence of N_1 and N_2. Also, $N(a,b]$ being the sum of two independent Poisson random variables of mean $\lambda_1(b-a)$ and $\lambda_2(b-a)$ is a Poisson variable of mean $\lambda(b-a)$ where $\lambda = \lambda_1 + \lambda_2$, and therefore, condition (β) of Definition 10.1.2 is satisfied. This proves (ii). But N is simple, and therefore (i) is true.

The above result can be extended to several – possibly infinitely many – homogeneous Poisson processes as follows:

Theorem 10.1.4 *Let $\{N_i\}_{i\geq 1}$ be a family of independent* HPP*s with respective positive intensities $\{\lambda_i\}_{i\geq 1}$. Then,*

(i) *two distinct* HPP*s of this family have no points in common, and*

(ii) *if $\lambda := \sum_{i=1}^{\infty} \lambda_i < \infty$, then $N(t) := \sum_{i=1}^{\infty} N_i(t)$ $(t \geq 0)$ defines the counting process of an* HPP *with intensity λ.*

Proof. Assertion (ii) has already been proven. Observe that for all $t \geq 0$, $N(t)$ is almost surely finite since

$$E[N(t)] = \sum_{i=1}^{\infty} E[N_i(t)] = \left(\sum_{i=1}^{\infty} \lambda_i\right) t < \infty\,.$$

In particular, $N(a,b]$ is almost surely finite for all $(a,b] \subset \mathbb{R}_+$. The proof of lack of memory of N is the same as in the case of two superposed Poisson processes. Finally, $N(a,b]$ is a Poisson random variable of mean $\lambda(b-a)$ since

$$\begin{aligned} P(N(a,b]=k) &= \lim_{n\uparrow\infty} P\left(\sum_{i=1}^{n} N_i(a,b] = k\right) \\ &= \lim_{n\uparrow\infty} \mathrm{e}^{-(\sum_{i=1}^{n}\lambda_i(b-a)} \frac{[\sum_{i=1}^{n}\lambda_i(b-a)]^k}{k!} \\ &= \mathrm{e}^{-\lambda(b-a)} \frac{[\lambda(b-a)]^k}{k!}\,. \end{aligned}$$

□

The next result is called the *competition theorem* because it features HPPs competing for the production of the first event.

Theorem 10.1.5 *In the situation of Theorem 10.1.4, where $\lambda := \sum_{i=1}^{\infty} \lambda_i < \infty$, denote by Z the first event time of $N = \sum_{i=1}^{\infty} N_i$ and by J the index of the* HPP *responsible for it (Z is the first event of N_J). Then*

$$P(J=i, Z \geq a) = P(J=i)P(Z \geq a) = \frac{\lambda_i}{\lambda}\mathrm{e}^{-\lambda a}\,. \tag{10.3}$$

In particular, J and Z are independent, $P(J=i) = \frac{\lambda_i}{\lambda}$ and Z is exponential with mean λ^{-1}.

Proof.

A. We first prove the result for a finite number of Poisson processes. We have to show that if $X_1, \ldots, X_K$ are K independent exponential variables with means

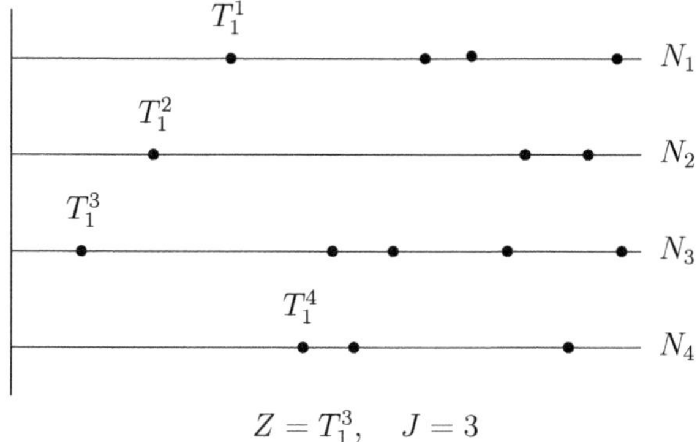

$Z = T_1^3, \quad J = 3$

Competition among four point processes

$\lambda_1^{-1}, \dots, \lambda_K^{-1}$ and if J_K is defined by $X_{J_K} = Z_K := Z_K := \inf(X_1, \dots, X_K)$, then

$$P(J_K = i, Z_K \geq a) = \frac{\lambda_i}{\lambda_1 + \cdots + \lambda_K} \exp\{-(\lambda_1 + \cdots + \lambda_K)a\}. \qquad (\star)$$

First observe that

$$P(Z_K \geq a) = P\left(\cap_{j=1}^K \{X_j \geq a\}\right) = \prod_{j=1}^K P(X_j \geq a) = \prod_{j=1}^K \mathrm{e}^{-\lambda_j a} = \mathrm{e}^{-(\lambda_1 + \cdots + \lambda_K)a}.$$

Letting $U := \inf(X_2, \dots, X_K)$, we have

$$\begin{aligned} P(J_K = 1, Z_K \geq a) &= P(a \leq X_1 < U) \\ &= \int_a^\infty P(U > x)\lambda_1 \mathrm{e}^{-\lambda_1 x} \mathrm{d}x = \int_a^\infty \mathrm{e}^{-(\lambda_2 + \cdots + \lambda_K)x} \lambda_1 \mathrm{e}^{-\lambda_1 x} \mathrm{d}x \\ &= \frac{\lambda_1}{\lambda_1 + \cdots + \lambda_K} \mathrm{e}^{-(\lambda_1 + \cdots + \lambda_K)a}. \end{aligned}$$

This gives $(\star)$. Letting $a = 0$ yields $P(J_K = 1) = \frac{\lambda_1}{\lambda_1 + \cdots + \lambda_K}$. This, together with $(\star)$ and the expression for $P(Z_K \geq a)$ gives (10.3), for $i = 1$, without loss of generality.

B. Suppose the result true for a finite number of HPPs. Since the event $\{J_K = 1, Z_K \geq a\}$ decreases to $\{J = 1, Z \geq a\}$ as $K \uparrow \infty$, we have

$$P(J = 1, Z \geq a) = \lim_{K \uparrow \infty} P(J_K = 1, Z_K \geq a),$$

from which (10.3) follows, using the result of part A of the proof. □

10.2 Generalities on Point Processes

A few definitions concerning general point processes are in order.

Let Δ be an arbitrary "dummy" element not in $\mathbb{R}^m$. Let ε_a be the Dirac measure at a if $a \in \mathbb{R}^m$, the null measure if $a = \Delta$.

Definition 10.2.1 *Let $\{X_n\}_{n\in\mathbb{N}}$ is a sequence of random variables with values in $\mathbb{R}^m \cup \{\Delta\}$. The collection $\{X_n\}_{n\in\mathbb{N}}$ is called a* point process *on $\mathbb{R}^m$, and the X_n's are the* points *of this point process.*

This point process may be represented by the (random) measure

$$N := \sum_{n\in\mathbb{N}} \varepsilon_{X_n} \,. \tag{10.4}$$

In particular,

$$N(C) = \sum_{n\in\mathbb{N}} 1_C(X_n)$$

counts the number of points in $C \in \mathcal{B}(\mathbb{R}^m)$. The Δ element plays the role of ∞ ("a point that does not exist"). Note that it may occur that for some of the values in the list $\{X_n\}_{n\in\mathbb{N}}$ are the same, thus allowing for multiple points.

Definition 10.2.2 *The point process N is called* simple *if almost surely $N(\omega)(\{c\}) \le 1$ for all $c \in \mathbb{R}^m$.*

Definition 10.2.3 *The point process N is called* locally finite *if $P(N(C) < \infty) = 1$ for all* bounded *measurable sets $C \subset \mathbb{R}^m$.*

Definition 10.2.4 *The locally finite point process N is called a* first-order point process *if $E[N(C)] < \infty$ for all bounded measurable sets $C \in \mathbb{R}^m$.*

EXAMPLE 10.2.5: THE BINOMIAL POINT PROCESS. This point process on $\mathbb{R}^m$ has a (finite number) of points, T, where T is a binomial random variable of size n and parameter $p \in (0,1)$:

$$P(T = k) = \binom{n}{k} p^k (1-p)^{n-k} \quad (0 \le k \le n)\,.$$

If $T = k$, the k points are located independently of one another on $\mathbb{R}^m$ according to the same probability distribution Q. It is locally finite. It is simple if and only if Q is non-atomic. It is a first-order point process because T is integrable.

Definition 10.2.6 *Let N be a point process on $\mathbb{R}^m$ and let P be a probability measure on $(\Omega, \mathcal{F})$. The distribution of N consists of the probability distributions of the vectors $(N(C_1), \dots, N(C_K))$ $(K \geq 1,\ C_1, \dots, C_K \in \mathcal{B}(\mathbb{R}^m))$.*

EXAMPLE 10.2.7: THE POISSON PROCESS. Let ν be a σ-finite measure on $\mathbb{R}^m$. The point process N on $\mathbb{R}^m$ is called a *Poisson process* on $\mathbb{R}^m$ with *intensity measure* ν if

(i) for all finite families of mutually *disjoint* sets $C_1, \dots, C_K \in \mathcal{B}(\mathbb{R}^m)$, the random variables $N(C_1), \dots, N(C_K)$ are independent, and

(ii) for any set $C \in \mathcal{B}(\mathbb{R}^m)$ such that $\nu(C) < \infty$,

$$P(N(C) = k) = \mathrm{e}^{-\nu(C)} \frac{\nu(C)^k}{k!} \qquad (k \geq 0)\,.$$

If ν is of the form $\nu(C) = \int_C \lambda(x)\mathrm{d}x$ for some non-negative measurable function $\lambda : \mathbb{R}^m \to \mathbb{R}$, the Poisson process N is said to admit the *intensity function* $\lambda(x)$. If in addition $\lambda(x) \equiv \lambda$, N is called a *homogeneous* Poisson process (HPP) on $\mathbb{R}^m$ with *intensity* or *rate* λ.

Definition 10.2.8 *A point process N on $\mathbb{R}^m$ is called* stationary *if for all families of measurable sets $C_1, \dots, C_K$ of $\mathbb{R}^m$, $K \geq 1$, the distribution of the random vector $(N(C_1 + a), \dots, N(C_K + a))$ is independent of $a \in \mathbb{R}^m$.*

EXAMPLE 10.2.9: A STATIONARY GRID, TAKE 1. A grid on $\mathbb{R}^2$ is a deterministic point process on $\mathbb{R}^2$ whose points are

$$(nT_1, mT_2) \quad (n, m \in \mathbb{Z})\,,$$

where T_1 and T_2 are positive real numbers. It is not a stationary point process. However, the shifted version of it,

$$(nT_1 + V_1, mT_2 + V_2) \quad (n, m \in \mathbb{Z})\,,$$

where V_1 and V_2 are independent random variables uniformly distributed on $[0, T_1)$ and $[0, T_2)$ respectively, is stationary. This can be proved directly, or by using the Laplace functional (Example 10.2.21).

Independent Point Processes

Definition 10.2.10 *The family N_i $(i \in I)$ of point processes on $\mathbb{R}^m$, where I is an arbitrary index, is called independent if for all finite sets of distinct indices $i_1, \dots, i_K$ in I, all integers $\ell_{i_1}, \dots, \ell_{i_K}$, and all $C_{i_1}^1, \dots, C_{i_K}^{\ell_{i_K}}(\mathbb{R}^m)$, the random vectors*

$$(N_{i_1}(C_{i_1}^1), \dots, N_{i_1}(C_{i_1}^{\ell_{i_1}})),$$
$$\dots$$
$$(N_{i_K}(C_{i_K}^1), \dots, N_{i_K}(C_{i_K}^{\ell_{i_K}}))$$

are independent.

Marked Point Processes

Let N and $\{X_n\}_{n\in\mathbb{N}}$ be as in Definition 10.2.1. Let $(K, \mathcal{K})$ be some $(\mathbb{R}^d, \mathcal{B}(\mathbb{R}^d))$ and let $\{Z_n\}_{n\in\mathbb{N}}$ be a random sequence with values in K.

Definition 10.2.11 *The sequence $\{\widetilde{X}_n\}_{n\in\mathbb{N}}$, where*

$$\widetilde{X}_n := (X_n, Z_n) \quad (n \in \mathbb{N}),$$

defines a point process $\widetilde{N}$ on $\mathbb{R}^{m+d}$ called a marked point process *on $\mathbb{R}$ with marks in K; $\{Z_n\}_{n\in\mathbb{N}}$ is the* mark sequence *and N is the* basic point process.

The notation K for $\mathbb{R}^d$ is used for rendering the distinction between the marks and the original point process N more visual.

The random variable

$$\widetilde{N}(C \times L) := \sum_{n\in\mathbb{N}} 1_C(X_n) 1_L(Z_n) \quad (C \in \mathcal{B}(\mathbb{R}^m), L \in \mathcal{K}) \tag{10.5}$$

counts the number of points in the original point process N in C with marks in L. Note that since $\Delta \notin C$, the points $X_n \in \{\Delta\}$ do not appear in the sum above ("points at infinity are excluded"). We shall occasionally use the notation N_Z instead of $\overline{N}$.

The following phrases are then considered equivalent:

(i) "the marked point process $\{(X_n, Z_n)\}_{n\in\mathbb{N}}$",

(ii) "the marked point process $\widetilde{N}$",

(iii) "the marked point process (N, Z)".

(iv) "the marked point process (N_Z)".

Definition 10.2.12 *If in addition $\{Z_n\}_{n\in\mathbb{N}}$ is* IID *and independent of N, with common probability distribution Q_Z, then $\overline{N}$ is called a marked point process with* independent IID marks.

Point Process Integrals

Since a point process is a sum of Dirac measures, point process integrals are in fact sums.

Let μ be a measure on $(\mathbb{R}^m, \mathcal{B}(\mathbb{R}^m))$ and let $\varphi : (\mathbb{R}^m, \mathcal{B}(\mathbb{R}^m)) \to (\overline{\mathbb{R}}, \mathcal{B}(\overline{\mathbb{R}}))$ be a measurable function for which the integral $\int_{\mathbb{R}^m} \varphi \, \mathrm{d}\mu$ is well defined. This integral is also denoted by $\mu(\varphi)$.

When N is a point process, the following notations represent the same mathematical object (if it is well defined):

$$\sum_{n\in\mathbb{N}} \varphi(X_n) , \qquad \int_{\mathbb{R}}^{m} \varphi(x) N(\mathrm{d}x) , \qquad N(\varphi) .$$

In the first notation, we use the *convention* that the sum extends only to those indices n such that $X_n \in \mathbb{R}^m$, excluding the points "at infinity" (in fact, $\varphi(\Delta)$ is not defined). In the situation of Definition 10.2.11, observe that

$$\int_{\mathbb{R}^m \times K} \varphi(x, z) \widetilde{N}(\mathrm{d}x \times \mathrm{d}z) = \sum_{n\in\mathbb{N}} \varphi(X_n, Z_n),$$

with the same convention as the one just agreed upon concerning points at infinity.

The Intensity Measure

Definition 10.2.13 *Let N be a locally finite point process on $\mathbb{R}^m$. The set function*

$$\nu \mapsto \nu(C) := E\,[N(C)] \qquad (C \in \mathcal{B}(\mathbb{R}^m))$$

defines a measure on $(\mathbb{R}^m, \mathcal{B}(\mathbb{R}^m))$, called the mean measure *or the* intensity measure *of N.*

The intensity measure of a marked point process $\widetilde{N}$ (Definition 10.2.11) is the measure $\widetilde{N}$ on $(\mathbb{R}^m \times K, \mathcal{B}(\mathbb{R}^m) \otimes \mathcal{K})$ defined by

$$\widetilde{\nu}(\widetilde{C} := \left[\widetilde{N}(\widetilde{C}\right] \qquad (\widetilde{C} \in \mathcal{B}(\mathbb{R}^m) \otimes \mathcal{K}) .$$

EXAMPLE 10.2.14: THE INTENSITY MEASURE OF A MARKED POINT PROCESS WITH INDEPENDENT IID MARKS. Let $\widetilde{N}$ be as in Definition 10.2.11. Denoting by Q_Z the common distribution of the marks and by ν the intensity measure of the basic point process N, the intensity measure of $\widetilde{N}$ is the product measure $\widetilde{\nu}(\mathrm{d}x \times \mathrm{d}z) = \nu(\mathrm{d}x)Q_Z(\mathrm{d}z)$. The easy proof is left as an exercise.

Campbell's Formula

Theorem 10.2.15 *Let N be a point process on $\mathbb{R}^m$ with intensity measure ν. Then, for all measurable functions $\varphi : \mathbb{R}^m \to \mathbb{R}$ which are either non-negative or ν-integrable, the integral $N(\varphi)$ is well defined (possibly infinite when φ is only assumed to be non-negative) and*

$$E\left[N(\varphi)\right] = \nu(\varphi)\,. \tag{10.6}$$

In particular, $N(\varphi)$ is a.s. finite if φ is ν-integrable.

Proof. First, suppose that φ is a *simple* non-negative measurable function, that is, of the form

$$\sum_{h=1}^{L} \alpha_h 1_{C_h}\,,$$

where $L \in \mathbb{N}_+$, $\alpha_h \in \mathbb{R}_+$ and $C_1, \ldots, C_L$ are disjoint measurable subsets of $\mathbb{R}^m$. Then

$$E[N(\varphi)] = E\left[\sum_{h=1}^{L} a_h N(C_h)\right] = \sum_{h=1}^{L} a_h \nu(C_h) = \nu(\varphi)\,.$$

Now let φ be a non-negative measurable function and let $\{\varphi_n\}_{n\in\mathbb{N}}$ be a non-decreasing sequence of simple non-negative measurable functions with limit φ. Letting $n \uparrow \infty$ in

$$E[N(\varphi_n)] = \nu(\varphi_n)$$

yields the announced result, by monotone convergence. In the case where $\varphi \in L^1_{\mathbb{R}}(\nu)$, since $E\left[N(\varphi^{\pm})\right] = \nu(\varphi^{\pm}) < \infty$, the random variables $N(\varphi^{\pm})$ are P-a.s. finite, and therefore $N(\varphi) = N(\varphi^+) - N(\varphi^-)$ is well defined and finite, and

$$E[N(\varphi)] = E[N(\varphi^+)] - E[N(\varphi^-)] = \nu(\varphi^+) - \nu(\varphi^-) = \nu(\varphi)\,.$$

□

EXAMPLE 10.2.16: CAMPBELL'S FORMULA FOR MARKED POINT PROCESSES WITH INDEPENDENT IID MARKS. Let $\widetilde{N}$ be as in Definition 10.2.11. Campbell's theorem then reads as follows. If the measurable function $\varphi : \mathbb{R}^m \times K \to \mathbb{R}$ is either non-negative or in $L^1_{\mathbb{R}}(\nu \times Q_Z)$, then the sum

$$\sum_{n\in\mathbb{N}} \varphi(X_n, Z_n)$$

is P-a.s. well defined (possibly infinite if φ is only assumed non-negative) and

$$E\left[\sum_{n\in\mathbb{N}} \varphi(X_n, Z_n)\right] = \int_{\mathbb{R}^m} E\left[\varphi(x, Z)\right] \nu(\mathrm{d}x),$$

where Z is a K-valued random variable with distribution Q_Z.

The Laplace Functional

This functional plays for point processes a role analogous to that of the usual Laplace transform for random vectors.

Definition 10.2.17 *Let N be a point process on $\mathbb{R}^m$. The* Laplace functional *of N is the mapping L_N associating with a non-negative measurable function $\varphi : \mathbb{R}^m \to \mathbb{R}_+$ the non-negative real number*

$$L_N(\varphi) := E\left[\mathrm{e}^{-N(\varphi)}\right].$$

EXAMPLE 10.2.18: THE LAPLACE FUNCTIONAL OF A POISSON PROCESS. Anticipating a later result (Theorem 10.3.7 thereof), the Laplace functional of a Poisson process on $\mathbb{R}^m$ with intensity measure ν is

$$L_N(\varphi) = \exp\left\{\int_{\mathbb{R}^m} \left(\mathrm{e}^{-\varphi(x)} - 1\right) \nu(\mathrm{d}x)\right\}.$$

Theorem 10.2.19 *The Laplace functional of a locally finite random measure N on E characterizes its distribution.*

Proof. It suffices to show that the Laplace functional of a point process N characterizes its finite-dimensional distributions. For this, just observe that for all $K \geq 1$ and all disjoint measurable sets $C_1, \ldots, C_K$ in $\mathcal{B}(\mathbb{R}^m)$, the Laplace transform of the vector $(N(C_1), \ldots, N(C_K))$, that is, the function

$$(t_1, \ldots, t_K) \in \mathbb{R}_+^K \mapsto E\left[\mathrm{e}^{-t_1 N(C_1) - \cdots - t_K N(C_K)}\right],$$

is of the form $E\left[\mathrm{e}^{-N(\varphi)}\right]$, where $\varphi = t_1 1_{C_1} + \cdots + t_K 1_{C_K}$. □

Corollary 10.2.20 *A point process N on $\mathbb{R}^m$ is stationary if and only if its Laplace functional L_N is such that*

$$L_N(\varphi) = L_N(S_a\varphi)$$

for all non-negative functions φ from $\mathbb{R}^m$ to $\mathbb{R}$ and all $a \in \mathbb{R}^m$, where $(S_a\varphi)(t) := \varphi(t-a)$.

EXAMPLE 10.2.21: A STATIONARY GRID, TAKE 2. In order to prove the stationarity of the shifted grid of Example 10.2.9, it suffices to show that for any non-negative function φ from $\mathbb{R}^2$ to $\mathbb{R}$, the quantity

$$E\left[e^{\sum_{n,m\in\mathbb{Z}} \varphi(nT_1+V_1+\alpha, nT_2+V_2+\beta)}\right]$$

is independent of α, $\beta \in \mathbb{R}$. This quantity equals

$$\int_0^{T_1}\left\{\int_0^{T_2} e^{\sum_{n,m\in\mathbb{Z}} \varphi(nT_1+v_1+\alpha, nT_2+v_2+\beta)}\, dv_2\right\} dv_1 .$$

The conclusion follows from the fact that for any non-negative function $\psi : \mathbb{R} \to \mathbb{R}$,

$$\int_0^T \psi(nT+u+\alpha)\, du = \int_0^T \psi(nT+u)\, du$$

for all $\alpha \in \mathbb{R}$, by the shift-invariance of the Lebesgue measure.

Theorem 10.2.22 *The family N_i $(i \in I)$ of point processes on $\mathbb{R}^m$, where I is an arbitrary index set, is an independent family if and only if for any finite subset $J \subseteq I$, and any collection φ_i $(i \in J)$ of non-negative measurable functions from $\mathbb{R}^m$ to $\mathbb{R}$,*

$$E\left[e^{-\sum_{i\in J} N_i(\varphi_i)}\right] = \prod_{i\in J} E\left[e^{-N_i(\varphi_i)}\right] . \tag{10.7}$$

Proof. The sufficiency follows immediately from the definition of independence for point processes. The necessity is left as an exercise. □

EXAMPLE 10.2.23: THE LAPLACE FUNCTIONAL OF THINNED POINT PROCESSES. Let N be a simple point process on $\mathbb{R}^m$ with point sequence $\{X_n\}_{n\in\mathbb{N}}$. Let $\{Z_n\}_{n\in\mathbb{N}}$ be an IID sequence of independent marks of N, each Z_n taking its values in $\{0,1\}$, with probability $p \in (0,1)$ for the value 1. The point process $N_{thin,p}$ defined by

$$N_{thin,p}(C) := \sum_{n\in\mathbb{N}} 1_C(X_n) Z_n$$

is called the *p-thinning* of N. Each point of N is retained in $N_{thin,p}$ with probability p, independently of everything else. We compute the Laplace functional of the thinned point process:

$$\begin{aligned} L_{N_{thin,p}}(\varphi) &= E\left[\exp\left\{-\sum_{n\in\mathbb{N}}\varphi(X_n)Z_n\right\}\right] \\ &= \lim_{k\uparrow\infty} E\left[\exp\left\{-\sum_{n=1}^{k}\varphi(X_n)Z_n\right\}\right]. \end{aligned}$$

But

$$\begin{aligned} E\left[\exp\left\{-\sum_{n=1}^{k}\varphi(X_n)Z_n\right\}\right] &= E\left[\prod_{n=1}^{k}\exp\left(-\varphi(X_n)Z_n\right)\right] \\ &= E\left[E\left[\prod_{n=1}^{k}\exp\left(-\varphi(X_n)Z_n\right)|X_1,\dots,X_k\right]\right] \\ &= E\left[\prod_{n=1}^{k}E\left[\exp\left(-\varphi(X_n)Z_n\right)|X_1,\dots,X_k\right]\right] \\ &= E\left[\prod_{n=1}^{k}\{p\exp(-\varphi(X_n))+(1-p)\}\right] \\ &= E\left[\exp\left(\sum_{n=1}^{k}\log\left(p\exp(-\varphi(X_n))+(1-p)\right)\right)\right]. \end{aligned}$$

Therefore finally, after letting $k\uparrow\infty$,

$$L_{N_{thin,p}}(\varphi) = L_N\left(-\log\left(p\mathrm{e}^{-\varphi(\cdot)}+1-p\right)\right).$$

For future reference, we record the intermediary result obtained in the line before the last one in the above calculation:

$$L_{N_{thin,p}}(\varphi) := E\left[\exp\left\{\int_{\mathbb{R}^m}\log\left(1-p(1-\mathrm{e}^{-\varphi(x)})\right)\, N(\mathrm{d}x)\right\}\right]. \tag{10.8}$$

10.3 Spatial Poisson Processes

Recall the definition given in Example 10.2.7.

Definition 10.3.1 *Let ν be a σ-finite measure on $\mathbb{R}^m$. The point process N on $\mathbb{R}^m$ is called a* Poisson process *on $\mathbb{R}^m$ with* intensity measure *ν if*

(i) *for all finite families of mutually* disjoint *sets $C_1, \dots, C_K \in \mathcal{B}(\mathbb{R}^m)$, the random variables $N(C_1), \dots, N(C_K)$ are independent, and*

(ii) *for any set $C \in \mathcal{B}(\mathbb{R}^m)$ such that $\nu(C) < \infty$,*

$$P(N(C) = k) = \mathrm{e}^{-\nu(C)} \frac{\nu(C)^k}{k!} \qquad (k \geq 0).$$

If ν is of the form $\nu(C) = \int_C \lambda(x)\mathrm{d}x$ for some non-negative measurable function $\lambda : \mathbb{R}^m \to \mathbb{R}$, the Poisson process N is said to admit the *intensity function* $\lambda(x)$. If in addition $\lambda(x) \equiv \lambda$, N is called a *homogeneous* Poisson process (HPP) on $\mathbb{R}^m$ with *intensity* or *rate* λ.

We now construct the Poisson process. The basic result is the following:

Theorem 10.3.2 *Let T be a Poisson random variable of mean θ. Let $\{Z_n\}_{n\geq 1}$ be an* IID *sequence of random elements with values in $\mathbb{R}^m$ and common distribution Q. Assume that T is independent of $\{Z_n\}_{n\geq 1}$. The point process N on $\mathbb{R}^m$ defined by*

$$N(C) = \sum_{n=1}^{T} 1_C(Z_n) \quad (C \in \mathcal{B}(\mathbb{R}^m))$$

is a Poisson process with intensity measure $\nu(\cdot) := \theta \times Q(\cdot)$.

Proof. It suffices to show that for any finite family $C_1, \dots, C_K$ of pairwise disjoint measurable sets of $\mathbb{R}^m$ with finite ν-measure and all non-negative reals $t_1, \dots, t_K$,

$$E[\mathrm{e}^{-\sum_{j=1}^K t_j N(C_j)}] = \Pi_{j=1}^K \exp\left\{\nu(C_j)(\mathrm{e}^{-t_j} - 1)\right\}.$$

We have

$$\sum_{j=1}^K t_j N(C_j) = \sum_{j=1}^K t_j \left(\sum_{n=1}^T 1_{C_j}(Z_n)\right) = \sum_{n=1}^T \left(\sum_{j=1}^K t_j 1_{C_j}(Z_n)\right) = \sum_{n=1}^T Y_n,$$

where $Y_n = \sum_{j=1}^K t_j 1_{C_j}(Z_n)$. By Theorem 3.2.22,

$$E[\mathrm{e}^{-\sum_{n=1}^T Y_n}] = g_T(E[\mathrm{e}^{-Y_1}]),$$

where g_T is the generating function of T. Here, since T is Poisson mean θ,

$$g_T(z) = \exp\left\{\theta(z - 1)\right\}.$$

The random variable Y_1 takes the values $t_1, \dots, t_K$ and 0 with the respective probabilities $Q(C_1), \dots, Q(C_K)$ and $1 - \sum_{j=1}^{K} Q(C_j)$. Therefore

$$E[\mathrm{e}^{-Y_1}] = \sum_{j=1}^{K} \mathrm{e}^{-t_j} Q(C_j) + 1 - \sum_{j=1}^{K} Q(C_j) = 1 + \sum_{j=1}^{K} \left(\mathrm{e}^{-t_j} - 1\right) Q(C_j),$$

from which we obtain the announced result. □

The above is a special case of what is to be done, that is, to construct a Poisson process on $\mathbb{R}^m$ with an intensity measure ν that is *σ-finite* (not just finite). Such a measure can be decomposed as

$$\nu(\cdot) = \sum_{j=1}^{\infty} \theta_j \times Q_j(\cdot),$$

where the θ_j's are positive real numbers and the Q_j's are probability distributions on $\mathbb{R}^m$. One can construct independent Poisson processes N_j on E with respective intensity measures $\theta_j Q_j(\cdot)$. The result then follows from the following:

Theorem 10.3.3 *Let ν be a σ-finite measure on $\mathbb{R}^m$ of the form $\nu = \sum_{i=1}^{\infty} \nu_i$, where the ν_i's $(i \geq 1)$ are σ-finite measures on $\mathbb{R}^m$. Let N_i $(i \geq 1)$ be a family of independent Poisson processes on E with respective intensity measures ν_i $(i \geq 1)$. Then the point process*

$$N = \sum_{j=1}^{\infty} N_j$$

is a Poisson process with intensity measure ν.

Proof. For mutually disjoint measurable sets $C_1, \dots, C_K$ of finite ν-measures, and non-negative reals $t_1, \dots, t_K$,

$$\begin{aligned} E\left[\mathrm{e}^{-\sum_{\ell=1}^{K} t_\ell N(C_\ell)}\right] &= E\left[\mathrm{e}^{-\sum_{\ell=1}^{K} t_\ell \left(\sum_{j=1}^{\infty} N_j(C_\ell)\right)}\right] \\ &= E\left[\mathrm{e}^{-\lim_{n\uparrow\infty} \sum_{\ell=1}^{K} t_\ell \left(\sum_{j=1}^{n} N_j(C_\ell)\right)}\right] \\ &= \lim_{n\uparrow\infty} E\left[\mathrm{e}^{-\sum_{\ell=1}^{K} t_\ell \left(\sum_{j=1}^{n} N_j(C_\ell)\right)}\right], \end{aligned}$$

by dominated convergence. But

$$E\left[\mathrm{e}^{-\sum_{\ell=1}^{K} t_\ell\left(\sum_{j=1}^{n} N_j(C_\ell)\right)}\right] = E\left[\mathrm{e}^{-\sum_{j=1}^{n}\left(\sum_{\ell=1}^{K} t_\ell N_j(C_\ell)\right)}\right]$$

$$= \prod_{j=1}^{n} E\left[\mathrm{e}^{-\sum_{\ell=1}^{K} t_\ell N_j(C_\ell)}\right] = \prod_{j=1}^{n}\prod_{\ell=1}^{K} \mathrm{e}^{-t_\ell N_j(C_\ell)}$$

$$= \prod_{j=1}^{n}\prod_{\ell=1}^{K} \exp\left\{(\mathrm{e}^{-t_\ell} - 1)\nu_j(C_\ell)\right\}$$

$$= \prod_{j=1}^{n} \exp\left\{\sum_{\ell=1}^{K} \left(\mathrm{e}^{-t_\ell} - 1\right) \nu_j(C_\ell)\right\}$$

$$= \exp\left\{\sum_{\ell=1}^{K} \left(\mathrm{e}^{-t_\ell} - 1\right)\left(\sum_{j=1}^{n} \nu_j(C_\ell))\right)\right\}.$$

Letting $n \uparrow \infty$ we obtain, by dominated convergence,

$$E\left[\mathrm{e}^{-\sum_{\ell=1}^{K} t_\ell N(C_\ell)}\right] = \exp\left\{\sum_{\ell=1}^{K} \left(\mathrm{e}^{-t_\ell} - 1\right) \nu(C_\ell)\right\}.$$

Therefore $N(C_1)$, ..., $N(C_K)$ are independent Poisson random variables with respective means $\nu(C_1)$, ..., $\nu(C_K)$. □

Theorem 10.3.4 *Let N be a Poisson process on $\mathbb{R}^m$ with intensity measure ν.*

(a) *If ν is locally finite, then N is locally finite.*

(b) *If ν is locally finite and non-atomic, then N is simple.*

Proof. (a) If C is a bounded measurable set, it is of finite ν-measure, and therefore $E[N(C)] = \nu(C) < \infty$, which implies that $N(C) < \infty$, P-almost surely.

(b) It suffices to show this for a finite intensity measure $\nu(\cdot) = \theta(\cdot)\, Q$, where θ is a positive real number and Q is a non-atomic probability measure on $\mathbb{R}^m$, and then use the construction of Theorem 10.3.2. In turn, it suffices to show that for each $n \geq 1$, $P(Z_i = Z_j$ for some pair $(i,j)\,(1 \leq i < j \leq n) \mid N(\mathbb{R}^m) = n) = 0$. This is the case because for IID vectors $Z_1, \ldots, Z_n$ with a non-atomic probability distribution, $P(Z_i = Z_j$ for some pair $(i,j)\,(1 \leq i < j \leq n)) = 0$. □

EXAMPLE 10.3.5: THINNED POISSON PROCESS. If the initial point process is a Poisson process with the locally integrable intensity measure ν,

$$L_{N_{thin,p}}(\varphi) = L_N(\psi) = \mathrm{e}^{-\int_{\mathbb{R}^m}\left(\mathrm{e}^{-\psi(x)}-1\right)\nu(\mathrm{d}x)},$$

where $\psi(x) := -\log\left(p\mathrm{e}^{-\varphi(x)} + 1 - p\right)$. Therefore $\mathrm{e}^{-\psi(x)} = p\mathrm{e}^{-\varphi(x)} + 1 - p = p(\mathrm{e}^{-\varphi(x)} - 1) + 1$ and finally

$$L_{N_{thin,p}}(\varphi) = \mathrm{e}^{-\int_{\mathbb{R}^m}\left(\mathrm{e}^{-\varphi(x)}-1\right)p\nu(\mathrm{d}x)}.$$

We therefore retrieve the standard result: p-thinning a Poisson process of intensity measure $\nu(\cdot)$ results in a Poisson process of intensity measure $\nu_p(\cdot) = p\nu(\cdot)$.

Doubly Stochastic Poisson Processes

Doubly stochastic Poisson processes are also called *Cox processes.*

Let $\{\lambda(x)\}_{x\in\mathbb{R}^m}$ be a real-valued non-negative stochastic process such that almost surely

$$\int_C \lambda(x)\,\mathrm{d}x < \infty \text{ for all bounded } C \in \mathcal{B}(\mathbb{R}^m).$$

A point process is constructed as follows: first generate the *stochastic intensity process* $\{\lambda(x)\}_{x\in\mathbb{R}^m}$ and, having done so, generate a Poisson process N with this intensity. The resulting point process is called a *doubly stochastic Poisson process* (or *Cox process*) with the (stochastic) intensity function $\{\lambda(x)\}_{x\in\mathbb{R}^m}$.

In the case where

$$\lambda(x) = \Lambda \quad (x \in \mathbb{R}^m),$$

where Λ is a non-negative finite random variable, the corresponding Cox process is also called a *mixed Poisson process.*

The Covariance Formula

Let N be a Poisson process on $\mathbb{R}^m$, with intensity measure ν. Recall Campbell's theorem (Theorem 10.2.15). Let $\varphi : \mathbb{R}^m \to \bar{\mathbb{R}}$ be a ν-integrable measurable function. Then $N(\varphi)$ is a well-defined *integrable* random variable, and

$$E\left[\int_{\mathbb{R}^m} \varphi(x)\,N(\mathrm{d}x)\right] = \int_{\mathbb{R}^m} \varphi(x)\,\nu(\mathrm{d}x). \tag{10.9}$$

Theorem 10.3.6 *Let N be as above. Let $\varphi, \psi : E \to \mathbb{C}$ be two ν-integrable measurable functions such that moreover $|\varphi|^2$ and $|\psi|^2$ are ν-integrable. Then $N(\varphi)$ and $N(\psi)$ are well-defined* square-integrable *random variables and*

$$\mathrm{cov}\left(\int_{\mathbb{R}^m} \varphi(x)\,N(\mathrm{d}x), \int_{\mathbb{R}^m} \psi(x)\,N(\mathrm{d}x)\right) = \int_{\mathbb{R}^m} \varphi(x)\psi(x)^*\,\nu(\mathrm{d}x). \tag{10.10}$$

Proof. It is enough to consider the case of real functions. First suppose that φ and ψ are *simple* non-negative Borel functions. We can always assume that

$$\varphi := \sum_{h=1}^{K} a_h 1_{C_h} \,, \quad \psi := \sum_{h=1}^{K} b_h 1_{C_h} ,$$

where $C_1, \dots, C_K$ are *disjoint* measurable subsets of $\mathbb{R}^m$. In particular, $\varphi(x)\psi(x) = \sum_{h=1}^{K} a_h b_h 1_{C_h}(x)$. Using the facts that if $i \neq j$, $N(C_i)$ and $N(C_j)$ are independent, and that a Poisson random variable with mean θ has variance θ,

$$\begin{aligned}
E[N(\varphi)N(\psi)] &= \sum_{h,l=1}^{K} a_h b_l E[N(C_h)N(C_l)] \\
&= \sum_{\substack{h,l=1\\ h\neq l}}^{K} a_h b_l E[N(C_h)N(C_l)] + \sum_{l=1}^{K} a_l b_l E[N(C_l)^2] \\
&= \sum_{\substack{h,l=1\\ h\neq l}}^{K} a_h b_l E[N(C_h)]E[N(C_l)] + \sum_{l=1}^{K} a_l b_l E[N(C_l)^2] ,
\end{aligned}$$

and therefore

$$\begin{aligned}
E[N(\varphi)N(\psi)] &= \sum_{\substack{h,l=1\\ h\neq l}}^{K} a_h b_l \nu(C_h)\nu(C_l) + \sum_{l=1}^{k} a_l b_l [\nu(C_l) + \nu(C_l)^2] \\
&= \sum_{h,l=1}^{k} a_h b_l \nu(C_h)\nu(C_l) + \sum_{l=1}^{k} a_l b_l \nu(C_l) \\
&= \nu(\varphi)\nu(\psi) + \nu(\varphi\psi) .
\end{aligned}$$

Let now φ, ψ be non-negative and let $\{\varphi_n\}_{n\geq 1}$, $\{\psi_n\}_{n\geq 1}$ be non-decreasing sequences of simple non-negative functions, with respective limits φ and ψ. Letting n go to ∞ in the equality

$$E[N(\varphi_n)N(\psi_n)] = \nu(\varphi_n\psi_n) + \nu(\varphi_n)\nu(\psi_n)$$

yields the announced results, by monotone convergence.

We have that for any ν-integrable function $\varphi : E \to \mathbb{C}$

$$E\left[N(\varphi)\right] = E\left[N(\varphi^+)\right] - E\left[N(\varphi^-)\right] = \nu(\varphi^+) - \nu(\varphi^-) = \nu(\varphi) .$$

Also by the result in the non-negative case, $E\left[N(|\varphi|)^2\right] = \nu(|\varphi|^2) + \nu(|\varphi|)^2 < \infty$. Therefore, since $|N(\varphi)| \leq N(|\varphi|)$, $N(\varphi)$ is a square-integrable variable, as well as $N(\psi)$ for the same reasons. Therefore, by Schwarz's inequality, $N(\varphi)N(\psi)$ is integrable. We have

$$\begin{aligned}
E[N(\varphi)N(\psi)] &= E\left[\left(N(\varphi^+) - N(\varphi^-)\right)\left(N(\psi^+) - N(\psi^-)\right)\right] \\
&= E\left[N(\varphi^+)N(\psi^+)\right] + E\left[N(\varphi^-)N(\psi^-)\right] \\
&\quad - E\left[N(\varphi^+)N(\psi^-)\right] - E\left[N(\varphi^-)N(\psi^+)\right] \\
&= \left(\nu(\varphi^+\psi^+) + \nu(\varphi^+)\nu(\psi^+)\right) + \left(\nu(\varphi^-\psi^-) + \nu(\varphi^-)\nu(\psi^-)\right) \\
&\quad - \left(\nu(\varphi^+\psi^-) + \nu(\varphi^+)\nu(\psi^-)\right) - \left(\nu(\varphi^-\psi^+) + \nu(\varphi^-)\nu(\psi^+)\right) \\
&= \nu(\varphi\psi) + \nu(\varphi)\nu(\psi),
\end{aligned}$$

from which (10.10) follows. □

The Exponential Formula

We now turn to the *exponential formula* for Poisson processes.

Theorem 10.3.7 *Let N be a Poisson process on $\mathbb{R}^m$ with intensity measure ν. Let $\varphi : \mathbb{R}^m \to \overline{\mathbb{R}}$ be a* non-negative *measurable function. Then,*

$$E[\mathrm{e}^{-\int_{\mathbb{R}^m} \varphi(x)\, N(\mathrm{d}x))}] = \exp\left\{\int_{\mathbb{R}^m} (\mathrm{e}^{-\varphi(x)} - 1)\, \nu(\mathrm{d}x)\right\}$$

and

$$E[\mathrm{e}^{\int_{\mathbb{R}^m} \varphi(x)\, N(\mathrm{d}x))}] = \exp\left\{\int_{\mathbb{R}^m} (\mathrm{e}^{\varphi(x)} - 1)\, \nu(\mathrm{d}x)\right\}.$$

Proof. We prove the first formula, the proof of the second being similar. Suppose that φ is simple and non-negative: $\varphi = \sum_{h=1}^{K} a_h 1_{C_h}$ where $C_1, \dots, C_K$ are mutually disjoint measurable subsets of $\mathbb{R}^m$. Then

$$\begin{aligned}
E[\mathrm{e}^{-N(\varphi)}] &= E\left[\mathrm{e}^{-\sum_{h=1}^{K} a_h N(C_h))}\right] = E\left[\prod_{h=1}^{K} \mathrm{e}^{-a_h N(C_h)}\right] \\
&= \prod_{h=1}^{K} E\left[\mathrm{e}^{-a_h N(C_h)}\right] = \prod_{h=1}^{K} \exp\left\{(\mathrm{e}^{-a_h} - 1)\nu(C_h)\right\} \\
&= \exp\left\{\sum_{h=1}^{K} (\mathrm{e}^{-a_h} - 1)\nu(C_h)\right\} = \exp\left\{\nu(\mathrm{e}^{-\varphi} - 1)\right\}.
\end{aligned}$$

The formula is therefore true for non-negative simple functions. Take now a non-decreasing sequence $\{\varphi_n\}_{n\geq 1}$ of such functions converging to φ. For all $n \geq 1$,

$$E[\mathrm{e}^{-N(\varphi_n)}] = \exp\left\{\nu(\mathrm{e}^{-\varphi_n} - 1)\right\}.$$

By monotone convergence, the limit as n tends to ∞ of $N(\varphi_n)$ is $N(\varphi)$. Consequently, by dominated convergence, the limit of the left-hand side is $E[\mathrm{e}^{-N(\varphi)}]$. The function $g_n = -(\mathrm{e}^{-\varphi_n} - 1)$ is a non-negative function increasing to $g = -(\mathrm{e}^{-\varphi} - 1)$, and therefore, by monotone convergence, $\nu(\mathrm{e}^{-\varphi_n} - 1) = -\nu(g_n)$ converges to $\nu(\mathrm{e}^{-\varphi} - 1) = -\nu(g)$, which in turn implies that the right-hand side of the last displayed equality tends to $\exp\{\nu(\mathrm{e}^{-\varphi} - 1)\}$ as n tends to ∞. □

The covariance formula can of course be obtained from the exponential formula by differentiation of $t \mapsto E\left[\mathrm{e}^{-tN(\varphi)}\right]$.

EXAMPLE 10.3.8: THE MAXIMUM FORMULA. Let N be a simple Poisson process on $\mathbb{R}^m$ with intensity measure ν and let $\varphi : E \to \mathbb{R}$. Then

$$P(\sup_{n\in\mathbb{N}} \varphi(X_n) \le a) = \exp\left\{-\int_{\mathbb{R}^m} 1_{\{\varphi(x)>a\}}\nu(\mathrm{d}x)\right\}.$$

A direct proof based on the construction of Poisson processes in Section 10.3 is possible (Exercise 10.5.21). We take another path and first prove that

$$\lim_{\theta\uparrow\infty} E\left[\mathrm{e}^{-\theta\sum_{n\in\mathbb{N}} 1_{\{\varphi(X_n)>a\}}}\right] = P(\sup_{n\in\mathbb{N}} \varphi(X_n) \le a). \tag{$\star$}$$

Indeed, the sum $\sum_{n\in\mathbb{N}} 1_{\{\varphi(X_n)>a\}}$ is strictly positive, except when $\sup_{n\in\mathbb{N}} \varphi(X_n) \le a$, in which case it is null. Therefore

$$\lim_{\theta\uparrow\infty} \mathrm{e}^{-\theta\sum_{n\in\mathbb{N}} 1_{\{\varphi(X_n)>a\}}} = 1_{\{\sup_{n\in\mathbb{N}} \varphi(X_n)\le a\}}.$$

Taking expectations yields ($\star$), by dominated convergence. Now, by Theorem 10.3.7,

$$\begin{aligned} E\left[\mathrm{e}^{-\theta\sum_{n\in\mathbb{N}} 1_{\{\varphi(X_n)>a\}}}\right] &= \exp\left\{\int_{\mathbb{R}^m} \left(\mathrm{e}^{-\theta 1_{\{\varphi(x)>a\}}} - 1\right)\nu(\mathrm{d}x)\right\} \\ &= \exp\left\{\int_{\mathbb{R}^m} \left(\mathrm{e}^{-\theta} - 1\right) 1_{\{\varphi(x)>a\}}\nu(\mathrm{d}x)\right\} \end{aligned}$$

and the limit of the latter quantity as $\theta \uparrow \infty$ is $\exp\left\{-\int_{\mathbb{R}^m} 1_{\{\varphi(x)>a\}}\nu(\mathrm{d}x)\right\}$.

EXAMPLE 10.3.9: THE LAPLACE FUNCTIONAL OF A POISSON PROCESS. According to Theorem 10.3.7, the Laplace functional of a Poisson process N on $\mathbb{R}^m$ with intensity measure ν is

$$L_N(\varphi) = \exp\left\{\nu\left(\mathrm{e}^{-\varphi} - 1\right)\right\}.$$

Theorem 10.3.10 *Let N_i $(i \in J)$ be a finite collection of simple point processes on $\mathbb{R}^m$. If for any collection $\varphi_i : E \to \mathbb{R}_+$ $(i \in J)$ of non-negative measurable functions,*

$$E\left[e^{-\sum_{i\in J} N_i(\varphi_i)}\right] = \prod_{i\in J} \exp\left\{\int_{\mathbb{R}^m} \left(e^{-\varphi_i(x)} - 1\right) \nu_i(dx)\right\}, \tag{10.11}$$

where ν_i, $i \in J$, is a collection of σ-finite measures on $\mathbb{R}^m$, then N_i, $i \in J$, is a family of independent Poisson processes with respective intensity measures ν_i, $i \in J$.

Proof. Taking all the φ_i's identically null except the first one, we have

$$E\left[e^{-N_1(\varphi_1)}\right] = \exp\left\{\int_{\mathbb{R}^m} \left(e^{-\varphi_1(x)} - 1\right) \nu_1(dx)\right\},$$

and therefore N_1 is a Poisson process with intensity measure ν_1. Similarly, for any $i \in J$, N_i is a Poisson process with intensity measure ν_i. Independence follows from Theorem 10.2.22. □

Marked Spatial Poisson Processes

Let

(α) N be a simple and locally finite process on $\mathbb{R}^m$, with point sequence $\{X_n\}_{n\in\mathbb{N}}$, and

(β) $\{Z_n\}_{n\in\mathbb{N}}$ be a sequence of random elements taking their values in the measurable space $(K, \mathcal{K}) := (\mathbb{R}^d, \mathcal{B}(\mathbb{R}^d))$ for some integer $d \geq 1$.

The sequence $\{X_n, Z_n\}_{n\in\mathbb{N}}$ is a marked point process, with the interpretation that Z_n is the *mark* associated with the *point* X_n. N is the *base* point process of the marked point process, and $\{Z_n\}_{n\in\mathbb{N}}$ is the associated *sequence of marks*. One also calls N a simple and locally finite point process on $\mathbb{R}^m$ with marks $\{Z_n\}_{n\in\mathbb{N}}$ in K. If moreover

(1) N is a Poisson process with intensity measure ν,

(2) $\{Z_n\}_{n\in\mathbb{N}}$ is an IID sequence, and

(3) $\{Z_n\}_{n\in\mathbb{N}}$ and N are independent,

the corresponding marked point process is called a Poisson process on $\mathbb{R}^m$ *with independent* IID *marks*. This model can be slightly generalized by allowing the mark distribution to depend on the location of the marked point. More precisely, we replace (2) and (3) by

(2') $\{Z_n\}_{n\in\mathbb{N}}$ is, conditionally on N, an independent sequence,

(3') given X_n, the random vector Z_n is independent of X_k ($k \in \mathbb{N}, k \neq n$), and

(4') for all $n \in \mathbb{N}$ and all $L \in \mathcal{K}$,

$$P(Z_n \in L \mid X_n) = Q(X_n, L),$$

where $Q(\cdot,\cdot)$ is a stochastic kernel from $(\mathbb{R}^m, \mathcal{B}(\mathbb{R}^m))$ to $(K, \mathcal{K})$, that is, Q is a function from $\mathbb{R}^m \times \mathcal{K}$ to $[0,1]$ such that for all $L \in \mathcal{K}$ the map $x \mapsto Q(x, L)$ is measurable, and for all $x \in \mathbb{R}^m$, $Q(x,\cdot)$ is a probability measure on $(K, \mathcal{K})$.

Theorem 10.3.11 *Let $\{X_n, Z_n\}_{n\in\mathbb{N}}$ be as in (α) and (β) above, and define the point process $\widetilde{N}$ on $\mathbb{R}^m \times K$ by*

$$\widetilde{N}(A) = \sum_{n\in\mathbb{N}} 1_A(X_n, Z_n) \quad (A \in \mathcal{B}(\mathbb{R}^m) \otimes \mathcal{K}). \tag{10.12}$$

If conditions (1), (2'), (3'), and (4') above are satisfied, then $\widetilde{N}$ is a simple Poisson process with intensity measure $\widetilde{\nu}$ given by

$$\widetilde{\nu}(C \times L) = \int_C Q(x, L)\, \nu(\mathrm{d}x) \qquad (C \in \mathcal{B}(\mathbb{R}^m),\, L \in \mathcal{K}).$$

Proof. In view of Theorem 10.2.19, it suffices to show that the Laplace functional of $\widetilde{N}$ has the appropriate form, that is, for any non-negative measurable function $\widetilde{\varphi} : E \times K \to \mathbb{R}$,

$$E\left[\mathrm{e}^{-\widetilde{N}(\widetilde{\varphi})}\right] = \exp\left\{\int_{\mathbb{R}^m}\int_K \left(\mathrm{e}^{-\widetilde{\varphi}(t,z)} - 1\right) \widetilde{\nu}(\mathrm{d}t \times \mathrm{d}z)\right\}.$$

By dominated convergence,

$$E\left[\mathrm{e}^{-\widetilde{N}(\widetilde{\varphi})}\right] = E\left[\mathrm{e}^{-\sum_{n\in\mathbb{N}} \widetilde{\varphi}(X_n, Z_n)}\right] = \lim_{L\uparrow\infty} E\left[\mathrm{e}^{-\sum_{n\leq L} \widetilde{\varphi}(X_n, Z_n)}\right].$$

For the time being, fix a positive integer L. Then, taking into account assumptions (2') and (3'),

$$
\begin{aligned}
E\left[\mathrm{e}^{-\sum_{n\le L}\widetilde{\varphi}(X_n,Z_n)}\right] &= E\left[\prod_{n\le L}\mathrm{e}^{-\widetilde{\varphi}(X_n,Z_n)}\right] \\
&= E\left[E\left[\prod_{n\le L}\mathrm{e}^{-\widetilde{\varphi}(X_n,Z_n)} \mid X_j, j\le L\right]\right] \\
&= E\left[\mathrm{e}^{-\sum_{n\le L}\psi(X_n)}\right],
\end{aligned}
$$

where $\psi(x) := -\log\int_K \mathrm{e}^{-\widetilde{\varphi}(x,z)}Q(x,\mathrm{d}z)$, a non-negative function. Letting $L\uparrow\infty$, we have, by dominated convergence,

$$
\begin{aligned}
E\left[\mathrm{e}^{-\widetilde{N}(\widetilde{\varphi})}\right] &= E\left[\mathrm{e}^{-\sum_{n\in\mathbb{N}}\psi(X_n)}\right] = E\left[\mathrm{e}^{-N(\psi)}\right] \\
&= \exp\left\{\int_{\mathbb{R}^m}\left(\mathrm{e}^{-\psi(x)}-1\right)\nu(\mathrm{d}x)\right\} \\
&= \exp\left\{\int_{\mathbb{R}^m}\left[\int_K \mathrm{e}^{-\widetilde{\varphi}(x,z)}Q(x,\mathrm{d}z)-1\right]\nu(\mathrm{d}x)\right\} \\
&= \exp\left\{\int_{\mathbb{R}^m}\left[\int_K \left(\mathrm{e}^{-\widetilde{\varphi}(x,z)}-1\right)Q(x,\mathrm{d}z)\right]\nu(\mathrm{d}x)\right\} \\
&= \exp\left\{\int_{\mathbb{R}^m}\int_K \left(\mathrm{e}^{-\widetilde{\varphi}(x,z)}-1\right)\widetilde{\nu}(\mathrm{d}x\times\mathrm{d}z)\right\}.
\end{aligned}
$$

□

EXAMPLE 10.3.12: THE M/GI/∞ MODEL, TAKE 1. The model of this example is of interest in queueing theory and in the traffic analysis of communications networks. We adopt the queueing interpretation. Let N be an HPP on $\mathbb{R}$ with intensity λ, and $\{\sigma_n\}_{n\in\mathbb{Z}}$ be a sequence of random vectors taking their values in $\mathbb{R}_+$ with probability distribution Q. Assume moreover that $\{\sigma_n\}_{n\in\mathbb{Z}}$ and N are independent. The n-th event time of N, T_n, is the arrival time of the n-th customer, and σ_n is her service time request. Define the point process $\widetilde{N}$ on $\mathbb{R}\times\mathbb{R}_+$ by

$$
\widetilde{N}(C) = \sum_{n\in\mathbb{Z}} 1_C(T_n,\sigma_n)
$$

for all $C\in\mathcal{B}(\mathbb{R})\otimes\mathcal{B}(\mathbb{R}_+)$. According to Theorem 10.3.11, $\widetilde{N}$ is a simple Poisson process with intensity measure

$$
\widetilde{\nu}(\mathrm{d}t\times\mathrm{d}z) = \lambda\mathrm{d}t\times Q(\mathrm{d}z)\,.
$$

In the M/GI/∞ model,[2] a customer arriving at time T_n is immediately served, and therefore departs from the "system" at time $T_n + \sigma_n$. The number $X(t)$ of customers present in the system at time t is therefore given by the formula

$$X(t) = \sum_{n\in\mathbb{Z}} 1_{(-\infty,t]}(T_n) 1_{(t,\infty)}(T_n + \sigma_n)\,.$$

(The n-th customer is in the system at time t if and only if she arrived at time $T_n \le t$ and departed at time $T_n + \sigma_n > t$.)

Assume that the service times have finite expectation: $E[\sigma_1] < \infty$. Then, for all $t \in \mathbb{R}$, $X(t)$ is a Poisson random variable with mean $\lambda E[\sigma_1]$.

Proof. Observe that

$$X(t) = \widetilde{N}(C(t))\,,$$

where $C(t) := \{(s,\sigma);\ s \le t,\ s+\sigma > t\} \subset \mathbb{R} \times \mathbb{R}_+$. In particular, $X(t)$ is a Poisson random variable with mean

$$\begin{aligned}
\widetilde{\nu}(C(t)) &= \int_{\mathbb{R}}\int_{\mathbb{R}_+} 1_{\{s+\sigma>t\}} 1_{\{s\le t\}} \widetilde{\nu}(\mathrm{d}s \times d\sigma) \\
&= \int_{\mathbb{R}}\int_{\mathbb{R}_+} 1_{\{s+\sigma>t\}} 1_{\{s\le t\}} \lambda\,\mathrm{d}s \times Q(d\sigma) \\
&= \int_{\mathbb{R}} \left(\int_{\mathbb{R}_+} 1_{\{s+\sigma>t\}} Q(d\sigma) \right) 1_{\{s\le t\}} \lambda\,\mathrm{d}s \\
&= \lambda \int_{-\infty}^{t} Q((t-s, +\infty))\,\mathrm{d}s \\
&= \lambda \int_{0}^{\infty} Q((s, +\infty))\,\mathrm{d}s = \lambda \int_{0}^{\infty} P(\sigma_1 > s)\mathrm{d}s = \lambda E[\sigma_1]\,.
\end{aligned}$$

□

It can be shown that the *departure* process D of departure times, defined by

$$D(C) := \sum_{n\in\mathbb{Z}} 1_C(T_n + \sigma_n)\,,$$

is an HPP of intensity λ (Exercise 10.5.16).

[2] "∞" represents the number of servers. This model is sometimes called a "queueing" system, although in reality there is no queueing, since customers are served immediately upon arrival and without interruption. It is in fact a "pure delay" system.

Formulas such as Campbell's first formula and the Poisson exponential formula are straightforwardly extended to *marked* point processes.

In the situation prevailing in Theorem 10.3.11, consider sums of the type

$$\widetilde{N}(\widetilde{\varphi}) := \sum_{n\in\mathbb{N}} \widetilde{\varphi}(X_n, Z_n)\,, \tag{10.13}$$

for functions $\widetilde{\varphi} : \mathbb{R}^m \times K \to \mathbb{R}$. Note that, denoting by $Z_1(x)$ any random element of K with the distribution $Q(x, \mathrm{d}z)$,

$$\widetilde{\nu}(\widetilde{\varphi}) = \int_{\mathbb{R}^m}\int_K \widetilde{\varphi}(x,z)Q(x,\mathrm{d}z)\,\nu(\mathrm{d}x) = \int_{\mathbb{R}^m} E\left[\widetilde{\varphi}(x, Z_1(x))\right]\,\nu(\mathrm{d}x)\,,$$

whenever the quantities involved have a meaning. Using this observation, the formulas obtained in the previous subsection can be applied in terms of marked point processes. The corollaries below do not require proofs, since they are *reformulations* of previous results, namely Theorem 10.3.6 and Theorem 10.3.7.

Let $0 < p < \infty$. Recall that a measurable function $\widetilde{\varphi} : E \times K \to \mathbb{R}$ (resp. $\to \mathbb{C}$) is said to be in $L^p_{\mathbb{R}}(\widetilde{\nu})$ (resp. $L^p_{\mathbb{C}}(\widetilde{\nu})$) if

$$\int_{\mathbb{R}^m}\int_K |\widetilde{\varphi}(x,z)|^p\,\nu(\mathrm{d}x)\,Q(x,\mathrm{d}z) < \infty\,.$$

Corollary 10.3.13 *Suppose that $\widetilde{\varphi} \in L^1_{\mathbb{C}}(\widetilde{\nu})$. Then the sum (10.13) is well defined, and moreover*

$$E\left[\sum_{n\in\mathbb{N}} \widetilde{\varphi}(X_n, Z_n)\right] = \int_{\mathbb{R}^m} E\left[\widetilde{\varphi}(x, Z_1(x))\right]\nu(\mathrm{d}x)\,.$$

Let $\widetilde{\varphi}, \widetilde{\psi} : \mathbb{R} \times E \to \mathbb{C}$ be two measurable functions in $L^1_{\mathbb{C}}(\widetilde{\nu}) \cap L^2_{\mathbb{C}}(\widetilde{\nu})$. Then

$$\begin{aligned}\mathrm{cov}&\left(\sum_{n\in\mathbb{N}} \widetilde{\varphi}(X_n, Z_n), \sum_{n\in\mathbb{N}} \widetilde{\psi}(X_n, Z_n)\right)\\ &= \int_{\mathbb{R}^m} E\left[\widetilde{\varphi}(x, Z_1(x))\widetilde{\psi}(x, Z_1(x))^*\right]\,\nu(\mathrm{d}x)\,.\end{aligned}$$

Corollary 10.3.14 *Let $\widetilde{\varphi}$ be a non-negative function from $\mathbb{R}^m \times K$ to $\mathbb{R}$. Then,*

$$E\left[\mathrm{e}^{-\sum_{n\in\mathbb{N}} \widetilde{\varphi}(X_n, Z_n)}\right] = \exp\left\{\int_{\mathbb{R}^m} E\left[\mathrm{e}^{-\widetilde{\varphi}(x, Z_1(x))} - 1\right]\,\nu(\mathrm{d}x)\right\}\,.$$

10.4 Operations on Poisson Processes

The framework and the results concerning marked Poisson processes is especially convenient to study the effects of various operations on Poisson processes, such as thinning, coloring, transportation, translation and filtering.

Thinning and Coloring

Thinning is the operation of randomly erasing points of a Poisson process. It is a particular case of the independent coloring operation whereby the points of a Poisson process are independently colored with the result of obtaining independent Poisson processes, each one corresponding to a different color.

Theorem 10.4.1 *Consider the situation depicted in Theorem 10.3.11. Let I be an arbitrary index set and let $\{L_i\}_{i\in I}$ be a family of disjoint measurable sets of K. Define for each $i \in I$ the simple point process N_i on $\mathbb{R}^m$ by*

$$N_i(C) = \sum_{n\in\mathbb{N}} 1_C(X_n)1_{L_i}(Z_n)\,.$$

Then the family N_i $(i \in I)$ is an independent family of Poisson processes with respective intensity measures ν_i $(i \in I)$, where

$$\nu_i(\mathrm{d}x) = Q(x, L_i)\,\nu(\mathrm{d}x)\,.$$

Proof. According to the definition of independence, it suffices to consider a *finite* index set I. Define the simple point process $\widetilde{N}$ on $\mathbb{R}^m \times K$ as in (10.12). Then $\widetilde{N}$ is a Poisson process with intensity measure $\widetilde{\nu}(C \times L) = \int_C Q(x, L)\nu(\mathrm{d}x)$. Defining $\widetilde{\varphi}(x, z) = \sum_{i\in I} \varphi_i(x)1_{L_i}(z)$, we have $\sum_{i\in I} N_i(\varphi_i) = \widetilde{N}(\widetilde{\varphi})$. Therefore

$$\begin{aligned}
E\left[\mathrm{e}^{-\sum_{i\in I} N_i(\varphi_i)}\right] &= E\left[\mathrm{e}^{-\widetilde{N}(\widetilde{\varphi})}\right] \\
&= \exp\left\{\int_{\mathbb{R}^m}\int_K \left(\mathrm{e}^{-\widetilde{\varphi}(x,z)} - 1\right)\widetilde{\nu}(\mathrm{d}x \times \mathrm{d}z)\right\} \\
&= \exp\left\{\int_{\mathbb{R}^m}\int_K \left(\mathrm{e}^{-\widetilde{\varphi}(x,z)} - 1\right)Q(x, \mathrm{d}z)\nu(\mathrm{d}x)\right\} \\
&= \exp\left\{\int_{\mathbb{R}^m}\int_K \left(\mathrm{e}^{-\sum_{i\in I}\varphi_i(x)1_{L_i}(z)} - 1\right)Q(x, \mathrm{d}z)\nu(\mathrm{d}x)\right\} \\
&= \exp\left\{\int_{\mathbb{R}^m}\int_K \sum_{i\in I}\left(\mathrm{e}^{-\varphi_i(x)} - 1\right)1_{L_i}(z)Q(x, \mathrm{d}z)\nu(\mathrm{d}x)\right\}
\end{aligned}$$

$$= \exp\left\{\int_{\mathbb{R}^m} \sum_{i\in I} \left(e^{-\varphi_i(x)} - 1\right) Q(x, L_i)\nu(dx)\right\}$$

$$= \prod_{i\in I} \exp\left\{\int_{\mathbb{R}^m} \left(e^{-\varphi_i(x)} - 1\right) Q(x, L_i)\nu(dx)\right\}.$$

Therefore,

$$E\left[e^{-\sum_{i\in I} N_i(\varphi_i)}\right] = \prod_{i\in I} \exp\left\{\int_{\mathbb{R}^m} \left(e^{-(\varphi_i)} - 1\right) \nu_i(dx)\right\}$$

and the result follows from Theorem 10.3.10. □

The above theorem is indeed about thinning. For instance the point process N_1 is obtained by thinning of N, each point x of which being saved with probability $Q(x, L_1)$.

EXAMPLE 10.4.2: ERASURES. Let in this special case $K = \{0, 1\}$ and

$$P(Z_n = 1 \mid X_n = x) = p(x).$$

We shall now define the point processes N^p and $\overline{N}^p$ on $\mathbb{R}^m$ by

$$N^p(C) = \sum_{n\geq 1} Z_n 1_C(X_n) \text{ and } \overline{N}^p(C) = \sum_{n\geq 1} (1 - Z_n) 1_C(X_n).$$

The interpretation is that $\overline{N}^p$ is obtained from N by erasing points, a point of the original point process N located at x being erased with probability $p(x)$ independently of everything else.

By Theorem 10.4.1,his point t process is a Poisson process with intensity measure

$$\overline{\nu}^p(dx) := p(x)\nu(dx).$$

Transportation

This is the operation of moving the points of a Poisson process.

More precisely, consider the situation depicted in Theorem 10.3.11. Form a point process N^* on K by associating to a point $X_n \in \mathbb{R}^m$ a point $Z_n \in K$:

$$N^*(L) := \sum_{n\in\mathbb{N}} 1_L(Z_n),$$

where $L \in \mathcal{B}(\mathbb{R}^m)$. We then say that N^* is obtained by transporting N *via the stochastic kernel* $Q(x, \cdot)$.

Theorem 10.4.3 *N^* is a Poisson process on K with intensity measure ν^* given by*

$$\nu^*(L) = \int_{\mathbb{R}^m} \nu(\mathrm{d}x) Q(x, L) .$$

Proof. Let $\varphi^* : K \to \mathbb{R}$ be a non-negative measurable function. We have

$$\begin{aligned} E\left[\mathrm{e}^{-N^*(\varphi^*)}\right] &= E\left[\mathrm{e}^{-\sum_{n\in\mathbb{N}} \varphi^*(Z_n)}\right] \\ &= \exp\left\{\int_{\mathbb{R}^m}\int_K \left(\mathrm{e}^{-\varphi^*(z)} - 1\right) \nu(\mathrm{d}x) Q(x, \mathrm{d}z)\right\} \\ &= \exp\left\{\int_K \left(\mathrm{e}^{-\varphi^*(z)} - 1\right) \int_{\mathbb{R}^m} \nu(\mathrm{d}x) Q(x, \mathrm{d}z)\right\}. \end{aligned}$$

□

EXAMPLE 10.4.4: TRANSLATION. Let N be a Poisson process on $\mathbb{R}^m$ with intensity measure ν and let $\{V_n\}_{n\in\mathbb{N}}$ be an IID sequence random vectors of $\mathbb{R}^m$ with common distribution Q. Form the point process N^* on $\mathbb{R}^m$ by translating each point X_n of N by V_n. Formally,

$$N^*(C) = \sum_{n\in\mathbb{N}} 1_C(X_n + V_n).$$

We are in the situation of Theorem 10.4.3 with $Z_n = X_n + V_n$. In particular, $Q(x, A) = Q(A - x)$. It follows that N^* is a Poisson process on $\mathbb{R}^m$ with intensity measure

$$\nu^*(L) = \int_{\mathbb{R}^m} Q(L - x)\, \nu(\mathrm{d}x) ,$$

the convolution of ν and Q.

Poisson Shot Noise

Let N be a simple and locally finite point process on $\mathbb{R}^m$ with point sequence $\{X_n\}_{n\in\mathbb{N}}$ and with marks $\{Z_n\}_{n\in\mathbb{N}}$ in the measurable space $(K, \mathcal{K})$. Let $h : \mathbb{R}^m \times K \to \mathbb{C}$ be a measurable function. The complex-valued spatial stochastic process $\{X(y)\}_{y\in\mathbb{R}^m}$ given by

$$X(y) := \sum_{n\in\mathbb{N}} h(y - X_n, Z_n) , \tag{10.14}$$

where the right-hand side is assumed well defined (for instance, when h takes real non-negative values), is called a *spatial shot noise with random impulse response*. If N is a simple and locally finite Poisson process on $\mathbb{R}^m$ with independent IID marks $\{Z_n\}_{n\in\mathbb{N}}$, $\{X(y)\}_{y\in\mathbb{R}^m}$ is called a Poisson spatial shot noise *with random impulse response and independent* IID *marks*.

The following result is a direct application of Theorems 10.3.6 and 10.3.11.

Theorem 10.4.5 *Consider the above Poisson spatial shot noise with random impulse response and independent* IID *marks. Suppose that for all $y \in \mathbb{R}^m$,*

$$\int_{\mathbb{R}^m} E\left[|h(y-x, Z_1)|\right] \nu(\mathrm{d}x) < \infty$$

and

$$\int_{\mathbb{R}^m} E\left[|h(y-x, Z_1)|^2\right] \nu(\mathrm{d}x) < \infty .$$

Then the complex-valued spatial stochastic process $\{X(y)\}_{y\in\mathbb{R}^m}$ given by (10.14) is well defined, and for any $y, \xi \in \mathbb{R}^m$, we have

$$E\left[X(y)\right] = \int_{\mathbb{R}^m} E\left[h(y-x, Z_1)\right] \nu(\mathrm{d}x)$$

and

$$\operatorname{cov}\left(X(y+\xi), X(y)\right) = \int_{\mathbb{R}^m} E\left[h(y-x, Z_1)h^*(y+\xi-x, Z_1)\right] \nu(\mathrm{d}t) .$$

In the case where the base point process N is an HPP with intensity λ, we find that

$$E\left[X(y)\right] = \lambda \int_{\mathbb{R}^m} E\left[h(x, Z_1)\right] \mathrm{d}x$$

and

$$\operatorname{cov}\left(X(y+\xi), X(y)\right) = \lambda \int_{\mathbb{R}^m} E\left[h(x, Z_1)h^*(\xi+x, Z_1)\right] \mathrm{d}x .$$

Observe that these quantities do not depend on $y \in \mathbb{R}^m$. The process $\{X(y)\}_{y\in\mathbb{R}^m}$ is therefore a *wide-sense stationary process* (see Chapter 12 for a definition).

10.5 Exercises

Exercise 10.5.1. BACKWARD AND FORWARD RECURRENCE TIMES
Let $\{T_n\}_{n\in\mathbb{Z}}$ be the sequence of event times of an HPP on $\mathbb{R}$ with the intensity $\lambda > 0$.

For fixed $t \in \mathbb{R}$, define the backward and forward recurrence times respectively by

$$B(t) = \inf\{t - T_n\,;\, T_n \le t\}$$
$$F(t) = \inf\{T_n - t\,;\, T_n > t\}$$

What is the distribution of the vector $(B(t), F(t))$? Compute $E[B(t) + F(t)]$.

Exercise 10.5.2. POISSON AND MULTINOMIAL
Let N be a HOMOGENEOUS Poisson process on $\mathbb{R}^m$ with intensity λ. Let $C_1, \ldots,$ C_K be disjoint bounded measurable sets of $\mathbb{R}^m$, and call C their union. Let n be an integer. What is the conditional distribution of the vector $(N(C_1), \ldots, N(C_K))$ given that $N(C) = n$?

Exercise 10.5.3. POISSON UNDER THE LINE
A. Let $\tilde{N}$ be an HPP on $\mathbb{R}^2$ with intensity 1. Let $\lambda : \mathbb{R} \to \mathbb{R}_+$ be a non-negative locally integrable function. Define a point process N on $\mathbb{R}$ as follows. The point $t \in N$ if and only if there exists a $z \in \mathbb{R}$ such that $0 \le z \le \lambda(t)$ and $(t, z) \in \tilde{N}$. Prove that N is a Poisson process on $\mathbb{R}$ with intensity function $\lambda(t)$.

B. Let N be a Poisson process on $\mathbb{R}$ with intensity function $\lambda(t)$. Denoting by T_n the n-th point of N strictly to the right of the origin, prove that T_n is an absolutely continuous random variable and give its probability density. Give an expression for the joint density of (T_n, S_{n+1}).

Exercise 10.5.4. POISSONIAN DISKS
Let N be a homogeneous Poisson process on $\mathbb{R}^2$, of intensity λ. Draw around each point $x \in N$ a closed disk of radius a. Let $X\,(y)$ be the number of disks covering $y \in \mathbb{R}^2$.

1) Compute for $y \in \mathbb{R}^2$, $\theta \in \mathbb{R}_+$

$$E\left[e^{-\theta X(y)}\right];$$

2) Deduce from this result the probability distribution of $X\,(y)$;

3) Give the average area inside the square $[0, T] \times [0, T]$ that is not covered by a disk;

4) This area is delimited by a curve. Give its average length (excluding the parts on the boundaries of $[0, T] \times [0, T]$).

Exercise 10.5.5. LINE OF SIGHT
Consider a Poisson N on $\mathbb{R}^2$ with diffuse and locally finite mean measure ν. There is a random shape centered around each of its points. Let the generic shape S

be distributed according to some probability distribution Q_S. Now consider two arbitrary points A, B. We say that A and B can communicate if the line connecting A and B does not intersect any of the existing shapes around the points of the point process (for all $n \geq 1$, the "existing shape around" $X_n \in N$ is $X_n + S_n$, that is S_n translated by X_n, where S_n is distributed according to Q_S). We assume that $\{S_n\}_{n\geq 1}$ is an IID sequence independent of N. What is the probability that A and B can successfully communicate? Keep the calculations as general as possible, and then, give the explicit result when N is an HPP of intensity λ and when the shape is (1): a circle of fixed radius a; (2) a circle of random radius uniformly distributed on $[0, 1]$.

Exercise 10.5.6. CELLPHONES
Consider two independent Poisson processes N_1 and N_2 on $\mathbb{R}^m$ with respective mean measures ν_1 and ν_2. Assume that $\nu_i(\mathbb{R}^m) < \infty$, $i = 1, 2$. Compute the average number of elements in N_1 that see no point of N_2 with distance a.

Exercise 10.5.7. MUTUALLY SINGULAR
Let N be a point process on $\mathbb{R}$ defined on a measurable space $(\Omega, \mathcal{F})$. Let P_1 and P_2 be two probability measures on $(\Omega, \mathcal{F})$ that make of N an HPP of intensity $\lambda_1 > 0$ and $\lambda_2 > 0$ respectively.

Show that if $\lambda_1 \neq \lambda_2$, P_1 and P_2 are mutually singular, that is to say, that there exists a set $A \in \mathcal{F}$ such that $P_1(A) = 1$ and $P_2(\overline{A}) = 0$.

Exercise 10.5.8. COUPLED HPPS
Let for $i = 1, 2$, $\lambda_i : \mathbb{R}_+ \to \mathbb{R}$ be a non-negative measurable function, locally integrable. Suppose that

$$\int_0^\infty |\lambda_1(t) - \lambda_2(t)|\, dt < \infty .$$

Show that one can construct *on the same probability space* $(\Omega, \mathcal{F}, P)$ two Poisson processes on $\mathbb{R}_+$, with respective intensity functions $\lambda_1(t)$ and $\lambda_2(t)$, with the following coupling property:

There exists an almost surely finite random variable τ such that

$$P(N_1(C \cap [\tau, \infty)) = N_2(C \cap [\tau, \infty)), \text{ for all } C \in \mathcal{B}(\mathbb{R}_+)) = 1 .$$

What is the probability distribution of the last point Z of either N_1 or N_2 that is not a shared point?

Exercise 10.5.9. GAUSSIAN LIMIT OF A SHOT NOISE
Consider the shot noise process $\{X(t)\}_{t\in\mathbb{R}}$ given by

$$X(t) = \sum_{n\in\mathbb{Z}} h(t - T_n),$$

where $\{T_n\}_{n\in\mathbb{Z}}$ is an HPP on $\mathbb{R}$ with intensity $\lambda = n\lambda_0$ and $h(t) = \frac{1}{\sqrt{n}}h_0(t)$ for some integrable function $h_0(t)$ such that $\int_{\mathbb{R}} h_0(t)\,\mathrm{d}t = 0$. Show that the finite distributions of $\{X(t)\}_{t\in\mathbb{R}}$ converge as $n \uparrow \infty$ to the finite distributions of a centered Gaussian process $\{Y(t)\}_{t\in\mathbb{R}}$ with covariance function $E[Y(t+\tau)Y(t)] = \lambda_0 \int_{\mathbb{R}} h_0(s+\tau)h_0(t)\,\mathrm{d}t$.

Exercise 10.5.10. PROVE THEOREM 10.2.22
Prove Theorem 10.2.22.

Exercise 10.5.11. DROPPING HUMANITARIAN PARCELS
Parcels are dropped on the plane $\mathbb{R}^2$. The impact times $\{T_n\}_{n\in\mathbb{Z}}$ form a simple Poisson process of mean measure ν, the impact locations $\{Z_n\}_{n\in\mathbb{Z}}$ are IID and independent of the impact times, and their common probability distribution is Q. A "shape" moves on $\mathbb{R}^2$ in order to collect the parcels as they impact on it. More precisely, there is for each time $t \in \mathbb{R}$ a measurable subset $S(t) \in \mathcal{B}(\mathbb{R}^2)$ and the point process $\hat{N}$ counting the parcels falling on the shape is defined by

$$\hat{N}(C) = \sum_{n\in\mathbb{Z}} 1_C(T_n)1_{S(t)}(Z_n).$$

Prove that $\hat{N}$ is a Poisson process and give its mean measure.

Exercise 10.5.12. POISSONIAN DISC CLUTTERS
Let N be a homogeneous Poisson process on $\mathbb{R}^2$, of intensity λ. Draw around each point $x \in N$ a closed disk of radius a. Let $X(y)$ be the number of disks covering $y \in \mathbb{R}^2$.

1) Compute for $y \in \mathbb{R}^2$, $\theta \in \mathbb{R}_+$

$$E\left[\mathrm{e}^{-\theta X(y)}\right];$$

2) Deduce from this result the probability distribution of $X(y)$;

3) Give the average surface inside the square $[0,T] \times [0,T]$ that is not covered by a disk;

4) This surface is delimited by a curve. Give its average length (excluding the parts on the boundaries of $[0,T] \times [0,T]$).

Exercise 10.5.13. RANDOM POINTS UNIFORMLY DISTRIBUTED ON $[0,1]$
Construct a point process N on $\mathbb{R}$ in the following way. First draw a finite integer-valued random variable T, and then an IID sequence $\{U_n\}_{n\geq 1}$ uniformly distributed on $[0,1]$, independent of T. Define $\alpha_k := P(T=k)$ $(k \geq 0)$. Finally, let $N = \sum_{k=1}^{T} \varepsilon_{U_k}$, where ε_a is the Dirac measure at a, and where $\sum_{k=1}^{0} \varepsilon_{U_k}$ is the null measure by convention. What is the Laplace functional of N? What about the case where T is a Poisson variable of mean θ?

Exercise 10.5.14. LAPLACE FUNCTIONAL OF A CONTRACTED POINT PROCESS
Let N be a simple point process on $\mathbb{R}^m$ with point sequence $\{X_n\}_{n\in\mathbb{N}}$ and let $\alpha > 0$. Define the "contracted"[3] point process $N_{c,\alpha}$ defined by its sequence of points $\{\alpha X_n\}_{n\in\mathbb{N}}$. Prove that its Laplace functional is

$$L_{N_{c,\alpha}}(\varphi) = E\left[\exp\left(-\sum_{n\in\mathbb{Z}} \varphi(\alpha X_n)\right)\right] = L_N(\varphi(\alpha\,\cdot)).$$

Exercise 10.5.15. DISTRIBUTION OF THE MAXIMUM INTERFERENCE
Let N be a homogeneous Poisson process on $\mathbb{R}^m$ of positive intensity λ and with point sequence $\{X_n\}_{n\geq 1}$. Let $\{Z_n\}_{n\geq 1}$ be an IID sequence of real non-negative random variables with common distribution Q, and independent of N. Compute the distribution of the random variable

$$\max_{n\geq 1} Z_n \mathrm{e}^{-\beta\|X_n\|} \qquad (\beta > 0).$$

(The title of the exercise refers to mobile communications: Z_n is the noise intensity generated at point X_n, and $\mathrm{e}^{-\beta\|X_n\|}$ is an attenuation factor for a receiver located at 0.)

Exercise 10.5.16. THE M/GI/∞ MODEL, TAKE 2
In Example 10.3.12,

(i) prove that the departure process is a homogeneous Poisson process with intensity λ,

(ii) compute $\mathrm{cov}(X(t), X(t+\tau))$ for all $t, \tau \in \mathbb{R}$, $\tau \geq 0$, and

(iii) interpret the process $\{X(t)\}_{t\in\mathbb{R}}$ as a shot noise in order to obtain the results of Example 10.3.12, and of (i) and (ii), from the general results of Section 10.3 (subsection *Marked Spatial Poisson Processes*, page 403).

[3] Of course, if $\alpha > 1$, it is in fact dilated...

Exercise 10.5.17. LIFTING

Let N be a Poisson process on $\mathbb{R}$ with (locally integrable) intensity function $\lambda : \mathbb{R} \to \mathbb{R}$. Let $\{T_n\}_{n\in\mathbb{Z}}$ be its sequence of points, and let $\{U_n\}_{n\in\mathbb{Z}}$ be an IID sequence of random variables uniformly distributed on $[0, 1]$. Let $\widehat{N}$ be an HPP on $\mathbb{R} \times \mathbb{R}_+$, with intensity 1, independent of N and of $\{U_n\}_{n\in\mathbb{Z}}$. Define a point process $\widetilde{N}$ on $\mathbb{R} \times \mathbb{R}_+$ by

$$\widetilde{N}(C) := \sum_{n\in\mathbb{Z}} 1_C((T_n, U_n\lambda(T_n))) + \widehat{N}(C \cap \bar{H}),$$

where

$$H := \{(t, z) \in \mathbb{R} \times \mathbb{R}_+ ; 0 \le z \le \lambda(t)\}.$$

Show that $\widetilde{N}$ is an HPP on $\mathbb{R} \times \mathbb{R}_+$ with intensity 1.

Exercise 10.5.18. WATER BOMBS

You are initially located at the origin $(0, 0)$ of the plane at which is centered a disk D of radius R. You run in a straight line from the origin to the "shelter point" $(0, R)$ at constant speed v. The reason why you are running is that water bombs are being dropped on the disk D. The times of impact form an HPP of intensity λ, and each impact is located independently of all the rest, uniformly on the disk. You will get wet if the impact of the bomb is within distance a of your position at the time of impact. Once arrived at the shelter point $(0, R)$, the bombing stops. What are your chances of not getting wet? Given that you did get wet, what is the expected time that you remained dry?

Exercise 10.5.19. SMOKING POT AT SAINT MARY-JANE'S

Smoking pot was recently banned on the Saint Mary-Jane's college campus. The authorities noticed that the violators of the ban make use of a restroom in a secluded wing of the campus. They consequently devised a strategy to send "cops" to capture the culprits. Assume that the schoolboys' arrival times in the restroom premises form a Poisson process with independent IID marks. Let τ_n denote the n-th arrival time of a schoolboy in the pot sanctuary (the restrooms) and let σ_n be the time he spends smoking. Cops also form a Poisson process with independent IID marks. Denote the k-th arrival time of a cop on the potential crime scene by T_k and by S_k the lingering time there of the corresponding representative of the college authority. The probability distribution of σ is Q_s and that of S is Q_c. Assuming the point processes of students and of the cops to be HPPs with respective intensities $\lambda_s > 0$ and $\lambda_c > 0$, compute the average number of students caught per unit of time.

Exercise 10.5.20. LAPLACE FUNCTIONAL OF A HOMOGENEOUS COX PROCESS
Let N be a Cox process on $\mathbb{R}^m$ with constant intensity process, that is, $\nu(\mathrm{d}x) := \Lambda\ell^m(\mathrm{d}x)$, where ℓ^m is the Lebesgue measure on $\mathbb{R}^m$ and Λ is a non-negative random variable with Laplace transform $L_\Lambda(t) := E\left[\mathrm{e}^{-t\Lambda}\right]$. Show that its Laplace functional is

$$L_N(\varphi) = L_\Lambda\left(\int_{\mathbb{R}^m} (1 - \mathrm{e}^{-\varphi(x)})\,\mathrm{d}x\right).$$

Exercise 10.5.21. THE MAXIMUM FORMULA
Give a direct proof of the result of Example 10.3.8 based on the construction of Section 10.3.

Exercise 10.5.22. LIKELIHOOD RATIO
Let N be an HPP on $\mathbb{R}$ of intensity 1, and let Λ be a non-negative random variable, both defined on the same probability space $(\Omega, \mathcal{F}, P)$. Let $T > 0$ be a fixed real number. Define

$$L(T) = \Lambda^{N(T)} \exp(-(\Lambda - 1)T).$$

(1) Show that $E[L(T)] = 1$

(2) Define a probability Q on $(\Omega, \mathcal{F})$ by $Q(A) = E[L(T)1_A]$. Show that under Q, N restricted to the interval $[0, T]$ is a Cox process with intensity $\lambda(t) = \Lambda$.

(3) Show that for $t \in [0, T]$,

$$E_Q[\Lambda|\mathcal{F}_t^N] = \frac{\varphi(N(t) + 1, t)}{\varphi(N(t), t)}$$

where

$$\varphi(n, t) = \int_{\mathbb{R}_+} \lambda^n \mathrm{e}^{-(\lambda-1)t}\,\mathrm{d}F(\lambda),$$

is the CDF of Λ.

Chapter 11

Brownian Motion

Brownian motion owes its name to the botanist Robert Brown who observed the chaotic motion of pollen grains in a liquid. From the mathematical point of view, it received attention from Albert Einstein and Louis Bachelier. The latter was motivated by his interest in finance, finding that the model could serve to describe the fluctuations of the stock market, and nowadays, its role in mathematical finance is well established. Brownian motion is also called the Wiener process, after Norbert Wiener, who introduced it in the theory of stochastic systems driven by white noise, a notion that we shall discuss in the next chapter.

11.1 Continuous-time Stochastic Processes

Some generalities on continuous stochastic processes are necessary before addressing the central topic of this chapter.

Definition 11.1.1 *A* stochastic process (*or* random process) *is a family* $\{X(t)\}_{t\in\mathbf{T}}$ *of random variables taking their values in some measurable space* $(E,\mathcal{E})$ *and defined on the same probability space* $(\Omega,\mathcal{F},P)$.

(The spaces E of interest in this chapter are $\mathbb{R}^m$ ($m\geq 1$), $\mathbb{C}$, $\mathbb{Z}$ and $\mathbb{N}$.)

It is called a *real* (resp., *complex*) stochastic process if it takes real (resp., complex) values, a *continuous-time* stochastic process when the index set $\mathbf{T}$ is $\mathbb{R}$ or $\mathbb{R}_+$, and a *discrete-time* stochastic process when it is $\mathbb{N}$ or $\mathbb{Z}$. When the index set is $\mathbb{N}$ or $\mathbb{Z}$, we also use the notation n instead of t for the time index, and write X_n instead of $X(t)$.

For each $\omega\in\Omega$, the function $t\mapsto X(t,\omega)$ is called a *trajectory* (more precisely, the ω-trajectory). This is why a stochastic process is sometimes called a *random function.*

P. Brémaud, *An Introduction to Applied Probability*, Texts in Applied Mathematics 77,
https://doi.org/10.1007/978-3-031-49306-5_11

EXAMPLE 11.1.2: RANDOM SINUSOID. Let A be some real non-negative random variable, let $\nu_0 \in \mathbb{R}$ be a positive constant and let Φ be a random variable with values in $[0, 2\pi]$. The formula

$$X(t) = A \sin(2\pi\nu_0 t + \Phi)$$

defines a stochastic process. For each sample $\omega \in \Omega$, the function $t \mapsto X(t, \omega)$ is a sinusoid with frequency ν_0, random amplitude $A(\omega)$ and random phase $\Phi(\omega)$.

One way of describing the probabilistic behavior of a stochastic process is by means of its finite-dimensional distribution.

Definition 11.1.3 *The* finite-dimensional (fidi) distribution *of a stochastic process $\{X(t)\}_{t\in\mathbf{T}}$ is the collection of probability distributions of the random vectors*

$$(X(t_1), \dots, X(t_k)) \quad (k \geq 1, t_1, \dots, t_k \in \mathbf{T}).$$

Definition 11.1.4 *A stochastic process $\{X(t)\}_{t\in\mathbb{R}}$ is said to be* stationary *iff for all $k \geq 1$ and all $t_1, \dots, t_k \in \mathbb{R}$ the probability distribution of the random vector*

$$(X(t_1 + \tau), \dots, X(t_k + \tau))$$

is independent of τ.

Definition 11.1.5 *A complex stochastic process $\{X(t)\}_{t\in\mathbb{R}}$ is said to have* independent increments *if for all $n \geq 2$ and for all mutually disjoint intervals $(a_1, b_1]$,..., $(a_n, b_n]$ of $\mathbb{R}$, the random variables*

$$X(b_1) - X(a_1), \dots, X(b_n) - X(a_n)$$

are independent.

It is sometimes useful to view a stochastic process as a *mapping* $X : \mathbf{T}\times\Omega \to E$, defined by $(t, \omega) \mapsto X(t, \omega)$.

Definition 11.1.6 *The stochastic process $\{X(t)\}_{t\in\mathbb{R}}$ is said to be* measurable *iff the mapping from $\mathbb{R} \times \Omega$ into E defined by $(t, \omega) \mapsto X(t, \omega)$ is measurable with respect to $\mathcal{B}(\mathbb{R}) \otimes \mathcal{F}$ and $\mathcal{E}$.*

In particular, by the Fubini–Tonelli theorem (Theorem 4.4.7), for any $\omega \in \Omega$ the mapping $t \mapsto X(t,\omega)$ is measurable with respect to the σ-fields $\mathcal{B}(\mathbb{R})$ and $\mathcal{E}$. Also, if $E = \mathbb{R}$ and if $X(t)$ is non-negative, one can define the Lebesgue integral

$$\int_{\mathbb{R}} X(t,\omega)\,\mathrm{d}t$$

for each $\omega \in \Omega$, and also apply Tonelli's theorem to obtain

$$E\left[\int_{\mathbb{R}} X(t)\mathrm{d}t\right] = \int_{\mathbb{R}} E\left[X(t)\right]\mathrm{d}t\,.$$

By Fubini's theorem, the last equality also holds true for measurable stochastic processes of arbitrary sign such that $\int_{\mathbb{R}} E\left[|X(t)|\right]\mathrm{d}t < \infty$.

The next theorem tells us that the stochastic processes occurring in applications are measurable.

Theorem 11.1.7 *A right-continuous (resp., left-continuous) stochastic process $\{X(t)\}_{t\in\mathbb{R}}$ taking its values in $\mathbb{R}^m$ is measurable.*

Proof. For all $n \geq 0$ and all $t \geq 0$, let

$$X_n(t) := \sum_{k=-n(2^n-1)}^{n(2^n-1)} X((k+1)/2^{-n})\, 1_{\{[k2^{-n},(k+1)2^{-n})\}}(t)\,.$$

The stochastic process $\{X_n(t)\}_{t\in\mathbb{R}}$ is measurable. If $t \mapsto X(t,\omega)$ is right-continuous, $X(t,\omega)$ is the limit of $X_n(t,\omega)$ for all $(t,\omega) \in \mathbb{R}\times\Omega$, and therefore $(t,\omega) \mapsto X(t,\omega)$ is measurable. The case of a left-continuous process is treated in a similar manner. □

Second-order Stochastic Processes

Definition 11.1.8 *A complex stochastic process $\{X(t)\}_{t\in\mathbf{T}}$ satisfying the condition*

$$E[|X(t)|^2] < \infty \quad (t \in \mathbf{T})$$

is called a second-order *stochastic process.*

In particular, the *mean* function $m : \mathbf{T} \to \mathbb{C}$ and the *covariance* function $\Gamma : \mathbf{T}\times\mathbf{T} \to \mathbb{C}$ are well defined by

$$m(t) := E[X(t)]$$

and

$$\Gamma(t,s) := \text{ cov } (X(t), X(s)) = E[X(t)X(s)^*] - m(t)m(s)^* .$$

When the mean function is the null function, the stochastic process is said to be *centered.*

Theorem 11.1.9 *Let $\{X(t)\}_{t\in\mathbf{T}}$ be a second-order complex stochastic process with mean function m and covariance function Γ. Then, for all $s,t \in \mathbf{T}$,*

$$E\left[|X(t) - m(t)|\right] \le \Gamma(t,t)^{\frac{1}{2}}$$

and

$$|\Gamma(t,s)| \le \Gamma(t,t)^{\frac{1}{2}}\Gamma(s,s)^{\frac{1}{2}} .$$

Proof. Apply Schwarz's inequality

$$E\left[|X|\,|Y|\right] \le E\left[|X|^2\right]^{\frac{1}{2}} E\left[|Y|^2\right]^{\frac{1}{2}}$$

with $X := X(t) - m(t)$ and $Y := 1$ for the first inequality, and with $X := X(t) - m(t)$ and $Y := X(s) - m(s)$ for the second one. □

Theorem 11.1.10 *Let $\{X(t)\}_{t\in\mathbb{R}}$ be a second-order complex-valued measurable stochastic process with mean function m and covariance function Γ. Let $f : \mathbb{R} \to \mathbb{C}$ be a measurable function such that*

$$\int_{\mathbb{R}} |f(t)| E\left[|X(t)|\right] \mathrm{d}t < \infty . \tag{11.1}$$

Then the integral $\int_{\mathbb{R}} f(t)X(t)\,\mathrm{d}t$ is almost surely well defined and

$$E\left[\int_{\mathbb{R}} f(t)X(t)\,\mathrm{d}t\right] = \int_{\mathbb{R}} f(t)m(t)\,\mathrm{d}t .$$

Suppose in addition that f satisfies the condition

$$\int_{\mathbb{R}} |f(t)||\Gamma(t,t)|^{\frac{1}{2}}\,\mathrm{d}t < \infty \tag{11.2}$$

and let $g : \mathbb{R} \to \mathbb{C}$ be a function with the same properties as f. Then $\int_{\mathbb{R}} f(t)X(t)\,\mathrm{d}t$ is square-integrable and

$$\text{cov}\left(\int_{\mathbb{R}} f(t)X(t)\,\mathrm{d}t, \int_{\mathbb{R}} g(t)X(t)\,\mathrm{d}t\right) = \int_{\mathbb{R}}\int_{\mathbb{R}} f(t)g^*(s)\Gamma(t,s)\,\mathrm{d}t\,\mathrm{d}s .$$

Remark: Since $E[|X(t)|] \le E[1 + |X(t)|^2] = 1 + \Gamma(t,t)$, condition (11.1) is satisfied as soon as f is an integrable function such that $\int_{\mathbb{R}} |f(t)|\Gamma(t,t)\,\mathrm{d}t < \infty$.

Proof. By Tonelli's theorem

$$E\left[\int_{\mathbb{R}} |f(t)||X(t)|\,\mathrm{d}t\right] = \int_{\mathbb{R}} |f(t)|E\left[|X(t)|\right]\,\mathrm{d}t < \infty$$

and therefore $\int_{\mathbb{R}} |f(t)||X(t)|\,\mathrm{d}t < \infty$ almost surely, so that almost surely the integral $\int_{\mathbb{R}} f(t)X(t)\,\mathrm{d}t$ is well defined and finite. Also (Fubini)

$$E\left[\int_{\mathbb{R}} f(t)X(t)\,\mathrm{d}t\right] = \int_{\mathbb{R}} E\left[f(t)X(t)\right]\,\mathrm{d}t = \int_{\mathbb{R}} f(t)E\left[X(t)\right]\,\mathrm{d}t.$$

Suppose now (without loss of generality) that the process is centered. By Tonelli's theorem

$$E\left[\left(\int_{\mathbb{R}} |f(t)||X(t)|\,\mathrm{d}t\right)\left(\int_{\mathbb{R}} |g(t)||X(t)|\,\mathrm{d}t\right)\right]$$
$$= \int_{\mathbb{R}}\int_{\mathbb{R}} |f(t)||g(s)|E\left[|X(t)||X(s)|\right]\,\mathrm{d}t\,\mathrm{d}s\,.$$

But (Schwarz's inequality) $E\left[|X(t)||X(s)|\right] \le \Gamma(t,t)|^{\frac{1}{2}}\Gamma(s,s)|^{\frac{1}{2}}$, and therefore the right-hand side of the last equality is bounded by

$$\left(\int_{\mathbb{R}} |f(t)|\Gamma(t,t)|^{\frac{1}{2}}\,\mathrm{d}t\right)\left(\int_{\mathbb{R}} |g(s)|\Gamma(s,s)|^{\frac{1}{2}}\,\mathrm{d}s\right) < \infty\,.$$

One may therefore apply Fubini's theorem to obtain

$$E\left[\left(\int_{\mathbb{R}} f(t)X(t)\,\mathrm{d}t\right)\left(\int_{\mathbb{R}} g(t)X(t)\,\mathrm{d}t\right)\right] = \int_{\mathbb{R}}\int_{\mathbb{R}} f(t)g^*(s)E\left[X(t)X(s)\right]\,\mathrm{d}t\,\mathrm{d}s\,.$$

□

Obviously, for a stationary second-order complex stochastic process $\{X(t)\}_{t\in\mathbb{R}}$, for all $s, t \in \mathbb{R}$,

$$m(t) \equiv m\,, \tag{11.3}$$

where $m \in \mathbb{C}$ and

$$\Gamma(t,s) = C(t-s) \tag{11.4}$$

for some function $C : \mathbb{R} \to \mathbb{C}$, also called the *covariance function* of the process. The complex number m is called the *mean* of the process.

Definition 11.1.11 *A second-order stochastic process $\{X(t)\}_{t\in\mathbb{R}}$ is said to have* orthogonal increments *if for all $n \ge 2$ and for all mutually disjoint intervals $(a_1, b_1], \dots, (a_n, b_n]$ of $\mathbb{R}$, the random variables*

$$X(b_1) - X(a_1), \dots, X(b_n) - X(a_n)$$

are mutually orthogonal.

Clearly, a *centered* second-order stochastic process with independent increments has *a fortiori* orthogonal increments.

Wide-sense Stationarity

Let $\mathbf{T} = \mathbb{R}$, $\mathbb{R}_+$, $\mathbb{Z}$ or $\mathbb{N}$, and let $\{X(t)\}_{t\in\mathbf{T}}$ be a second-order stochastic process.

Definition 11.1.12 *If conditions (11.3) and (11.4) are satisfied for all* $s, t \in \mathbf{T}$, *the complex second-order stochastic process* $\{X(t)\}_{t\in\mathbf{T}}$ *is called* wide-sense stationary. *In continuous time (*$\mathbf{T} = \mathbb{R}$ *or* $\mathbb{R}_+$*) this appellation is reserved for wide-sense stationary processes that have in addition a continuous covariance function.*

There exist stochastic processes that are wide-sense stationary but not strictly stationary (Exercise 11.6.1).

Note that $C(0) = \sigma_X^2$, the variance of any of the random variables $X(t)$.

As an immediate corollary of Theorem 11.1.9, we have:

Corollary 11.1.13 *Let* $\{X(t)\}_{t\in\mathbf{T}}$ *be a wide-sense stationary stochastic process with mean* m *and covariance function* C. *Then*

$$E\left[|X(t) - m|\right] \leq C(0)^{\frac{1}{2}}$$

and

$$|C(\tau)| \leq C(0)\,.$$

Recall the definition of the *correlation coefficient* ρ between two non-trivial real square-integrable random variables X and Y with respective means m_X and m_Y and respective variances σ_X^2 and σ_Y^2:

$$\rho := \frac{\text{cov}\,(X, Y)}{\sigma_X \sigma_Y}\,.$$

The variable $aX + b$ that minimizes the function $F(a, b) := E\left[(Y - aX - b)^2\right]$ is

$$\widehat{Y} = m_Y + \frac{\text{cov}\,(X, Y)}{\sigma_X^2}(X - m_X)$$

and moreover

$$E\left[\left(\widehat{Y} - Y\right)^2\right] = \left(1 - \rho^2\right)\sigma_Y^2$$

(Theorem 3.3.9). This random variable is called the *best linear-quadratic estimate* of Y given X, or the *linear regression* of Y on X.

For a WSS stochastic process with covariance function C, the function

$$\rho(\tau) = \frac{C(\tau)}{C(0)}$$

is called the *autocorrelation function.* It is in fact, for any t, the correlation coefficient between $X(t)$ and $X(t+\tau)$. In particular, the best linear-quadratic estimate of $X(t+\tau)$ given $X(t)$ is

$$\widehat{X}(t+\tau|t) := m + \rho(\tau)(X(t) - m)\,.$$

The estimation error is then, according to the above,

$$E\left[\left(\widehat{X}(t+\tau|t) - X(t+\tau)\right)^2\right] = \sigma_X^2\left(1 - \rho(\tau)^2\right)\,.$$

In the continuous time case, this shows that if the support of the covariance function is concentrated around $\tau = 0$, the process tends to be "unpredictable". We shall come back to this when we discuss the notion of white noise.

EXAMPLE 11.1.14: HARMONIC PROCESS. Let $\{U_k\}_{k\geq 1}$ be square-integrable centered random variables that are mutually uncorrelated. Let $\{\Phi_k\}_{k\geq 1}$ be completely random phases, that is, real random variables uniformly distributed on $[0, 2\pi]$. Suppose moreover that the U variables are independent of the Φ variables. Finally, suppose that $\sum_{k=1}^{\infty} E[|U_k|^2] < \infty$. For all $t \in \mathbb{R}$, the series on the right-hand side of

$$X(t) = \sum_{k=1}^{\infty} U_k \cos(2\pi\nu_k t + \Phi_k)\,,$$

where the ν_k's are arbitrary real numbers (frequencies), is convergent in the quadratic mean and defines a centered WSS stochastic process with covariance function

$$C(\tau) = \sum_{k=1}^{\infty} \frac{1}{2} E[|U_k|^2] \cos(2\pi\nu_k\tau)\,.$$

(This stochastic process is called a *harmonic process.*)

We first do the proof for a finite number N of terms, that is with $X(t) = \sum_{k=1}^{N} U_k \cos(2\pi\nu_k t + \Phi_k)$. We then have

$$\begin{aligned}
E[X(t)] &= E\left[\sum_{k=1}^{N} U_k \cos(2\pi\nu_k t + \Phi_k)\right] \\
&= \sum_{k=1}^{N} E[U_k \cos(2\pi\nu_k t + \Phi_k)] = \sum_{k=1}^{N} E[U_k] E[\cos(2\pi\nu_k t + \Phi_k)] = 0
\end{aligned}$$

and

$$\begin{aligned}
E[X(t+\tau)X(t)^*] &= E\left[\sum_{k=1}^{N}\sum_{\ell=1}^{N} U_k U_\ell^* \cos(2\pi\nu_k(t+\tau)+\Phi_k)\cos(2\pi\nu_\ell t+\Phi_\ell)\right] \\
&= \sum_{k=1}^{N}\sum_{\ell=1}^{N} E[U_k U_\ell^* \cos(2\pi\nu_k(t+\tau)+\Phi_k)\cos(2\pi\nu_\ell t+\Phi_\ell)] \\
&= \sum_{k=1}^{N}\sum_{\ell=1}^{N} E[U_k U_\ell^*]E[\cos(2\pi\nu_k(t+\tau)+\Phi_k)\cos(2\pi\nu_\ell t+\Phi_\ell)] \\
&= \sum_{k=1}^{N} E[|U_k|^2]E[\cos(2\pi\nu_k(t+\tau)+\Phi_k)\cos(2\pi\nu_k t+\Phi_k)] \\
&= \sum_{k=1}^{N} E[|U_k|^2]E\left[\frac{1}{2}\left(\cos(2\pi\nu_k(2t+\tau)+2\Phi_k)+\cos(2\pi\nu_k\tau)\right)\right].
\end{aligned}$$

The announced result then follows since

$$E[\cos(2\pi\nu_k(2t+\tau)+2\Phi_k)] = \frac{1}{2\pi}\int_0^{2\pi}\cos(2\pi\nu_k(2t+\tau)+2\varphi)\,\mathrm{d}\varphi = 0\,.$$

The extension of this result to an infinite sum of complex exponentials is a straightforward consequence of the result of Example 6.4.8.

11.2 Gaussian Processes

Brownian motion is a particular type of Gaussian process, which we now introduce.

Gaussian processes are important for at least three reasons:

(1) because of their mathematical tractability due in particular to the stability of the Gaussianity of stochastic processes: (α) by linear transformations (Theorem 3.4.5) and (β) by limits in the quadratic mean (see Theorem 7.4.5),

(2) because of their ubiquity due to the many forms of the central limit theorem (Theorem 7.2.1), and

(3) because the most important Gaussian process, Brownian motion, plays a fundamental role in the noise theory in communications and in mathematical finance.

Let $\mathbf{T}$ be an arbitrary index.

Definition 11.2.1 *The real-valued stochastic process* $\{X(t)\}_{t\in\mathbf{T}}$ *is called a* Gaussian process *if for all* $n \geq 1$ *and for all* $t_1, \ldots, t_n \in \mathbf{T}$, *the random vector* $(X(t_1), \ldots, X(t_n))$ *is Gaussian.*

In particular, its characteristic function is given by the formula

$$E\left[\exp\left\{i\sum_{j=1}^{n} u_j X(t_j)\right\}\right] = \exp\left\{i\sum_{j=1}^{n} u_j m(t_j) - \frac{1}{2}\sum_{j=1}^{n}\sum_{k=1}^{n} u_j u_k \Gamma(t_j, t_k)\right\}, \tag{11.5}$$

where $u_1, \ldots, u_n \in \mathbb{R}$ and where m and Γ are the mean and covariance functions respectively.

Theorem 11.2.2 *For a Gaussian process with index set* $\mathbf{T} = \mathbb{R}$ *or* $\mathbb{Z}$ *to be stationary, it is necessary and sufficient that* $m(t) = m$ *and* $\Gamma(t, s) = C(t - s)$ *for all* $s, t \in \mathbf{T}$.

Proof. The necessity is obvious, whereas the sufficiency is proven by replacing the t_ℓ's in (11.5) by $t_\ell + h$ to obtain the characteristic function of

$$(X(t_1 + h), \ldots, X(t_n + h))$$

namely,

$$\exp\left\{i\sum_{j=1}^{n} u_j m - \frac{1}{2}\sum_{j=1}^{n}\sum_{k=1}^{n} u_j u_k C(t_j - t_k)\right\},$$

and then observing that this quantity is independent of h. □

EXAMPLE 11.2.3: CLIPPED GAUSSIAN PROCESS, I. Let $\{X(t)\}_{t\in\mathbb{R}}$ be a centered stationary Gaussian process with covariance function $C_X(\tau)$. Define the *clipped* (or *hard-limited*) process

$$Y(t) = \text{sign } X(t),$$

with the convention sign $X(t) = 0$ if $X(t) = 0$ (note however that this occurs with null probability if $C_X(0) = \sigma_X^2 > 0$, which is henceforth assumed). Clearly this stochastic process is centered. Moreover, it is unchanged when $\{X(t)\}_{t\in\mathbb{R}}$ is multiplied by a positive constant. In particular, we may assume that the variance $C_X(0)$ equals 1, so that the covariance matrix of the vector $(X(0), X(\tau))^T$ is

$$\Gamma(\tau) = \begin{pmatrix} 1 & \rho_X(\tau) \\ \rho_X(\tau) & 1 \end{pmatrix},$$

where $\rho_X(\tau)$ is the correlation coefficient of $X(0)$ and $X(\tau)$. We assume that $\Gamma(\tau)$ is invertible, that is, $|\rho_X(\tau)| < 1$.

We then have the *Van Vleck–Middleton formula:*

$$C_Y(\tau) = \frac{2}{\pi} \sin^{-1}\left(\frac{C_X(\tau)}{C_X(0)}\right).$$

Proof. Since for each t the random variable $Y(t)$ takes the values ± 1 and 0, the latter with null probability, we can express the autocovariance function of the clipped process as

$$C_Y(\tau) = 2\{P(X(0) > 0, X(\tau) > 0) - P(X(0) > 0, X(\tau) < 0)\},$$

where it was noted that

$$P(X(0) < 0, X(\tau) < 0) = P(X(0) > 0, X(\tau) > 0)$$

and that

$$P(X(0) < 0, X(\tau) > 0) = P(X(0) > 0, X(\tau) < 0).$$

The result then follows from that of Exercise 3.6.33 with $\rho = \rho_X(\tau)$. □

The Wiener Process

Definition 11.2.4 *By definition, a* standard Brownian motion, *or* standard Wiener process, *is a continuous centered Gaussian process* $\{W(t)\}_{t\in\mathbb{R}}$ *with independent increments, such that* $W(0) = 0$, *and such that for any interval* $[a, b] \subset \mathbb{R}$, *the variance of* $W(b) - W(a)$ *is equal to* $b - a$.

In particular, the vector $(W(t_1), \ldots, W(t_k))$ with $0 < t_1 < \ldots < t_k$ admits the probability density function

$$\frac{1}{(\sqrt{2\pi})^k \sqrt{t_1\,(t_2 - t_1)\cdots(t_k - t_{k-1})}}\, e^{-\frac{1}{2}\left(\frac{x_1^2}{t_1} + \frac{(x_1+x_2)^2}{t_2-t_1} + \cdots + \frac{(x_1+\cdots+x_k)^2}{t_k-t_{k-1}}\right)}.$$

Note for future reference that for $s, t \in \mathbb{R}_+$,

$$E[W(t)W(s)] = t \wedge s. \tag{11.6}$$

In fact, for $0 \le s \le t$,

$$\begin{aligned} E[W(t)W(s)] &= E[(W(t) - W(s))W(s)] + E[W(s)^2] \\ &= E[(W(t) - W(s))(W(s) - W(0))] + E[(W(s) - W(0))^2] \\ &= 0 + s = t \wedge s\,. \end{aligned}$$

We now give the description of the Wiener process as limit of a properly rescaled (both in time and amplitude) symmetric random walk. Let $\{X_n\}_{n\ge 0}$ be a symmetric random walk on $\mathbb{Z}$ starting from 0, of the form

$$X_n = \sum_{k=1}^{n} Z_k\,,$$

where $\{Z_n\}_{n\ge 1}$ is an IID sequence of $\{-1, +1\}$-valued random variables with $P(Z_n = -1) = P(Z_n = 1) = \frac{1}{2}$. Construct a continuous time stochastic process $\{X(t)\}_{t\ge 0}$ from this sequence as follows:

$$X(t) = \delta X_{\lfloor t/\Delta \rfloor} = \delta \sum_{k=1}^{\lfloor t/\Delta \rfloor} Z_k\,.$$

(Recall the notation $\lfloor a \rfloor = \sup\{k \in \mathbb{N}; k \le a\}$.) Since the Z_k's are centered and of variance 1, we have that

$$E\left[X(t)\right] = 0\,, \qquad \text{Var}\,\left(X(t)\right) = (\delta)^2 \times \lfloor t/\Delta \rfloor\,.$$

Let Δ and δ tend to 0 in such a way that the limit is not trivial. With respect to this goal, the choice $\Delta = \delta$ is not satisfactory since $E\left[X(t)\right] = 0$ and $\lim_{t\downarrow 0} \text{Var}\,\left(X(t)\right) = 0$, leading to a null process. If we take $\delta^2 = \Delta$, we have $E\left[X(t)\right] = 0$ and $\lim_{\Delta\downarrow 0} \text{Var}\,\left(X(t)\right) = t$. We show that in this case, for all $t_1, \dots, t_m$ in $\mathbb{R}_+$ forming an increasing sequence, the limit distribution of the vector $(X(t_1), \dots, X(t_m))$ is that corresponding to a Wiener process.

We consider the case $m = 1$, the general case being an easy adaptation. Let $t_1 = t$. In this case, since by the central limit theorem

$$\frac{\sum_{k=1}^{n} Z_k}{\sqrt{n}} \to \mathcal{N}(0, 1)$$

we have, by Slutsky's lemma (Theorem 7.1.6),

$$\frac{X(t)}{\sqrt{t}} = \frac{\sum_{k=1}^{\lfloor t/\Delta \rfloor} Z_k}{\sqrt{\lfloor \frac{t}{\Delta} \rfloor}} \frac{\sqrt{\lfloor \frac{t}{\Delta} \rfloor}}{\sqrt{\frac{t}{\Delta}}} \to \mathcal{N}(0, 1)\,.$$

Therefore, at the limit (in distribution), $X(t)$ is a centered Gaussian variable with variance $\sqrt{t}$.

Pathology

Definition 11.2.4 of the Wiener process does not say much about the qualitative behavior of this process. Although the trajectories of the Brownian motion are, almost surely, continuous functions, their behavior is rather chaotic. First of all we observe that, for fixed $t_0 > 0$, the random variable

$$\frac{W(t_0 + h) - W(t_0)}{h} \sim \mathcal{N}\left(0, h^{-1}\right)$$

and therefore it cannot converge in distribution as $h \downarrow 0$ since the limit of its characteristic function is the null function, which is not a characteristic function. In particular, it does not converge almost surely to any random variable. Therefore, for any $t_0 > 0$,

$$P\left(t \mapsto W(t) \text{ is not differentiable at } t_0\right) = 1\,.$$

But the situation is even more dramatic:

Theorem 11.2.5 *Almost all the paths of the Wiener process are nowhere differentiable.*

We shall not prove this result here,[1] but state one of its consequences.

Corollary 11.2.6 *Almost all the paths of the Wiener process are of unbounded variation on finite intervals.*

Proof. This is because any function of bounded variation is differentiable almost everywhere (with respect to Lebesgue measure).[2] □

The Brownian Bridge

This is the process $\{X(t)\}_{t\in[0,1]}$ obtained from the standard Brownian motion $\{W(t)\}_{t\in[0,1]}$ by

$$X(t) := W(t) - tW(1) \quad (t \in [0,1])\,.$$

It is a Gaussian process since for all $t_1, \dots, t_k \in [0,1]$, $(X(t_1), \dots, X(t_k))$ is Gaussian vector, being a linear function of the Gaussian vector $(W(t_1), \dots, W(t_k), W(1))$. In particular, since it is a centered Gaussian process, its distribution is entirely characterized by its covariance function and a simple calculation (Exercise 11.6.5) gives

$$\operatorname{cov}\left(X(t), X(s)\right) = s(1-t) \quad (0 \le s \le t \le 1)\,.$$

[1] See for instance [7], Theorem 11.2.8.
[2] See for instance Corollary 6, section 5.2 of [15].

In particular, $X(0) = X(1) = 0$.

The Brownian bridge $\{X(t)\}_{t\in[0,1]}$ is distributionwise a Wiener process $\{W(t)\}_{t\in[0,1]}$ conditioned by $W(1) = 0$. This statement is problematic in that the conditioning event has a null probability. However, it is true "at the limit":

Theorem 11.2.7 *Let $f : \mathbb{R}^k \to \mathbb{R}$ be a bounded and continuous function. Then, for any $0 \le t_1 < t_2 < \cdots < t_k \le 1$,*

$$\lim_{\varepsilon\downarrow 0} \mathrm{E}\left[f(W(t_1),\dots,W(t_k)) \mid |W(1)| \le \varepsilon\right] = \mathrm{E}\left[f(X(t_1),\dots,X(t_k))\right] .$$

Proof.

$$\begin{aligned}
&\mathrm{E}\left[f(W(t_1),\dots,W(t_k)) \mid |W(1)| \le \varepsilon\right] \\
&\quad = \mathrm{E}\left[f(X(t_1)+t_1W(1),\dots,X(t_k)+t_kW(1)) \mid |W(1)| \le \varepsilon\right] \\
&\quad = \frac{\mathrm{E}\left[f(X(t_1)+t_1W(1),\dots,X(t_k)+t_kW(1))1_{|W(1)|\le\varepsilon}\right]}{P(|W(1)| \le \varepsilon)} .
\end{aligned}$$

In view of the independence of $\{X(t)\}_{t\in[0,1]}$ and $W(1)$ (Exercise 11.6.7), this last quantity equals

$$\frac{\int_{-\varepsilon}^{+\varepsilon} \mathrm{e}^{-\frac{1}{2}x^2} \mathrm{E}\left[f(X(t_1)+t_1x,\dots,X(t_k)+t_kx)\right] \mathrm{d}x}{\int_{-\varepsilon}^{+\varepsilon} \mathrm{e}^{-\frac{1}{2}x^2}\,\mathrm{d}x} ,$$

which tends to $\mathrm{E}\left[f(X(t_1),\dots,X(t_k))\right]$ as $\varepsilon \downarrow 0$. □

Gauss–Markov Processes

We now investigate another type of Gaussian processes, those having the additional property of being Markovian. We first give the general definition of a *Markov process*:

Definition 11.2.8 *Let $\mathbf{T}$ be $\mathbb{R}_+$ or $\mathbb{N}$. A real-valued stochastic process $\{X(t)\}_{t\in\mathbf{T}}$ is called a Markov process if for $f : \mathbb{R} \to \mathbb{R}$ that is non-negative or such that $E\left[|f(X(t))|\right] < \infty$ $(t \in \mathbf{T})$,*

$$E\left[f(X(t))|X(s), X(t_1),\dots,X(t_k)\right] = E\left[f(X(t))|X(s)\right] \tag{11.7}$$

for all $0 \le t_1 \le t_2 \le \cdots \le t_n \le s \le t$.

Of course this definition fits the special case of Markov chains.

EXAMPLE 11.2.9: WIENER IS GAUSS–MARKOV. The Wiener process is a Gauss–Markov process (Exercise 11.6.9).

EXAMPLE 11.2.10: A DISCRETE-TIME GAUSS–MARKOV PROCESS. A discrete-time stochastic process $\{X_n\}_{n\geq 0}$ defined by $X_{n+1} = aX_n + \varepsilon_{n+1}$ $(n \geq 0)$, where $\{\varepsilon_n\}_{n\geq 1}$ is an IID centered Gaussian sequence and X_0 is a Gaussian random variable independent of this sequence, is a Gauss–Markov process (Exercise 11.6.9).

The stochastic processes that are Gaussian and Markovian are in fact Wiener processes with a different time scale. The proof starts with a simple lemma.

Lemma 11.2.11 *Let $\{X(t)\}_{t\geq 0}$ be a centered Gaussian process with covariance function Γ such that $\Gamma(t,t) > 0$ for all $t \geq 0$. If in addition it is Markov, then for all $t > s > t_0 \geq 0$,*

$$\Gamma(t, t_0) = \frac{\Gamma(t,s)\Gamma(s,t_0)}{\Gamma(s,s)}. \tag{11.8}$$

Proof. By the Gaussian property, the linear regression of $X(t)$ on $X(t_0)$ is equal to the conditional expectation of $X(t)$ given $X(t_0)$:

$$E[X(t)|X(t_0)] = \frac{\Gamma(t,t_0)}{\Gamma(t_0,t_0)} X(t_0). \tag{$\star$}$$

Using this remark and the Markov property,

$$\begin{aligned} E[X(t)|X(t_0)] &= E[E[X(t)|X(t_0), X(s)]\,|X(t_0)] \\ &= E[E[X(t)|X(s)]\,|X(t_0)] = E\left[\frac{\Gamma(t,s)}{\Gamma(s,s)} X(s)|X(t_0)\right] \\ &= \frac{\Gamma(t,s)}{\Gamma(s,s)} E[X(s)|X(t_0)] = \frac{\Gamma(t,s)}{\Gamma(s,s)} \frac{\Gamma(s,t_0)}{\Gamma(t_0,t_0)} X(t_0). \end{aligned}$$

Comparing with the right-hand side of ($\star$), and since $P(X(t_0) \neq 0) > 0$ (in fact $= 1$), we obtain (11.8). □

Theorem 11.2.12 *Let $\{X(t)\}_{t\geq 0}$ be a centered Gaussian process with continuous covariance function Γ such that $\Gamma(t,t) > 0$ for all $t \geq 0$. It is Markov if and only there exist functions $f, g : \mathbb{R}_+ \to \mathbb{R}_+$ such that for all $s, t \geq 0$*

$$\Gamma(t,s) = f(t \vee s)g(t \wedge s). \tag{11.9}$$

Proof.

Necessity. Suppose the process is Gauss–Markov. Let

$$\rho(t,s) := \frac{\Gamma(t,s)}{(\Gamma(t,t))^{\frac{1}{2}}\,(\Gamma(s,s))^{\frac{1}{2}}}$$

denote its autocorrelation function. By Eqn. (11.8), for all $t > s > t_0 \geq 0$,

$$\rho(t,t_0) = \rho(t,s)\rho(s,t_0)\,. \tag{$\star\star$}$$

We show that $\rho(t,s) > 0$ for all $t, s \geq 0$. Indeed, assuming $s > t$ and using $(\star\star)$ repeatedly, for all $n \geq 1$,

$$\rho(t,s) = \prod_{k=0}^{n-1} \rho\left(t + \frac{k(s-t)}{n}, t + \frac{(k+1)(s-t)}{n}\right),$$

and therefore, using the facts that $\rho(u,u) = 1$ for all u and that ρ is uniformly continuous on bounded rectangles, we can choose n large enough as to make all the elements in the above product positive. Therefore, we may divide by $\rho(t,t_0)$ and write $(\star\star)$ as

$$\rho(t,s) = \frac{\rho(t,t_0)}{\rho(s,t_0)}$$

or

$$\Gamma(t,s) = \rho(t,t_0)\Gamma(t,t)^{\frac{1}{2}} \times \frac{\Gamma(s,s)^{\frac{1}{2}}}{\rho(s,t_0)},$$

from which we obtain the desired conclusion (here $s = t \wedge s$ and $t = t \vee s$).

Sufficiency. Suppose that the process is Gaussian and that (11.8) holds true. Assume $t > s$. Therefore $\Gamma(t,s) = f(t)g(s)$. By Schwarz's inequality, $\Gamma(t,s) \leq \Gamma(t,t)^{\frac{1}{2}}\,(\Gamma(s,s))^{\frac{1}{2}}$ or, equivalently, $f(t)g(s) \leq (f(t)g(t)f(s)g(s))^{\frac{1}{2}}$. Therefore, the function

$$\tau(t) = \frac{g(t)}{f(t)}$$

is monotone non-decreasing. In particular, the centered Gaussian process

$$Y(t) = f(t)W(\tau(t))$$

is Markov (because the Wiener process is Markov). Its covariance function is

$$\begin{aligned} E\left[Y(t)Y(s)\right] &= f(t)f(s)E\left[W(\tau(t))W(\tau(s))\right] \\ &= f(t)f(s)(\tau(t) \wedge \tau(s)) \\ &= f(t)f(s)\tau(s) = f(t)g(s)\,. \end{aligned}$$

Since it has the same covariance as $\{X(t)\}_{t\geq 0}$ and since both processes are centered and Gaussian, they have the same distribution. In particular $\{X(t)\}_{t\geq 0}$ is Markov. □

Theorem 11.2.13 *A* WSS *Gaussian stochastic process* $\{X(t)\}_{t\geq 0}$ *is Markov if and only if its covariance function has the form*

$$C(\tau) = C(0)\mathrm{e}^{-\lambda|\tau|}$$

for some $\lambda \geq 0$.

Proof. If $\{X(t)\}_{t\geq 0}$ is WSS, $\Gamma(t,s) = C(t-s)$ and therefore, with $\rho(t) := \frac{C(t)}{C(0)}$,

$$\rho(t+s) = \rho(t)\rho(s)\,,$$

which implies that $\rho(t) = c\mathrm{e}^{\alpha t}$ for some $\alpha \in \mathbb{R}$. Here $c = 1$ since $\rho(0) = 1$. Now $\rho(1) = \frac{C(1)}{C(0)} = \mathrm{e}^{\alpha}$. But (Schwarz's inequality) $C(1) \leq 1$ so that $\alpha \leq 0$. □

11.3 The Wiener–Doob Integral

The *Doob stochastic integral*, a special case of which is the *Wiener stochastic integral*

$$\int_{\mathbb{R}} f(t)\,\mathrm{d}W(t) \qquad (\star)$$

that is defined for a certain class of measurable functions f, is not of the usual types. For instance, it cannot be defined pathwise as a Stieltjes–Lebesgue integral since the trajectories of the Brownian motion are of unbounded variation (Corollary 11.2.6). Nor can this integral be interpreted as $\int_{\mathbb{R}} f(t)\dot{W}(t)\,\mathrm{d}t$ (where the dot denotes derivation), since the Brownian motion sample paths are not differentiable (Theorem 11.2.5).

The integral in $(\star)$ will therefore be defined in a radically different way. In fact, the Doob stochastic integral will be defined more generally, with respect to a process with centered and uncorrelated increments.

Definition 11.3.1 *Let* $\{Z(t)\}_{t\in\mathbb{R}}$ *be a complex stochastic process such that for all intervals* $[t_1,t_2] \subset \mathbb{R}$ *the increments* $Z(t_2) - Z(t_1)$ *are in* $L^2_{\mathbb{C}}(P)$, *centered and such that for some locally finite measure* μ *on* $(\mathbb{R}, \mathcal{B}(\mathbb{R}))$:

$$E[(Z(t_2) - Z(t_1))(Z(t_4) - Z(t_3))^*] = \mu((t_1,t_2] \cap (t_3,t_4])$$

for all $[t_1, t_2] \subset \mathbb{R}$ *and all* $[t_3, t_4] \subset \mathbb{R}$. *Such stochastic process* $\{Z(t)\}_{t\in\mathbb{R}}$ *is called a stochastic process with* centered *and* uncorrelated increments, *and* μ *is its* structural measure.

In particular, if the intervals $(t_1, t_2]$ and $(t_3, t_4]$ are disjoint, $Z(t_2) - Z(t_1)$ and $Z(t_4) - Z(t_3)$ are orthogonal elements of the Hilbert space $L^2_{\mathbb{C}}(P)$.

EXAMPLE 11.3.2: WIENER PROCESS. The Wiener process $\{W(t)\}_{t\in\mathbb{R}}$ is a process with centered and uncorrelated increments whose structural measure is the Lebesgue measure.

Gaussian Subspaces

Before proceeding to the construction of the Wiener–Doob integral, the definition of a Gaussian subspace is necessary.

Definition 11.3.3 *Let* $\{X_i\}_{i\in I}$ *be an arbitrary collection of complex (resp., real) random variables in* $L^2_{\mathbb{C}}(P)$ *(resp.,* $L^2_{\mathbb{R}}(P)$*). The Hilbert subspace of* $L^2_{\mathbb{C}}(P)$ *(resp.,* $L^2_{\mathbb{R}}(P)$*) consisting of the closure of the vector space of finite linear complex (resp., real) combinations of elements of* $\{X_i\}_{i\in I}$ *is called the complex (resp., real)* Hilbert subspace generated by $\{X_i\}_{i\in I}$, *and is denoted by* $H_{\mathbb{C}}(X_i, i \in I)$ *(resp.,* $H_{\mathbb{R}}(X_i, i \in I)$*).*

More explicitly, in the complex case for instance: the Hilbert subspace $H_{\mathbb{C}}(X_i, i \in I) \subseteq L^2_{\mathbb{C}}(P)$ consists of all complex square-integrable random variables that are limits in the quadratic mean (that is, limits in $L^2_{\mathbb{C}}(P)$) of some sequence of finite complex linear combinations of elements in the set $\{X_i\}_{i\in I}$.

Definition 11.3.4 *A collection* $\{X_i\}_{i\in I}$ *of real random variables defined on the same probability space, where* I *is an arbitrary index set, is called a* Gaussian family *if for all finite set of indices* $i_1, \ldots, i_k \in I$, *the random vector* $(X_{i_1}, \ldots, X_{i_k})$ *is Gaussian. A Hilbert subspace* G *of the real Hilbert space* $L^2_{\mathbb{R}}(P)$ *is called a* Gaussian (Hilbert) subspace *if it is a Gaussian family.*

Theorem 11.3.5 *Let* $\{X_i\}_{i\in I}$ *be a Gaussian family of random variables of* $L^2_{\mathbb{R}}(P)$. *Then the Hilbert subspace* $H_{\mathbb{R}}(X_i, i \in I)$ *generated by* $\{X_i\}_{i\in I}$ *is a Gaussian subspace of* $L^2_{\mathbb{R}}(P)$.

Proof. By definition, the Hilbert subspace $H_{\mathbb{R}}(X_i, i \in I)$ consists of all the random variables in $L^2_{\mathbb{R}}(P)$ that are limits in the quadratic mean of finite linear combinations of elements of the family $\{X_i\}_{i\in I}$. The result follows from that in Example 7.4.5. □

Construction of the Wiener–Doob Integral

This integral is defined for all integrands $f \in L^2_{\mathbb{C}}(\mu)$ in the following manner. First, we define it for all $f \in \mathcal{L}$, the vector subspace of $L^2_{\mathbb{C}}(\mu)$ formed by the finite complex linear combinations of interval indicator functions

$$f(t) = \sum_{i=1}^{N} \alpha_i 1_{(a_i,b_i]}(t)\,.$$

For such functions, *by definition*,

$$\int_{\mathbb{R}} f(t)\,\mathrm{d}Z(t) := \sum_{i=1}^{N} \alpha_i (Z(b_i) - Z(a_i))\,.$$

Observe that this random variable belongs to the Hilbert subspace $H_{\mathbb{C}}(Z)$ of $L^2_{\mathbb{C}}(P)$ generated by $\{Z(t)\}_{t\in\mathbb{R}}$. One easily verifies that the linear mapping

$$\varphi : f \in \mathcal{L} \mapsto \int_{\mathbb{R}} f(t)\,\mathrm{d}Z(t) \in L^2_{\mathbb{C}}(P)$$

is an isometry, that is,

$$\int_{\mathbb{R}} |f(t)|^2\,\mu(\mathrm{d}t) = E\left[\left|\int_{\mathbb{R}} f(t)\,\mathrm{d}Z(t)\right|^2\right].$$

Since $\mathcal{L}$ is a dense subset of $L^2_{\mathbb{C}}(\mu)$ [3], φ can be uniquely extended to an isometric linear mapping of $L^2_{\mathbb{C}}(\mu)$ into $H_{\mathbb{C}}(Z)$ (Theorem A.0.6). We continue to call this extension φ and then define, for all $f \in L^2_{\mathbb{C}}(\mu)$, the Doob integral of f with respect to $\{Z(t)\}_{t\in\mathbb{R}}$ by

$$\int_{\mathbb{R}} f(t)\,\mathrm{d}Z(t) := \varphi(f)\,.$$

The fact that φ is an isometry is expressed by the *Doob isometry formula*

$$E\left[\left(\int_{\mathbb{R}} f(t)\,\mathrm{d}Z(t)\right)\left(\int_{\mathbb{R}} g(t)\,\mathrm{d}Z(t)\right)^*\right] = \int_{\mathbb{R}} f(t)g^*(t)\,\mu(\mathrm{d}t)\,, \tag{11.10}$$

[3] The proof is not obvious. See for instance Theorem 9.4 of *Théorie de l'intégration*, M. Biane and G. Pagès, Vuibert, Paris, 2004.

where f and g are in $L^2_{\mathbb{C}}(\mu)$. Note also that for all $f \in L^2_{\mathbb{C}}(\mu)$:

$$E\left[\int_{\mathbb{R}} f(t)\,\mathrm{d}Z(t)\right] = 0\,, \tag{11.11}$$

since the Doob integral is the limit in $L^2_{\mathbb{C}}(\mu)$ of random variables of the type $\sum_{i=1}^{N} \alpha_i(Z(b_i) - Z(a_i))$ that have mean 0 (use the continuity of the inner product in $L^2_{\mathbb{C}}(P)$).

A Formula of Integration by Parts

Theorem 11.3.6 *Let $\{W(t)\}_{t\in\mathbb{R}_+}$ be a standard Wiener process and denote by $H_{\mathbb{R}}(W)$ the Gaussian real Hilbert space that it generates. The Wiener integral $Y = \int_{\mathbb{R}_+} f(t)\,\mathrm{d}W(t)$, where $f \in L^2_{\mathbb{R}}(\mathbb{R}_+)$, is characterized by the following two properties:*

(a) *$Y \in H_{\mathbb{R}}(W)$;*

(b) *$E[YW(s)] = \int_0^s f(t)\,\mathrm{d}t$ for all $s \geq 0$.*

Proof. Necessity: We have already noted that, by construction, $\int_{\mathbb{R}_+} f(t)\,\mathrm{d}W(t) \in H_{\mathbb{R}}(W)$. As for (b), this is just the isometry formula

$$E\left[\int_{\mathbb{R}_+} f(t)\,\mathrm{d}W(t) \int_{\mathbb{R}_+} 1_{\{s\leq t\}}\,\mathrm{d}W(t)\right] = \int_0^s f(t)\,\mathrm{d}t\,.$$

Sufficiency: Since $Y - \int_0^t f(s)\,\mathrm{d}W(s)$ is in $H_{\mathbb{R}}(W)$, it suffices to show that this random variable is orthogonal to the generators $W(s)$, $s \in \mathbb{R}_+$, of $H_{\mathbb{R}}(W)$ and therefore is the null element of $H_{\mathbb{R}}(W)$, and therefore $Y = \int_0^t f(s)\,\mathrm{d}W(s)$, P-a.s. But, by (b) and, again, by the isometry formula,

$$E\left[\left(Y - \int_{\mathbb{R}_+} f(u)\,\mathrm{d}W(u)\right)W(s)\right] = \int_0^s f(t)\,\mathrm{d}t - \int_0^s f(t)\,\mathrm{d}t = 0\,.$$

□

Theorem 11.3.7 *Let $\{W(t)\}_{t\in\mathbb{R}}$ be a standard Wiener process. Let T be a positive real number and let $f : [0,T] \to \mathbb{R}$ be a continuously differentiable function. In particular $f \in L^2_{\mathbb{R}}([0,T])$ and therefore the integral $\int_0^T f(t)\,\mathrm{d}W(t)$ is well defined. Then:*

$$\int_0^T f(t)\,\mathrm{d}W(t) + \int_0^T f'(t)W(t)\,\mathrm{d}t = f(T)W(T)\,.$$

Proof. By Theorem 11.3.6, it suffices to prove that for all $s \in [0,T]$,

$$E\left[\left(f(T)W(T) - \int_0^T f'(t)W(t)\,\mathrm{d}t\right)W(s)\right] = \int_0^s f(t)\,\mathrm{d}t\,,$$

which, using the equality $E\left[W(a)W(b)\right] = \min(a,b)$, reduces to

$$f(T)s - E\left[\left(\int_0^T f'(t)W(t)\,\mathrm{d}t\right)W(s)\right] = \int_0^s f(t)\mathrm{d}t\,.$$

By Fubini:

$$\begin{aligned} E\left[\left(\int_0^T f'(t)W(t)\,\mathrm{d}t\right)W(s)\right] &= \int_0^T f'(t)E\left[W(t)W(s)\right]\,\mathrm{d}t \\ &= \int_0^T f'(t)\min(t,s)\,\mathrm{d}t\,. \end{aligned}$$

We therefore have to check that

$$f(T)s - \int_0^s f'(t)t\,\mathrm{d}t - s\int_s^T f'(t)\,\mathrm{d}t = \int_0^s f(t)\mathrm{d}t\,,$$

or

$$f(T)s - \int_0^s f'(t)t\,\mathrm{d}t - s(f(T) - f(s)) = \int_0^s f(t)\mathrm{d}t\,.$$

But this is

$$-\int_0^s f'(t)t\,\mathrm{d}t + sf(s) = \int_0^s f(t)\mathrm{d}t\,,$$

which follows by integration by parts. □

11.4 Two Applications

Langevin's Equation

Definition 11.4.1 *Let $\{W(t)\}_{t\in\mathbb{R}}$ be a standard Wiener process, and let for all $t \in \mathbb{R}$*

$$X(t) = (2\alpha)^{\frac{1}{2}} \int_{-\infty}^t \mathrm{e}^{-\alpha(t-s)}\,\sigma\,\mathrm{d}W(s)\,,$$

where $\alpha > 0$ and $\sigma > 0$. The process $\{X(t)\}_{t\in\mathbb{R}}$ defined in this way is called the Ornstein–Uhlenbeck process.

Since for all $t \in \mathbb{R}$, $X(t)$ belongs to $H_{\mathbb{R}}(W)$, it is a Gaussian process (Theorem 11.3.5). It is centered, with covariance function

$$\Gamma(t,s) = \mathrm{e}^{-\alpha|t-s|},$$

as follows directly from the isometry formula (11.10).

Definition 11.4.2 *The* Langevin equation *is, by definition, the equation*

$$\mathrm{d}V(t) + \alpha V(t)\,\mathrm{d}t = \sigma\,\mathrm{d}W(t)$$

to be interpreted as

$$V(t) - V(0) + \alpha \int_0^t V(s)\,\mathrm{d}s = \sigma\,W(t)\,.$$

Theorem 11.4.3 *The unique solution of the Langevin equation with initial value $V(0)$ is*

$$V(t) = \mathrm{e}^{-\alpha t}V(0) + \int_0^t \mathrm{e}^{-\alpha(t-s)}\sigma\,\mathrm{d}W(s)\,.$$

In particular, with the choice $V(0) = \int_{-\infty}^0 \mathrm{e}^{\alpha s}\,\mathrm{d}W(s)$,

$$V(t) = \int_{-\infty}^t \mathrm{e}^{-\alpha(t-s)}\sigma\,\mathrm{d}W(s)$$

is the Ornstein–Uhlenbeck process.

Proof. Using the integration by parts formula of Theorem 11.3.7, the Langevin equation is found to be equivalent to

$$V(t) = \mathrm{e}^{-\alpha t}V(0) + \sigma W(t) - \int_0^t \alpha\mathrm{e}^{-\alpha(t-s)}\sigma W(s)\,\mathrm{d}s\,. \qquad (\star)$$

By the (classical) formula of integration by parts,

$$\mathrm{e}^{-\alpha u}\int_0^u \mathrm{e}^{+\alpha s}\sigma W(s)\,\mathrm{d}s = -\int_0^u \mathrm{e}^{-\alpha t}\left(\int_0^t \mathrm{e}^{+\alpha s}\sigma W(s)\,\mathrm{d}s\right)\mathrm{d}t + \sigma\int_0^u W(t)\,\mathrm{d}t\,.$$

Therefore, integrating both sides of $(\star)$ from 0 to u

$$\alpha\int_0^u V(t)\,\mathrm{d}t = (1-\mathrm{e}^{-\alpha u})V(0) + \int_0^u \alpha\mathrm{e}^{-\alpha(u-s)}\sigma W(s)\,\mathrm{d}s$$

and finally:

$$
\begin{aligned}
V(u) - V(0) + \alpha \int_0^u V(t)\,\mathrm{d}t \\
&= V(u) - \mathrm{e}^{-\alpha u} V(0) + \int_0^u \alpha \mathrm{e}^{-\alpha(u-s)} \sigma W(s)\,\mathrm{d}s = \sigma W(u)\,.
\end{aligned}
$$

We now prove unicity. Let V' be another solution with the same initial value. With $U := V - V'$, we therefore have

$$U(t) = \alpha \int_0^t U(s)\,\mathrm{d}s\,,$$

whose unique solution is the null function, by Gronwall's lemma:

Lemma 11.4.4 *Let $x : \mathbb{R}^+ \to \mathbb{R}$ be a positive locally integrable real function such that*

$$x(t) \le a + b \int_0^t x(s) \qquad (\star)$$

for some $a \ge 0$ and $b > 0$. Then

$$x(t) \le a\mathrm{e}^{bt}\,.$$

□

Proof. Multiplying $(\star)$ by e^{-bt}, we have

$$\mathrm{e}^{-bt} x(t) \le a\mathrm{e}^{-bt} + b\mathrm{e}^{-bt} \int_0^t x(s)\,\mathrm{d}s$$

or, equivalently

$$
\begin{aligned}
a\mathrm{e}^{-bt} &\ge \mathrm{e}^{-bt} x(t) - b\mathrm{e}^{-bt} \int_0^t x(s)\,\mathrm{d}s \\
&= \frac{d}{\mathrm{d}t} \mathrm{e}^{-bt} \int_0^t x(s)\,\mathrm{d}s\,.
\end{aligned}
$$

Integrating this inequality:

$$\frac{a}{b}(1 - \mathrm{e}^{-bt}) \ge \mathrm{e}^{-bt} \int_0^t x(s)\,\mathrm{d}s\,.$$

Substituting this into $(\star)$:

$$x(t) \le a + b\mathrm{e}^{-bt} \frac{a}{b}(1 - \mathrm{e}^{-bt}) = a\mathrm{e}^{bt}\,.$$

□

The Cameron–Martin Formula

This result is of interest in communications and detection theory. One will recognize the likelihood ratio associated with the hypothesis "signal plus white Gaussian noise" against the hypothesis "white Gaussian noise only".

Theorem 11.4.5 *Let $\{X(t)\}_{t\geq 0}$ be, with respect to probability P, a Wiener process with variance σ^2 and let $\gamma : \mathbb{R} \to \mathbb{R}$ be in $L^2_{\mathbb{R}}(\ell)$. For any $T \in \mathbb{R}_+$, the formula*

$$\frac{\mathrm{d}Q}{\mathrm{d}P} = \mathrm{e}^{\frac{1}{\sigma^2}\left\{\int_0^T \gamma(t)\mathrm{d}X(t) - \frac{1}{2}\int_0^T \gamma^2(t)\mathrm{d}t\right\}} \tag{11.12}$$

defines a probability measure Q on $(\Omega, \mathcal{F})$ with respect to which

$$X(t) - \int_0^t \gamma(s)\,\mathrm{d}s$$

is, on the interval $[0, T]$, a Wiener process with variance σ^2.

The proof of Theorem 11.4.5 is based on the following preliminary result.

Lemma 11.4.6 *Let $\{X(t)\}_{t\geq 0}$ be a Wiener process with variance σ^2 and let $\varphi : \mathbb{R} \to \mathbb{R}$ be in $L^2_{\mathbb{R}}(\ell)$. Then, for any $T \in \mathbb{R}_+$,*

$$\mathrm{E}\left[\mathrm{e}^{\int_0^T \varphi(t)\mathrm{d}X(t)}\right] = \mathrm{e}^{\frac{1}{2}\sigma^2\int_0^T \varphi^2(t)\mathrm{d}t}\,. \tag{11.13}$$

Proof. First consider the case

$$\varphi(t) = \sum_{k=1}^{N} \alpha_k 1_{(a_k, b_k]}(t)\,, \tag{11.14}$$

where $\alpha_k \in \mathbb{R}$ and the intervals $(a_k, b_k]$ are disjoint. For this special case, formula (11.13) reduces to

$$\mathrm{E}\left[\mathrm{e}^{\sum_{k=1}^N \alpha_k (X(b_k) - X(a_k))}\right] = \mathrm{e}^{\frac{1}{2}\sigma^2\sum_{k=1}^N \alpha_k^2 (b_k - a_k)}\,,$$

and therefore follows directly from the independence of the increments of a Wiener process and from the Gaussian property of these increments, in particular, the formula giving the Laplace transform of the centered Gaussian variable $X(b) - X(a)$ with variance $\sigma^2(b-a)$:

$$\mathrm{E}\left[\mathrm{e}^{\alpha(X(b) - X(a))}\right] = \mathrm{e}^{\frac{1}{2}\sigma^2\alpha^2(b-a)}\,.$$

Let now $\{\varphi_n\}_{n\geq 1}$ be a sequence of functions of type (11.14) converging in $L^2_{\mathbb{R}}(\ell)$ to φ (in particular, $\lim_{n\uparrow\infty}\int_0^T \varphi_n^2(t)\mathrm{d}t = \int_0^T \varphi^2(t)\mathrm{d}t$). Therefore,

$$\lim_{n\uparrow\infty}\int_0^T \varphi_n(t)\mathrm{d}X(t) = \int_0^T \varphi(t)\mathrm{d}X(t),$$

where the latter convergence is in $L^2_{\mathbb{R}}(P)$. This convergence can be assumed to take place almost surely by taking if necessary a subsequence. From the equality

$$\mathrm{E}\left[\mathrm{e}^{\int_0^T \varphi_n(t)\mathrm{d}X(t)}\right] = \mathrm{e}^{\sigma^2\int_0^T \varphi_n^2(t)\mathrm{d}t}$$

we can then deduce (11.13), at least if the sequence of random variables in the left-hand side is uniformly integrable. This is the case because the quantity

$$\mathrm{E}\left[\left|\mathrm{e}^{\int_0^T \varphi_n(t)\mathrm{d}X(t)}\right|^2\right] = \mathrm{E}\left[\mathrm{e}^{2\int_0^T \varphi_n(t)\mathrm{d}X(t)}\right] = \mathrm{e}^{2\sigma^2\int_0^T \varphi_n^2(t)\mathrm{d}t}$$

is uniformly bounded, and therefore the uniform integrability claim follows from Theorem 6.5.5, with $G(t) = t^2$. □

We may now turn to the proof of Theorem 11.4.5.

Proof. The fact that (11.12) properly defines a probability Q, that is, that the expectation of the right-hand side of (11.12) equals 1, follows from Lemma 11.4.6 with $\varphi(t) = \frac{1}{\sigma^2}\gamma(t)$.

Letting

$$Y(t) := X(t) - \int_0^t \gamma(s)\mathrm{d}s\,,$$

we have to prove that this centered stochastic process is Gaussian. To do this, we must show that

$$\mathrm{E}_Q\left[\mathrm{e}^{\sum_{k=1}^N \alpha_k(Y(b_k)-Y(a_k))}\right] = \mathrm{e}^{\frac{1}{2}\sigma^2\sum_{k=1}^N \alpha_k^2(b_k-a_k)}\,,$$

where $\alpha_k \in \mathbb{R}$ and the intervals $(a_k, b_k] \subseteq [0,T]$ are disjoint, that is, letting $\psi(t) = \sum_{k=1}^N \alpha_k 1_{(a_k,b_k]}(t)$,

$$\mathrm{E}_Q\left[\mathrm{e}^{\int_0^T \psi(t)\mathrm{d}Y(t)}\right] = \mathrm{e}^{\frac{1}{2}\sigma^2\int_0^T \psi^2(t)\mathrm{d}t}\,,$$

or equivalently,

$$\mathrm{E}_P\left[\frac{\mathrm{d}Q}{\mathrm{d}P}\mathrm{e}^{\int_0^T \psi(t)(\mathrm{d}X(t)-\gamma(t)\mathrm{d}t)}\right] = \mathrm{e}^{\frac{1}{2}\sigma^2\int_0^T \psi^2(t)\mathrm{d}t}\,,$$

that is,

$$\mathrm{E}_P\left[\mathrm{e}^{\frac{1}{\sigma^2}\left\{\int_0^T \gamma(t)\mathrm{d}X(t)-\frac{1}{2}\int_0^T \gamma^2(t)\mathrm{d}t\right\}}\mathrm{e}^{\int_0^T \psi(t)(\mathrm{d}X(t)-\gamma(t)\mathrm{d}t)}\right] = \mathrm{e}^{\frac{1}{2}\sigma^2\int_0^T \psi^2(t)\mathrm{d}t}.$$

Simplifying:

$$\mathrm{E}_P\left[\mathrm{e}^{\int_0^T \left(\psi(t)+\frac{1}{\sigma^2}\gamma(t)\right)\mathrm{d}X(t)-\int_0^T(\gamma(t)\psi(t))\mathrm{d}t-\frac{1}{2}\int_0^T\frac{\gamma^2(t)}{\sigma^2}\mathrm{d}t}\right] = \mathrm{e}^{\frac{1}{2}\sigma^2\int_0^T \psi^2(t)\mathrm{d}t},$$

and using (11.13) with $\varphi(t) = \psi(t) + \frac{1}{\sigma^2}\gamma(t)$, the left-hand side is equal to

$$\mathrm{E}_P\left[\mathrm{e}^{\frac{1}{2}\sigma^2\int_0^T \left(\psi(t)+\frac{1}{\sigma^2}\gamma(t)\right)^2\mathrm{d}t-\int_0^T(\gamma(t)\psi(t))\mathrm{d}t-\frac{1}{2}\int_0^T\frac{\gamma^2(t)}{\sigma^2}\mathrm{d}t}\right].$$

The proof is completed since

$$\begin{aligned}\frac{1}{2}\sigma^2\int_0^T \left(\psi(t)+\frac{1}{\sigma^2}\gamma(t)\right)^2\mathrm{d}t - \int_0^T(\gamma(t)\psi(t))\,\mathrm{d}t - \frac{1}{2}\int_0^T\frac{\gamma^2(t)}{\sigma^2}\\ = \frac{1}{2}\sigma^2\int_0^T \psi^2(t)\mathrm{d}t.\end{aligned}$$

□

11.5 Fractal Brownian Motion

The Wiener process $\{W(t)\}_{t\geq 0}$ has the following property. If c is a positive constant, the process $\{W_c(t)\}_{t\geq 0} := \{c^{-\frac{1}{2}}W(ct)\}_{t\geq 0}$ is also a Wiener process. It is indeed a centered Gaussian process with independent increments, null at the time origin, and for $0 < a < b$,

$$\mathrm{E}\left[|W_c(b) - W_c(a)|^2\right] = c^{-1}\mathrm{E}\left[|W(cb) - W(ca)|^2\right] = c^{-1}(cb - ca) = b - a.$$

This is a particular instance of a self-similar stochastic process.

Definition 11.5.1 *A real-valued stochastic process $\{Y(t)\}_{t\geq 0}$ is called* self-similar *with* (Hurst) self-similarity parameter H *if for any $c > 0$,*

$$\{Y(t)\}_{t\geq 0} \stackrel{\mathcal{D}}{\sim} \{c^{-H}Y(ct)\}_{t\geq 0},$$

where the symbol $\stackrel{\mathcal{D}}{\sim}$ means "have the same distribution", or "have the same finite-dimensional distribution", depending on the context.

The Wiener process is therefore self-similar with similarity parameter $H = \frac{1}{2}$.

It follows from the definition that $Y(t) \stackrel{\mathcal{D}}{\sim} t^H Y(1)$, and therefore, if $P(Y(1) \neq 0) > 0$:

If $H < 0$, $Y(t) \to 0$ in distribution as $t \to \infty$ and $Y(t) \to \infty$ in distribution as $t \to 0$.

If $H > 0$, $Y(t) \to \infty$ in distribution as $t \to 0$ and $Y(t) \to 0$ in distribution as $t \to \infty$.

If $H = 0$, $Y(t)$ has a distribution independent of t.

In particular, when $H \neq 0$, a self-similar process cannot be stationary (strictly or in the wide sense).

We shall be interested in self-similar processes that have *stationary increments.*

Theorem 11.5.2 *Let $\{Y(t)\}_{t\geq 0}$ be a non-negative self-similar stochastic process with stationary increments and self-similarity parameter $H > 0$ (in particular, $Y(0) = 0$). Its covariance function is given by*

$$\Gamma(s,t) := \operatorname{cov}(Y(s), Y(t)) = \frac{1}{2}\sigma^2 \left[t^{2H} - |t-s|^{2H} + s^{2H}\right],$$

where $\sigma^2 = \mathrm{E}\left[(Y(t+1) - Y(t))^2\right] = \mathrm{E}\left[Y(1)^2\right]$.

Proof. Assume without loss of generality that the process is centered. Let $0 \leq s \leq t$. Then

$$\begin{aligned}\mathrm{E}\left[(Y(t) - Y(s))^2\right] &= \mathrm{E}\left[(Y(t-s) - Y(0))^2\right] \\ &= \mathrm{E}\left[(Y(t-s))^2\right] = \sigma^2 (t-s)^{2H}\end{aligned}$$

and

$$2\mathrm{E}\left[Y(t)Y(s)\right] = \mathrm{E}\left[Y(t)^2\right] + \mathrm{E}\left[Y(s)^2\right] - \mathrm{E}\left[(Y(t) - Y(s))^2\right],$$

hence the result. □

The fractal Brownian motion is a Gaussian process that in a sense generalizes the Wiener process.

Definition 11.5.3 *A* fractal Brownian motion *on $\mathbb{R}_+$ with Hurst parameter $H \in (0,1)$ is a centered Gaussian process $\{B_H(t)\}_{t\geq 0}$ with continuous paths such that $B_H(0) = 0$, and with covariance function*

$$\mathrm{E}[B_H(t)B_H(s)] = \frac{1}{2}\left(|t|^{2H} + |s|^{2H} - |t-s|^{2H}\right). \tag{11.15}$$

We shall prove the existence of the fractal Brownian motion by constructing it as a Wiener integral. More precisely, define for $0 < H < 1$, $w_H(t,s) := 0$ for $t \le s$,

$$w_H(t,s) := (t-s)^{H-\frac{1}{2}} \text{ for } 0 \le s \le t$$

and

$$w_H(t,s) := (t-s)^{H-\frac{1}{2}} - (-s)^{H-\frac{1}{2}} \text{ for } s < 0.$$

Observe that for any $c > 0$

$$w_H(ct,s) = c^{H-\frac{1}{2}} w_H(t, sc^{-1}).$$

Define

$$B_H(t) := \int_{\mathbb{R}} w_H(t,s)\, \mathrm{d}W(s)\,.$$

The Wiener integral of the right-hand side is, more explicitly,

$$A - B := \int_0^t (t-s)^{H-\frac{1}{2}}\, \mathrm{d}W(s) - \int_{-\infty}^0 \left((t-s)^{H-\frac{1}{2}} - (-s)^{H-\frac{1}{2}}\right) \mathrm{d}W(s). \quad (11.16)$$

It is well defined and with the change of variable $u = c^{-1}s$ it becomes

$$c^{H-\frac{1}{2}} \int_{\mathbb{R}} w_H(t,u)\, \mathrm{d}W(cu)\,.$$

Using the self-similarity of the Brownian motion, the process defined by the last display has the same distribution as the process defined by

$$c^{H-\frac{1}{2}} c^{\frac{1}{2}} \int_{\mathbb{R}} w_H(t,u)\, \mathrm{d}W(u).$$

Therefore $\{B_H(t)\}_{t\ge 0}$ is self-similar with similarity parameter H.

It is tempting to rewrite (11.16) as $Z(t) - Z(0)$, where

$$Z(t) = \int_{-\infty}^t (t-s)^{H-\frac{1}{2}}\, \mathrm{d}W(s).$$

However this last integral is not well defined as a Doob integral since for all $H > 0$, the function $s \to (t-s)^{H-\frac{1}{2}} 1_{\{s \le t\}}$ is not in $L^2_{\mathbb{R}}(\mathbb{R})$.

11.6 Exercises

Exercise 11.6.1. WIDE-SENSE STATIONARY, BUT NOT STATIONARY
Give a simple example of a discrete-time stochastic process that is wide-sense stationary, but not strictly stationary. Do the same for a continuous-time wide-sense stationary process.

Exercise 11.6.2. CLOSE RELATIVES OF THE BROWNIAN MOTION
Let $\{W(t)\}_{t\geq 0}$ be a standard Brownian motion. What can you say about the process $\{X(t)\}_{t\in[0,1]}$, where:

A. $X(t) = t\,W\left(\frac{1}{t}\right)$ with $X(0) := 0$? (You will admit continuity of the process at time 0.)

B. $X(t) = W(1) - W(1-t)$?

Exercise 11.6.3. SQUARED BROWNIAN MOTION
1. Show that for a Brownian motion $\{W(t)\}_{t\geq 0}$,

$$E\left[|W(t) - W(s)|^4\right] = 3\,|t-s|^2 .$$

2. Let $\{X(t)\}_{t\in\mathbb{R}}$ be a centered wide-sense stationary Gaussian process with covariance function C_X. Compute the probability that $X(t_1) > X(t_2)$ where $t_1, t_2 \in \mathbb{R}$ are fixed times.

3. Give the mean function and the covariance function of the process $\{X(t)^2\}_{t\in\mathbb{R}}$.

Exercise 11.6.4. CONTINUITY OF THE COVARIANCE FUNCTION
Prove that for the covariance function of a complex wide-sense stationary process $\{X(t)\}_{t\geq 0}$ to be continuous, it suffices that it be continuous at the origin, and that this is in turn equivalent to continuity in the quadratic mean of the stochastic process, that is, for all $\mathbb{R}$,

$$\lim_{h\to 0} E\left[|X(t+h) - X(t)|^2\right] = 0 .$$

Show that in fact, the covariance function is then uniformly continuous on $\mathbb{R}$.

Exercise 11.6.5. A BASIC FORMULA
Let $\{W(t)\}_{t\geq 0}$ be a standard Wiener process. Prove that for $s, t \in \mathbb{R}_+$,

$$\mathrm{E}[W(t)W(s)] = t \wedge s .$$

Let $\{Y(t)\}_{t\geq 0}$ be a Brownian bridge. Prove that

$$\mathrm{cov}\left(X(t), X(s)\right) = s(1-t) \quad (0 \leq s \leq t \leq 1) .$$

Exercise 11.6.6. A REPRESENTATION OF THE BROWNIAN BRIDGE
Let $\{W(t)\}_{t\geq 0}$ be a standard Brownian motion. Let for $t \in [0,1)$,

$$Y(t) := (1-t)\int_0^t \frac{\mathrm{d}W(s)}{1-s}\,\mathrm{d}s\,.$$

(i) Prove that the integral in the right-hand side is well defined on $[0,1)$ as a Wiener integral.

(ii) Prove that as $t \downarrow 0$, $Y(t) \to 0$ in quadratic mean.

(iii) Define $Y(0) := 0$. Show that $\{Y(t)\}_{t\in[0,1]}$ is a Gaussian process.

(iv) Show that $\{Y(t)\}_{t\in[0,1]}$ is (has the same distribution as) a Brownian bridge.

Exercise 11.6.7. BROWNIAN BRIDGE
Let $\{W(t)\}_{t\in[0,1]}$ be a Wiener process. Show that the Brownian bridge

$$\{X(t) := W(t) - tW(1)\}_{t\in[0,1]}$$

is a Gaussian process independent of $W(1)$ and compute its autocovariance function. Show that the process $\{X(1-t)\}_{t\in[0,1]}$ is a Brownian bridge.

Exercise 11.6.8. STRUCTURAL MEASURE
Let $\{Z(t)\}_{t\in\mathbb{R}}$ be a second-order real-valued centered stochastic process, right-continuous in the quadratic mean, such that $Z(0) = 0$ and with uncorrelated increments (for all $a \leq b \leq c \leq d$, we have that $E\left[(Z(b) - Z(a))(Z(d) - Z(c))\right] = 0$). Show that there exists a locally finite measure μ on $(\mathbb{R}, \mathcal{B}(\mathbb{R}))$ such that

$$E\left[(Z(b) - Z(a))^2\right] = \mu((a,b])\,.$$

Exercise 11.6.9. SOME GAUSS–MARKOV PROCESSES

A. Show that the Wiener process is a Gauss–Markov process.

B. Show that a discrete-time stochastic process $\{X_n\}_{n\geq 1}$ defined by $X_{n+1} = aX_n + \varepsilon_{n+1}$ $(n \geq 0)$, where $\{\varepsilon_n\}_{n\geq 1}$ is an IID centered Gaussian sequence and X_0 is a Gaussian random variable independent of this sequence, is a Gauss–Markov process.

C. For each $t \geq 0$, let $X(t) = a(t)W(\tau(t))$, where $\{W(t)\}_{t\geq 0}$ is a standard Wiener process, $a : \mathbb{R}_+ \to \mathbb{R}$ and $\tau : \mathbb{R}_+ \to \mathbb{R}$ are measurable functions, and moreover

τ is strictly increasing, with $\tau(0) = 0$. Prove that $\{X(t)\}_{t\geq 0}$ is a Gauss–Markov stochastic process and give explicitly the functions f and g of Theorem 11.2.12.

Exercise 11.6.10. THE ORNSTEIN–UHLENBECK PROCESS.
Let

$$X(t) := \mathrm{e}^{-\alpha t} W(\mathrm{e}^{2\alpha t}) \quad (t \geq 0)$$

where $\{W(t)\}_{t\in\geq 0}$ is a standard Wiener process and α is a positive real number. Prove that $\{X(t)\}_{t\in\geq 0}$ is an Ornstein–Uhlenbeck process.

Exercise 11.6.11. ORNSTEIN–UHLENBECK IS GAUSS–MARKOV
Show that the Ornstein–Uhlenbeck process is a Gauss–Markov process. Describe the functions f and τ in its representation as

$$X(t) = f(t)W(\tau(t)).$$

Exercise 11.6.12. MICROPULSES AND FRACTAL BROWNIAN MOTION.
Let $\bar{N}_\epsilon$ be a Poisson process on $\mathbb{R} \times \mathbb{R}_+$ with the mean measure $\nu(\mathrm{d}t \times \mathrm{d}z) = \frac{1}{2\epsilon^2} z^{-1-\theta}\,\mathrm{d}t \times \mathrm{d}z$, where $0 < \theta < 1$ and $\epsilon > 0$. For all $t \geq 0$, let $S^+_{0,t} = \{(s,z) : 0 < s < t,\, t - s < z\}$ and $S^-_{0,t} = \{(s,z) : -\infty < s < 0,\, -s < z < t - s\}$, and define[4]

$$X_\epsilon(t) = \epsilon \left\{ \bar{N}_\epsilon(S^+_{0,t}) - \bar{N}_\epsilon(S^-_{0,t}) \right\}.$$

(1) Show that $X_\epsilon(t)$ is well defined for all $t \geq 0$.

(2) Compute for all $0 \leq t_1 \leq t_2 \leq \cdots \leq t_n$ the characteristic function of $(X_\epsilon(t_1), \ldots, X_\epsilon(t_n))$.

(3) Show that for all $0 \leq t_1 \leq t_2 \leq \cdots \leq t_n$, $(X_\epsilon(t_1), \ldots, X_\epsilon(t_n))$ converges in distribution to $(B_H(t_1), \ldots, B_H(t_n))$ as $\epsilon \downarrow 0$, where $\{B_H(t)\}_{t\geq 0}$ is a fractal Brownian motion (fBm) with Hurst parameter $H = \frac{1-\theta}{2}$ and variance $E\left[B_H(1)^2\right] = \theta^{-1}(1-\theta)^{-1}$. Recall that $\{B_H(t)\}_{t\geq 0}$ is called an fBm with Hurst parameter H, $0 < H < \frac{1}{2}$, if it is a centered Gaussian process such that $B_H(0) = 0$ with covariance function

$$E\left[B_H(t)B_H(s)\right] = \frac{1}{2}\left(|s|^{2H} + |t|^{2H} - |s-t|^{2H}\right) E\left[B_H(1)^2\right].$$

[4] R. Cioczek-Georges and B.B. Mandelbrot, A class of micropulses and antipersistent fractal Brownian motion, *Stochastic Processes and their Applications*, 60, pp. 1–18, (1995).

Chapter 12

Wide-sense Stationary Processes

This chapter concerns a topic of interest in many fields of application, most notably signal processing and communications theory, as well as econometrics and the earth sciences. The main notion here is that of power spectrum (power spectral measure).

12.1 The Power Spectral Measure

As we shall now see, the classical Fourier analysis of square-integrable (with respect to Lebesgue measure) functions has a counterpart in the theory of wide-sense stationary processes.

Consider first a WSS stochastic process $\{X(t)\}_{t\in\mathbb{R}}$ with *integrable and continuous* covariance function C. The Fourier transform f of this covariance function is therefore well defined by

$$f(\nu) := \int_{\mathbb{R}} \mathrm{e}^{-2i\pi\nu\tau} C(\tau)\, \mathrm{d}\tau \,. \tag{12.1}$$

It is called the *power spectral density* (PSD). It turns out that it is non-negative and integrable, as we shall soon see. Since it is integrable, the Fourier inversion formula

$$C(\tau) = \int_{\mathbb{R}} \mathrm{e}^{2i\pi\nu\tau} f(\nu)\, \mathrm{d}\nu \tag{12.2}$$

holds almost everywhere, and in fact everywhere since both sides of the equality are continuous (Example 4.1.24).[1] In the context of WSS stochastic processes, (12.2) is called the *Bochner formula*. Letting $\tau = 0$ in this formula, we obtain,

[1] See for instance [5], Theorem 1.1.8.

P. Brémaud, *An Introduction to Applied Probability*, Texts in Applied Mathematics 77,
https://doi.org/10.1007/978-3-031-49306-5_12

since $C(0) = \text{Var}\,(X(t) := \sigma^2$,

$$\sigma^2 = \int_{\mathbb{R}} f(\nu)\mathrm{d}\nu\,. \tag{12.3}$$

EXAMPLE 12.1.1: THE ORNSTEIN–UHLENBECK PROCESS. Let $\{X(t)\}_{t\in\geq 0}$ be an Ornstein–Uhlenbeck process. It is a centered Gaussian process, and, using (11.6), we have for $t \geq s$,

$$\begin{aligned}
E[X(t)X(s)] &= E[\mathrm{e}^{-\alpha t}W(\mathrm{e}^{2\alpha t})\mathrm{e}^{-\alpha s}W(\mathrm{e}^{2\alpha s})] \\
&= \mathrm{e}^{-\alpha(t+s)}E[W(\mathrm{e}^{2\alpha t})W(\mathrm{e}^{2\alpha s})] \\
&= \mathrm{e}^{-\alpha(t+s)}\min(\mathrm{e}^{2\alpha t}\mathrm{e}^{2\alpha s}) \\
&= \mathrm{e}^{-\alpha(t+s)}\mathrm{e}^{2\alpha s} = \mathrm{e}^{-\alpha(t-s)}\,,
\end{aligned}$$

and therefore, for all $s, t \in \mathbb{R}_+$

$$E[X(t)X(s)] = \mathrm{e}^{-\alpha|t-s|}\,.$$

It is therefore a WSS stochastic process with integrable covariance function, and its power spectral density is then the Fourier transform of the covariance function:

$$f(\nu) = \int_{\mathbb{R}} \mathrm{e}^{-2i\pi\nu\tau}\mathrm{e}^{-\alpha|\tau|}\mathrm{d}\tau = \frac{2\alpha}{\alpha^2 + 4\pi^2\nu^2}\,.$$

Not all WSS stochastic processes admit a power spectral density. For instance, consider a wide-sense stationary process with a covariance function of the form

$$C(\tau) = \sum_{k\in\mathbb{Z}} P_k \mathrm{e}^{2i\pi\nu_k\tau}, \tag{12.4}$$

where

$$P_k \geq 0 \text{ and } \sum_{k\in\mathbb{Z}} P_k < \infty \tag{12.5}$$

(say, the harmonic process of Example 11.1.14). Clearly, this covariance function is not integrable, and in fact there does not exist a power spectral density. In particular, a representation of the covariance function such as (12.2) is not available, at least if the function f is interpreted in the ordinary sense. However, there is a formula such as (12.2) if we consent to define the PSD in this case to be the *pseudo-function*

$$f(\nu) = \sum_{k\in\mathbb{Z}} P_k\,\delta(\nu - \nu_k)\,, \tag{12.6}$$

where $\delta(\nu - a)$ is the delayed Dirac pseudo-function informally defined by

$$\int_{\mathbb{R}} \varphi(\nu)\,\delta(\nu - a)\,\mathrm{d}\nu = \varphi(a)\,.$$

Indeed, with such a convention,

$$\int_{\mathbb{R}} f(\nu)\mathrm{e}^{2i\pi\nu\tau} f(\nu)\,\mathrm{d}\nu = \sum_{k\in\mathbb{Z}} P_k \int_{\mathbb{R}} \mathrm{e}^{2i\pi\nu\tau}\delta(\nu - \nu_k)\,\mathrm{d}\nu = \sum_{k\in\mathbb{Z}} P_k \mathrm{e}^{2i\pi\nu_k\tau}\,.$$

The General Case

Remember that the characteristic function φ of a real random variable X has the following properties:

A. it is hermitian symmetric, that is, $\varphi(-u) = \varphi(u)^*$, and it is uniformly bounded: $|\varphi(u)| \le \varphi(0)$,

B. it is uniformly continuous on $\mathbb{R}$, and

C. it is definite non-negative, in the sense that for all integers n, all $u_1, \dots, u_n \in \mathbb{R}$, and all $z_1, \dots, z_n \in \mathbb{C}$,

$$\sum_{j=1}^{n}\sum_{k=1}^{n} \varphi(u_j - u_k) z_j z_k^* \ge 0$$

(just observe that the left-hand side equals $E\left[\left|\sum_{j=1}^{n} z_j \mathrm{e}^{iu_j X}\right|^2\right]$).

It turns out that Properties A , B and C characterize characteristic functions up to a multiplicative constant. This is the content of *Bochner's theorem* (Theorem 7.1.7), which is now recalled for easier reference:

Let $\varphi : \mathbb{R} \to \mathbb{C}$ be a function satisfying properties A, B and C. Then there exists a constant $0 \le \beta < \infty$ and a real random variable X such that for all $u \in \mathbb{R}$,

$$\varphi(u) = \beta E\left[\mathrm{e}^{iuX}\right]\,.$$

Bochner's theorem is all that is needed to define the power spectral measure of a wide-sense stationary stochastic process continuous in the quadratic mean.

Theorem 12.1.2 *Let $\{X(t)\}_{t\in\mathbb{R}}$ be a* WSS *stochastic process continuous in the quadratic mean, with covariance function C. Then, there exists a* unique *measure μ on $\mathbb{R}$ such that*

$$C(\tau) = \int_{\mathbb{R}} \mathrm{e}^{2i\pi\nu\tau}\mu(\mathrm{d}\nu)\,. \tag{12.7}$$

In particular, μ is a *finite* measure:

$$\mu(\mathbb{R}) = C(0) = \text{Var}\,(X(0)) < \infty\,. \tag{12.8}$$

Proof. It suffices to observe that the covariance function of a WSS stochastic process that is continuous in the quadratic mean shares the properties A, B and C of the characteristic function of a real random variable. Indeed,

(a) it is hermitian symmetric, and $|C(\tau)| \le C(0)$ (Schwarz's inequality),

(b) it is uniformly continuous, and

(c) it is definite non-negative, in the sense that for all integers n, all $\tau_1, \ldots, \tau_n \in \mathbb{R}$, and all $z_1, \ldots, z_n \in \mathbb{C}$,

$$\sum_{j=1}^{n}\sum_{k=1}^{n} C(\tau_j - \tau_k) z_j z_k^* \ge 0$$

(just observe that the left-hand side is equal to $E\left[\left|\sum_{j=1}^{n} z_j X(t_j)\right|^2\right]$).

Therefore, by Theorem 7.1.7, the covariance function C is up to a multiplicative constant a characteristic function. This is exactly what (12.7) says, since μ thereof is a finite measure, that is, up to a multiplicative constant, a probability distribution.

Uniqueness of the power spectral measure follows from the fact that a finite measure (up to a multiplicative constant: a probability) on $\mathbb{R}$ is characterized by its Fourier transform (Theorem 5.3.2). □

Special Cases

The case of an absolutely continuous spectrum corresponds to the situation where μ admits a density f with respect to Lebesgue measure: $\mu(\mathrm{d}\nu) = f(\nu)\,\mathrm{d}\nu$. (In particular, f is non-negative and integrable with respect to Lebesgue measure.) As we saw before, we then say that the WSS stochastic process in question admits the *power spectral density* (PSD) f.

The case of a "*line spectrum*" corresponds to a spectral measure that is a weighted sum of Dirac measures:

$$\mu(\mathrm{d}\nu) = \sum_{k\in\mathbb{Z}} P_k\,\varepsilon_{\nu_k}(\mathrm{d}\nu)\,.$$

Since μ is a measure, the P_k's are non-negative, and since μ is a finite measure, they have a finite sum.

12.2 Filtering of WSS Stochastic Processes

We recall a few standard results concerning the (convolutional) filtering of deterministic functions.

Let $f, g : (\mathbb{R}, \mathcal{B}(\mathbb{R})) \to (\mathbb{R}, \mathcal{B}(\mathbb{R}))$ be integrable functions with respective Fourier transforms $\widehat{f}$ and $\widehat{g}$. Then (Exercise 4.5.12),

$$\int_{\mathbb{R}} \int_{\mathbb{R}} |f(t-s)g(s)| \,\mathrm{d}t \,\mathrm{d}s < \infty \,,$$

and therefore, for almost all $t \in \mathbb{R}$, the function $s \mapsto f(t-s)g(s)$ is Lebesgue integrable. In particular, the convolution

$$(f * g)(t) := \int_{\mathbb{R}} f(t-s)g(s) \,\mathrm{d}s$$

is almost everywhere well defined. For all t such that the last integral is not defined, set $(f * g)(t) = 0$. Then $f * g$ is Lebesgue integrable and its Fourier transform is $\widehat{f * g} = \widehat{f}\widehat{g}$, where $\hat{f}, \hat{g}$ are the Fourier transforms of f and g, respectively (Exercise 4.5.13).

Let $h : (\mathbb{R}, \mathcal{B}(\mathbb{R})) \to (\mathbb{R}, \mathcal{B}(\mathbb{R}))$ be an integrable function. The operation that associates to the integrable function $x : (\mathbb{R}, \mathcal{B}(\mathbb{R})) \to (\mathbb{R}, \mathcal{B}(\mathbb{R}))$ the integrable function

$$y(t) := \int_{\mathbb{R}} h(t-s)x(s) \,\mathrm{d}s$$

is called a stable *convolutional filter*. The function h is called the *impulse response* of the filter, and x and y are respectively the *input* and the *output* of this filter. The Fourier transform $\widehat{h}$ of the impulse response is the *transmittance* of the filter.

Let now $\{X(t)\}_{t\in\mathbb{R}}$ be a WSS stochastic process with continuous covariance function C_X. We examine the effect of filtering on this process. The output process is the process defined by

$$Y(t) := \int_{\mathbb{R}} h(t-s)X(s)\mathrm{d}s \,. \tag{12.9}$$

Note that the integral (12.9) is well defined under the integrability condition for the impulse response h. This follows from Theorem 11.1.10 according to which the integral

$$\int_{\mathbb{R}} f(s)X(s,\omega) \,\mathrm{d}s$$

is well defined for P-almost all ω when f is integrable (in the special case of WSS stochastic processes, $m(t) = m$ and $\Gamma(t,t) = C(0) + |m|^2$, and therefore the conditions on f and g thereof reduce to integrability of these functions). Referring to the same theorem, we have

$$E[\int_{\mathbb{R}} f(t)X(t)\,\mathrm{d}t] = \int_{\mathbb{R}} f(t)E[X(t)]\,\mathrm{d}t = m\int_{\mathbb{R}} f(t)\,\mathrm{d}t\,. \tag{12.10}$$

Let now $f, g : \mathbb{R} \to \mathbb{C}$ and be integrable functions. As a special case of Theorem 11.1.10, we have

$$\operatorname{cov}\left(\int_{\mathbb{R}} f(t)X(t)\,\mathrm{d}t\,,\,\int_{\mathbb{R}} g(s)X(s)\,\mathrm{d}s\right) = \int_{\mathbb{R}}\int_{\mathbb{R}} f(t)g^*(s)C(t-s)\,\mathrm{d}t\,\mathrm{d}s\,. \tag{12.11}$$

We shall see that, in addition,

$$\operatorname{cov}\left(\int_{\mathbb{R}} f(t)X(t)\,\mathrm{d}t\,,\,\int_{\mathbb{R}} g(s)X(s)\,\mathrm{d}s\right) = \int_{\mathbb{R}}\int_{\mathbb{R}} \widehat{f}(-\nu)\widehat{g}^*(-\nu)\mu(\mathrm{d}\nu)\,. \tag{12.12}$$

Proof. Assume without loss of generality that $m = 0$. From Bochner's representation of the covariance function, we obtain for the last double integral in (12.11)

$$\begin{aligned}\int_{\mathbb{R}}\int_{\mathbb{R}} f(t)g^*(s)\left(\int_{\mathbb{R}} \mathrm{e}^{+2j\pi\nu(t-s)}\,\mu(\mathrm{d}\nu)\right)\mathrm{d}t\,\mathrm{d}s \\ = \int_{\mathbb{R}}\left(\int_{\mathbb{R}} f(t)\mathrm{e}^{+2j\pi\nu t}\mathrm{d}t\right)\left(\int_{\mathbb{R}} g(s)\mathrm{e}^{+2j\pi\nu s}\,\mathrm{d}s\right)^*\mu(\mathrm{d}\nu)\,.\end{aligned}$$

Here again we have to justify the change of order of integration using Fubini's theorem. For this, it suffices to show that the function

$$(t,s,\nu) \mapsto \left|f(t)g^*(s)\mathrm{e}^{+2j\pi\nu(t-s)}\right| = |f(t)|\,|g(s)|\,1_{\mathbb{R}}(\nu)$$

is integrable with respect to the product measure $\ell \times \ell \times \mu$. This is indeed true, the integral being equal to $(\int_{\mathbb{R}} |f(t)|\,\mathrm{d}t) \times (\int_{\mathbb{R}} |g(t)|\,\mathrm{d}t) \times \mu(\mathbb{R})$. □

In view of the above results, the right-hand side of formula (12.9) is well defined. Moreover

Theorem 12.2.1 *When the input process* $\{X(t)\}_{t\in\mathbb{R}}$ *is a* WSS *stochastic process with power spectral measure* μ_X*, the output* $\{Y(t)\}_{t\in\mathbb{R}}$ *of a* stable *convolutional filter of transmittance* $\widehat{h}$ *is a* WSS *stochastic process with the power spectral measure*

$$\mu_Y(\mathrm{d}\nu) = |\widehat{h}(\nu)|^2\mu_X(\mathrm{d}\nu)\,. \tag{12.13}$$

This formula will be referred to as the *fundamental filtering formula*.

Proof. Just apply formulas (12.10) and (12.12) with the functions

$$f(u) := h(t-u), \qquad g(v) := h(s-v),$$

to obtain

$$E[Y(t)] = m \int_{\mathbb{R}} h(t)\mathrm{d}t\,,$$

and

$$E[(Y(t)-m)(Y(s)-m)^*] = \int_{\mathbb{R}} |\widehat{h}(\nu)|^2 \mathrm{e}^{+2j\pi\nu(t-s)} \mu(\mathrm{d}\nu)\,.$$

□

EXAMPLE 12.2.2: TWO SPECIAL CASES. In particular, if the input process admits a PSD f_X, the output process also admits a PSD given by

$$f_Y(\nu) = |\widehat{h}(\nu)|^2 f_X(\nu)\,\mathrm{d}\nu\,.$$

When the input process has a line spectrum, the power spectral measure of the output process takes the form

$$\mu_Y(\mathrm{d}\nu) = \sum_{k=1}^{\infty} P_k |\widehat{h}(\nu_k)|^2 \varepsilon_{\nu_k}(\mathrm{d}\nu)\,.$$

White Noise

By analogy with Optics, one calls *white noise* any centered WSS stochastic process $\{B(t)\}_{t\in\mathbb{R}}$ with constant power spectral density $f_B(\nu) = 1$. Such a definition presents a theoretical difficulty, because

$$\int_{-\infty}^{+\infty} f_B(\nu)\,\mathrm{d}\nu = +\infty\,,$$

which contradicts the finite power property of wide-sense stationary processes. We have therefore to find other ways to deal with white noise.

Heuristics I: The Large Flat Spectrum Approach

From a pragmatic point of view, one could define a white noise to be a centered WSS stochastic process whose PSD is constant over a "large", yet bounded, range of frequencies $[-A, +A]$. The calculations below show what happens as A tends to infinity. Let therefore $\{X(t)\}_{t\in\mathbb{R}}$ be a centered WSS stochastic process with PSD

$$f(\nu) = 1_{[-A,+A]}(\nu) .$$

Let φ_1, $\varphi_2 : \mathbb{R} \to \mathbb{C}$ be two functions in $L^1_{\mathbb{C}}(\mathbb{R}) \cap L^2_{\mathbb{C}}(\mathbb{R})$ with Fourier transforms $\widehat{\varphi}_1$ and $\widehat{\varphi}_2$, respectively. Then

$$\begin{aligned}\lim_{A\uparrow\infty} E\left[\left(\int_{\mathbb{R}} \varphi_1(t)X(t)\,\mathrm{d}t\right)\left(\int_{\mathbb{R}} \varphi_2(t)X(t)\,\mathrm{d}t\right)^*\right] &= \int_{\mathbb{R}} \varphi_1(t)\varphi_2^*(t)\,\mathrm{d}t \\ &= \int_{\mathbb{R}} \widehat{\varphi}_1(\nu)\widehat{\varphi}_2^*(\nu)\,\mathrm{d}\nu .\end{aligned}$$

Proof. We have

$$E\left[\left(\int_{\mathbb{R}} \varphi_1(t)X(t)\,\mathrm{d}t\right)\left(\int_{\mathbb{R}} \varphi_2(t)X(t)\,\mathrm{d}t\right)^*\right] = \int_{\mathbb{R}}\int_{\mathbb{R}} \varphi_1(u)\varphi_2(v)^* C_X(u-v)\,\mathrm{d}u\,\mathrm{d}v .$$

The latter quantity is equal to

$$\begin{aligned}&\int_{-\infty}^{+\infty} \varphi_1(u)\varphi_2(v)^* \left(\int_{-A}^{+A} \mathrm{e}^{2i\pi\nu(u-v)}\,\mathrm{d}\nu\right)\mathrm{d}u\,\mathrm{d}v \\ &= \int_{-A}^{+A} \left(\int_{-\infty}^{+\infty} \varphi_1(u)\mathrm{e}^{2i\pi\nu u}\,\mathrm{d}u\right)\left(\int_{-\infty}^{+\infty} \varphi_2(v)^*\mathrm{e}^{-2i\pi\nu v}\,\mathrm{d}v\right)\mathrm{d}\nu \\ &= \int_{-A}^{+A} \widehat{\varphi}_1(-\nu)\widehat{\varphi}_2(-\nu)^*\,\mathrm{d}\nu ,\end{aligned}$$

and the limit of this quantity as $A \uparrow \infty$ is:

$$\int_{-\infty}^{+\infty} \widehat{\varphi}_1(\nu)\widehat{\varphi}_2^*(\nu)\,\mathrm{d}\nu = \int_{-\infty}^{+\infty} \varphi_1(t)\varphi_2(t)^*\,\mathrm{d}t ,$$

where the last equality is the Plancherel–Parseval identity. □

Let now $h : \mathbb{R} \to \mathbb{C}$ be in $L^1_{\mathbb{C}}(\mathbb{R}) \cap L^2_{\mathbb{C}}(\mathbb{R})$, and define

$$Y(t) = \int_{\mathbb{R}} h(t-s)X(s)\,\mathrm{d}s .$$

Applying the above result with $\varphi_1(u) = h(t-u)$ and $\varphi_2(v) = h(t+\tau-v)$, we find that the covariance function C_Y of this WSS stochastic process is such that

$$\lim_{A\uparrow\infty} C_Y(\tau) = \int_{\mathbb{R}} \mathrm{e}^{2i\pi\nu\tau}|\widehat{h}(\nu)|^2\,\mathrm{d}\nu .$$

The limit is finite since $\widehat{h} \in L^2_{\mathbb{C}}(\mathbb{R})$ and is a covariance function corresponding to a *bona fide* (that is, integrable) PDF $f_Y(\nu) = |\widehat{h}(\nu)|^2$. With $f(\nu) \equiv 1$, we formally retrieve the usual filtering formula,

$$f_Y(\nu) = |\widehat{h}(\nu)|^2 f(\nu).$$

Heuristics II: The Approximate Derivative Approach

Here, we consider the white Gaussian noise. The heuristic approach in this case substitutes for $\{B(t)\}_{t\in\mathbb{R}}$ the "finitesimal" derivative of the Brownian motion

$$B_h(t) = \frac{W(t+h) - W(t)}{h}.$$

For fixed $h > 0$ this defines a proper WSS stochastic process centered, with covariance function

$$C_h(\tau) = \frac{(h - |\tau|)^+}{h^2}$$

and (Exercise 12.5.6) power spectral density

$$f_h(\nu) = \left(\frac{\sin \pi\nu h}{\pi\nu h}\right)^2. \tag{12.14}$$

Note that, as $h \downarrow 0$, the power spectral density tends to the constant function 1, the power spectral density of the "white noise". At the same time, the covariance function "tends to the Dirac function" and the energy $C_h(0) = \frac{1}{h}$ tends to infinity. This is another feature of white noise: unpredictability. Indeed, for $\tau \geq h$, the value $B_h(t+\tau)$ cannot be predicted from the value $B_h(t)$, since both are independent random variables.

One then lets

$$\int_{\mathbb{R}_+} f(t)B(t)\,\mathrm{d}t := \int_{\mathbb{R}_+} f(t)B_h(t)\,\mathrm{d}t.$$

The Wiener Approach to White Noise

The third approach to white noise differs from the previous ones, involving limits, in that it consists in working right away "at the limit".

In this approach, one does not attempt to define the white noise $\{B(t)\}_{t\in\mathbb{R}}$ directly (for good reasons since it does not exist as a *bona fide* WSS stochastic process, as we noted earlier). Instead, the symbolic integral $\int_{\mathbb{R}} f(t)B(t)\,\mathrm{d}t$ is defined, for integrands f to be described below, by

$$\int_{\mathbb{R}} f(t)B(t)\,\mathrm{d}t := \int_{\mathbb{R}} f(t)\,dZ(t), \tag{12.15}$$

where $\{Z(t)\}_{t\in\mathbb{R}}$ is a centered stochastic process with uncorrelated increments. One then says that $\{B(t)\}_{t\in\mathbb{R}}$ is a *white noise* and that $\{Z(t)\}_{t\in\mathbb{R}}$ is an *integrated white noise*.

When $\{Z(t)\}_{t\in\mathbb{R}} \equiv \{W(t)\}_{t\in\mathbb{R}}$, a standard Brownian motion, $\{B(t)\}_{t\in\mathbb{R}}$ is called a *Gaussian white noise*.

In the Gaussian white noise case, we have that for all $f, g \in L^2_{\mathbb{C}}(\mathbb{R})$,

$$E\left[\int_{\mathbb{R}} f(t)\, B(t)\, \mathrm{d}t\right] = 0\,,$$

and by the isometry formulas for the Doob–Wiener integral,

$$E\left[\left(\int_{\mathbb{R}} f(t)\, B(t)\, \mathrm{d}t\right)\left(\int_{\mathbb{R}} g(t)\; B(t)\, \mathrm{d}t\right)^*\right] = \int_{\mathbb{R}} f(t)g(t)^*\, \mathrm{d}t\,,$$

which can be formally rewritten, using the Dirac symbolism:

$$\int_{\mathbb{R}} f(t)g(s)^*\, E\left[B(t)B^*(s)\right]\, \mathrm{d}t\, \mathrm{d}s = \int_{\mathbb{R}} f(t)g(s)^*\, \delta(t-s)\, \mathrm{d}t\, \mathrm{d}s\,.$$

Hence "the covariance function of the white noise $\{B(t)\}_{t\in\mathbb{R}}$ is a Dirac pseudo-function: $C_B(\tau) = \delta(\tau)$".

Let $\{B(t)\}_{t\in\mathbb{R}}$ be a white noise with structural measure 1, for example the Gaussian white noise. Let $h : \mathbb{R} \to \mathbb{C}$ be in $L^1_{\mathbb{C}} \cap L^2_{\mathbb{C}}$ and define the output of a filter with impulse response h when the white noise $\{B(t)\}_{t\in\mathbb{R}}$ is the input, by

$$Y(t) = \int_{\mathbb{R}} h(t-s)B(t)\, \mathrm{d}s.$$

By the isometry formula for the Wiener–Doob integral,

$$E[Y(t)Y(s)^*] = \int_{\mathbb{R}} h(t-s-u)h^*(u)\, \mathrm{d}u\,,$$

and therefore (Plancherel–Parseval equality)

$$C_Y(\tau) = \int_{\mathbb{R}} \mathrm{e}^{2i\pi\nu\tau} |\widehat{h}(\nu)|^2\, \mathrm{d}\nu\,.$$

The stochastic process $\{Y(t)\}_{t\in\mathbb{R}}$ is therefore centered and WSS, with power spectral density

$$f_Y(\nu) = |\widehat{h}(\nu)|^2 f_B(\nu)\,,$$

where

$$f_B(\nu) := 1\,.$$

We therefore once more recover formally the fundamental equation of linear filtering of WSS continuous-time stochastic processes.

The connection with the approximate derivative approach is the following: For all $f \in L^2_{\mathbb{C}}(\mathbb{R}) \cap L^1_{\mathbb{C}}(\mathbb{R})$,

$$\lim_{h \downarrow 0} \int_{\mathbb{R}} f(t) B_h(t)\, \mathrm{d}t = \int_{\mathbb{R}} f(t)\, dW(t)$$

in the quadratic mean. The proof is omitted.

12.3 The Cramér–Khinchin Decomposition

Almost surely, a trajectory of a stationary stochastic process is neither in $L^1_{\mathbb{C}}(\ell)$ nor in $L^2_{\mathbb{C}}(\ell)$, unless it is identically null. The formal argument will not be given here[2], but the examples show this convincingly. Therefore such trajectory does not have a Fourier transform in the usual senses. There exists however, in some particular sense, a kind of Fourier spectral decomposition of the trajectories of a WSS stochastic process, as we shall now see.

Theorem 12.3.1 *Let $\{X(t)\}_{t\in\mathbb{R}}$ be a centered* WSS *stochastic process, continuous in the quadratic mean, and let μ be its power spectral measure. There exists a* unique *(more precision below the theorem) centered stochastic process $\{x(\nu)\}_{\nu\in\mathbb{R}}$ with uncorrelated increments and with structural measure μ, such that for all $t \in \mathbb{R}$, P-a.s.,*

$$X(t) = \int_{\mathbb{R}} \mathrm{e}^{2i\pi\nu t}\, \mathrm{d}x(\nu)\,, \tag{12.16}$$

where the integral on the right-hand side is a Doob integral.

The decomposition (12.16) is unique in the following sense: If there exists another centered stochastic process $\{\tilde{x}(\nu)\}_{\nu\in\mathbb{R}}$ with uncorrelated increments, and with finite structural measure $\tilde{\mu}$, such that for all $t \in \mathbb{R}$, we have P-a.s., $X(t) = \int_{\mathbb{R}} \mathrm{e}^{2i\pi\nu t}\, d\tilde{x}(\nu)$, then for all $a, b \in \mathbb{R}$, $a \le b$, $\tilde{x}(b) - \tilde{x}(a) = x(b) - x(a)$, P-a.s.

We shall say: "$\mathrm{d}x(\nu)$ is the *(Cramer–Khinchin) spectral decomposition*" of the WSS stochastic process.

---

[2] See Remark 12.1.1 of [7].

Proof. 1. Denote by $\mathcal{H}(X)$ the vector subspace of $L^2_{\mathbb{C}}(P)$ formed by the finite complex linear combinations of the type

$$Z = \sum_{k=1}^{K} \lambda_k X(t_k)$$

and let us denote by φ the mapping of $\mathcal{H}(X)$ into $L^2_{\mathbb{C}}(\mu)$ defined by

$$\varphi : Z \mapsto \sum_{k=1}^{K} \lambda_k \mathrm{e}^{2i\pi\nu t_k} .$$

Using Bochner's theorem, we verify that it is a linear isometry of $\mathcal{H}(X)$ into $L^2_{\mathbb{C}}(\mu)$:

$$\begin{aligned}
E\left[\left|\sum_{k=1}^{K} \lambda_k X(t_k)\right|^2\right] &= \sum_{k=1}^{K}\sum_{\ell=1}^{K} \lambda_k \lambda_\ell^* E\left[X(t_k)X(t_\ell)^*\right] \\
&= \sum_{k=1}^{K}\sum_{\ell=1}^{K} \lambda_k \lambda_\ell^* C(t_k - t_\ell) = \sum_{k=1}^{K}\sum_{\ell=1}^{K} \lambda_k \lambda_\ell^* \int_R \mathrm{e}^{2i\pi\nu(t_k - t_\ell)}\,\mu(\mathrm{d}\nu) \\
&= \int_R \left(\sum_{k=1}^{K}\sum_{\ell=1}^{K} \lambda_k \lambda_\ell^* \mathrm{e}^{2i\pi\nu(t_k - t_\ell)}\right)\mu(\mathrm{d}\nu) = \int_R \left|\sum_{k=1}^{K} \lambda_k \mathrm{e}^{2i\pi\nu t_k}\right|^2 \mu(\mathrm{d}\nu) .
\end{aligned}$$

2. This isometric linear mapping can be uniquely extended to an isometric linear mapping (that we shall continue to call φ) from $H(X)$, the closure of $\mathcal{H}(X)$, into $L^2_{\mathbb{C}}(\mu)$ (Theorem A.0.6). As the combinations $\sum_{k=1}^{K} \lambda_k \mathrm{e}^{2i\pi\nu t_k}$ are dense in $L^2_{\mathbb{C}}(\mu)$ when μ is a finite measure[3], φ is *onto*. Therefore, it is a linear isometric bijection between $H(X)$ and $L^2_{\mathbb{C}}(\mu)$.

3. Let $x(\nu_0)$ be the random variable in $H(X)$ that corresponds in this isometry to the function $1_{(-\infty,\nu_0]}(\nu)$ of $L^2_{\mathbb{C}}(\mu)$. First, observe that

$$E[x(\nu_2) - x(\nu_1)] = 0$$

since $H(X)$ is the closure in $L^2_{\mathbb{C}}(P)$ of a family of centered random variables. Also, by isometry,

$$\begin{aligned}
E[(x(\nu_2) - x(\nu_1))(x(\nu_4) - x(\nu_3))^*] &= \int_{\mathbb{R}} 1_{(\nu_1,\nu_2]}(\nu) 1_{(\nu_3,\nu_4]}(\nu)\,\mu(\mathrm{d}\nu) \\
&= \mu((\nu_1, \nu_2] \cap (\nu_3, \nu_4]) .
\end{aligned}$$

[3] This will be admitted.

One can therefore define the Doob integral $\int_{\mathbb{R}} f(\nu)\,\mathrm{d}x(\nu)$ for all $f \in L^2_{\mathbb{C}}(\mu)$.

4. Let now

$$Z_n(t) := \sum_{k\in\mathbb{Z}} \mathrm{e}^{2i\pi t(k/2^n)} \left(x\left(\frac{k+1}{2^n}\right) - x\left(\frac{k}{2^n}\right)\right) .$$

We have

$$\lim_{n\to\infty} Z_n(t) = \int_{\mathbb{R}} \mathrm{e}^{2i\pi\nu t}\,\mathrm{d}x(\nu)$$

(limit in $L^2_{\mathbb{C}}(P)$). In fact,

$$Z_n(t) = \int_{\mathbb{R}} f_n(t,\nu)\,\mathrm{d}x(\nu) ,$$

where

$$f_n(t,\nu) := \sum_{k\in\mathbb{Z}} \mathrm{e}^{2i\pi t(k/2^n)} 1_{(k/2^n,(k+1)/2^n]}(\nu) ,$$

and therefore, by isometry,

$$E\left| Z_n(t) - \int_{\mathbb{R}} \mathrm{e}^{2i\pi\nu t}\,\mathrm{d}x(\nu)\right|^2 = \int_{\mathbb{R}} |\mathrm{e}^{2i\pi\nu t} - f_n(t,\nu)|^2\,\mu(\mathrm{d}\nu) ,$$

a quantity which tends to zero when n tends to infinity (by dominated convergence, using the fact that μ is a bounded measure). On the other hand, by definition of φ,

$$Z_n(t) \overset{\varphi}{\mapsto} f_n(t,\nu) .$$

Since, for fixed t, $\lim_{n\to\infty} Z_n(t) = \int_{\mathbb{R}} \mathrm{e}^{2i\pi\nu t}\,\mathrm{d}x(\nu)$ in $L^2_{\mathbb{C}}(P)$ and $\lim_{n\to\infty} f_n(t,\nu) = \mathrm{e}^{2i\pi\nu t}$ in $L^2_{\mathbb{C}}(\mu)$,

$$\int_{\mathbb{R}} \mathrm{e}^{2i\pi\nu t}\,\mathrm{d}x(\nu) \overset{\varphi}{\mapsto} \mathrm{e}^{2i\pi\nu t} .$$

But, by definition of φ,

$$X(t) \overset{\varphi}{\mapsto} \mathrm{e}^{2i\pi\nu t} .$$

Therefore $X(t) = \int_{\mathbb{R}} \mathrm{e}^{2i\pi\nu t}\,\mathrm{d}x(\nu)$.

5. We now prove uniqueness. Suppose that there exists another spectral decomposition $\mathrm{d}\tilde{x}(\nu)$. Denote by $\mathcal{G}$ the set of finite linear combinations of complex exponentials. Since by hypothesis

$$\int_{\mathbb{R}} \mathrm{e}^{2i\pi\nu t}\,\mathrm{d}x(\nu) = \int_{\mathbb{R}} \mathrm{e}^{2i\pi\nu t}\,\mathrm{d}\tilde{x}(\nu) \quad (= X(t))$$

we have

$$\int_{\mathbb{R}} f(\nu)\,\mathrm{d}x(\nu) = \int_{\mathbb{R}} f(\nu)\,\mathrm{d}\tilde{x}(\nu)$$

for all $f \in \mathcal{G}$, and therefore, for all $f \in L^2_{\mathbb{C}}(\mu) \cap L^2_{\mathbb{C}}(\tilde{\mu}) \subseteq L^2_{\mathbb{C}}(\frac{1}{2}(\mu + \tilde{\mu}))$ because $\mathcal{G}$ is dense in $L^2_{\mathbb{C}}(\frac{1}{2}(\mu + \tilde{\mu}))$. In particular, with $f = 1_{(a,b]}$,

$$x(b) - x(a) = \tilde{x}(b) - \tilde{x}(a).$$

□

More details can be obtained as to the continuity properties (in the quadratic mean) of the increments of the spectral decomposition. For instance, it is right-continuous in the quadratic mean, and it admits a left-hand limit in the quadratic mean at any point $\nu \in \mathbb{R}$. If such limit is denoted by $x(\nu-)$, then, for all $a \in \mathbb{R}$,

$$E[|x(a) - x(a-)|^2] = \mu(\{a\}).$$

Proof. The right-continuity follows from the continuity of the (finite) measure μ:

$$\lim_{h \downarrow 0} E[|x(a+h) - x(a)|^2] = \lim_{h \downarrow 0} \mu((a, a+h]) = \mu(\varnothing) = 0.$$

As for the existence of left-hand limits, it is guaranteed by the Cauchy criterion, since for all $a \in \mathbb{R}$,

$$\lim_{h,h' \downarrow 0, h < h'} E[|x(a-h) - x(a-h')|^2] = \lim_{h,h' \downarrow 0, h < h'} \mu((a-h', a-h]) = 0.$$

Finally,

$$E[|x(a) - x(a-)|^2] = \lim_{h \downarrow 0} E[|x(a) - x(a-h)|^2] = \lim_{h \downarrow 0} \mu((a-h, a]) = \mu(\{a\}).$$

□

Theorem 12.3.2 *Let $\{X(t)\}_{t \in \mathbb{R}}$ be a* WSS *stochastic process continuous in the quadratic mean. It is real if and only if its spectral decomposition is hermitian symmetric, that is, for all $[a, b] \subset \mathbb{R}$,*

$$x(b) - x(a) = (x(-a_-) - x(-b_-))^*.$$

Proof. If the stochastic process is real,

$$\begin{aligned} X(t) &= \int_{\mathbb{R}} e^{2i\pi\nu t} \, dx(\nu) = \left(\int_{\mathbb{R}} e^{2i\pi\nu t} \, dx(\nu) \right)^* \\ &= \int_{\mathbb{R}} e^{-2i\pi\nu t} \, dx^*(\nu) = \int_{\mathbb{R}} e^{2i\pi\nu t} \, dx^*(-\nu), \end{aligned}$$

and therefore, by uniqueness of the spectral decomposition, $\mathrm{d}x(\nu) = \mathrm{d}x^*(-\nu)$. Similarly, if $\mathrm{d}x(\nu) = \mathrm{d}x^*(-\nu)$,

$$\begin{aligned} X(t) &= \int_{\mathbb{R}} \mathrm{e}^{2i\pi\nu t}\, \mathrm{d}x(\nu) \\ &= \int_{\mathbb{R}} \mathrm{e}^{2i\pi\nu t}\, \mathrm{d}x^*(-\nu) = \left(\int_{\mathbb{R}} \mathrm{e}^{2i\pi\nu t}\, \mathrm{d}x(\nu) \right)^* = X(t)^*, \end{aligned}$$

and therefore the process is real. □

Theorem 12.3.3 *Let* $\{X(t)\}_{t\in\mathbb{R}}$ *be a centered* WSS *stochastic process continuous in the quadratic mean. Then*

$$H_C(x(\nu); \nu \in \mathbb{R}) = H_C(X(t); t \in \mathbb{R})$$

and both Hilbert subspaces are identical with

$$\{Z = \int_{\mathbb{R}} g(\nu)\, \mathrm{d}x(\nu);\ g \in L^2_{\mathbb{C}}(\mu)\}.$$

Proof. 1. For all $\nu \in \mathbb{R}$, $x(\nu) \in H_{\mathbb{C}}(X(t); t \in \mathbb{R})$ (by definition of $x(\nu)$; see the proof of Theorem 12.3.1). Therefore,

$$H_C(x(\nu); \nu \in \mathbb{R}) \subseteq H_{\mathbb{C}}(X(t); t \in \mathbb{R}).$$

On the other hand, for all $t \in \mathbb{R}$, $X(t) = \int_{\mathbb{R}} \mathrm{e}^{-2i\pi\nu t}\, \mathrm{d}x(\nu) \in H_{\mathbb{C}}(x(\nu); \nu \in \mathbb{R})$. Therefore

$$H_{\mathbb{C}}(X(t); t \in \mathbb{R}) \subseteq H_{\mathbb{C}}(x(\nu); \nu \in \mathbb{R}).$$

2. Defining $H := \{Z = \int_{\mathbb{R}} g(\nu)\, \mathrm{d}x(\nu);\ g \in L^2_{\mathbb{C}}(\mu)\}$, then $H \subseteq H_C(x(\nu))$. Moreover, since H contains all the $X(t) = \int_{\mathbb{R}} \mathrm{e}^{-2i\pi\nu t}\, \mathrm{d}x(\nu)$, $H_{\mathbb{C}}(X(t); t \in \mathbb{R}) \subseteq H$. Therefore

$$H_{\mathbb{C}}(X(t); t \in \mathbb{R}) \subseteq H \subseteq H_C(x(\nu))$$

and the conclusion follows from Part 1 of the proof. □

A Plancherel–Parseval Formula

The following result is the analog of the Plancherel–Parseval formula of classical Fourier analysis.

Theorem 12.3.4 *Let $f : \mathbb{R} \to \mathbb{C}$ be in $L^1_{\mathbb{C}}(\mathbb{R})$ with Fourier transform $\hat{f}$. Let $\{X(t)\}_{t\in\mathbb{R}}$ be a centered* WSS *stochastic process with power spectral measure μ and Cramér–Khinchin spectral decomposition $\mathrm{d}x(\nu)$. Then:*

$$\int_{\mathbb{R}} \hat{f}(\nu)^* \mathrm{d}x(\nu) = \int_{\mathbb{R}} f(t)^* X(t)\, \mathrm{d}t\,. \tag{12.17}$$

Proof. Since $\hat{f}$ is bounded and continuous (as the Fourier transform of an integrable function), and since μ is a finite measure, we have that $\hat{f} \in L^2_{\mathbb{C}}(\mu)$, and

$$\sum_n \hat{f}\left(\frac{k}{2^n}\right) 1_{\left(\frac{k}{2^n}, \frac{k+1}{2^n}\right]} \to \hat{f} \text{ in } L^2_{\mathbb{C}}(\mu)$$

and therefore (all limits in the following sequence of equalities are in $L^2_{\mathbb{C}}(P)$):

$$\begin{aligned}
\int_{\mathbb{R}} \hat{f}(\nu)^* \,\mathrm{d}x(\nu) &= \lim_{n\to\infty} \sum_{-n2^n}^{n2^n-1} \hat{f}\left(\frac{k}{2^n}\right)^* \left(x\left(\frac{k+1}{2^n}\right) - x\left(\frac{k}{2^n}\right)\right) \\
&= \lim_{n\to\infty} \sum_{-n2^n}^{n2^n-1} \left(\int_{\mathbb{R}} f^*(t) \mathrm{e}^{+2i\pi(k/2^n)t}\,\mathrm{d}t\right)\left(x\left(\frac{k+1}{2^n}\right) - x\left(\frac{k}{2^n}\right)\right) \\
&= \lim_{n\to\infty} \int_{\mathbb{R}} f^*(t) \sum_{-n2^n}^{n2^n-1} \left[\mathrm{e}^{+2i\pi(k/2^n)t}\left(x\left(\frac{k+1}{2^n}\right) - x\left(\frac{k}{2^n}\right)\right)\right] \mathrm{d}t \\
&= \lim_{n\to\infty} \int_{\mathbb{R}} f^*(t) X_n(t)\,\mathrm{d}t\,,
\end{aligned}$$

where

$$X_n(t) = \sum_{-n2^n}^{n2^n-1} \mathrm{e}^{+2i\pi(k/2^n)t}\left(x\left(\frac{k+1}{2^n}\right) - x\left(\frac{k}{2^n}\right)\right) \to X(t) \text{ in } L^2_{\mathbb{C}}(P)\,.$$

The announced result will then follow once we prove that

$$\lim_{n\to\infty} \int_{\mathbb{R}} f^*(t) X_n(t)\,\mathrm{d}t = \int_{\mathbb{R}} f^*(t) X(t)\,\mathrm{d}t\,,$$

where the limit is in $L^2_{\mathbb{C}}(P)$. In fact, with $Y_n(t) = X(t) - X_n(t)$,

$$E\left[\left|\int_{\mathbb{R}} f(t) Y_n(t)\,\mathrm{d}t\right|^2\right] = \int_{\mathbb{R}}\int_{\mathbb{R}} f(t) f(s)^* E\left[Y_n(t) Y_n(s)^*\right] \mathrm{d}t\,\mathrm{d}s\,.$$

But $\lim_{n\uparrow\infty} Y_n(t) = 0$ (in $L^2_{\mathbb{C}}(P)$) and therefore $\lim_{n\uparrow\infty} E[Y_n(t)Y_n(s)^*] = 0$. Moreover $E[Y_n(t)Y_n(s)^*]$ is uniformly bounded in n. Therefore, by dominated convergence,

$$\lim_{n\uparrow\infty} \int_{\mathbb{R}} \int_{\mathbb{R}} f(t)f(s)^* E[Y_n(t)Y_n(s)^*] \, \mathrm{d}t \, \mathrm{d}s = 0 \, .$$

□

EXAMPLE 12.3.5: CONVOLUTIONAL FILTERING. Let $h \in L^1_{\mathbb{C}}(\mathbb{R})$ and let $\hat{h}$ be its Fourier transform. Then

$$\int_{\mathbb{R}} h(t-s)X(s) \, \mathrm{d}s = \int_{\mathbb{R}} \hat{h}(\nu) \mathrm{e}^{2i\pi\nu t} \, \mathrm{d}x(\nu) \, . \tag{12.18}$$

Proof. It suffices to apply (12.17) to the function $s \mapsto h^*(t-s)$, whose Fourier transform is $\hat{h}(\nu)^* \mathrm{e}^{-2i\pi\nu t}$. □

Linear Operations on WSS Stochastic Processes

A function $g : \mathbb{R} \to \mathbb{C}$ in $L^2_{\mathbb{C}}(\mu)$ defines a linear operation on the centered WSS stochastic process $\{X(t)\}_{t\in\mathbb{R}}$ (called the *input*) by associating with it the centered stochastic process (called the *output*)

$$Y(t) = \int_{\mathbb{R}} \mathrm{e}^{2i\pi\nu t} g(\nu) \, \mathrm{d}x(\nu) \, . \tag{12.19}$$

On the other hand, the calculation of the covariance function

$$C_Y(\tau) = E[Y(t)Y(t+\tau)^*]$$

of the output gives, by isometry,

$$C_Y(\tau) = \int_{\mathbb{R}} \mathrm{e}^{2i\pi\nu\tau} |g(\nu)|^2 \, \mu_X(\mathrm{d}\nu) \, ,$$

where μ_X is the power spectral measure of the input. The power spectral measure of the output process is then

$$\mu_Y(\mathrm{d}\nu) = |g(\nu)|^2 \, \mu_X(\mathrm{d}\nu) \, . \tag{12.20}$$

This is similar to the formula obtained when $\{Y(t)\}_{t\in\mathbb{R}}$ is the output of a stable convolutional filter with impulse response h and transmittance $\hat{h}$: $\mu_Y(\mathrm{d}\nu) =$

$|\hat{h}(\nu)|^2 \, \mu_X(\mathrm{d}\nu)$. We therefore say that g is the *transmittance* of the "filter" (12.19). Note however that this filter is not necessarily of the convolutional type, since g may well not be the Fourier transform of an integrable function (for instance it may be unbounded, as the next example shows).

EXAMPLE 12.3.6: DIFFERENTIATION. Let $\{X(t)\}_{t\in\mathbb{R}}$ be a WSS stochastic processes with spectral measure μ_X such that

$$\int_{\mathbb{R}} |\nu|^2 \, \mu_X(\mathrm{d}\nu) < \infty . \tag{12.21}$$

Then

$$\lim_{h\to 0} \frac{X(t+h) - X(t)}{h} = \int_{\mathbb{R}} (2i\pi\nu) \mathrm{e}^{2i\pi\nu t} \mathrm{d}x(\nu) ,$$

where the limit is in the quadratic mean. The linear operation corresponding to the transmittance $g(\nu) = 2i\pi\nu$ is therefore the *differentiation in quadratic mean.*

Proof. Let $h \in \mathbb{R}$. From the equality

$$\frac{X(t+h) - X(t)}{h} - \int_{\mathbb{R}} (2i\pi\nu) \mathrm{e}^{2i\pi\nu t} \mathrm{d}x(\nu) = \int_{\mathbb{R}} \mathrm{e}^{2i\pi\nu t} \left(\frac{\mathrm{e}^{2i\pi\nu h} - 1}{h} - 2i\pi\nu \right) \mathrm{d}x(\nu)$$

we have, by isometry,

$$\lim_{h\to 0} E\left[\left| \frac{X(t+h) - X(t)}{h} - \int_{\mathbb{R}} (2i\pi\nu) \mathrm{e}^{2i\pi\nu t} \mathrm{d}x(\nu) \right|^2 \right]$$
$$= \lim_{h\to 0} \int_{\mathbb{R}} \left| \frac{\mathrm{e}^{2i\pi\nu h} - 1}{h} - 2i\pi\nu \right|^2 \mu_X(\mathrm{d}\nu) .$$

The latter limit is 0, by dominated convergence, since $\left| \frac{\mathrm{e}^{2i\pi\nu h} - 1}{h} - 2i\pi\nu \right|^2 \leq 4\pi^2\nu^2$ and in view of the hypothesis (12.21). □

"A line spectrum corresponds to a combination of sinusoids." More precisely:

Theorem 12.3.7 *Let $\{X(t)\}_{t\in\mathbb{R}}$ be a centered* WSS *stochastic processes with spectral measure*

$$\mu_X(\mathrm{d}\nu) = \sum_{k\in\mathbb{Z}} P_k\, \epsilon_{\nu_k}(\mathrm{d}\nu)\,,$$

where ϵ_{ν_k} is the Dirac measure at $\nu_k \in \mathbb{R}$, $P_k \in \mathbb{R}_+$ and $\sum_{k\in\mathbb{Z}} P_k < \infty$. Then

$$X(t) = \sum_{k\in\mathbb{Z}} U_k \mathrm{e}^{2i\pi\nu_k t}\,,$$

where $\{U_k\}_{k\in\mathbb{Z}}$ is a sequence of centered uncorrelated square-integrable complex variables, and $E[|U_k|^2] = P_k$.

Proof. Let

$$g(\nu) = \sum_{k\in\mathbb{Z}} 1_{\{\nu_k\}}(\nu)\,.$$

It is in $L^2_{\mathbb{C}}(\mu_X)$, as is $1 - g(\nu)$. Also $\int_{\mathbb{R}} |1 - g(\nu)|^2\, \mu_X(\mathrm{d}\nu) = 0$, and in particular $\int_{\mathbb{R}} (1 - g(\nu))\mathrm{e}^{2i\pi\nu t}\, \mathrm{d}x(\nu) = 0$. Therefore

$$\begin{aligned} X(t) &= \int_{\mathbb{R}} g(\nu)\mathrm{e}^{2i\pi\nu t}\, \mathrm{d}x(\nu) \\ &= \sum_{k\in\mathbb{Z}} \mathrm{e}^{2i\pi\nu_k t}(x(\nu_k) - x(\nu_k-))\,. \end{aligned}$$

We conclude by defining $U_k = x(\nu_k) - x(\nu_k-)$. □

Linear Transformations of Gaussian Processes

We call a *linear transformation* of the WSS stochastic process $\{X(t)\}_{t\in\mathbb{R}}$ a transformation of it into the second-order process (not WSS in general)

$$Y(t) = \int_{\mathbb{R}} g(\nu, t)\, \mathrm{d}x(\nu)\,, \tag{12.22}$$

where

$$\int_{\mathbb{R}} |g(t, \nu)|^2\, \mu_X(\mathrm{d}\nu) < \infty \quad \text{for all } t \in \mathbb{R}\,.$$

Theorem 12.3.8 *Every linear transformation of a Gaussian* WSS *stochastic process yields a Gaussian stochastic process.*

Proof. Let $\{X(t)\}_{t\in\mathbb{R}}$ be centered, Gaussian, WSS, with Cramer–Khinchin decomposition $\mathrm{d}x(\nu)$. For each $\nu \in \mathbb{R}$, the random variable $x(\nu)$ is in $H_{\mathbb{R}}(X)$, by construction. Now, if $\{X(t)\}_{t\in\mathbb{R}}$ is a Gaussian process, $H_{\mathbb{R}}(X)$ is a Gaussian subspace. But (Theorem 12.3.3) $H_{\mathbb{R}}(X) = H_{\mathbb{R}}(x)$. Therefore the process (12.22) is in $H_{\mathbb{C}}(X)$, hence Gaussian. □

EXAMPLE 12.3.9: CONVOLUTIONAL FILTERING OF A WSS GAUSSIAN PROCESS. In particular, if $\{X(t)\}_{t\in\mathbb{R}}$ is a Gaussian WSS process with Cramer–Khinchin decomposition $\mathrm{d}x(\nu)$, and if $g \in L^2_{\mathbb{C}}(\mu_X)$, the process

$$Y(t) = \int_{\mathbb{R}} \mathrm{e}^{2i\pi\nu t} g(\nu)\,\mathrm{d}x(\nu)$$

is a Gaussian process.

A particular case is when $g = \hat{h}$, the Fourier transform of a filter with integrable impulse response h; the signal $\{Y(t)\}_{t\in\mathbb{R}}$ is the one obtained by convolutional filtering of $\{X(t)\}_{t\in\mathbb{R}}$ with this filter.

12.4 Multivariate WSS Stochastic Processes

Let $\{X(t)\}_{t\in\mathbb{R}}$ be a stochastic process with values in $E := \mathbb{C}^L$, where L is an integer greater than or equal to 2: $X(t) = (X_1(t), \ldots, X_L(t))$. This process is assumed to be of the second order, that is:

$$E[||X(t)||^2] < \infty \quad \text{for all} \quad t \in \mathbb{R},$$

and centered. Furthermore, it will be assumed that it is wide-sense stationary, in the sense that the mean vector of $X(t)$ and the cross-covariance matrix of the vectors $X(t+\tau)$ and $X(t)$ do not depend upon t. The matrix-valued function C defined by

$$C(\tau) = \mathrm{cov}\,(X(t+\tau), X(t)) \tag{12.23}$$

is called the (matrix) *covariance function* of the stochastic process. Its general entry is

$$C_{ij}(\tau) = \mathrm{cov}(X_i(t), X_j(t+\tau)).$$

Therefore, each of the processes $\{X_i(t)\}_{t\in\mathbb{R}}$ is a WSS stochastic process, but, furthermore, they are *stationarily correlated* or "jointly WSS". The vector-valued stochastic process $\{X(t)\}_{t\in\mathbb{R}}$ is then called a *multivariate* WSS *stochastic process*.

EXAMPLE 12.4.1: SIGNAL PLUS NOISE. The following model frequently appears in signal processing:

$$Y(t) = S(t) + B(t),$$

where $\{S(t)\}_{t\in\mathbb{R}}$ and $\{B(t)\}_{t\in\mathbb{R}}$ are two *uncorrelated* centered WSS stochastic processes with respective covariance functions C_S and C_B. Then, $\{(Y(t), S(t))^T\}_{t\in\mathbb{R}}$ is a bivariate WSS stochastic process. In fact, by the assumption of non-correlation:

$$C(\tau) = \begin{pmatrix} C_S(\tau) + C_B(\tau) & C_S(\tau) \\ C_S(\tau) & C_S(\tau) \end{pmatrix}.$$

We shall need at this point a minor extension of the notion of measure.

Definition 12.4.2 *A* finite complex measure *on the measurable space* $(X, \mathcal{X})$ *is, by definition, a mapping* $\mu : \mathcal{X} \to \mathbb{C}$ *of the form*

$$\mu = \mu_R + i\mu_I,$$

where μ_R *and* μ_I *are finite measures on* $(X, \mathcal{X})$. *The integral of a measurable function* $f : (X, \mathcal{X}) \to (\mathbb{R}, \mathcal{B}(\mathbb{R}))$ *with respect to such measure is defined by*

$$\int_X f(x)\mu(\mathrm{d}x) := \int_X f(x)\mu_R(\mathrm{d}x) + i\int_X f(x)\mu_I(\mathrm{d}x)$$

whenever f *is integrable with respect to both* μ_R *and* μ_I.

Theorem 12.4.3 *Let* $\{X(t)\}_{t\in\mathbb{R}}$ *be an L-dimensional multivariate* WSS *stochastic process. For all* r, s $(1 \le r, s \le L)$ *there exists a finite complex measure* μ_{rs} *such that*

$$C_{rs}(\tau) = \int_{\mathbb{R}} \mathrm{e}^{2i\pi\nu\tau} \mu_{rs}(\mathrm{d}\nu). \tag{12.24}$$

Proof. (The case $r = 1$, $s = 2$). Let us consider the stochastic processes

$$Y(t) = X_1(t) + X_2(t), \qquad Z(t) = iX_1(t) + X_2(t).$$

These are WSS stochastic processes with respective covariance functions

$$C_Y(\tau) = C_1(\tau) + C_2(\tau) + C_{12}(\tau) + C_{21}(\tau),$$
$$C_Z(\tau) = -C_1(\tau) + C_2(\tau) + iC_{12}(\tau) - iC_{21}(\tau).$$

¿From these two equalities we deduce

$$C_{12}(\tau) = \frac{1}{2}\{[C_Y(\tau) - C_1(\tau) - C_2(\tau)] - i[C_Z(\tau) - C_1(\tau) + C_2(\tau)]\},$$

from which the result follows with

$$\mu_{12} = \frac{1}{2}\{[\mu_Y - \mu_1 - \mu_2] - i[\mu_Z - \mu_1 + \mu_2]\}.$$

□

The matrix

$$M := \{\mu_{ij}\}_{1\le i,j\le k}$$

(whose entries are finite complex measures) is the *interspectral power measure matrix* of the multivariate WSS stochastic process $\{X(t)\}_{t\in\mathbb{R}}$. It is clear that for all $z = (z_1, \dots, z_k) \in \mathbb{C}^k$, $U(t) = z^T X(t)$ defines a WSS stochastic process with spectral measure $\mu_U = z\, M\, z^\dagger$ (recall that † means transpose conjugate).

The link between the interspectral measure μ_{12} and the Cramer–Khinchin decompositions $\mathrm{d}x_1(\nu)$ and $\mathrm{d}x_2(\nu)$ is the following:

$$E[x_1(\nu_2) - x_1(\nu_1))(x_2(\nu_4) - x_2(\nu_3))^*] = \mu_{12}((\nu_1, \nu_2] \cup (\nu_3, \nu_4]).$$

This is a particular case of the following: for all functions $g_i : \mathbb{R} \to \mathbb{C}$, $g_i \in L^2_{\mathbb{C}}(\mu_i)$ $(i = 1, 2)$

$$E\left[\left(\int_{\mathbb{R}} g_1(\nu)\,\mathrm{d}x_1(\nu)\right)\left(\int_{\mathbb{R}} g_2(\nu)\,\mathrm{d}x_2(\nu)\right)^*\right] = \int_{\mathbb{R}} g_1(\nu) g_2(\nu)^*\,\mu_{12}(\mathrm{d}\nu). \quad (12.25)$$

Indeed, equality (12.25) is true for $g_1(\nu) = \mathrm{e}^{2i\pi t_1\nu}$, $g_2(\nu) = \mathrm{e}^{2i\pi t_2\nu}$, since it then reduces to

$$E[X_1(t)X_2(t)^*] = \int_{\pi} \mathrm{e}^{2i\pi(t_1-t_2)\nu}\,\mu_{12}(\mathrm{d}\nu).$$

This is therefore verified for $g_1, g_2 \in \mathcal{E}$, the set of finite linear combinations of functions of the type $\nu \mapsto \mathrm{e}^{2i\pi t\nu}$ $(t \in \mathbb{R})$. But $\mathcal{E}$ is dense in $L^2_{\mathbb{C}}(\mu_i)$ $(i = 1, 2)$,[4] and therefore the equality (12.25) is true for all $g_i \in L^2_{\mathbb{C}}(\mu_i)$ $(i = 1, 2)$.

Theorem 12.4.4 *The interspectral measure μ_{12} is absolutely continuous with respect to each of the spectral measures μ_1 and μ_2.*

Proof. This means that $\mu_{12}(A) = 0$ whenever $\mu_1(A) = 0$ or $\mu_2(A) = 0$. Indeed,

$$\mu_{12}(A) = E\left[\left(\int_A \mathrm{d}Z_1\right)\left(\int_A \mathrm{d}Z_2\right)^*\right]$$

[4] This will be admitted.

and $\mu_1(A) = 0$ implies $\int_A \mathrm{d}Z_1 = 0$ since

$$E\left[\left|\int_A \mathrm{d}Z_1\right|^2\right] = \mu_1(A).$$

□

Therefore, each of the spectral measures μ_{ij} is absolutely continuous with respect to the trace

$$\mathrm{Tr}\, M := \sum_{j=1}^{k} \mu_j$$

of the power spectral measure matrix. By the Radon–Nikodym theorem there exists a function $g_{ij} : \mathbb{R} \to \mathbb{C}$ such that

$$\mu_{ij}(A) = \int_A g_{ij}(\nu)\, \mathrm{Tr}\ M(\mathrm{d}\nu)\,.$$

We say that the matrix

$$g(\nu) = \{g_{ij}(\nu)\}_{1\le i,j\le k}$$

is the *canonical spectral density* matrix of $\{X(t)\}_{t\in\mathbb{R}}$. One should insist that it is not required that the stochastic processes $\{X_i(t)\}_{t\in\mathbb{R}}$, $1 \le i \le k$, admit power spectral densities.

The correlation matrix $C(\tau)$ has, with the above notations, the representation

$$C(\tau) = \int_{\mathbb{R}} \mathrm{e}^{2i\pi\nu\tau} g(\nu)\, \mathrm{Tr}\ M(\mathrm{d}\nu)\,.$$

If each of the WSS stochastic processes $\{X_i(t)\}_{t\in\mathbb{R}}$ admits a spectral density, $\{X(t)\}_{t\in\mathbb{R}}$ admits an interspectral density matrix

$$f(\nu) = \{f_{ij}(\nu)\}_{1\le i,j\le k}\,,$$

that is:

$$C_{ij}(\tau) = \mathrm{cov}\ (X_i(t+\tau), X_j(t)) = \int_{\mathbb{R}} \mathrm{e}^{2i\pi\nu\tau} f_{ij}(\nu)\, \mathrm{d}\nu\,.$$

EXAMPLE 12.4.5: INTERFERENCES. Let $\{X(t)\}_{t\in\mathbb{R}}$ be a centered WSS stochastic process with power spectral measure μ_X. Let $h_1, h_2 : \mathbb{R} \to \mathbb{C}$ be integrable functions with respective Fourier transforms $\widehat{h}_1$ and $\widehat{h}_2$. Define for $i = 1, 2$,

$$Y_i(t) = \int_{\mathbb{R}} h_i(t-s) X(s)\, \mathrm{d}s\,.$$

The WSS stochastic processes $\{Y_1(t)\}_{t\in\mathbb{R}}$ and $\{Y_2(t)\}_{t\in\mathbb{R}}$ are stationarily correlated. In fact (assuming that they are centered, without loss of generality),

$$\begin{aligned}E[Y_1(t+\tau)Y_2(t)^*] &= E\left[\left(\int_{\mathbb{R}} h_1(t+\tau-s)X(s)\,\mathrm{d}s\right)\left(\int_{\mathbb{R}} h_2(t-s)X(s)\,\mathrm{d}s\right)^*\right] \\ &= \int_{\mathbb{R}}\int_{\mathbb{R}} h_1(t+\tau-u)h_2^*(t-v)C_X(u-v)\,\mathrm{d}u\,\mathrm{d}v \\ &= \int_{\mathbb{R}}\int_{\mathbb{R}} h_1(\tau-u)h_2^*(-v)C_X(u-v)\,\mathrm{d}u\,\mathrm{d}v\,,\end{aligned}$$

and this quantity depends only upon τ. Replacing $C_X(u-v)$ by its expression in terms of the spectral measure μ_X, one obtains

$$C_{Y_1Y_2}(\tau) = \int_{\mathbb{R}} \mathrm{e}^{2i\pi\nu\tau} T_1(\nu)T_2^*(\nu)\,\mu_X(\mathrm{d}\nu)\,.$$

The power spectral matrix of the bivariate process $\{Y_1(t), Y_2(t)\}_{t\in\mathbb{R}}$ is therefore

$$\mu_Y(\mathrm{d}\nu) = \begin{pmatrix} |T_1(\nu)|^2 & T_1(\nu)T_2^*(\nu) \\ T_1^*(\nu)T_2(\nu) & |T_2(\nu)|^2 \end{pmatrix} \mu_X(\mathrm{d}\nu)\,.$$

Band-pass Stochastic Processes

Let $\{X(t)\}_{t\in\mathbb{R}}$ be a centered WSS stochastic process with power spectral measure μ_X and Cramér–Khinchin decomposition $\mathrm{d}x(\nu)$. This process is assumed real, and therefore

$$\mu_X(-\mathrm{d}\nu) = \mu_X(\mathrm{d}\nu), \qquad \mathrm{d}x(-\nu) = \mathrm{d}x(\nu)^*\,.$$

Definition 12.4.6 *The above* WSS *stochastic process is called* band-pass (ν_0, B), *where* $\nu_0 > B > 0$, *if the support of* μ_X *is contained in the frequency band* $[-\nu_0 - B, -\nu_0 + B] \cup [\nu_0 - B, \nu_0 + B]$. *It is called* base-band (B) *in if in addition* $\nu_0 = 0$.

Our purpose is to show that such a band-pass stochastic process admits the following *quadrature decomposition*

$$X(t) = M(t)\cos 2\pi\nu_0 t - N(t)\sin 2\pi\nu_0 t\,, \tag{12.26}$$

where $\{M(t)\}_{t\in\mathbb{R}}$ and $\{N(t)\}_{t\in\mathbb{R}}$, called the *quadrature components*, are real base-band (B) WSS stochastic process. To prove this, let $G(\nu) := -i\,\mathrm{sign}(\nu)$ (= 0 if

$\nu = 0$). The function G is the so-called *Hilbert filter* transmittance. The *quadrature process* associated with $\{X(t)\}_{t\in\mathbb{R}}$ is defined by

$$Y(t) = \int_{\mathbb{R}} G(\nu) e^{2i\pi\nu t} \, dx(\nu) .$$

The right-hand side of the preceding equality is well defined since $\int_{\mathbb{R}} |G(\nu)|^2 \mu_X(d\nu) = \mu_X(\mathbb{R}) < \infty$. Moreover, this stochastic process is real, since its spectral decomposition is hermitian symmetric. The *analytic process* associated with $\{X(t)\}_{t\in\mathbb{R}}$ is, by definition, the stochastic process

$$Z(t) = X(t) + iY(t) = \int_{\mathbb{R}} (1 + iG(\nu)) e^{2i\pi\nu t} \, dx(\nu) = 2 \int_{(0,\infty)} e^{2i\pi\nu t} \, dx(\nu) .$$

Taking into account that $|G(\nu)|^2 = 1$, the preceding expressions and the Wiener isometry formulas lead to the following properties:

$$\mu_Y(d\nu) = \mu_X(d\nu), \qquad C_Y(\tau) = C_X(\tau), \qquad C_{XY}(\tau) = -\,C_{YX}(\tau),$$

$$\mu_Z(d\nu) = 4\, 1_{\mathbb{R}_+}(\nu)\, \mu_X(d\nu), \qquad C_Z(\tau) = 2\left\{C_X(\tau) + iC_{YX}(\tau)\right\},$$

and

$$E[Z(t+\tau)Z(t)] = 0 . \qquad (\star)$$

Defining the *complex envelope* of $\{X(t)\}_{t\in\mathbb{R}}$ by

$$U(t) = Z(t) e^{-2i\pi\nu_0 t} , \qquad (\star\star)$$

it follows from this definition that

$$C_U(\tau) = e^{-2i\pi\nu_0\tau} C_Z(\tau), \qquad \mu_U(d\nu) = \mu_Z(d\nu + \nu_0), \qquad (\dagger)$$

whereas $(\star)$ and $(\star\star)$ give

$$E[U(t+\tau)U(t)] = 0 . \qquad (\dagger\dagger)$$

The quadrature components $\{M(t)\}_{t\in\mathbb{R}}$ and $\{N(t)\}_{t\in\mathbb{R}}$ of $\{X(t)\}_{t\in\mathbb{R}}$ are the *real* WSS stochastic processes defined by

$$U(t) = M(t) + iN(t) .$$

Since

$$X(t) = \mathrm{Re}\{Z(t)\} = \mathrm{Re}\{U(t) e^{2i\pi\nu_0 t}\} ,$$

we have the decomposition (12.26). Taking (††) into account we obtain:

$$C_M(\tau) = C_N(\tau) = \frac{1}{4}\{C_U(\tau) + C_U(\tau)^*\},$$

and

$$C_{MN}(\tau) = C_{NM}(\tau) = \frac{1}{4i}\{C_U(\tau) - C_U(\tau)^*\}, \tag{$\diamondsuit$}$$

and the corresponding relations for the spectra

$$\mu_M(\mathrm{d}\nu) = \mu_N(\mathrm{d}\nu) = \{\mu_X(\mathrm{d}\nu - \nu_0) + \mu_X(\mathrm{d}\nu + \nu_0)\}\, 1_{[-B,+B]}(\nu).$$

From ($\diamondsuit$) and the observation that $C_U(0) = C_U(0)^*$ (since $C_U(0) = E[|U(0)|^2]$ is real), we deduce $C_{MN}(0) = 0$, that is to say,

$$E[M(t)N(t)] = 0. \tag{12.27}$$

If, furthermore, the original process has a power spectral measure that is symmetric about ν_0 in the band $[\nu_0 - B, \nu_0 + B]$, the same holds for the spectrum of the analytic process and, by (†), the complex envelope has a spectral measure symmetric about 0, which implies $C_U(\tau) = C_U(\tau)^*$ and then, by ($\diamondsuit$),

$$E[M(t)N(t+\tau)] = 0. \tag{12.28}$$

In summary:

Theorem 12.4.7 *Let $\{X(t)\}_{t\in\mathbb{R}}$ be a centered real band-pass (ν_0, B)* WSS *stochastic process. The values of its quadrature components at a given time are uncorrelated. Moreover, if the original stochastic process has a power spectral measure symmetric about ν_0, the quadrature component processes are uncorrelated.*

More can be said when the original process is Gaussian. In this case, the quadrature component processes are jointly Gaussian (being obtained from the original Gaussian process by linear operations). In particular, for all $t \in \mathbb{R}$, $M(t)$ and $N(t)$ are jointly Gaussian and uncorrelated, and therefore independent.

If moreover the original process has a spectrum symmetric about ν_0, then, by (12.28), $M(t_1)$ and $N(t_2)$ ($t_1, t_2 \in \mathbb{R}$) are uncorrelated jointly Gaussian variables, and therefore independent. In other words, the quadrature component processes are two independent centered Gaussian WSS stochastic processes.

12.5 Exercises

Exercise 12.5.1. APPROXIMATE DERIVATIVE OF THE BROWNIAN MOTION
Prove Formula 12.14.

Exercise 12.5.2. STATIONARIZATION OF A CYCLIC STOCHASTIC PROCESS
Let $\{Y(t)\}_{t\geq 0}$ be the stochastic process taking its values in $\{-1,+1\}$ defined by

$$Y(t) := Z \times (-1)^n \text{ on } (nT, (n+1)T] \quad (n \geq 0)\,,$$

where T is a positive real number and Z is a random variable equidistributed on $\{-1,+1\}$.

(1) Show that $\{Y(t)\}_{t\geq 0}$ is not a stationary (neither strictly nor in the wide sense) stochastic process.

(2) Let now U be a random variable uniformly distributed on $[0,T]$ and independent of Z. Define for all $t \geq 0$,

$$X(t) = Y(t-U)^+\,.$$

Show that $\{X(t)\}_{t\geq 0}$ is a strictly stationary stochastic process and compute its covariance function.

Exercise 12.5.3. AN ERGODIC PROPERTY
Let $\{X(t)\}_{t\geq 0}$ be a wide-sense stationary stochastic process with mean m and covariance function $C(\tau)$. Prove that in order that

$$\lim_{T\uparrow\infty} \frac{1}{T}\int_0^T X(s)\,\mathrm{d}s = m_X$$

holds in the quadratic mean, it is necessary and sufficient that

$$\lim_{T\uparrow\infty} \frac{1}{T}\int_0^T \left(1-\frac{u}{T}\right) C(u)\,\mathrm{d}u = 0\,. \tag{12.29}$$

Show that this condition is satisfied in particular when the covariance function is integrable.

Exercise 12.5.4. SYMMETRIC POWER SPECTRAL MEASURE
Show that the power spectral measure of a real WSS stochastic process is symmetric.

Exercise 12.5.5. PRODUCTS OF INDEPENDENT WSS STOCHASTIC PROCESSES
Let $\{X(t)\}_{t\in\mathbb{R}}$ and $\{Y(t)\}_{t\in\mathbb{R}}$ be two independent centered WSS stochastic processes of respective covariance functions $C_X(\tau)$ and $C_Y(\tau)$.

1. Show that $Z(t) := X(t)Y(t)$ $(t \in \mathbb{R})$ is a WSS stochastic process. Give its mean and covariance function.

2. Assume in addition that $\{X(t)\}_{t\in\mathbb{R}}$ is the harmonic process of Example 11.1.14. Suppose that $\{Y(t)\}_{t\in\mathbb{R}}$ admits a power spectral density $f_Y(\nu)$. Give the power spectral density $f_Z(\nu)$ of $\{Z(t)\}_{t\in\mathbb{R}}$.

Exercise 12.5.6. THE APPROXIMATE DERIVATIVE OF A WIENER PROCESS
Let $\{W(t)\}_{t\geq 0}$ be a Wiener process. Show that for $a > 0$, the stochastic process

$$X_a(t) := \frac{W(t+a) - W(t)}{a} \quad (t \in \mathbb{R})$$

is a WSS stochastic process. Compute its mean, its covariance function and its power spectral density.

Exercise 12.5.7. THE SQUARE OF A BAND-LIMITED WHITE NOISE
Let $\{X(t)\}_{t\in\mathbb{R}}$ be a wide-sense stationary *centered* Gaussian process with covariance function $C_X(\tau)$ and with the power spectral density

$$f_X(\nu) = \frac{N_0}{2} 1_{[-B,+B]}(\nu),$$

where $N_0 > 0$ and $B > 0$.

1. Let $Y(t) = X(t)^2$. Show that $\{Y(t)\}_{t\in\mathbb{R}}$ is a wide-sense stationary process.

2. Give its power spectral density $f_Y(\nu)$.

Exercise 12.5.8. PROJECTION OF WHITE NOISE ONTO AN ORTHONORMAL BASE
Let the set of square-integrable functions $\varphi : [0,T] \to \mathbb{R}$ $(1 \leq i \leq N)$ be such that

$$\int_0^T \varphi_i(t)\varphi_j(t)\,\mathrm{d}t = \delta_{ij} \quad (1 \leq i, j \leq N),$$

and let $\{B(t)\}_{t\in\mathbb{R}}$ be a Gaussian white noise with PSD 1. Show that the vector $B = (B_1, \ldots, B_N)^T$ defined by

$$B_i = \int_0^T B(t)\varphi_i(t)\,\mathrm{d}t \quad (1 \leq i \leq N)$$

is a centered Gaussian vector with covariance matrix $\Gamma_B = I$, the identity matrix of size N (In particular, the components $B_1, \ldots, B_N$ are identically distributed, independent, and centered Gaussian random variables with common variance 1.)

Exercise 12.5.9. AN IID SEQUENCE CARRIED BY AN HPP
Let N be a homogeneous Poisson process on $\mathbb{R}_+$ of intensity $\lambda > 0$, and let $\{Z_n\}_{n\geq 0}$ be an IID sequence of integrable real random variables, centered, with finite variance σ^2, and independent of N.

1) Show that $\{Z_{N((0,t])}\}_{t\geq 0}$ is a wide-sense stationary stochastic process and give its covariance function.

2) Give its power spectral density.

3) Compute $P(X(t_1) = X(t_2))$ and $P(X(t_1) > X(t_2))$.

Exercise 12.5.10. POISSON SHOT NOISES
Let N_1, N_2 and N_3 be three independent homogeneous Poisson processes on $\mathbb{R}$ with respective intensities $\theta_1 > 0$, $\theta_2 > 0$ and $\theta_3 > 0$. Let $\{X_1(t)\}_{t\in\mathbb{R}}$ be the shot noise constructed on $N_1 + N_3$ with an impulse function $h : \mathbb{R} \to \mathbb{R}$ that is bounded and with compact support (null outside a finite interval). Let $\{X_2(t)\}_{t\in\mathbb{R}}$ be the shot noise constructed on $N_2 + N_3$ with the same impulse function h.

Compute the power spectral density of the wide-sense stationary process $\{X(t)\}_{t\in\mathbb{R}}$, where $X(t) = X_1(t) + X_2(t)$.

Exercise 12.5.11. FLIP-FLOP
Let N be an HPP on $\mathbb{R}_+$ with intensity λ. Define the (*telegraph* or *flip-flop*) process $\{X(t)\}_{t\geq 0}$ with state space $E = \{+1, -1\}$ by

$$X(t) = Z(-1)^{N(t)},$$

where $X(0) = Z$ is an E-valued random variable independent of the counting process N. (Thus the telegraph process switches between -1 and $+1$ at each event of N.) The probability distribution of Z is arbitrary.

1. Compute $P(X(t+s) = j | X(s) = i)$ for all $t,\ s \geq 0$ and all $i,\ j \in E$.

2. Give, for all $i \in E$, the limit of $P(X(t) = i)$ as t tends to ∞.

3. Show that when $P(Z = 1) = \frac{1}{2}$, the process is a stationary process and give its power spectral measure.

Exercise 12.5.12. FLIP-FLOP WITH LIMITED MEMORY
Let N be a HPP on $\mathbb{R}$ with intensity $\lambda > 0$. Define for all $t \in \mathbb{R}$

$$X(t) = (-1)^{N((t,t+a])}.$$

1. Show that $\{X(t)\}_{t\in\mathbb{R}}$ is a WSS stochastic process.
2. Compute its power spectral density.
3. Give the best affine estimate of $X(t+\tau)$ in terms of $X(t)$, that is, find α, β minimizing

$$\mathrm{E}\left[|X(t+\tau) - (\alpha + \beta X(t))|^2\right], \quad \text{when } \tau > 0.$$

Exercise 12.5.13. JUMPING PHASE
Define for each $t \in \mathbb{R}$, $t \geq 0$,

$$X(t) = \mathrm{e}^{i\Phi_{N(t)}},$$

where $\{N(t)\}_{t\geq 0}$ is the counting process of a homogeneous Poisson process on $\mathbb{R}_+$ with intensity $\lambda > 0$, and $\{\Phi_n\}_{n\geq 0}$ is an IID sequence of random variables uniformly distributed on $[0, 2\pi]$, and independent of the Poisson process.

Show that $\{X(t)\}_{t\geq 0}$ is a wide-sense stationary process, give its covariance function $C_X(\tau)$ and its power spectral measure.

Appendix A

A Review of Hilbert Spaces

Basic Definitions

Let H be a vector space with scalar field $K = \mathbb{C}$ or $\mathbb{R}$, endowed with a map $(x, y) \in H \times H \to \langle x, y \rangle \in K$ such that for all $x, y, z \in H$ and all $\lambda \in K$,

1. $\langle y, x \rangle = \langle x, y \rangle^*$,
2. $\langle \lambda y, x \rangle = \lambda \langle y, x \rangle$,
3. $\langle x, y + z \rangle = \langle x, y \rangle + \langle x, z \rangle$,
4. $\langle x, x \rangle \geq 0\,;$ and $\langle x, x \rangle = 0$ if and only if $x = 0$.

Then H is called a *pre-Hilbert space* over K and $\langle x, y \rangle$ is called the *inner product* of x and y. For any $x \in E$, define

$$\|x\|^2 = \langle x, x \rangle\,.$$

The *parallelogram identity*

$$\|x\|^2 + \|y\|^2 = \frac{1}{2}(\|x + y\|^2 + \|x - y\|^2)$$

is obtained by expanding the right-hand side and using the equality

$$\|x + y\|^2 = \|x\|^2 + \|y\|^2 + 2\mathrm{Re}\,\{\langle x, y \rangle\}\,.$$

The *polarization identity*

$$\langle x, y \rangle = \frac{1}{4}\left\{\|x + y\|^2 - \|x - y\|^2 + i\|x + iy\|^2 - i\|x - iy\|^2\right\}$$

is checked by expanding the right-hand side. It shows in particular that two inner products $\langle \cdot, \cdot \rangle_1$ and $\langle \cdot, \cdot \rangle_2$ on E such that $\| \cdot \|_1 = \| \cdot \|_2$ are identical.

P. Brémaud, *An Introduction to Applied Probability*, Texts in Applied Mathematics 77,
https://doi.org/10.1007/978-3-031-49306-5

Schwarz's Inequality

Theorem A.0.1 *For all $x, y \in H$,*

$$|\langle x, y\rangle| \leq \|x\| \times \|y\|.$$

Equality occurs if and only if x and y are colinear.

Proof. Say $K = \mathbb{C}$. If x and y are colinear, that is, $x = \lambda y$ for some $\lambda \in \mathbb{C}$, the inequality is obviously an equality. If x and y are linearly independent, then for all $\lambda \in \mathbb{C}$, $x + \lambda y \neq 0$. Therefore

$$\begin{aligned} 0 < \|x + \lambda y\|^2 &= \|x\|^2 + |\lambda y|^2 \|\lambda y\|^2 + \lambda^* \langle x, y\rangle + \lambda \langle x, y\rangle^* \\ &= \|x\|^2 + |\lambda|^2 \|y\|^2 + 2\text{Re}(\lambda^* \langle x, y\rangle). \end{aligned}$$

Take $u \in \mathbb{C}$, $|u| = 1$, such that $u^*\langle x, y\rangle = |\langle x, y\rangle|$. Take any $t \in \mathbb{R}$ and put $\lambda = tu$. Then

$$0 < \|x\|^2 + t^2 \|y\|^2 + 2t|\langle x, y\rangle|.$$

This being true for all $t \in \mathbb{R}$, the discriminant of the second degree polynomial in t of the right-hand side must be strictly negative, that is, $4|\langle x, y\rangle|^2 - 4\|x\|^2 \times \|y\|^2 < 0$. □

Theorem A.0.2 *The mapping $x \to \|x\|$ is a* norm *on E, that is to say, for all $x, y \in E$, and all $\alpha \in \mathbb{C}$,*

(a) $\|x\| \geq 0$; *and* $\|x\| = 0$ *if and only if* $x = 0$,

(b) $\|\alpha x\| = |\alpha| \, \|x\|$, *and*

(c) $\|x + y\| \leq \|x\| + \|y\|$ (triangle inequality).

Proof. The proof of (a) and (b) is immediate. For (c) write

$$\|x + y\|^2 = \|x\|^2 + \|y\|^2 + \langle x, y\rangle + \langle y, x\rangle$$

and

$$(\|x\| + \|y\|)^2 = \|x\|^2 + \|y\|^2 + 2\|x\|\|y\|.$$

It therefore suffices to prove

$$\langle x, y\rangle + \langle y, x\rangle = 2\text{Re}(\langle x, y\rangle) \leq 2\,\|x\|\|y\|,$$

which follows from Schwarz's inequality. □

The norm $\|\cdot\|$ induces a *metric* $d(\cdot, \cdot)$ on H by

$$d(x, y) = \|x - y\|.$$

Recall that a mapping $d : E \times E \to \mathbb{R}_+$ is called a *metric* on E if, for all $x, y, z \in E$,

(a') $d(x,y) \geq 0$; and $d(x,y) = 0$ if and only if $x = y$,

(b') $d(x,y) = d(y,x)$, and

(c') $d(x,y) \geq d(x,z) + d(z,y)$.

The above properties are immediate consequences of (a), (b), and (c) of Theorem A.0.2. When endowed with a metric, a space H is called a *metric space.*

Definition A.0.3 *A pre-Hilbert space H is called a* Hilbert space *if it is a* complete *metric space with respect to the metric d.*

By this, the following is meant: If $\{x_n\}_{n\geq 1}$ is a Cauchy sequence in H, that is, if $\lim_{m,n\uparrow\infty} d(x_m, x_n) = 0$, then there exists an $x \in H$ such that $\lim_{n\uparrow\infty} d(x_n, x) = 0$.

Theorem A.0.4 *Let $\{x_n\}_{n\geq 1}$ and $\{y_n\}_{n\geq 1}$ be sequences in a Hilbert space H that converge to x and y, respectively. Then,*

$$\lim_{m,n\uparrow\infty} \langle x_n, y_m \rangle = \langle x, y \rangle .$$

In other words, the inner product of a Hilbert space is *bicontinuous.* In particular, the norm $x \mapsto \|x\|$ is a continuous function from H to $\mathbb{R}_+$.

Proof. We have for all h_1, h_2 in H,

$$|\langle x + h_1, y + h_2 \rangle - \langle x, y \rangle| = |\langle x, h_2 \rangle + \langle h_1, y \rangle + \langle h_1, h_2 \rangle| .$$

By Schwarz's inequality $|\langle x, h_2 \rangle| \leq \|x\|\|h_2\|$, $|\langle h_1, y \rangle| \leq \|y\|\|h_1\|$, and $|\langle h_1, h_2 \rangle| \leq \|h_1\|\|h_2\|$. Therefore

$$\lim_{\|h_1\|,\|h_2\|\downarrow 0} |\langle x + h_1, y + h_2 \rangle - \langle x, y \rangle| = 0 .$$

□

Isometric Extension

Definition A.0.5 *Let H and K be two Hilbert spaces with inner products denoted by $\langle \cdot, \cdot \rangle_H$ and $\langle \cdot, \cdot \rangle_K$, respectively, and let $\varphi : H \mapsto K$ be a linear mapping such that for all $x, y \in H$*

$$\langle \varphi(x), \varphi(y) \rangle_K = \langle x, y \rangle_H .$$

Then, φ is called a linear isometry *from H into K. If, moreover, φ is from H* onto *K, then H and K are said to be* isomorphic.

Note that a linear isometry is necessarily injective, since $\varphi(x) = \varphi(y)$ implies $\varphi(x - y) = 0$, and therefore

$$0 = \|\varphi(x-y)\|_K = \|x-y\|_H ,$$

which implies $x = y$. In particular, if the linear isometry is *onto*, it is bijective.

Recall that a subset $A \in E$, where (E, d) is a metric space, is said to be *dense* in E if, for all $x \in E$, there exists a sequence $\{x_n\}_{n\geq 1}$ in A converging to x.

Theorem A.0.6 *Let H and K be two Hilbert spaces with inner products denoted by $\langle \cdot, \cdot \rangle_H$ and $\langle \cdot, \cdot \rangle_K$, respectively. Let V be a vector subspace of H that is dense in H, and $\varphi : V \mapsto K$ be a linear isometry from V to K. Then, there exists a unique linear isometry $\widetilde{\varphi} : H \mapsto K$ whose restriction to V is φ.*

Proof. We shall first define $\widetilde{\varphi}(x)$ for $x \in H$. Since V is dense in H, there exists a sequence $\{x_n\}_{n\geq 1}$ in V converging to x. Since φ is isometric,

$$\|\varphi(x_n) - \varphi(x_m)\|_K = \|x_n - x_m\|_H \quad \text{for all } m, n \geq 1 .$$

In particular, $\{\varphi(x_n)\}_{n\geq 1}$ is a Cauchy sequence in K and therefore it converges to some element of K, which we denote by $\widetilde{\varphi}(x)$.

The definition of $\widetilde{\varphi}(x)$ is independent of the sequence $\{x_n\}_{n\geq 1}$ converging to x. Indeed, for another such sequence $\{y_n\}_{n\geq 1}$,

$$\lim_{n\uparrow\infty} \|\varphi(x_n) - \varphi(y_n)\|_K = \lim_{n\uparrow\infty} \|x_n - y_n\|_H = 0 .$$

The mapping $\widetilde{\varphi} : H \mapsto K$ so constructed is clearly an extension of φ (for $x \in V$ one can take for an approximating sequence of x the sequence $\{x_n\}_{n\geq 1}$ such that $x_n \equiv x$).

The mapping $\widetilde{\varphi}$ is linear. Indeed, let $x, y \in H$, $\alpha, \beta \in \mathbb{C}$, and let $\{x_n\}_{n\geq 1}$ and $\{y_n\}_{n\geq 1}$ be two sequences in V converging to x and y, respectively. Then $\{\alpha x_n + \beta y_n\}_{n\geq 1}$ converges to $\alpha x + \beta y$. Therefore

$$\lim_{n\uparrow\infty} \varphi(\alpha x_n + \beta y_n) = \widetilde{\varphi}(\alpha x + \beta y) .$$

But

$$\varphi(\alpha x_n + \beta y_n) = \alpha\varphi(x_n) + \beta\varphi(y_n) \to \alpha\widetilde{\varphi}(x) + \beta\widetilde{\varphi}(y)$$

tends to $\widetilde{\varphi}(\alpha x + \beta y) = \alpha\widetilde{\varphi}(x) + \beta\widetilde{\varphi}(y)$.

The mapping $\widetilde{\varphi}$ is isometric since, in view of the bicontinuity of the inner product and of the isometricity of φ, if $\{x_n\}_{n\geq 1}$ and $\{y_n\}_{n\geq 1}$ are two sequences in

V converging to x and y, respectively, then

$$\begin{aligned}\langle\widetilde{\varphi}(x),\widetilde{\varphi}(y)\rangle_K &= \lim_{n\uparrow\infty}\langle\varphi(x_n),\varphi(y_n)\rangle_K\\ &= \lim_{n\uparrow\infty}\langle x_n,y_n\rangle_H = \langle x,y\rangle_H\,.\end{aligned}$$

□

Orthogonal Projection

A subset G of a Hilbert space H is said to be closed in H if every convergent sequence of G has a limit in G.

Theorem A.0.7 *Let $G \subseteq H$ be a vector subspace of the Hilbert space H. Endow G with the inner product which is the restriction to G of the inner product on H. Then, G is a Hilbert space if and only if G is closed in H.*

G is then called a *Hilbert subspace* of H.

Proof. (i) Assume that G is closed. Let $\{x_n\}_{n\in\mathbb{N}}$ be a Cauchy sequence in G. It is a fortiori a Cauchy sequence in H, and therefore it converges in H to some x, and this x must be in G, because it is a limit of elements of G and G is closed.

(ii) Assume that G is a Hilbert space with the inner product induced by the inner product of H. In particular every convergent sequence $\{x_n\}_{n\in\mathbb{N}}$ of elements of G converges to some element of G. Therefore G is closed. □

Definition A.0.8 *Two elements x, y of the Hilbert space H are said to be* orthogonal *if $\langle x, y\rangle = 0$. Let G be a Hilbert subspace of the Hilbert space H. The* orthogonal complement *of G in H, denoted $G^\perp$, is defined by*

$$G^\perp = \{z \in H : \langle z, x\rangle = 0 \text{ for all } x \in G\}\,.$$

Clearly, $G^\perp$ is a vector space over $\mathbb{C}$. Moreover, it is closed in H since if $\{z_n\}_{n\geq 1}$ is a sequence of elements of $G^\perp$ converging to $z \in H$ then, by continuity of the inner product,

$$\langle z, x\rangle = \lim_{n\uparrow\infty}\langle z_n, x\rangle = 0 \quad \text{for all } x \in H\,.$$

Therefore $G^\perp$ is a Hilbert subspace of H.

Note that a decomposition $x = y + z$ where $y \in G$ and $z \in G^\perp$ is necessarily unique. Indeed, let $x = y' + z'$ be another such decomposition. Then, letting $a = y - y'$, $b = z - z'$, we have that $0 = a + b$ where $a \in G$ and $b \in G^\perp$. Therefore, in particular, $0 = \langle a, a\rangle + \langle a, b\rangle$. But $\langle a, b\rangle = 0$, and therefore $\langle a, a\rangle = 0$, which implies that $a = 0$. Similarly, $b = 0$.

Theorem A.0.9 *Let G be a Hilbert subspace of H. For all $x \in H$, there exists a unique element $y \in G$ such that $x - y \in G^{\perp}$. Moreover,*

$$\|y - x\| = \inf_{u \in G} \|u - x\| . \tag{A.1}$$

Proof. Let $d(x, G) = \inf_{z \in G} d(x, z)$ and let $\{y_n\}_{n \geq 1}$ be a sequence in G such that

$$d(x, G)^2 \leq d(x, y_n)^2 \leq d(x, G)^2 + \frac{1}{n} . \tag{$\star$}$$

The parallelogram identity gives, for all $m, n \geq 1$,

$$\|y_n - y_m\|^2 = 2(\|x - y_n\|^2 + \|x - y_m\|^2) - 4\|x - \frac{1}{2}(y_m + y_n)\|^2 .$$

Since $\frac{1}{2}(y_n + y_m) \in G$,

$$\|x - \frac{1}{2}(y_m + y_n)\|^2 \geq d(x, G)^2,$$

and therefore

$$\|y_n - y_m\|^2 \leq 2\left(\frac{1}{n} + \frac{1}{m}\right) .$$

The sequence $\{y_n\}_{n \geq 1}$ is therefore a Cauchy sequence in G and consequently it converges to some $y \in G$ since G is closed. Passing to the limit in ($\star$) gives (A.1).

Uniqueness of y satisfying (A.1): Let $y' \in G$ be another such element. Then

$$\|x - y'\| = \|x - y\| = d(x, G) ,$$

and from the parallelogram identity

$$\begin{aligned}\|y - y'\|^2 &= 2\|y - x\|^2 + 2\|y' - x\|^2 - 4\|x - \frac{1}{2}(y + y')\|^2 \\ &= 4d(x, G)^2 - 4\|x - \frac{1}{2}(y + y')\|^2 .\end{aligned}$$

Since $\frac{1}{2}(y + y') \in G$,

$$\|x - \frac{1}{2}(y + y')\|^2 \geq d(x, G)^2 ,$$

and therefore $\|y - y'\|^2 \leq 0$, which implies $\|y - y'\|^2 = 0$ and therefore $y = y'$.

It now remains to show that $x - y$ is orthogonal to G, that is, $\langle x - y, z \rangle = 0$ for all $z \in G$. Since this is trivially true if $z = 0$, we may assume $z \neq 0$. Because $y + \lambda z \in G$ for all $\lambda \in \mathbb{R}$,

$$\|x - (y + \lambda z)\|^2 \geq d(x, G)^2 ,$$

that is,

$$\|x - y\|^2 + 2\lambda \text{Re}\{\langle x - y, z \rangle\} + \lambda^2 \|z\|^2 \geq d(x, G)^2 .$$

Since $\|x - y\|^2 = d(x, G)^2$, we have

$$-2\lambda \text{Re}\{\langle x - y, z \rangle\} + \lambda^2 \|z\|^2 \geq 0 \quad \text{for all } \lambda \in \mathbb{R},$$

which implies $\text{Re}\{\langle x - y, z \rangle\} = 0$. The same type of calculation with $\lambda \in i\mathbb{R}$ (pure imaginary) leads to $\Im\{\langle x - y, z \rangle\} = 0$. Therefore $\langle x - y, z \rangle = 0$.

That y is the unique element of G such that $y - x \in G^\perp$ follows from the remark preceding Theorem A.0.9. □

Definition A.0.10 *The element y in Theorem A.0.9 is called the* orthogonal projection *of x on G and is denoted by $P_G(x)$.*

The projection theorem states, in particular, that for any $x \in G$ there is a unique decomposition

$$x = y + z, \quad y \in G, \ z \in G^\perp,$$

and that $y = P_G(x)$, the (unique) element of G closest to x. Therefore

Theorem A.0.11 *The orthogonal projection $y = P_G(x)$ is characterized by the two following properties:*

(1) $y \in G$;

(2) $\langle y - x, z \rangle = 0$ for all $z \in G$.

This characterization is known as the *projection principle* of Hilbert spaces.

Let C be a collection of vectors in the Hilbert space H. The linear span of C, denoted span(C) is, by definition, the set of all finite linear combinations of vectors of C. This is a vector space. The closure of this vector space, $\overline{\text{span}(C)}$, is called the Hilbert subspace generated by C. By definition, x belongs to this subspace if and only if there exists a sequence of vectors $\{x_n\}_{n \geq 1}$ such that

(i) for all $n \geq 1$, x_n is a finite linear combination of vectors of C, and

(ii) $\lim_{n \uparrow \infty} x_n = x$.

Theorem A.0.12 *An element $\widehat{x} \in H$ is the projection of x onto $G = \overline{\text{span}(C)}$ if and only if*

(α) $\widehat{x} \in G$, and

(β) $\langle x - \widehat{x}, z \rangle = 0$ for all $z \in C$.

Note that we have to satisfy requirement not for all $z \in G$, but only for all $z \in C$.

The proof is easy. We have to show that $\langle x - \widehat{x}, z \rangle = 0$ for all $z \in G$. But $z = \lim_{n\uparrow\infty} z_n$, where $\{z_n\}_{n\geq 1}$ is a sequence of vectors of span(C) such that $\lim_{n\uparrow\infty} z_n = z$. By hypothesis, for all $n \geq 1$, $\langle x - \widehat{x}, z_n \rangle = 0$. Therefore, by continuity of the inner product,

$$\langle x - \widehat{x}, z \rangle = \lim_{n\uparrow\infty} \langle x - \widehat{x}, z_n \rangle = 0 \,.$$

Bibliography

[1] Bauer, H., *Measure and Integration Theory*, de Gruyter (2001).

[2] Billingsley, P., *Convergence of Probability Measures*, Wiley (1968).

[3] —, *Probability and Measure*, Anniversary ed., Wiley (2012).

[4] Brémaud, P., *Markov Chains*, 2nd ed., Springer (2020).

[5] —, *Fourier Analysis and Stochastic Processes*, Springer (2014).

[6] —, *Discrete Probability Models and Methods*, Springer (2017).

[7] —, *Probability Theory and Stochastic Processes*, Springer (2020).

[8] —, *Point Process Calculus on the Line and in Space*, Springer (2020).

[9] Dembo, A., and O. Zeitouni, *Large Deviations Techniques and Applications*, 2nd ed., Springer (2010).

[10] Durrett, R., *Probability and Examples*, Duxbury Press (1996).

[11] Kallenberg, O., *Foundations of Modern Probability*, 3rd ed., vol. 1 and 2, Springer (2021).

[12] Klenke, A., *Probability Theory; A Comprehensive Course*, Springer (2008).

[13] Levin, D.A., Peres, Y., and E.L. Wilmer, *Markov Chains and Mixing Times*, American Mathematical Society (2009).

[14] Lindvall, T., *Lectures on the Coupling Method*, Wiley (1992).

[15] Royden, H.L., *Real Analysis*, Macmillan (1988).

[16] Shiryaev, A.N., *Probability*, 2nd ed., Springer (2013).

[17] Williams, D., *Probability and Martingales*, Cambridge University Press (1991).

P. Brémaud, *An Introduction to Applied Probability*, Texts in Applied Mathematics 77,
https://doi.org/10.1007/978-3-031-49306-5

Index

P. Brémaud, *An Introduction to Applied Probability*, Texts in Applied Mathematics 77,
https://doi.org/10.1007/978-3-031-49306-5

GPSR Compliance

The European Union's (EU) General Product Safety Regulation (GPSR) is a set of rules that requires consumer products to be safe and our obligations to ensure this.

If you have any concerns about our products, you can contact us on ProductSafety@springernature.com

In case Publisher is established outside the EU, the EU authorized representative is:

Springer Nature Customer Service Center GmbH
Europaplatz 3
69115 Heidelberg, Germany

Batch number: 10365511

Printed by Printforce, the Netherlands